Microbiology of Foods

A Series of Books in Food and Nutrition

Editor: B. S. Schweigert

Microbiology of Foods

John C. Ayres
The University of Georgia

J. Orvin Mundt
The University of Tennessee

William E. Sandine
Oregon State University

W. H. Freeman and Company

San Francisco

Sponsoring Editor: Gunder Hefta
Manuscript Editor: Kevin Gleason
Designers: Gary A. Head and Sharon H. Smith
Production Coordinator: Linda Jupiter
Illustration Coordinator: Batyah Janowski
Artists: Evan Gillespie and Tim Keenan
Compositor: Graphic Typesetting Service
Printer and Binder: The Maple-Vail Book Manufacturing Group

Library of Congress Cataloging in Publication Data

Ayres, John Clifton, 1913–
  Microbiology of foods.

  (A Series of books in food and nutrition)
  Includes bibliographies and index.
  1. Food—Microbiology. 2. Food poisoning.
I. Mundt, John Orvin, 1912–   joint author.
II. Sandine, William E., 1928–   joint author.
III. Title.
QR115.A88   664'.001'576   79-16335
ISBN 0-7167-1049-8

Printed in the United States of America

9 8 7 6 5 4 3 2

Contents

Part IV   Foodborne Illnesses

Preface

The purpose of *Microbiology of Foods* is to serve as a text for the advanced undergraduate and the beginning graduate student preparing to enter the foods industries or involved with allied disciplines. Rather than present the reader with exhaustive source data, we have sought to present the material in sufficient detail for the beginning microbiologist and, to some extent, the lay reader to grasp. For those readers interested in more exhaustive treatment, we provide at the end of each chapter a list of the references cited in the chapter text.

Our text represents the first serious attempt since that of Tanner's *Microbiology of Foods,* first published in the early 1930s, to comprehensively embrace the available information about food microbiology. Tanner's text, authoritative as it was during the first half of the century, is obsolete today. Our text updates the treatment of food microbiology to make it useful in this latter part of the century. We also examine areas little touched upon by other food microbiology texts—for example, the importance of molds in the production of mycotoxins and antibiotics and the involvement of yeasts and molds in various fermentations and spoilage. We also essay—as other general references do not—to evaluate or interpret the confusing data on the role of chemicals, radiation, and microorganisms in benefiting or harming foods and their users.

Microbiology of Foods is divided into four parts. In Part I, Chapter 1 presents the classification of bacteria, yeasts, and molds. The remaining five chapters describe the major non-microbiological methods of preserving foods. Part II, composed of four chapters, describes the desirable alterations of cereals, milk, and vegetables by fermentation, to produce beverages and foods for human enjoyment and to effect preservation. Chapter 10, the final chapter of the section, identifies fermentations, most employing fungi, that are little known in the United States but much used in other parts of the world, particularly in the Far East. The 10 chapters of Part III treat the microbiology of specific food products: in the first five chapters, those obtained from spices, sugars and starches, cereals, flours, fruits and vegetables; in the last five chapters, those of animal origin—milk and fermented milk products, meats, fish and shellfish, poultry, and eggs. The three chapters of the last part describe the nonmicrobial and microbial foodborne infections and the bacterial and some fungal toxemias.

Although the principles of microbial spoilage and of nonmicrobiological methods of preservation are concentrated in the opening chapters, they are also interwoven throughout the remainder of the text. In other respects, however, each chapter is independent of preceding chapters, and may be rearranged into any sequence desired in a teaching program. Thus, if the time constraints of academic quarters must be accommodated, chapters deemed less essential to the teaching of food microbiology may be omitted without sacrificing the overall principles.

The authors wish to thank Kevin Gleason and Ruth Allen for their help during the preparation of this volume. Also, we extend our appreciation to Dr. B. S. Schweigert and Professor E. B. Collins for useful suggestions and to the anonymous reviewers for helpful advice and criticism. Finally, we are indebted to our many colleagues and friends who have graciously permitted us to use their figures and tables.

August 1979

John C. Ayres
J. Orvin Mundt
William E. Sandine

Part I
General Considerations

1

Classification of Microorganisms

Every form of life can be assigned membership in one of five kingdoms: **Plantae** (plants), **Animalia** (animals), **Protista, Fungi,** and **Monera.** The basic structural unit of living things is the cell, in which genetic information is replicated and stored, cellular components are synthesized, energy is generated, and, in some cells, a mechanism for movement is located. Both the plants and animals are multicellular. Plants have cell walls composed of cellulose, whereas animal cells do not have cell walls. Plants perform photosynthesis, whereas animals ingest and then digest their foods. To perform a variety of specialized functions the cellular matter of every plant and animal is differentiated into tissues, organs, and organ systems. All members of the kingdom Protista are single-cell life forms that perform all functions without specialization. (Some members, however, occur in well-developed colonial aggregates that resemble multicellular structures.) The kingdom Fungi is made up of single-cell forms, such as the yeasts, and other forms composed of masses of filaments. The kingdom Monera comprises single-cell life forms: the bacteria and the blue-green algae (see Table 1–1).

Table 1-1
Protista

Eucaryotes (having nucleus and nuclear membrane)
   Kingdom: Animalia
              Plantae
              Fungi
              Protista
Procaryotes (lacking nucleus and nuclear membrane)
   Kingdom: Monera

   Division I. Phototrophic procaryotes (photobacteria)
      Class I.   Blue-green photobacteria
      Class II.   Red photobacteria }
      Class III.   Green photobacteria } Part 1. Phototropic
   Division II. Procaryotes indifferent to light (scotobacteria)
      Class I.   The bacteria, Parts 2 to 17; chemotropic
      Class II.   Obligate intracellular scotobacteria in eucaryotic
                  cells, rickettsias, Part 18
      Class III.   Scotobacteria without cell walls, Mollicutes, Part 19

Sources: After Haeckel, 1894; Stanier and van Niel, 1962. Eucaryotae, after Allsopp, 1969.

These five kingdoms can be divided into two groups: the plants, animals, Fungi, and Protista are **eucaryotes;** the Monera are **procaryotes.** The two groups differ by type of nucleus. In the eucaryotic cell the nucleus is a discrete body surrounded by a membrane; the nucleus divides according to the classical Mendelian description. Each cell possesses spindles, golgi apparatus, and mitochondria. Cells of plants and some protists also contain plastic bodies. Eucaryotic ribosomes are of the large 80S type (Taylor and Storck, 1964).

The procaryotic cell contains none of these structures, having instead a long, double-stranded nucleus that weaves throughout the cell. This nucleus is not a discrete body and is not surrounded by a nuclear membrane. During cell division the double-stranded nucleus separates, each strand becoming a template for the formation of new, complementary strands. Procaryotic ribosomes are of the small 70S type.

Among the procaryotes the organisms of most concern to the food microbiologist are the bacteria. Among the eucaryotes the organisms of most concern are the fungi. Of far less importance are the algae and protozoa.

Except for a few siliceous forms, such as the diatoms, the Algae have cell walls composed of cellulose. Being chlorophyllous, the algae are photosynthetic—although occasionally achlorophyllous algae are isolated from untreated waters and foods washed or conveyed in such waters. The achlorophyllous forms are easily mistaken for yeasts. In the past the algae have played no role in the microbiology of foods, but they may be a future source of food for humans and animals. The Protozoa lack cell walls. They ingest and digest complex organic matter. They thrive in moist or wet environments. Some protozoa are food- and water-borne pathogens.

Fungi

Many fungi are multicellular; others, such as the yeasts, are unicellular; and some, called dimorphic, are unicellular in some conditions or stages of their existence but become multicellular in other conditions or stages.

In many varieties of fungi the cells grow end-to-end to form filaments, or **hyphae,** which in turn branch and intertwine into a network, termed the **mycelium.** In other varieties hyphae produce a compact stalk and aerial spore-forming structure, as in the fruiting body of mushrooms. In some fungi the hyphal strands have no crosswalls, or **septa,** and the cytoplasm can flow as the fungus grows. Although these nonseptate, or **coenocytic,** structures are multinucleate, they remain unicellular even when attaining heroic size. Since the same habit also occurs in some of the green algae, this group of fungi is called **Phycomycetes** (*phykos,* seaweed; *mykes,* fungus). The phycomycetes constitute one of the five classes of the true Fungi (Table 1-2). The remaining classes of fungi have a septate, or cellular mycelium; at certain stages of development the cells of some members may contain two or more nuclei.

Fungi are differentiated into classes by their reproductive habits. All except the members of the class **Mycelia Sterila** are able to produce spores asexually, by simple cell division without fusion of nuclei. The members of the classes Phycomycetes, Basidiomycetes, and Ascomycetes also reproduce sexually, by fusion of two similar or dissimilar cells, but asexual spores are produced much more commonly than are the sexual spores.

Table 1-2
Key to the classes and orders of fungi.

I. Reproduce by sexual mechanisms
  A. Mycelium, when present, coenocytic;
    zygotes produced

                                    Class: Phycomycetes
    1. Mycelium scanty or lacking   Subclass: Archimycetes
    2. Mycelium present
      a. Sex cells heterogamic   Subclass: Oomycetes
      b. Sex cells isogamic   Subclass: Zygomycetes
  B. Mycelium septate, or reproduces
    by budding
    1. Sexual spores borne
      externally   Class: Basidiomycetes
      a. Basidiospores not septate   Subclass: Homobasidio-
        or deeply divided   mycetes
      b. Basidiospores deeply   Subclass: Heterobasidio-
        divided or septate   mycetes
    2. Sexual spores borne within
      a membrane or sac   Class: Ascomycetes
      a. Unicellular or rarely
        filamentous   Subclass: Hemiascomycetes
      b. Loosely filamentous to
        fleshy structures   Subclass: Euascomycetes
         i. Produce cleistothecia
           Colonial structures
           filamentous   Order: Plectascales
           Colonial structures
           compact   Order: Tuberales
        ii. Produce perithecia
           Peridium dark in color   Order: Sphaeriales
           Peridium bright in color   Order: Hypocreales
        iii. Produce apothecia   Order: Pezizales
II. Do not reproduce by sexual mechanisms
  A. Produce neither sexual cells nor
    vegetative spores

                                      Class: Mycelia Sterila
  B. Produce conidia (conidia may be
    lacking in laboratory cultures)   Class: Fungi Imperfecti
    1. Conidia produced in pycnidia   Order: Sphaeropsidales
    2. Conidia produced openly, not
      in enclosed structures   Order: Moniales

Table 1-2, *continued*

a. Conidiophores scattered, not in a sheath or a globose structure	
i. Basal mycelium light in color; reverse not dark or darkening with age	Family: Moniliaceae
ii. Basal mycelium may be dark in color; reverse dark or darkening with age	Family: Dematiceae
b. Conidiophores gathered into sheath	Family: Stilbaceae
c. Conidiophores globose and sessile	Family: Tuberculariaceae

Source: Adapted from Gilman, 1957.

Because it is extremely difficult to induce the sexual, or **perfect,** stage in some fungi, these organisms are often identified by their vegetative growth and by their asexual fruiting structure. Such organisms are placed in an artificial class called the Fungi Imperfecti, or Deuteromycetes, discussed further below.

Phycomycetes

This class of fungi is generally divided into three subclasses: Archimycetes, Oomycetes, and Zygomycetes.

Archimycetes comprises the most primitive of the fungi, being rudimentary aquatic parasites composed of single or irregular masses of cells (**thalli;** singular, **thallus**) or forming, at most, only a rudimentary mycelium. The organisms produce both sexual and asexual spores. The latter form at the tips of hyphae, where they are separated from the rest of the filament by a septum, forming a **sporangium,** a sac containing one or more vegetatively produced spores. After the septum is formed, rapid mitotic divisions produce many nuclei around which the spore coats form to leave an anucleated sporangium. The numerous **sporangiospores** borne within the cell of an Archimycete have a distinguishing feature: they are genetically alike; those of the other two subclasses are not. Since each spore develops two flagella and is able to swim freely when released, it is called a **zoospore.** Few members of this subclass are encountered by the food microbiologist.

The members of the subclass **Oomycetes** also are primarily aquatic fungi, but some live in the soil and a few are highly specialized parasites of land plants. On rare occasions oomycetes are isolated from frozen fruits and other plant foods exposed to untreated well waters during preparation for market. Three genera, *Pythium, Phytophthora,* and *Plasmopara,* are of special interest to the food microbiologist, since they cause disease in cultivated plants, sometimes occasioning sizable economic losses. *Pythium* spp. cause "damping off" of seedlings, bean sprouts, and germinating barley; *Phytophthora* spp. cause late blight of potatoes; and *Plasmopara* spp. cause downy mildew of grapes.

Members of the genus *Saprolegnia,* ordinarily saprophytic, may become parasitic on fish such as aquarium and hatchery stock, and occasionally cause extensive kills of fingerlings in fresh waters. The fungus grows on the mouthparts, gills, fins, and tail, and unless the growth is halted quickly by antibiotics or other medication, death soon occurs. In *Saprolegnia* (see Figure 1-1) well-defined differentiated sex cells, or **gametes,** are formed. The larger of the cells, the **oogonium,** contains several eggs or single nucleated cells. The **antheridia,** delicate fertilization tubes that arise nearby, penetrate the oogonium. A nucleus from an antheridium fuses with that of an egg cell to form the **oospore** (Figure 1-2).

Most of the phycomycetes that occur in foods are members of the subclass **Zygomycetes.** These produce a large, dark-brown-to-black, thick-walled, warty-to-spiny spore termed the **zygote.** Two identical, or **isogamic,** hyphae come into contact; septa are formed in each hypha to isolate the gamete; the cell walls at the point of contact dissolve; and **plasmogamy** and **caryogamy** occur. Both the zygote and the suspending hyphae (termed **suspensor cells**) enlarge. *Zygorrhynchus* spp., common soil fungi, are **homothallic;** that is, both gametes arise from a single colony. Because the common members of the order Mucorales are **heterothallic,** zygotes are rare (see Figure 1-3). They can be induced by placing mating types, often designated as (+) and (–) strains, into the same petri dish. Upon germination, the zygotes produced by members of the family Mucoraceae give rise to a **sporangiophore,** a stalk bearing the

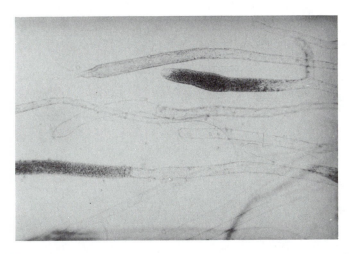

Figure 1-1
Photomicrograph of cells of the oomycete genus *Saprolegnia* (×950), showing the formation of sporangia.

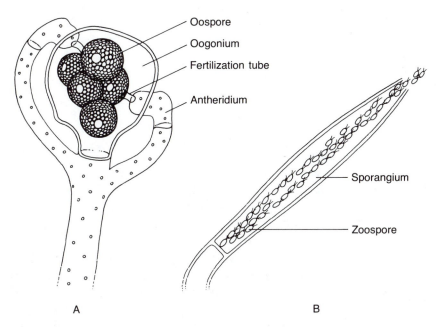

Oospore

Oogonium

Fertilization tube

Antheridium

Sporangium

Zoospore

A

B

Figure 1-2
A, sexual and, B, asexual reproduction of members of the genus *Saprolegnia*.

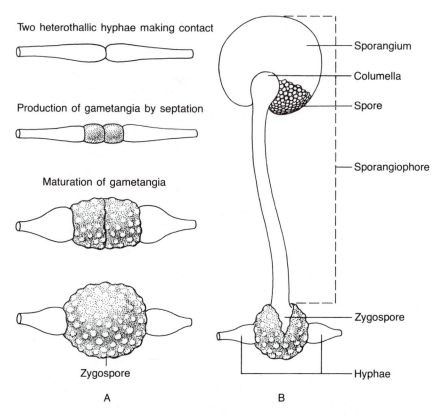

Two heterothallic hyphae making contact

Production of gametangia by septation

Maturation of gametangia

Zygospore

A

Sporangium

Columella

Spore

Sporangiophore

Zygospore

Hyphae

B

Figure 1-3
A, zygospore and, B, sporangiospore production of *Mucorales.*

sporangium. This stalk becomes inflated to form the **columella,**
on which is borne the **sporangium,** a cell or sac bearing vegetatively induced spores (see Figure 1-4).

The class Phycomycetes is divided into several orders (Gilman,
1957), but only members of the order Mucorales occur in
foods. Because zygote formation is rarely observed, these organisms are identified exclusively by their vegetative structure
and their mode of sporangiospore production. All members
produce prolific and characteristic growth on culture media,
with mycelia often completely filling the petri dish or culture
tube. The mycelium is white to gray. Members of the families

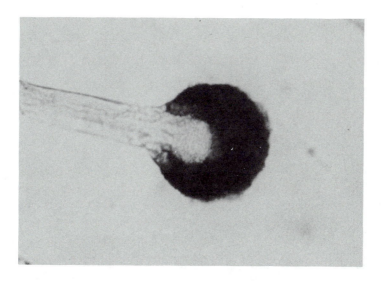

A

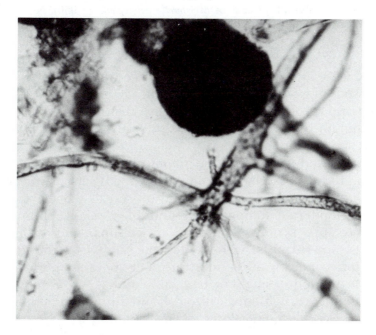

B

Figure 1-4
Photomicrograph of mold of *Rhizopus nigricans* (×950). A, sporangiophore; B, rhizoids and zygospore.

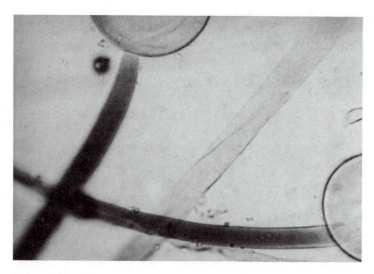

Figure 1-5
Photomicrograph of cells of *Mortierella polycephala* (×950). Note
constricted sporangiophore and lack of membrane.

Mucoraceae and Mortierellaceae form conspicuous, brown-to-
black sporangia embedded throughout the mycelial mass. *Mortierella* spp. produce a sporangiophore that tapers toward the
sporangium (Figure 1-5), whereas the mucors produce an inflated sporangiophore.

Rhizopus and *Mucor* spp. are among the most commonly occurring fungi on saccharine and starchy foods such as strawberries, sweet potatoes, and bread, and on cereals that have been
held at moderate temperatures and high relative humidity.
Rhizopus spp. produce runners, or **stolons,** and primitive
holdfasts, or **rhizoids,** at the **nodes,** the sites where the stolons
come into contact with the substrate. The sporangiophores also
arise at the nodes, singly or in small clusters. *Mucor* spp. produce neither stolons nor rhizoids, and the sporangiophores
may arise at any point along the mycelium.

Although not all members of the order Mucorales are commonly found on foods, species of *Thamnidium,* the sole genus in
the family Thamnidiaceae, are. They form the long "whiskers"
on meat. *Choanephora* and *Cunninghamella* (both of the family
Choanephoraceae; see Figure 1-6), and *Syncephalastrum* (of the

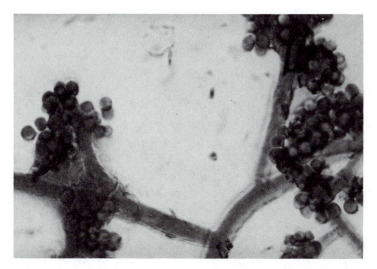

A

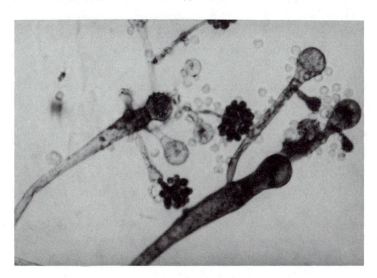

B

Figure 1-6
Members of the family Choanephoraceae (×950): A, cells of
Choanephora sporangioli; B, cells of *Cunninghamella elegans.*

family Cephalidaceae) occur on sweet potatoes, yams, and other vegetables and on aging meats held too long under moist or humid conditions.

Basidiomycetes

In the class Basidiomycetes the spores, usually four in number, result from reduction division within a terminal, swollen cell called the **basidium.** The class includes the **Homobasidiomycetes** (the mushrooms, puffballs, and bracket fungi), organisms having simple basidia, and the **Heterobasidiomycetes,** organisms having septate or deeply divided basidia. The latter group is divided into three orders: Tremallales, Uredinales, and Ustilaginales. **Tremallales,** which includes several yeastlike organisms, differs from the other two orders in that its members are saprophytic and have a well-developed **basidiocarp,** or fleshy fruiting structure. The members of **Uredinales,** or rusts, never reproduce by budding. Their spores are produced on a spiculelike extension, or **sterigma,** on the basidium. Basidiospores of the members of **Ustilaginales,** or smuts, lack such a stalk and are sessile. The smuts can reproduce vegetatively, by budding.

Ascomycetes

The **ascomycetes,** or sac fungi, constitute the largest class of fungi in number of genera and species. They range from the unicellular yeasts (Figure 1-7) and loosely filamentous forms commonly termed molds, to the highly organized fruiting structures of the fleshy fungi. The filamentous and fleshy forms initiate sexual reproduction by fusion of two like gametes. The fusion stimulates very rapid growth to produce a structure termed the **ascocarp,** the interior of which becomes lined with cells that ultimately become **asci** (singular, **ascus**), sacs in which the ascospores are borne. Within the asci reduction division occurs to produce, usually, eight ascospores. The **peridium,** or wall of the ascocarp, may be a loose, feltlike structure that resembles a mesh; or it may be compact and either membranous, or rigid, or leathery, or, occasionally, ligneous. Other than the yeasts (which will be treated later), the class Ascomycetes is

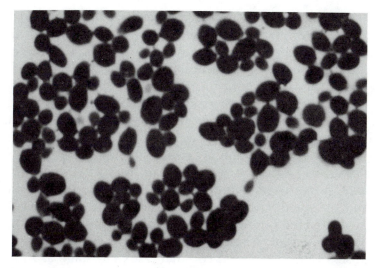

Figure 1-7
Saccharomyces ellipsoideus (× 950).

divided into two groups according to how the ascospores are borne in the ascocarp. The members of the group to which the orders **Plectascales** and **Perisporiales** belong produce structures termed **cleistothecia:** peridia that do not have openings, or ostioles. The filamentous fungi of the orders **Sphaeriales** and **Hypocreales** produce structures termed **perithecia:** scattered and hirsute or unadorned peridia that have ostioles. The fleshy structures of the fungi of the order **Pezizales** produce asci that are in **apothecia,** open-disc or cup-shaped ascocarps. Among the fleshy fungi only the members of the order **Tuberales** produce a structure that has no ostiole.

Fungi Imperfecti

The filamentous fungi manifest a peculiar characteristic: in many genera some members of a species produce ascospores and others of that species do not. The latter are assigned to the class Fungi Imperfecti. There are several possible causes of their asexuality: retrogression, mutation, or loss of essential genes (as through irradiation, for example). Among other possible causes are that a strain is self-sterile and that the strain of the opposite sex, termed the **mating type,** has not been discovered; or that there are peculiar requirements, still unknown, for the induction of sexual union; or that hybrids are sterile. Therefore, it is not unusual that when the same fungus is placed in the classes *Ascomycetes* and Fungi Imperfecti it receives two different names. This taxonomic convenience has resulted in confusing terminology. For example, a common fungus associated with herbarium specimens may be identified by its asexual characteristics as *Aspergillus glaucus* but by its cleistothecia as *Eurotium herbariorum.* Another fungus, occasionally found on peanuts, is called *Penicillium vermiculatum* in the schema of the Fungi Imperfecti but *Talaromyces vermiculatus* when sexual spores are evident. Similarly, one of the T-2 toxin-producing organisms is known in the conidial (asexual) stage as *Fusarium graminearum,* but when ascospores are formed it is called *Gibberella zeae.*

Members of the class Fungi Imperfecti are identified as to genus and species almost exclusively by their asexual fruiting structure or by the characteristics of the **conidiophore,** the aerial filament upon which the spores are borne. Or their genus can be assigned according to mode of spore formation. The references at the end of the chapter cite keys to both approaches. All members of the class produce loosely filamentous growth typical of the fungi bearing the fruiting structures. Members of one small order, Sphaeropsidales, produce vegetative spores within the **pycnidia,** structures that resemble the cleistothecia and perithecia of the Ascomycetes. Of the fungi that are not in the class Phycomycetes, most that are of interest to the food microbiologist (either as contaminants or as agents of spoilage) are either members of the order Moniliales or are identified through the keys to that order.

Members of the order Moniliales bear spores openly rather than in a sac such as the sporangium of the phycomycetes. The spores, termed **conidia,** may be produced directly, either upon the vegetative hyphae or upon the conidiophore, without intervention of an intermediary cell, or may instead be produced upon specialized intermediary cells termed **sterigmata.** Spores may be produced singly, in clusters, or in chains. Those fungi

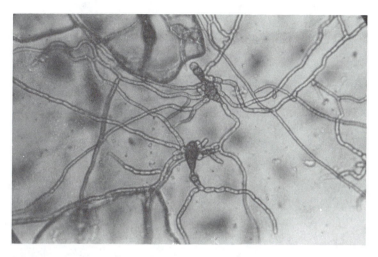

Figure 1-8
Alternaria spp. (× 220), of the family Dematiaceae, have muriform conidia.

that are clear in color (hyaline) and have scattered conidio-phores are placed in the family Moniliaceae. Except for members of the *Aspergillus niger* group, fungi whose spores are dark in color and whose scattered conidiophores often produce a smoky colony are placed into the family Dematiaceae (Figure 1-8). If the conidiophores are grouped to form a cylindrical sheaf, the **coremium,** the fungi are placed in the relatively small family Stilbaceae. The family Tuberculaceae, of which *Fusarium* spp. are commonly occurring members, produce globose bodies without stalks, or **sporodochia.** Figure 1-9 shows several typical fruiting structures.

Mycelia Sterila

Fungi of the order Mycelia Sterila produce only vegetative hyphae and neither vegetative nor sexually induced spores; they must be identified by colonial features. They are dominant in soils, such as those in rice lands, that have been underwater for a long time. If not given special care, some members of the

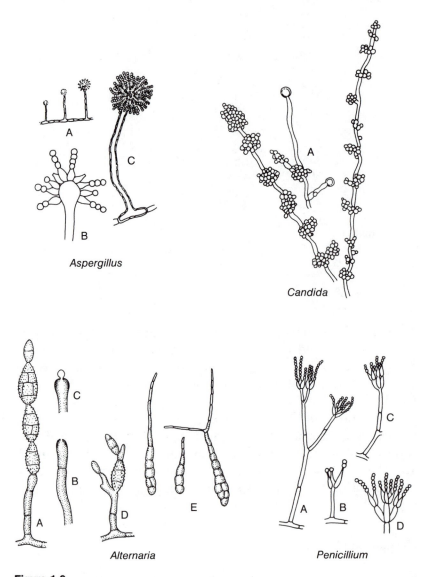

Aspergillus

Candida

Alternaria

Penicillium

Figure 1-9

Fruiting structures of representative members of *Moniliales*. A, E: representative conidiophores; B, C, D: formation of conidia.

Fungi Imperfecti, notably *Fusarium* and *Helminthosporium* spp., rapidly lose the ability to produce conidiospores in culture and can be mistakenly placed in the order Mycelia Sterila.

Characteristics of the Filamentous Fungi

From the primary habitat of all fungi, the soil, spores escape and disperse throughout the atmosphere and ultimately fall on organic material, including any that humans wish to preserve. Because soil is granular, the fungi usually occur as ramifying filaments. But on foods that provide solid surfaces, abundant nutrition, and sometimes little competition from other microorganisms, colonization is common. All fungi have preferred conditions for growth with respect to food, moisture, temperature, and acidity. Filamentous fungi are found on foods, but relatively few are found often; among the few are species of *Aspergillus, Penicillium, Rhizopus, Mucor, Absidia, Aureobasidium, Epicoccum, Alternaria, Helminthosporium, Monilia, Botrytis, Scopulariopsis, Cephalosporium,* and *Fusarium.* However, no one genus is encountered in foods under all conditions. The mucors prefer foods stored at moderate temperature and relatively high humidity. Species of *Penicillium* and *Aureobasidium* (also *Hormodendron* and *Cladosporium*) grow commonly on foods under refrigeration. Members of *Thamnidium* spp. grow readily on cold-stored cured meats. Some of the aspergilli, such as *A. niger,* grow in foods with fairly low water content or with water activity (a_w) of 0.65 (see Chapter 3).

Most fungi—with the notable exception of those in the order Mycelia Sterila—are prolific producers of spores. It has been estimated that a colony of *Aspergillus luchuensis* occupying half the diameter of a petri dish will produce 400 million spores within 4 days. Most spores are borne aerially, or on or above the surface of the substrate upon which the fungus is growing.

Yeasts

The true yeasts are unicellular members of the classes Ascomycetes and Fungi Imperfecti. Some fungi, notably mucors and members of the genus *Hormodendron,* if grown in or on foods under peculiar conditions, produce cells indistinguishable from yeasts when examined directly by microscope. But if they are grown in culture their filamentous nature is readily apparent. Several genera of yeasts are classified with the basidiomycetes, but none has been found in foods, and will not be discussed further.

The life cycles of the ascomycetous yeasts manifest three general patterns. **Haploid** yeasts spend the entire life cycle in the haploid state, except for one nuclear generation. At the time of mating, a fusion tube develops between two isogametes, plasmogamy and caryogamy occur, and the nucleus immediately undergoes reduction division to produce the characteristic number of spores. This type of reproduction occurs among members of the genus *Schizosaccharomyces,* which is also a fission yeast.

Diploid yeasts spend most of the vegetative growth period in the diploid state. Shortly after the haploid spore has germinated and its progeny have begun growing, mating, with plasmogamy and caryogamy, occurs. The yeast then normally continues for a prolonged period of budding, until ultimately reduction division spontaneously occurs, and material is formed around each daughter haploid nucleus to produce the ascospore. Two workers, Kruis and Sativa, had noted this mode of reproduction in 1918 but had published their report in an obscure, poorly circulated journal. The process was rediscovered by Winge and Laustsen in 1937. Their paper laid the foundation for the modern science of yeast genetics. It is not unusual to isolate diploid yeasts while the vegetative cell is in the haploid state. Because the vegetative progeny of such yeasts are self-sterile, they usually cannot be identified directly as ascomycetous yeasts. When identification is desired, mating types are sought on the basis of similar or identical physical and physiological properties, and cells of the known and unknown cultures are introduced into the same medium.

In the third general pattern of the ascomycetous life cycle, mating occurs within a closed system. There are many variations, but basically the nuclei of the mother and daughter cells mate and one or more ascospores are formed; or spores budding within the ascus mate with other spores there. Owing to the mother-daughter relationship within the closed ascus, it is impossible to introduce new genetic traits or to eliminate undesired traits, by genetic manipulation. While these yeasts are often encountered, none of those exhibiting these patterns is economically useful.

The class Fungi Imperfecti comprises two groups of yeast genera. Yeasts such as *Trigonopsis* and *Cryptococcus* spp. have physical and physiological characteristics that are not duplicated among the asomycetous yeasts. Others, such as species of *Torulopsis,* are unable to produce ascospores but are identical in all other respects with species of *Saccharomyces,* their counterparts in the class Ascomycetes. Further, some yeasts, such as one or more species in the genus *Candida,* have been kept in the class Fungi Imperfecti because they produce ascospores so rarely as to be laboratory curiosities.

Most yeasts occur normally as spherical, ellipsoidal, or elongated cells. Some yeasts produce a pseudomycelium from cells that elongate after budding; others produce a true mycelium, in which the cells appear to be cylindrical. They are often enumerated and cultivated on media adjusted to relatively low pH to inhibit the bacteria with which they may be associated, and thus they are described, erroneously, as being acidophilic. The ascomycetous, basidiomycetous, and imperfect yeasts compose a very large group of genera. Table 1-3 lists the genera of yeasts known to have species associated with foods. The list is intended to be indicative only, not all-inclusive or exhaustive, since members of other genera may someday be found in association with foods. The descriptive features given in this table are scarcely useful in generic identification. Yeast identification is an independent branch of the microbiological sciences, and interested students should consult manuals such as that by Lodder (1970).

Vegetative Reproduction. Most yeast reproduce vegetatively by bud formation; a few reproduce by fission and a few by blastospore formation. During budding the nucleus undergoes mitotic division, and at the same time a narrow portion of the cell wall softens and cytoplasm flows into the extrusion. One of the daughter nuclei migrates into the developing protuberance. When the bud reaches approximately one-third the size of the

Table 1-3
Genera of yeasts associated with foods.

Genus	Budding	Shape	Metabolism	Spores	Distinguishing feature
Citeromyces	Multipolar	Spher.-ellips.	Fermentative	1–2	Assimilates nitrate; negative 37°C
Debaromyces	Multipolar	Oval-elong.	Oxidative	1–2	May assimilate nitrite
Dekkera	Multipolar	Spher.-oval	Fermentative	1–4	Produces acetic acid
Hansenula	Multipolar	Spher.-oval	Fermentative	1–4	May produce fruity esters; assimilates nitrate
Kluyveromyces	Multipolar	Spher.-elong.	Fermentative	1–4	May produce red pigment
Lodderomyces	Multipolar	Oval-filam.	Weakly ferm.	2	
Pichia	Multipolar	Pseudofilam.	Variable	1–4	Pellicle on liquids
Saccharomyces	Multipolar	Spher.-oval	Fermentative	2–4	
Haniospora	Bipolar	Oval-oblong	Fermentative	2–4	
Saccharomycodes	Bipolar	Elongated	Fermentative	4	Buds across broad base
Schizosaccharo-myces	Fission	Sperh.-cylind.	Fermentative	4–8	May grow at 42°C
Endomycopsis	Blastospore	Pseudofilam.	Oxidative	1–4	May digest starch
Candida	Multipolar	Glob.-pseudo-mycelium	Weakly ferm. or none	None	May grow at 45°C

Genus	Budding	Shape	Metabolism	Spores	Distinguishing feature
Brettanomyces	Multipolar	Ogive dominant	Fermentative	None	Produces acetic acid
Cryptococcus	Multipolar	Spher.-glob.	Oxidative	None	Some produce pigments; digest gelatin
Kloekera	Bipolar	Oval-oblong	Fermentative	None	Buds across broad base
Rhodotorula	Multipolar	Spher.-oval	Oxidative	None	Red-orange pigments; NaCl-tolerant
Torulopsis	Multipolar	Globose	Fermentative	None	NaCl-tolerant
Trichosporon	Budding, arthrospores	Pseudomycelium	Weakly fermentative	None	
Trigonopsis	Budding on apices of tetrahedral cells	Tetrahedral	Oxidative		

mother cell, the cell wall closes. Species of the genera *Saccharo-mycodes* and *Kloeckera* differ slightly from this description in that the buds are produced over a broad base across the ends of elongated cells.

Bipolar yeasts bud only on apices of elongated cells. All other budding yeasts, variously termed multilateral or multipolar, produce buds at random locations on cells of spherical shape, or at the sites of greatest curvature on oval or ellipsoidal cells. Bud scars on the cell surface prevent further budding at these sites. A typical ellipsoidal yeast will produce 20 to 25 daughter cells. Cultures of such yeasts will contain a mixture of biologically old, and often enlarged, cells and biologically young, reproducing cells.

Fission yeasts undergo cell division, during which both the nucleus and the cellular material are equally divided between the two daughter cells. Biologically these yeasts are ever young. The mode of reproduction is reflected in the generic name *Schizosaccharomyces*. Fission yeasts are widespread and actively fermentative, especially in warmer climates. They are commonly found in saccharine materials undergoing spontaneous fermentation with the production of ethanol and carbon dioxide. Daughter cells are known technically as **arthrospores,** cells that are produced at the end of the filament.

Endomycopsis and *Trichosporon* are genera of filamentous yeasts with elongated cells upon which buds, or blastospores, are produced. *Trichosporon* spp. are sometimes considered pathogens, but their presence in foods seems to offer no pathogenic implications. Yeasts of the genus *Candida* are dimorphic; that is, they may occur as ellipsoidal, budding yeasts or they may be strongly filamentous, with buds termed **blastospores** borne on cells along the filament. The cells of *Trigonopsis* spp. are tetrahedral in shape, and buds are produced on the four apices. The cells of species of *Dekkera* and of its imperfect counterpart *Brettanomyces* often are **ogival** (pointed at one end, as a cathedral arch) in shape.

Oxidation and Fermentation. All yeasts draw energy from converting carbohydrates into carbon dioxide and water and, if ammonia is available, into yeast cell substance. Some yeasts, the **oxidative** yeasts, have only this mechanism for using sugars. Some oxidative yeasts oxidize to completion a number of organic acids, such as lactic and acetic. Such oxidation is detrimental to bacterial fermentations of some acids desired for preservation. Such yeasts, often termed **scum,** or pellicle, yeasts, tend to gather at the surfaces of liquids to form loose pellicles, probably because of the obligate demand for oxygen.

Fermentative yeasts decarboxylate pyruvic acid to produce acetaldehyde and carbon dioxide. The acetaldehyde becomes the hydrogen acceptor for the reduced diphosphopyridine nucleotide (which is active in internal hydrogen transport during respiration), and ethanol is produced. If the acetaldehyde becomes unavailable because sodium bisulfite has been added to the fermentation vessel, the yeasts use dihydroxyacetone phosphate as the hydrogen acceptor, and glycerol is produced. Among the yeasts most useful to humans are those of the fermentative genera *Saccharomyces* and *Torulopsis*. They have strong potential for the anaerobic production of ethanol and carbon dioxide and for the aerobic production of culture yeasts and dried yeasts as dietary supplements.

Bacteria

The bacteria and the closely related cyanobacteria constitute the kingdom Monera. All of the monerans are procaryotes. Their cells are bounded by a cell wall and an inner cytoplasmic membrane, and, in some, an outer capsule or slime layer. The cells of most procaryotes are enclosed in rigid, rarely flexible, cell walls composed of peptidoglycans in a three-dimensional linkage. The linkage provides extreme rigidity and extreme resistance to plasmoptysis. The peptidoglycans are composed of a variety of amino acids that are specific for various taxonomic groups and that are useful in determining taxonomic relationships among some groups of bacteria. The cell walls of gram-negative bacteria have a high content of lipopolysaccharide material, which is pyrogenic, or fever-producing; acts as an endotoxin when released upon autolysis of pathogenic bacteria such as shigellae; and has a specific chemical configuration important in antibody formation.

The cyanobacteria, commonly termed the blue-green algae, use water as the electron or hydrogen donor during photosynthesis, whereas the photosynthetic bacteria, which metabolize sulfide, obtain the hydrogen from H_2S. Several species of cyanobacteria produce paralyzing toxins, which concentrate in shellfish.

All bacteria associated with foods are heterotrophic. In their metabolism they make use of the common proteins, carbohydrates, fats, and the other components of foods. Some bacteria oxidize the molecules completely to CO_2 and H_2O and, if the food is an amino acid, liberate the alkalinizing ammonia in the process. Many bacteria ferment carbohydrates, and some ferment amino acids under anaerobic conditions to produce a variety of organic acids and simple, sometimes odorous, molecules. They grow over wide ranges of temperature and acidity.

When conditions permit, the majority of bacteria grow freely in foods, to bring about desirable or undesirable changes in texture, flavor, and odor. Some bacteria grow to very high numbers without producing pronounced organoleptic changes, whereas others bring about detectable changes when the microbial populations reach several million per unit quantity. A few bacteria elaborate toxins. Some bacteria, including some common pathogens, have very minimal nutritive requirements and grow in media containing only the essential minerals, a source of carbon such as a carbohydrate or acetic acid, and ammonia. Other microorganisms, including most fastidious pathogens, can be spread throughout foods and yet not normally grow because of low storage temperature or because preformed organic factors required for growth are missing.

Bacteria rarely reproduce sexually. When they do, only a portion of the bacterial chromosome moves from the donor to the recipient. Such properties of the bacteria as virulence and drug resistance are controlled by functional deoxyribonucleic acid (DNA) sequences on the bacterial chromosome, or by extrachromosomal elements called **episomes,** or **plasmids.** These plasmids may be transferred either in company of fragments of chromosomal material transmitted during genetic transfer, or independently of chromosomal material. Genetic transfer occurs frequently in members of the family Enterobacteriaceae but less frequently in some other bacteria, and then only under very specific conditions; and among still other bacteria the process has not yet been observed.

A phenomenon peculiar to bacteria is made possible by the double-stranded chromosome: temperate **bacteriophages,** viruses that parasitize bacteria, insert themselves into the bacterial chromosome; the bacterial reproduction mechanism then reproduces the bacteriophage. This process involves no evident harm to the host strain. Such bacterial cells are called **lysogenic.** At some future time the virus may leave the nuclear material and cause lysis of susceptible strains of bacteria. The cheese-making industry has suffered the loss of large quantities of milk when *Streptococcus lactis* has been destroyed by lysis. The bacteriophage, while resident, may also give its host unusual properties; the most spectacular example is provided by *Corynebacterium diphtheriae,* which can produce diphtherial toxin only when the cell is in the lysogenetic state.

Viruses

Viruses are the only group of organisms that remain without a clear position in the scheme of things. All viruses are parasites, incapable of independent existence outside their host. They are:

1. submicroscopic, or invisible in the light microscope;
2. composed of either DNA (deoxyribonucleic acid) or RNA (ribonucleic acid), but not both;
3. lacking in respiratory and other enzymes; and
4. reproduced by means of the host machinery.

A small number of viruses enter the body with food and drink, causing disease. By definition, viruses are nonliving; yet they must be considered as virulent agents, and they are cultivated much as are very fastidious bacteria.

Classification of the Bacteria

The present (eighth) edition of *Bergey's Manual of Determinative Bacteriology* (Buchanan and Gibbons, 1974) divides the bacteria into 19 "parts," or groups, which are not always in accordance with the standard assignments to order, family, and genus. Each "part" and its subdivisions are briefly characterized with both a taxonomically useful function and a behavioral role.

Table 1-4 lists those heterotrophic and pathogenic bacteria that are found in foods. Examples of "parts" that are omitted are members of the order Mycoplasmatales, which are pathogenic for poultry (as well as for other vertebrates), and members of the genus *Leptospira*. A number of bacteria are less competitive and for some reason do not thrive in foods. However, virtually any bacterium may be mechanically transferred to some type of food and be isolated from it. Following is a brief discussion of each of the "parts" important in food.

1. Gram-Negative Aerobic Rods and Cocci. All members of this very large group of bacteria are obligately aerobic. Some, such as members of the genera Rhizobiaceae and Azotobacteriaceae, are generally sluggish and may have long generation times. Most of those commonly found in foods are common residents of soil and water, and one is an animal pathogen. They are generally capable of rapid proliferation in foods.

Most of these bacteria oxidize sugars to the sugar acids, which are slightly bitter; but *Gluconobacter* and *Acetobacter* spp. oxidize ethanol to acetic acid. Amino acids are oxidized completely to CO_2 and H_2O, and the alkalinizing ammonia is liberated; the amino acids bearing the sulfhydryl group also liberate H_2S. Many of the bacteria grow freely on simple, unenriched media, and may grow on the simplest of media, composed only of the essential minerals, ammonia, and either a carbohydrate or acetate as the source of carbon. The optimal temperature for all except the pathogens is between 20 and 30°C, although some are capable of growth below 0°C.

The pseudomonads (genus *Pseudomonas*) are among the bacteria most common in soils (see Table 1-5). Most species are low-temperature mesophiles. Contrary to common belief, only a few species produce pigments. Members of the fluorescent group, typified by *Pseudomonas fluorescens*, produce the water-soluble, fluorescent pigment fluorescein. Members of the aeruginosa group, typified by *P. aeruginosa*, produce the soluble blue pigment pyocyanin; it is important to know, however, that a small percent of strains of *P. aeruginosa* do not produce a

Table 1-4
The "parts," families, and genera of bacteria common in foods.

1. Gram-negative aerobic rods and spheres
   Pseudomonadaceae   Halobacteriaceae
   Pseudomonas   *Halobacterium*
   Xanthomonas   *Halococcus*
   Gluconobacter

   Of uncertain position
   Alcaligenes
   Acetobacter
   Brucella

2. Gram-negative facultative anaerobic rods
   Enterobacteriaceae   Vibrionaceae
   Escherichia   *Vibrio*
   Edwardsiella   *Aeromonas*
   Salmonella   *Plesiomonas*
   Shigella   *Photobacterium*
   Proteus   *Lucibacterium*
   Klebsiella
   Enterobacter   Of uncertain position
   Hafnia   *Zymomonas*
   Serratia   *Chromobacterium*
   Erwinia   *Flavobacterium*

3. Gram-negative anaerobic bacteria
   Bacteroides

4. Gram-negative rods, coccobacilli
   Moraxella, Acinetobacter

5. Gram-positive spheres
   Micrococcaceae   Streptococcaceae
   Micrococcus   *Streptococcus*
   Staphylococcus   *Leuconostoc*
                             Pediococcus
                             Aerococcus

6. Gram-positive asporulating, catalase-negative rods
   Lactobacillus

7. Gram-positive endospore-forming rods
   Bacillus
   Clostridium
   Desulfatomaculum

8. Bacteria with branching cells
   Corynebacterium   *Propionibacterium*
   Brevibacterium   *Streptomyces*
   Microbacterium   *Streptoverticillum*
   Kurthia

Source: Adapted from *Bergey's Manual of Determinative Bacteriology*, 8th ed. 1974.

Table 1-5
Genera of bacteria known to occur in foods, according to families.

Family	Genus or genera	Of uncertain position
Pseudomonadaceae	*Pseudomonas* *Xanthomonas* *Gluconobacter*	*Alcaligenes* *Brucella* *Acetobacter*
Halobacteriaceae	*Halobacterium* *Halococcus*	
Enterobacteriaceae	*Escherichia* *Edwardsiella* *Citrobacter* *Salmonella* *Shigella* *Klebsiella* *Enterobacter* *Hafnia* *Serratia* *Proteus* *Yersinia* *Erwinia*	
Vibrionaceae	*Vibrio* *Aeromonas* *Plesiomonas* *Photobacterium* *Lucibacterium*	*Zymomonas* *Flavobacterium*
Bacteroidaceae	*Bacteroides*	*Desulfovibrio*
Neisseriaceae	*Moraxella* *Acinetobacter*	
Micrococcaceae	*Micrococcus* *Staphylococcus*	
Streptococcaceae	*Streptococcus* *Leuconostoc* *Pediococcus* *Aerococcus*	
Baccillaceae	*Bacillus* *Sporolactobacillus* *Clostridium* *Desulfotomaculum*	
Lactobacillaceae	*Lactobacillus*	
Coryneform group	*Corynebacterium* *Kurthia*	*Brevibacterium* *Microbacterium*
Mycobacteriaceae	*Mycobacterium*	
Streptomycetaceae[a]	*Streptomyces*[a]	
Rickettsiceae	*Coxiella*	
Spirochaetaceae	*Leptospira*	

[a]Indirect effects.

pigment. Unlike many pseudomonads, P. *aeruginosa* and P. *fluorescens* can grow at 37°C. Several species of *Pseudomonas* produce insoluble yellow or blue pigments. The remaining species are hyaline or colorless on agar media (see Box 1-1). All pseudomonads except the members of the phytopathogenic syringae group (typified by P. *syringae*), which are often isolated from vegetables, are **oxidase positive.** Only a few other genera of bacteria share this distinctive characteristic. When the dye dimethyl-*p*-phenylenediamine dihydrochloride is added, oxidase-positive bacteria produce a blue pigment instantaneously. Many species of *Pseudomonas* are very actively proteolytic or lipolytic, and some species are pectinolytic, especially at low temperatures. They produce flavors described as stale, oxidized, rancid, and bitter, or as resembling cabbage-water;

Box 1-1

Many gram-negative, nonpigmented, short, rod-shaped bacteria assigned to the genus *Achromobacter* in earlier editions of *Bergey's Manual* were reassigned to the genus *Pseudomonas* in the sixth edition. Brisou (1957) differentiated the genus *Acinetobacter* from other members of the order Pseudomonadaceae (Brisou, 1957) and Steel and Cowan (1964) redescribed it. Unfortunately, Steel and Cowan included a nonmotile pseudomonad, *Pseudomonas pseudomallei*, among the species they assigned to *Acinetobacter* (Redfearn et al., 1966). When the criterion of presence of cytochrome oxidase was adopted for the differentiation between *Moraxella* and *Acinetobacter* (Baumann et al., 1968) nonmotile organisms previously regarded as *Achromobacter* or *Pseudomonas* had to be reclassified. In many of their physiological and biochemical properties, the more nutritionally versatile moraxellas resemble pseudomonads. Morphologically, however, many of the moraxellas are difficult to distinguish from *Neisseria* spp. Further, since the oxidase-negative moraxellas are common inhabitants of soil and water and many use a wide range of carbon sources, they have been described under a bewildering variety of generic names, including *Bacterium, Neisseria, Micrococcus, Diplococcus, Alcaligenes, Achromobacter, Acinetobacter,* and *Pseudomonas*. Since the eighth edition of *Bergey's Manual* has eliminated the genus *Achromobacter,* the status of many of the organisms assigned to that genus is in doubt. Obviously, because most reference cultures are unavailable, these organisms cannot be reclassified at present.

some flavors, and odors as well, suggest old washcloths; some odors suggest boiled cabbage. Foods in which proteolysis and lipolysis proceed simultaneously develop soapy flavors.

Young colonies and cultures of members of the oxidase-negative *Xanthomonas* often have a weak orange-yellow, water-insoluble pigment. The color of the pigment intensifies with age; it is sensitive to the oxidation-reduction indicator 2,3,5-triphenyltetrazolium chloride. Media with the indicator are used for presumptive confirmation of suspected *Xanthomonas*. Although best known as a plant pathogen, it is one of the commonest bacteria on garden leaf lettuce, which has a surface pH of 5.7; on succulent yellow and zucchini squashes, with a surface pH of 5.9; and on asparagus.

Bacteria of the family Halobacteriaceae are halophilic, growing over a range of NaCl from 3.5 percent to saturation. They are found in the oceans and in solar salt. They are agents of spoilage of salted ocean fish that is improperly refrigerated.

2. The Gram-Negative Facultatively Anaerobic Rods. These bacteria thrive in both aerobic and anaerobic conditions. Aerobically they oxidize amino acids. In the absence of oxygen they become fermentative and derive energy by producing organic acids from sugars. Media designed to detect and enumerate coliform bacteria take advantage of the anaerobic properties when applied as an overlay. Nearly all species of bacteria in this group grow freely on simple media and over a wide range of pH and of temperature, from less than 10°C to more than 40°C. When conditions preclude growth they tend to persist in the dormant state for long periods of time. Of the group only members of *Erwinia* spp. are proteolytic or pectinolytic, and none is lipolytic.

Bacteria of the genera *Escherichia, Edwardsiella, Salmonella, Shigella,* and *Proteus* are primarily associated with humans and vertebrate animals (Table 1-6). *Escherichia coli,* the only species of its genus, is termed the **fecal coliform** bacterium. Two species of *Proteus* are generally considered to be saprophytic, and several species cause diarrhea in humans, particularly in children. Bacteria formerly placed in the genus *Providencia* are now known as *Proteus inconstans.*

Table 1-6

Presumptively confirmatory properties of the genera of the Enterobacteriaceae[a].

Genus	Glucose	Lactose	Indole	AMC (VP)	HSH	Motility	Lysine	Ornithine	Alanine	KCN	Citrate	Major residence
Escherichia	AG	AG	+	–	–	+	±	±	–	–	–	Human
Edwardsiella	AG	–	+	–	–	+	+	+	–	–	–	Human
Shigella	A	(A)	±	–	–	–	–	–	–	–	–	Human
Salmonella	AG	–	–	–	+	+	+	+	–	–	–	Human
Arizona	AG	(A)	–	–	+	+	+	+	–	–	+	Human
Citrobacter	AG	AG	±	–	+	+	–	±	–	–	+	Plant
Klebsiella	AG	AG	–	+	–	–	+	–	–	–	+	Human
Enterobacter	AG	AG	–	–	–	+	±	+	–	–	+	Plant (primarily)
Serratia	AG	V	–	+	–	+	+	+	–	–	+	Plant
Proteus	AG	–	–	–	V	+	V	V	+	–	–	Human
Providencia	A/AG	V	–	–	–	+	–	–	+	+	+	Human

[a] Abbreviations: AG: acid and gas; A: acid; (A): may ferment; V: variable; HSH: hydrogen sulfide produced; lysine, ornithine, alanine decarboxylated; KCN: growth in medium containing KCN; citrate utilized as sole source of carbon for energy.

Source: Modified from Edwards and Ewing, 1972.

Citrobacter, Klebsiella, Hafnia, Enterobacter, and *Serratia* spp. are residents of plants and soil. *Klebsiella* and *Enterobacter* spp. are readily established in the intestinal tract and are known as the **nonfecal coliform** bacteria. Members of *Hafnia, Citrobacter,* and *Serratia* are often isolated from natural sources. The isolation of *Klebsiella pneumoniae* from foods and sources in nature has caused some concerns regarding the role of raw plant foods in the spread of *Klebsiella* infections; however, not many of the serotypes of *K. pneumoniae* isolated from these sources have been incriminated in diseases of humans. Although students are told that *Serratia* produces a vivid red pigment, many strains isolated from foods and nature are white or hyaline. *Serratia* is heterofermentative but anaerogenic, since it lacks the enzyme formic acid dehydrogenase, which decarboxylates formic acid to produce CO_2 and H_2O.

Erwinia spp., best known as plant pathogens, are common plant residents and may be the dominant bacteria on some vegetables. Several species in each of the three subgroups of the genus grow at 37°C, and emerge on selective media used to detect coliform bacteria. Members of the *E. carotovora* group produce a small bubble of gas in insert vials; members of the *E. amylovora* group and the *E. herbicola* group produce an abundance of acid but no gas.

The polarly flagellated members of the family **Vibrionaceae** are oxidase-positive and fermentative and thus are easily separable from the pseudomonads and the enterobacteria. A cell cannot be reliably identified by its curvature, because not all cells lie on their sides, and the cells of some species are barely curved. *Vibrio cholerae* and *V. parahaemolyticus* are virulent human pathogens. *V. costicola* is a seldom-seen obligate halophile found in salted meats and brines. *Aeromonas* spp. are heterofermentative, nonpathogenic bacteria resembling *Enterobacter* spp.; they have been isolated from processed luncheon foods, particularly in Europe. *Plesiomonas,* with many synonyms and described as the aerogenic *Shigella,* is an agent of human infectious gastroenteritis. *V. fisheri* and *Photobacterium* and *Lucibacterium* spp. are luminescent because they contain the enzyme luciferinase, which oxidizes the reduced substrate **luciferin,** which produces light. (These bacteria are reputed to have been the source of great consternation in earlier times, when European housewives observed the salted fish in their storage cellars staring at them.)

Flavobacterium, a genus of oxidase-positive, yellow-pigmented, nonfermentative rods, once included both gram-positive and gram-negative members, but the former have been removed from the genus and have not yet been reclassified— nor have typical oxidase-negative flavobacteria. Flavobacteria grow poorly on simple nutrient media, but glucose and yeast extract enhance growth. Despite their nutritional fastidiousness, flavobacteria are common on fresh plant material.

Members of the genus *Zymomonas,* a highly unusual obligate anaerobe, produce catalase. The end products of fermentation are ethanol and CO_2, with traces of lactic acid. In tropical countries *Z. mobilis* cells reputedly replace yeasts during the fermentation of juices, and an ethanol content of 10% has been claimed for fermented juices. *Z. anaerobia* cells fail to reduce some or much of the acetaldehyde, and fruit beverages such as European ciders acquire a harsh, obnoxious flavor.

3. Gram-Negative, Anaerobic, Rod-Shaped Bacteria. Members of the genus *Bacteroides* are among the most numerous bacteria in the human intestinal tract. They have been isolated from meats, milk, and milk products. *Fusobacterium* spp. cause mouth infections and internal abscesses. They are communicated by unsanitary tableware and by kissing. *Desulfovibrio* spp. reduce sulfate to sulfide. In pickling brines the sulfide reacts with iron to form the dark ferrous sulfide and imparts an odor reminiscent of a manure pile.

4. Gram-Negative Rods and Coccobacilli. Bacteria in this group are rods, but some are so short as to look like spheres. Members of two of the four genera are commonly found in foods, but none of these bacteria produces pronounced changes in flavor, texture, or odor. *Moraxella osloensis,* the only saprophyte or nonpathogenic species, is oxidase-positive. *Acinetobacter calcoaceticus,* the only species in the genus, is oxidase-negative. Both species grow in simple broth media with acetate as the sole source of carbohydrate, while the pathogenic species of *Moraxella* require preformed organic factors for growth.

5. Gram-Positive Spherical Bacteria. The obligately aerobic, catalase-positive spheres that require oxygen are placed in the family Micrococcaceae, which includes the genera *Micrococcus* and *Staphylococcus.* The family Streptococcaceae comprises four genera of fermentative bacteria that tolerate oxygen but cannot use it and generally grow better in its absence.

The three species of the genus *Micrococcus* produce red and yellow pigments. They do not ferment sugars but may oxidize them to the sugar acids. The micrococci are weakly acid tolerant. They are found in milk, in refrigerated, low-acid salads, and on aging cured meats. They grow at 10°C but not at 46°C and thus are readily separable from the staphylococci.

Cells of species of *Staphylococcus* may be found singly or in clusters of varying size; the same microscopic preparation may contain them in pairs, in short chains, and in small to large clusters. Strains associated with raw milk often occur in tetrads and once were given the generic names *Gaffkya* and *Tetracoccus.* *Staphylococcus aureus,* the golden staphylococcus, is a virulent human and animal pathogen and many strains are toxigenic. *S. epidermidis* is a common, usually nonpathogenic, resident of the nares and skin. Several other species have been reported, one associated with hogs and another a halophilic species. These are not currently recognized in the classification.

Members of the family Streptococcaceae are catalase-negative, spherical bacteria that occur in chains (Figure 1-10) or in tetrads. Because they lack some or all components of the cytochrome system, they cannot use oxygen, and can make only limited, if any, use of amino acids for energy. All species are considered to be fastidious or difficult to cultivate, since they require many components in culture media. They obtain energy by fermenting sugars, and most species in the family produce an abundance of acid, reducing the pH to some value below 5.0. If the heme precursor is supplied in the form of heated blood, some strains and species produce catalase. Many strains of *Streptococcus faecium* and its yellow biotype are motile. Because some species in the family rapidly produce lactic acid under anaerobic conditions, they are among the most useful for preserving foods and for inhibiting the growth of pathogenic and toxigenic bacteria.

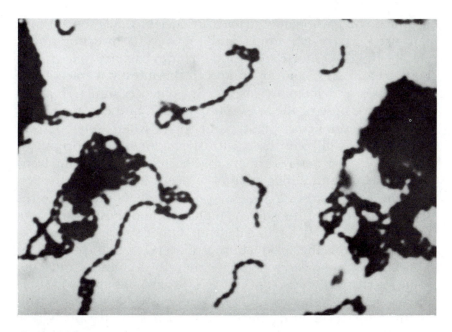

Figure 1-10
Chains of cocci of *Streptococcus faecalis* (× 950).

Species of the genus *Streptococcus* are homofermentative. The cells of most species are lanceolate rather than spherical and occur in short-to-long chains. *S. pyogenes* and related species in the pyogenic subgroup are hemolytic and pus-forming, and cause serious and often fatal diseases in humans and animals. Species in this group do not grow at either 10 or 45°C, whereas those in the remaining three subgroups grow at either or both temperatures.

Streptococcus bovis and *S. equinus* (which are members of the *S. viridans* group) are found in the intestinal tracts of domestic animals. *S. thermophilus*, found on plants in nature, is very useful in cheesemaking when the curd is heated to a high temperature. *S. salivarius* resides in the mouth, and is the cause of chronic infections of the teeth, tonsils, and heart. Species in the subgroup grow at 45°C but not at 10°C.

S. lactis and related species develop naturally in fermenting saccharine materials, and are useful in the dairy industry. They grow at 10°C but not at 45°C. No member of the subgroup is hemolytic or pathogenic; in fact, in ancient times soured milk was used to treat wounds and to prevent infections after childbirth. The **enteric** subgroup includes bacteria that grow at both 10 and 45°C and are found in the intestinal tracts of all vertebrates and in many insects. Streptococci similar but not identical to those of the *S. viridans, S. lactis,* and enteric groups are often resident on plants; and some, when isolated from plants, are readily confused with strains of human origin.

Leuconostoc is a heterofermentative bacterium producing lactic acid and CO_2 and either ethanol or acetic acid during the fermentation of hexoses. *L. mesenteroides* members metabolize the fructose portion of the sucrose molecule, and most strains convert the glucose portion to a dextran. Other species of *Leuconostoc* are useful in the production of diacetyl in milk products.

Aerococcus and *Pediococcus* spp. generally occur as tetrads, but some species of *Pediococcus* (Figure 1-11) grow as chains in some media. Selected species of *Pediococcus* prove useful as starter cultures for fermenting meats. *Aerococcus* spp. are plant residents that grow in meat brines and that cause a serious disease

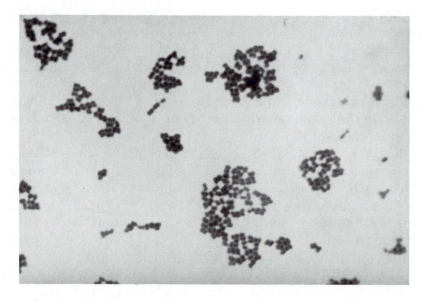

Figure 1-11
Tetrads and short chains of *Pediococcus acidilactici* ($\times 950$).

in lobsters. *P. acidilactici* and *P. pentosaceus* are plant residents that are active during early stages of vegetable fermentations. *P. cerevisiae* is an acidophilic species that tolerates hops in beers and is a cause of souring. *P. halophilus* is a halophile that grows well in 7% NaCl solutions and is commonly found in oriental fermentation of soybeans.

6. Gram-Positive, Asporogenous, Rod-Shaped Bacteria. These bacteria, in the genus *Lactobacillus* of the family Lactobacillaceae, resemble the streptococci in their nutritional requirements. They are catalase-negative and strongly fermentative. Species of the subgenus *Thermobacterium* are homofermentative, and grow best at 37°C and above. Species in the subgenus *Streptobacterium,* also homofermentative, grow well at 20 to 28°C. Species in the subgenus *Betabacterium* also grow well at the lower temperature and are heterofermentative, with end products of fermentation identical with those of *Leuconostoc* spp. The lactobacilli are very useful in vegetable fermentations, fermenting the sugars to exhaustion and thus preventing the growth of fermentative yeasts, and in dairy fermentations.

7. Spore-Forming Bacteria. All bacteria in the family Bacillaceae produce endospores. The endospore is a central or polar body in the cell that contains the essentials of life and is surrounded by a peptidoglycan and an outer spore coat (Figure 1-12 A, B). Some bacteria produce heat-resistant bodies that are not, however, like the bacillary spores.

Spores are exceedingly resistant to factors normally detrimental to the vegetative cell. They may remain dormant in the dried state for years. The natural residence of all spore-forming bacteria is the soil, but some species have adapted quite well to life in the intestinal tract. Sporulating bacteria that demand oxygen or grow best in its presence are placed in the genus *Bacillus.* Bacteria demanding the complete or nearly complete absence of oxygen are found in the genera *Clostridium* and *Desulfatomaculum.*

The 22 species of *Bacillus* are divided into 5 groups, and species of 4 groups are found in foods. Species in the *B. subtilis*

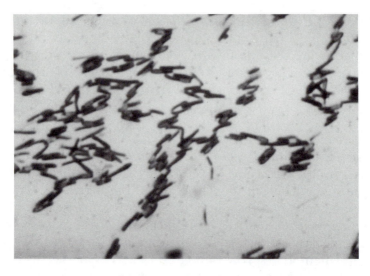

A

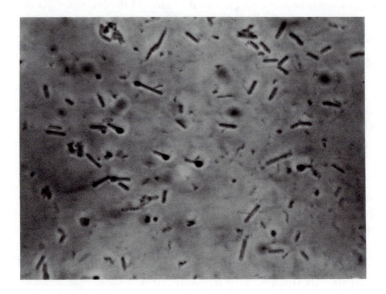

B

Figure 1-12

Members of the family Bacillaceae. Vegetative cells and spores of members of, A, *Bacillus subtilis* (× 950) and of, B, *Clostridium tetani* (× 450).

and *cereus* groups produce cylindrical, nonswollen spores; species in the *B. polymyxa* and *B. sphaericus* groups produce swollen spores. Species in the *B. subtilis* group produce slender spores, not over 0.9 μm in diameter (Figure 1-12 A). Species once known in food microbiology as *B. mesentericus, B. panis,* and *B. vulgatus* are now identified as variants of *B. subtilis.* Some strains of *B. subtilis* produce dark-brown-to-black melaninlike pigments.

Cells of the *Bacillus cereus* group are large, more than 0.9 μm in diameter. *B. cereus* and *B. megaterium* contain lipid granules of poly-β-hydroxybutyrate, which stain with fat stains but remain clear when simple stains such as methylene blue are used. Both species tend to form long, tortuous chains of cells. On agar, *B. cereus* var. *mycoides* grows as a thin, spreading colony with long, curling tendrils.

Under anaerobic conditions, species in the *Bacillus polymyxa* group ferment sugars, sometimes vigorously, and *B. polymyxa,* a fairly heat-resistant bacterium, develops in canned vegetables and in cheese. *B. stearothermophilus* is a thermophile that lacks the enzyme formic acid dehydrogenase and is termed a "flat sour" bacterium. Members of the *sphaericus* subgroup become gram-variable to gram-negative with age.

The genus *Clostridium* can be divided into several groups. The cellulose digesters, the pigment formers, the purine fermenters, and *C. kluyveri* occur in soil or water and appear to have no role in foods; the remaining groups are strongly fermentative or strongly proteolytic, or tend to have mixed characteristics. The butyric group, including *C. butyricum* and *C. acetobutylicum,* readily ferment carbohydrates. The amino acid fermenters are actively proteolytic and use amino acids preferentially to yield fatty acids, ammonia, gas, and a mixture of little-known, sometimes highly repulsive, volatile products.

The genus *Desulfatomaculum* was established recently to accommodate sporulating anaerobes that reduce sulfate to H_2S. Members of the genus occur chiefly in waste waters and effluents containing sulfate. One species, *D. nigrificans,* formerly *C. nigrificans,* is a thermophile that has been responsible for enormous economic loss through destruction of canned peas and corn.

8. Bacteria with Branching Cells. Whereas members of some genera produce branching cells, other bacteria do so, in culture or *in vivo,* only in certain conditions. *Corynebacterium diphtheriae* and the related human and animal pathogens produce club-shaped, banded, and granulated cells. No members of this group are encountered in foods. The plant-resident members of *Corynebacterium* are simple, unbranched, bacillary, slender cells 2 to 3 μm in length; they stain intensely blue with the Gram stain, and on slides form V- or Y-shaped or palisade arrangements. On nutrient agar they grow slowly, forming only small colonies. The colonies range in color from white through cream, from pale to deep yellow, salmon, pink, light to deep red, or orange. These bacteria are usually isolated from raw and frozen vegetables.

Brevibacterium and *Microbacterium* spp. form slender, short cells, sometimes less than 1 μm in length. They are frequently associated with milk and milk products, with batters, and with foods generally low in acid. *Brevibacterium linens* is associated with cheese of the limburger type. *Microbacterium* spp. are among the more heat-resistant asporulating bacteria known, and a laboratory test to confirm identification is to heat to 70°C for 15 minutes. *Kurthia zopfii* is a nonfermentative long rod occurring frequently in chains. Each of the bacteria grows to very high numbers in foods without noticeably altering texture, flavor, or color.

Propionibacterium spp., which are frequently branched, are seldom seen unless cultivated on highly nutritive media under anaerobic to microaerophilic conditions. The bacteria produce the propionic acid that gives cheeses of the Swiss type their characteristic flavor, and produce the CO_2 that gives the cheese its holes, or eyes. Pigmented species of *Propionibacterium* are a nuisance, producing yellow and dull red colors.

Streptomyces, which may branch irregularly, and *Streptoverticillium,* with sporophores in whorls, are not known to grow in foods. They produce volatile oils that are acquired by foods with a sensitive flavor and impart the odor and flavor of soil.

References

Allsopp, A. 1969. New Phytol. *68*, 591.

Barnett, H. L., and B. B. Hunter. 1972. *Illustrated Genera of Imperfect Fungi*. 3rd ed. Burgess Publishing Company, Minneapolis, Minn.

Barron, G. L. 1968. *The genera of Hyphomycetes from soil*. Williams and Wilkins Co., Baltimore, Md.

Baumann, P., M. Douderoff, and R. Y. Stanier. 1968. J. Bacteriol. *95*, 8; ibid. *95*,20.

Brisou, J. 1957. Ann. Inst. Pasteur *93*, 397.

Buchanan, R. E., and N. E. Gibbons, eds. 1974. *Bergey's Manual of Determinative Bacteriology*. 8th ed. Williams and Wilkins Co., Baltimore, Md.

Christensen, C. M. 1975. *Molds, Mushrooms, and Mycotoxins*. University of Minnesota Press, Minneapolis, Minn.

Edwards, P. R., and W. H. Ewing. 1972. *Identification of Enterobacteriaceae*. 3rd ed. Burgess Publishing Company, Minneapolis, Minn.

Gilman, J. C. 1957. *A Manual of Soil Fungi*. 2nd ed. Iowa State University Press, Ames, Iowa.

Gray, W. D. 1959. *The Relation of Fungi to Human Affairs*. Holt and Company, New York, N.Y.

Haeckel, E. H. 1894. *Systematische Phylogenie der Protisteu und Pflanzea*. I. G. Reimer, Berlin.

Lodder, J. 1970. *The Yeasts. A Taxonomic Study*. North-Holland Publishing Company, Amsterdam.

Phaff, H. J., M. W. Miller, and E. M. Mrak. 1966. *The Life of Yeasts*. Harvard University Press, Cambridge, Mass.

Raper, K. B., and D. I. Fennell. *The Genus Aspergillus*. Williams and Wilkins Co., Baltimore, Md.

Raper, K. B., and C. Thom. 1949. *A Manual of the Pencillia*. Williams and Wilkins Co., Baltimore, Md.

Redfearn, M. S., N. J. Palleroni, and R. Y. Stanier. 1966. J. Gen. Microbiol. *43*, 293.

Skerman, V. B. D. 1967. *A Guide to the Identification of the Genera of Bacteria*. 2nd ed. Williams and Wilkins Co., Baltimore, Md.

Stanier, R. Y., and C. B. van Niel, 1962. Arch. Mikrobiol. *42*, 17.

Steel, K. J., and S. T. Cowan, 1964. Ann. Inst. Pasteur *106*, 479.

Taylor, M. M., and R. Storck. 1964. Proc. Nat. Acad. Sci. *52*, *958*.

Winge, O., and O. Laustsen. 1937. Compt. rend. trav. lab. Carlsberg, Ser. physiol. *22*, 99.

2

Prevention of Food Spoilage

Spoilage

Food spoilage is any organoleptic change—that is, any tactile, visual, olfactory, or flavor change—that the consumer considers to be an unacceptable departure from the normal state. Spoilage can be microbial or mechanical in origin. Table 2-1 summarizes the visual and tactile defects of microbial origin; Table 2-2 summarizes textural and other organoleptic changes. Consumers reject foods that have suffered mechanical damage such as bruising, crushing, and abrasions, or that have absorbed the flavors of other foods. For example, meats may absorb the aroma of apples when stored in the same facility, or eggs left in contact with the ground may absorb the oil geosmin, produced by streptomycetes. Also, inherent pectinolytic and browning enzymes induce textural and color changes in stored fruits and vegetables, and inherent or catheptic enzymes induce textural changes in stored meats.

Table 2-1

Common visual and tactile food defects of microbial origin.

Defect	Nature or cause	General description
Color defects, bacterial	Pseudomonas spp.	Fluorescent yellow-green; pyocyanin blue; red or vivid green, as egg albumin
	Variants of Bacillus subtilis	Blackening from melanin formation
	Microbacterium thermosphactum	Surface darkening of raw meats, especially beef
	Propionibacterium spp.; Brevibacterium spp.; micrococci; flavobacteria	Red, yellow, brown discolorations of surfaces of meat, cheese, butter
	Lactobacillus viridescens	Microaerophilic greening of cooked and cured meats
Yeasts	Nonpigmented yeasts	White colonies on cured meats and luncheon meats; dried fruits, olives
	Pigmented yeasts; Rhodotorula spp.	Red-to-pink discolorations usually of protein, marine foods, oysters, over-salted cabbage in fermentation
Fungi	Nearly all fungi	Wide range of water-soluble and insoluble pigments, determined by species of mold
Sliminess	"Lardy" slime	Massive growth of bacteria and yeasts on surfaces
	"Mushy" slime	Pectinolysis of vegetables
	Viscous slime	Dextran formation by Leuconostoc mesenteroides; variants of Saccharomyces lactis

Table 2-2

Representative types of microbial spoilage of foods at low temperatures.

Type of spoilage	Microorganisms responsible	Microbial action involved
Bitterness	Pseudomonas spp.; Bacillus spp.; fungi	Hydrolysis of proteins; fungal oxidative degradation of fatty acids: aldonic acid formation from carbohydrates
Putrescence	Clostridia; fungi; gram-negative rods	Incomplete metabolism of amino acids
Alkalinization	Pseudomonas spp.; molds; clostridia	Formation of ammonia and alkanes from amino acids, in meats, milk, molded canned fruits

continued next page

Table 2-2, *continued*

Type of spoilage	Microorganisms responsible	Microbial action involved
Sulfide spoilage	Clostridia; *Proteus* spp.; sometimes members of the family Enterobacteriaceae	Liberation of H_2S from amino acids; reduction of sulfate by clostridia, *Desulfovibrio* spp.
Lipolytic rancidity	*Pseudomonas* spp.; fungi *Achromobacter* spp.; *Acinetobacter* spp.	Hydrolysis of triglycerides
Soapiness	*Pseudomonas* spp.; other proteolytic and lipolytic microorganisms	Reaction between freed ammonia and free fatty acids
Oxidative rancidity	Fungi, primarily	Incomplete metabolism of fatty acids, fragmenting to aldehydes, ketones, carboxyl groups
Pectinolysis	*Erwinia* spp.; sometimes clostridia	Digestion of pectins
Souring, lactic	*Streptococcus* spp.; *Lactobacillus* spp.; *Pediococcus* spp.; *Microbacterium* spp.	Homofermentation of sugars
Souring, gassy lactic	*Leuconostoc* spp.; heterofermentative *Lactobacillus* spp.	Heterofermentation of sugars yielding lactic and acetic acids, ethanol, and CO_2
Souring, butyric	Clostridia	Fermenation of sugars; may be associated with amino acid metabolism
Souring, acetic	*Acetobacter* spp.; *Gluconobacter* spp.	Oxidation of ethanol to acetic acid
"Unclean" flavor	Coliform bacteria; particularly *Escherichia coli*	Formation of indole and skatole from tryptophan
Souring of meat	Clostridia	Anaerobic metabolism of amino acids; may be coupled with butyric fermentation of sugars
Loss in flavor	*Pseudomonas* spp.	Destruction of flavor component such as diacetyl
Neutralization in flavor	Streptococci	Acid formation neutralizing effect of flavor component
Staling effects	Many bacteria	During induction period, before specific flavor changes become pronounced
Ethereal odors	Yeasts; *Pseudomonas fragi*	Ester formation by yeasts; strawberry aroma associated with *P. fragi*
Earthiness	Streptomycetes	Absorption of oil, geosmin, by foods
"Potato" aroma	*Pseudomonas taetrolens*	Growth of pseudomonads on lamb carcasses

Predictable and Unpredictable Spoilage

The kind of spoilage to be expected can often be predicted from the composition of the food. Hydrolysis of proteins, fats, and polysaccharides can cause changes in texture. Incomplete metabolism of the amino acids and fatty acids and fermentations of the simple sugars can cause changes in flavor. Foods of mixed composition frequently undergo several simultaneous changes in odor, flavor, and texture. Fruits and juices often exhibit a succession of changes, beginning with ethanolic fermentation and ending in acetification.

Unpredictable types of spoilage are changes that cannot be anticipated from knowledge of the composition of the food alone. The development of potato aroma on lamb carcasses by *Pseudomonas taetrolens* is one such example (Tompkin and Shaparis, 1972).

Changes in Flavor. Microbially induced changes in flavor are seldom abrupt. Usually, several successive steps are observed. Flavor can be lost through destruction of a flavor component, as when *Pseudomonas fragi* converts diacetyl to acetylmethylcarbinol in cottage cheese. Streptococci can produce minute amounts of lactic acid in chocolate milk. A resulting reduction of only 0.2 pH unit is sufficient to nullify the rich vanilla flavor.

Formation of components not normally present in a food can cause a gain in flavor. Consumers can readily notice the formation of lactic and acetic acids and ethanol in foods, as well as occurrences of lipolytic and oxidative rancidity. Less readily identifiable is the "unclean" (a euphemism for fecal) flavor, sometimes associated with milk and salt-rising breads, resulting when *Escherichia coli* produces indole and skatole from tryptophan. The actions of microorganisms on hydrolytic products of proteins and fats produce a complex array of substances that only now modern laboratory instrumentation is beginning to identify. Recently identified, for example, are the volatile substances produced by several species of *Pseudomonas* and *Achromobacter* (the problematic status of which was described in Box 1-1) that act in sterile fish muscle as methyl mercaptans and methyl mono-, di-, and tri-sulfides, and possibly butanone and the ethyl esters of acetic, butanoic, and hexanoic acids (Miller et al., 1973).

Flavor changes are not always associated with grossly proteolytic, fermentative, or lipolytic changes. Milk can acquire a
noticeably malty flavor from as little as 0.5 ppm 3-meth-
ylbutanal produced by *Streptococcus lactis* var. *maltigenes* (Jackson and Morgan, 1954). And a wide variety of metabolic products of microbial respiration give rise to a staling effect—that
is, a subtle loss in fresh flavor before definite products of microbial activity become identifiable by taste.

Estimating Microbial Loads. Generally microbial populations are estimated by conventional plating procedures, but dye reduction
tests and direct microscopic counts have also been used. The
operating principle of a **dye reduction** test is the change in
color of a reducible dye used as a hydrogen acceptor in lieu of
oxygen. The dairy industry once made much use of both
methylene blue and resazurin to test the quality of raw milk.
However, dyes are of limited use in determining the microbiological quality of foods, because microorganisms vary
greatly in their ability to reduce the dyes to leuco bases. The
bacteria most often found are slow to reduce the oxidation-
reduction potential, and changes in color are observed only if
large numbers of microorganisms are present or accumulate
through growth.

A direct microscopic count is a quantitative procedure: a
specified volume of food suspension is spread over a known
area of slide, is stained, and is then examined with a calibrated
microscope. The method has had few valuable uses, one being
the examination of fluid egg products. But it has no value when
applied to frozen fruits and vegetables, because particulate
materials resemble bacteria and cannot be accurately distinguished from them. Moreover, most vegetables possess a high
natural population prior to blanching, and many of the bacteria
are dislodged from the vegetable after freezing.

New Methodology. Several methods have been developed for identifying and enumerating bacteria:

A number of commercial kits incorporate multiple
minibiochemical testing procedures for identifying
microorganisms—in particular, members of the family Enterobacteriaceae.

The fluorescent antibody technique allows rapid presumptive identification of Enterobacteriaceae in foods. Fluorescein renders the antibody fluorescent. The tagged antibody is concentrated on the surface of cells that contain the complementary antigen. Thus, when the antibody is applied to a suspension of bacteria on a slide and the slide is examined under a fluorescent microscope, reacting cells are readily distinguished from nonfluorescing cells.

Bioluminescence is an indirect measure of adenosine triphosphate (ATP), a common component of microbial cells and the source of energy in the light-emitting luciferin-luciferinase reaction.

Several other, instrumental procedures have been developed to detect, enumerate, and identify bacteria (Goldschmidt and Fung, 1978):

Radiometry measures the rate of release of ^{14}C in CO_2 while bacteria are metabolizing specified organic molecules.

Microcalorimetry measures the heat released by bacteria during growth; it is also used to determine sterility of liquids.

The Coulter counter measures the changes in electrical conductivity in relation to the concentrations of microorganisms in suspension. Because different organisms have different resistances to electric current, several sophisticated systems have been developed to measure changes in **impedance,** which indicate increases or decreases in reaction.

Laser beam spectrophotometry measures the intensity of light that bacteria in suspension scatter at different angles. Specific species of bacteria have their own characteristic patterns of scattering.

Indicator Bacteria

Indicator bacteria are bacteria that most commonly occur if handling or processing is done unsanitarily, and thus their presence indicates such conditions. Coliform bacteria have been the most common indicator, although more recently enterococci have also been looked for. In theory, cooked foods should not contain staphylococci, coliform bacteria, or enterococci. But many foods are so easily contaminated that the theory cannot become practice.

In heated foods an abundance of staphylococci—that is, counts significantly over 100/g or /ml—signifies negligent personal habits, sanitation, or temperature control. Their significance lies not in the numbers imparted—which usually are low in relation to the total mass of the contaminated food—but in their potential for growth above 20°C to produce toxins. Like many of the animal microflora, they thrive in cooked meats, milk, and eggs and in foods made with these. They are killed at 85°C. *Streptococcus salivarius,* also a resident of the human mouth, dies so rapidly that it is a useful indicator of contamination only if worker hygiene lapses shortly (< 24 hours) before the test is made (Thatcher and Clark, 1968).

The analytical techniques for detecting and enumerating coliform bacteria are borrowed from practices in water and clinical microbiology. But only the techniques and interpretations for detecting the fecal coliform bacterium are valid. All others are only as effective as the food and the habitats and characteristics of associated bacteria allow.

Contamination by the fecal coliform bacterium *Escherichia coli* can be either direct (that is, from person to food) or indirect (from person or polluted water to environment, then to food). Direct contamination is rare in commercial processing of foods other than raw meats; indirect contamination appears to be the more common route. Direct contamination may result in no more than 25 mg fecal matter per 10 kg of food—a maximum density of 100 cells of any member of the family Enterobacteriaceae (Buttiaux and Mossel, 1961). *E. coli* may be considered a captive bacterium: it thrives only in the human and animal intestinal tracts. In nature it does not survive long or multiply well, and it is seldom airborne. It does, however, proliferate exceedingly rapidly on warm cooked poultry meats and in juices during processing. In certain settings it is an excellent indicator of personal sanitary practices.

The nonfecal coliform bacteria are members of the genera *Enterobacter, Klebsiella,* and *Citrobacter.* They grow freely in nature, particularly in the nectaries of flowers and on fruits such as bananas and peaches (Barber, 1955) and strawberries (Mundt et al., 1954). They travel freely by air from raw produce to post-heating processing areas and are natural residents, sometimes in high numbers, on raw seafoods. The controversy raging over how to interpret the presence of nonfecal coliform bacteria on seafoods caused Hunter (1939) to state a fundamental guideline and truism:

In the sanitary control of food production and handling there is, of necessity, considerable groping for bacteriological indices of pollution and potential danger to health. . . . A knowledge of the food product which permits an explanation of the sources of contamination is fully as important as a knowledge of the identity of the coliform organisms detected in it. This principle is elementary. . . . It is not unusual to encounter misconceptions and differences of opinion relating to the applicability of coliform bacteria as indices of pollution of food.

Today it is thought that significantly high populations of non-fecal coliform bacteria reflect faulty manufacturing practices.

Coliform bacteria are detectable in waters either by gas formation, if the liquid medium contains lactose, or by the formation of characteristic colonies upon selective agars. However, false presumptive tests are possible with liquid media, because foods contain sugars other than lactose that are fermented by other bacteria, which produce coliformlike colonies on the selective media. Lactose-fermenting yeasts also produce positive presumptive reactions in liquid media (Martinez and Appleman, 1949).

After Burton's report (1949) on the ubiquity of the enterococci in frozen vegetables, they were frequently proposed as indices of pollution of frozen and later other nonsterile foods. Before then they had been known only as residents of human and animal intestinal tracts. It then became known that the enterococci are widespread in nature and occur as plant residents during the warm months. They overwinter in insects and poikilothermic animals, and insects would seem to be the prominent vector (Martin and Mundt, 1972). Most of the identifiable homolactic spherical bacteria found on plants are members of *Streptococcus faecalis*, *S. faecium*, and *S. faecium* biot. *casseliflavus* (Mundt et al., 1966; Mundt and Graham, 1968).

Strains of *Streptococcus faecalis* found in plants differ little from those found in humans and animals. This organism produces a soft curd in milk that is digested in stratiform fashion. Few members of *S. faecium* conform to the classical description of the species, which is based upon studies of strains of human and animal origins. *S. bovis* and *S. equinus* are rarely isolated from plants or fresh and processed foods.

More than 30 variant groups of streptococci are known. Some superficially resemble the enterococci, some resemble *S. lactis,* and some remotely resemble *S. bovis* or *S. equinus.* Few of the streptococci impart a reaction in litmus milk. Most ferment sugars rapidly (within 48 hours) to pH 4.1 to 4.5 and grow in media containing sodium azide. For convenience the many groups have been condensed into seven plant streptococcal types (Mundt, 1976). None resembles the pyogenic or viridans groups of streptococci.

The enterococcal (streptococcal) types are found commonly in frozen and dried foods. Because many plant streptococci reactions on media are characteristic of the enterococci, using abbreviated procedures in presumptive detection and enumeration can arouse unwarranted concern about sanitation and safety. The enterococci and other streptococci in most foods are now generally thought to be established by environmental contamination (Virgilio et al., 1970), occurring far from the processing source (de Figueiredo, 1970).

Microbiological Standards

Microbiological standards are specified maximum microbial counts allowable on nonsterile foods. The standards specify either total viable populations or the tolerated numbers of specified groups, such as the coliform bacteria and the staphylococci, per unit quantity of food. The standards establish levels that are attainable with good manufacturing practice, and without undue hardship.

Microbial numbers are estimated by the use of solid or fluid culture media. The results are reported as numbers per gram of product when solid media are used, or as the most probable number (MPN) when multiple tubes over a series of dilutions are used. In discussing formulation of standards and handling codes Elliott and Michener (1961) proposed a set of eight statements that have since served as federal guidelines.

1. *A single set of microbial standards should not be applied to foods as a miscellaneous group such as "frozen foods" or "precooked foods."*

2. *Microbiological standards should be applied first to the more hazardous types of foods on an individual basis, after sufficient data are accumulated on expected bacterial levels, with consideration of variations in composition, processing procedures, and time of frozen storage.*

3. *When standards are chosen, there should be a definite relation between the standard and the hazard against which it is meant to protect the public.*

4. *Methods of sampling and analysis should be carefully studied for reliability and reproductibility among laboratories and chosen methods should be specified in detail as part of the standard.*

5. *Tolerance should be included in the standard to account for inaccuracies of sampling and analysis.*

6. *At first, the standard should be applied on a tentative basis to allow for voluntary compliance before becoming a strictly enforced regulation.*

7. *Microbiological standards will be expensive to enforce.*

8. *If standards are unwisely chosen, they will not stand in courts of law.*

Good Manufacturing Practices. Food processing plants are considered to be in conformity with good manufacturing practices if their products have low microbial content in comparison with products of similar plants that are considered marginal. Microbial quality may be assessed not only by total plate counts, but also by the numbers of bacteria of specific interest, such as the coliform bacteria and the staphylococci (Surkiewicz et al., 1973). Standards of good manufacturing practices may be applied during processing, when microbial counts may be high, but before final heat treatments destroy the bacteria (Surkiewicz et al., 1967). Surveys of a certain number of representative plants within an industry indicate what can be achieved by good management (Leininger et al., 1971) and establish a goal that marginally operated plants must, or should, strive to meet.

Refrigeration

Whether temperature is the most *important variable in the environment of living things some may debate, but none will doubt the cardinal role that temperature plays in the vital processes of living things. Man has learned to modify the normal temperature extremes of his environment, and in all warm-blooded animals (homiotherms) there has evolved a highly intricate mechanism for the control of body temperature within amazingly narrow limits. However, the vast majority of living things, including all lower animals and the whole plant kingdom as well as the entire array of microorganisms, can deviate little from the ambient temperature of the environment. (Cooney and Emerson, 1964)*

Probably throughout human history lowered temperatures have been used to prolong the acceptability of perishable foods to the consumer. Temperatures lower than ambient are found in natural recesses of earth, dugouts, root cellars, and basements; and refrigerating temperatures occur seasonally in the temperate and arctic regions of the earth. Any reduction in temperature normally reduces the activity of the self-contained enzymes in raw plant and animal foods, and the microbial activity in both raw and heated foods. Although lowering the temperature to 12.8°C would seem to have little effect in prolonging the acceptability of foods (Plateniu, 1939), decreases below this temperature have noticeable effect.

Determinants of Shelf Life

Refrigeration, however, is only one of many factors affecting the period of acceptability, or shelf life, of foods. It is affected also by the stage of maturity of fruits and vegetables at harvest, the condition of the animal at slaughter, the rapidity of temperature reduction, the cooling, mechanical cleaning, and sorting methods used, and the use of inhibitors and germicides to protect against contamination or recontamination.

Lower Temperature Limits for Growth. For convenience, micro-organisms are classified according to their preferred temperatures and those that limit growth (Table 2-3). The fastidious, high-temperature mesophiles and the thermophiles play no role in the low-temperature spoilage of foods. The **psychrophiles** are defined as true cold-loving microorganisms. These bacteria grow well at 0°C within two weeks (Ingraham and Stokes, 1959). Both in nature and in foods, true psychrophiles are rare; yet some microorganisms seem to effect a downward shift in temperature tolerance that causes them to mimic true psychrophiles.

Table 2-3
Classification of microorganisms according to temperature characteristics for growth.

Temperature group	Representative microorganisms	Temperature ranges for growth, °C		
		Minimum	Optimum range	Maximum
Psychrophiles		Freezing or below	5–15	±20
Low-temperature mesophiles, psychrotrophic and psychroduric microorganisms	Polarly flagellated gram-negative rods; gram-positive rods; cocci; most yeasts and molds; most lactic acid bacteria; bacilli; clostridia; streptomycetes	±10–+8	20–27	32–43
Nonfastidious high-temperature mesophiles	Members of the family Enterobacteriaceae; *Aeromonas* spp., staphylococci; some lactic acid bacteria	±8	35–43	43–45
Fastidious high-temperature mesophiles	*Brucella; Bacteroides* spp.; gram-positive anaerobic pathogens	20–25	37	?
Preferential thermophiles	*Bacillus coagulans; Lactobacillus bulgaricus;* few fungi	25–28	45–55	60–65
Obligate thermophiles	*Bacillus stearo-thermophilus;* thermophilic clostridia	37	50–60	60–75

A great variety of the low-temperature mesophiles grow at 5°C and less. Some grow exceedingly well—those that mimic psychrophiles. Two terms are commonly used: **psychrotrophic** organisms grow well at low temperatures but do not require them for growth; **psychroduric** organisms prefer moderate storage conditions but are able to persist at low temperature. At present, no psychrotrophic viruses are known.

Minimum Temperatures for Initiation of Growth. The precise minimum temperatures at which microbial growth is initiated are easily determined in the laboratory (Ingraham, 1969). For example, the freely growing, high-temperature mesophiles, the bacteria normally associated with warm-blooded vertebrate animals, are customarily said to require a minimum temperature of 8°C before initiating growth. However, species and strains within species differ in their minimum temperature requirements, and are further influenced by nutrient components, salts, other additives, water content, and the acidity of foods (see Chapter 22). Both cold-fitter and cold-sensitive mutant strains of bacterial species occur. **Cold-fitter** strains adapt to more rapid low-temperature growth than their parents while becoming less able to grow at high temperatures. Yeasts that adapt to growth in grape juice stored at 0°C may be cold-fitter mutants (Pedersen, 1959). At 20°C **cold-sensitive** strains cannot initiate DNA replication (Waskell and Glaser, 1974), or regenerate various amino acids, use various energy substrates, or synthesize functional ribosomes. It is generally agreed that the absolute minimum temperature for bacteria, yeasts, and molds is around −10°C.

In any genus some species and strains are low-temperature psychrotrophs, others are high-temperature psychrotrophs. *Pseudomonas aeruginosa* can grow at no lower than approximately 10°C, whereas the fluorescent pseudomonads grow well at 0°C and below. The minimum temperature among a collection of strains of *Aeromonas hydrophila, A. shigelloides,* and *A. salmonicida* varies between 0 and 15°C (Rouf and Rigney, 1971).

Every food or food type has properties that provide an environment favorable or unfavorable to growth of specific microorganisms. Thus, although apples and citrus fruits are both highly acidic, *Penicillium expansum* is the exclusive molding agent of apples, whereas *P. digitatum* and *P. italicum* are prominent spoilers of citrus fruits.

The presence or absence of associated organisms influences the nature of spoilage. *Bacillus cereus,* spores of which are activated to germinate rapidly by the heat of pasteurization, producing sweet curdling of milk and the formation of "bitty cream," is normally inhibited through acid formation by *Streptococcus lactis.* The surface salt content of dry-cured bacon, ham, and jowls favors the development of *Micrococcus* spp. In a processing plant heavily contaminated with *Bacillus mesentericus* the bacterium grew, producing slime and putrefraction on dressed ducks (Mallmann, 1932).

Refrigerated storage temperatures prevent the growth of bacteria and favor the growth of proteolytic fungi such as *Cladosporium, Cephalosporium,* and *Penicillium* on shell eggs. Treating dressed poultry with chlortetracyclene antibiotics eliminates bacterial spoilage but favors development of the yeast *Trichosporon.* The anaerobic conditions within cartons of refrigerated dough products favor the development of species of *Lactobacillus, Leuconostoc,* and streptococci, although species of *Micrococcus, Bacillus,* and gram-negative rods have been seen in some products (Hesseltine et al., 1969).

Growth along the Temperature Gradient. When plotted semilogarithmically, the generation time over a great portion of the temperature gradient increases or decreases at a rate suggesting a first-order reaction (Figures 2-1 and 2-2). When bacteria are exposed to a higher or lower temperature along this portion of the gradient, they immediately adopt the new growth rate, without lag or interruption. At some point along the gradient, the generation time becomes longer than would be expected at a given lower temperature. For the pseudomonads illustrated in Figure 2-1, the rate changes at about 12°C, confirming the observation made earlier, although much more crudely, by Plateniu. Ingraham (1961) reviews evidence that at some point on the gradient of decreasing temperature, cell damage occurs

58

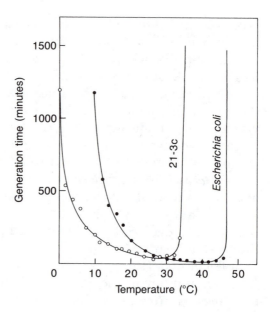

Figure 2-1
The effect of temperature
on generation time of a
psychrophilic
pseudomonad (21-3c) and
the mesophile *Escherichia
coli,* K-12 (Ingraham,
1958).

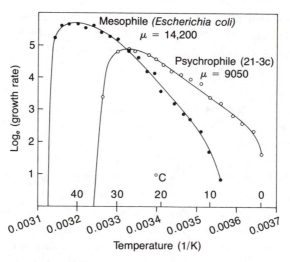

Figure 2-2
An Arrhenius plot of growth rate of a
psychrophilic pseudomonad (21-3c) as compared
with that of the mesophile *Escherichia coli,* K-12
(Ingraham, 1958).

and lengthens the generation time. At the minimum temperature growth ceases abruptly rather than asymptotically (Ingraham, 1958).

Several implications regarding temperature control become obvious to the producer, processor, vendor, and purchaser of perishable foods. First, on or in warm foods placed in refrigeration, microorganisms continue to proliferate at the successive rates dictated by the gradient. Thus, the rate of cooling directly determines the final populations and content of enzymes—a fact of extreme importance in the handling of fluid, semifluid, and plastic foods of high mass and a low rate of heat transfer. Temperature decreases inward from the surface of a container in proportion to the container's surface area and volume, a condition described as the "geometry of the mass" (Dickerson and Read, 1973).

The second implication is that, because each degree of temperature decrease at the lower end of the gradient markedly increases generation time, all cooked and pasteurized foods, raw and processed meats, milk and its fluid products, eggs, and some fruits and vegetables should be refrigerated at the lowest practical temperatures.

Achromobacter spp., Pseudomonads, and Enzymes

The nonproteolytic *Pseudomonas putida, Achromobacter* spp., and the proteolytic *P. fluorescens* are the dominant agents of spoilage of many heated and moist foods such as milks, proteins, gravies, and meat pies. All these bacteria are obligately aerobic: oxidation of the aldehyde moiety of the carbohydrate is the source of acidic reactions in aerobically incubated carbohydrate media.

The fluorescent group *Pseudomonas fluorescens* has been divided into seven subgroups (Stanier et al., 1966) that have absorbed many older species epithets, such as *graveolans, mucidolens,* and *putrefaciens.* All pseudomonads occurring in foods are oxidase-positive and highly motile. Despite the name, members of the fluorescent group seldom produce fluorescense in foods or culture media. *Achromobacter* spp. are oxidase-negative and nonmotile. Most strains, if incubated at 5°C, exhibit moderate

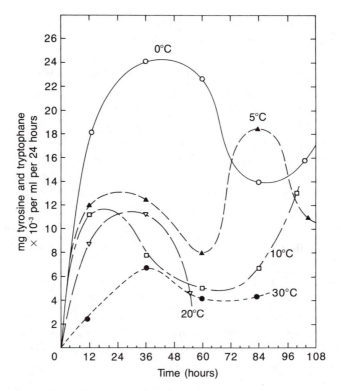

Figure 2-3
Elaboration of extracellular proteolytic enzymes by
Pseudomonas fluorscens when grown in TGE broth at various
temperatures (Peterson, 1962).

to copious growth on culture media within five to seven days.
Many strains initiate rapid growth at 0°C.

When strains of *Pseudomonas fluorescens* begin growth, they
elaborate protease, the maximum amount being elaborated
within 36 hours at 0°C (Figure 2-3) and less at higher temperatures of incubation. It is thought that a large quantity of protease is elaborated at 0°C to compensate for the slow rate of
enzymatic activity, thereby ensuring adequate nutrition for the
bacterium. The protease of *P. fragi* is excreted during the late
exponential phase of growth, and the enzyme is degraded dur-

ing the stationary phase (Tarrant et al., 1973). The optimum pH for elaboration of proteases is 6.5 to 7.0, with exceedingly slow growth at pH 5.5–5.6, the ultimate pH of red meat carcasses. For this reason, various authors have suggested that carcass surfaces be acidified to further retard the proteolytic bacteria (Elliott and Michener, 1965).

Pseudomonas fragi requires a mixture of 11 amino acids and 2 dipeptides for growth (Tarrant et al., 1973). Red meat carcasses have a paucity of free amino acids, and proteolytic pseudomonads do not grow until these are supplied by the animal's autolytic enzymes. Thereafter, the bacteria elaborate their own enzymes and become independent of the autolytic enzymes. The proteolytic enzymes are sensitive to heat; those of *P. fragi* are inactivated within five minutes at 40°C (Elliott and Michener, 1965).

Lipase is elaborated by *Pseudomonas fragi* in milk only at and below 15°C (Nashif and Nelson, 1953). Less than 50% of the enzyme is inactivated at 71.6°C in 30 minutes, and complete inactivation requires heating to 99°C. When produced during bacterial growth in raw milk, the enzyme persists after pasteurization, introducing a rancid flavor. The process of homogenization enhances the enzyme's activity in milk by reducing the fat droplets to diameters of 5 microns or less, greatly increasing surface area for the enzyme.

Enzymatic activity is slow at −12.2°C (Peterson, 1962). Undesirable enzymatically induced visual and other organoleptic changes occur in meat pies during defrost even when microbial counts are still well within acceptable limits (Figure 2-4). The flavor defects appear to be due to a combination of proteolysis and lipolysis, while the amylolytic defects are visible in deformations of the crust, uneven browning during baking, seepage of fluid through the crust, and the sticking of the bottom crust to the pan.

Preservation by Freezing

Foods were first mechanically frozen during the latter half of the last century. Frozen meats were first shipped from Australia to France in 1877, and to England in 1878. It was rather early

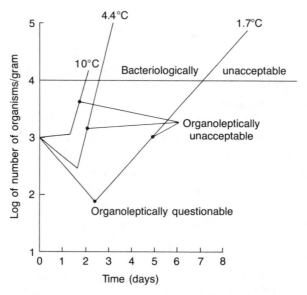

Figure 2-4
Relationship among time of defrost of chicken pies at
various temperatures, total number of bacteria in the
product, and organoleptic evaluation (Peterson,
1962).

observed that the ideal temperature during transit is –10°C,
and should not exceed –5.5°C (Hales, 1963). Fruits and eggs
for the baking industries were being frozen at the turn of the
century. Around 1929 Clarence Birdseye developed the commercial freezing of foods in packages for retail sale. Approximately 33% of all preserved foods consumed in the United
States are frozen foods.

Classification of Frozen Foods

Frozen foods are divided into four major classes according to
perishability in the unfrozen state:

1. *Perishable raw foods* include all red meats, poultry, fish,
   and raw breaded items; fruits and juices; a few vegetables; broken-out eggs.

2. *Perishable heated or cooked foods* include all blanched vegetables; all moist foods such as TV dinners, meat and
fruit pies, and deboned cooked meats such as turkey rolls and seafood items; some baked items; and
breaded foods such as french-fried onion rings.

3. *Semiperishable foods,* which have a moderately short but
stable shelf life at room or refrigerating temperature,
include breads, partially baked breads, and dry baked
pastries; unheated dough products; cheese; butter.

4. *Nonperishable foods* are dry foods, such as nuts, that are
frozen to prevent or retard oxidative or color changes
or loss of oils.

Perishable Raw Foods. The prefreezing microbiology of most
perishable raw foods is described in other chapters. There is
little available literature, suggesting that microorganisms play a
relatively inconsequential role during the preparation of foods
for freezing. Raw foods, whether animal or vegetable, have a
retarding effect on the growth of microorganisms occurring
during prefreezing processing. The major exceptions are certain juices and breaded items.

Breaded or battered foods pass through flowing sheets of a
flour batter. Batter not adhering to the food is returned to the
central supply well for repassage. During continuous, perhaps
hours-long processing in warm rooms, *Microbacterium* spp. become the dominant bacteria, reproducing to millions of cells
per ml of batter. Consumers do not notice the mildly acidic
condition, approximately pH 5.5, and the bacteria are killed
during the deep fat frying at around 175°C. However, partly
for aesthetic reasons and partly because at warm temperature
Staphylococcus spp. proliferate on some foods, such as shrimp,
batter mixes are now chilled to approximately 5°C.

Green and red peppers and onions (the centers of the onion
rings, diced for the hamburger trade) are frozen without
blanching. Fresh peppers usually have a low surface population
that is further reduced by washing. Onions are flame peeled to
burn the parchmentlike outer skins. Few microorganisms are
within the fleshy leaves of the onion proper. The cell juices of
both pepper and onion do not support the rapid growth of

bacteria on surfaces of processing machinery, even during long periods of continuous processing. *Leuconostoc mesenteroides* and plant-type streptococci are the dominant microflora of the frozen onion.

Fruits and juices are protected by the low pH that prevents the growth of all but a few aciduric bacteria and the relatively slow-growing yeasts and molds. Washing most small fruits reduces their surface loads of microorganisms. Some large fruits are peeled by hot water or skin-dissolving lye solutions; others are peeled mechanically. The exposed flesh becomes contaminated by microorganisms present on the equipment.

Juices become contaminated with yeasts and molds from the fruit surface. Grape juice is pasteurized and stabilized at 60°C and then filtered. Apple juice is filtered to provide the sparkling clear juice preferred by Americans, in contrast with the unfiltered, cloudy juice preferred by Europeans.

Orange juice is concentrated by evaporation under high vacuum. Species of *Lactobacillus, Leuconostoc,* plant-type enterococci, *Enterobacter, Xanthomonas,* and *Achromobacter* grow in the holding tanks, lines, and hot wells to produce acids and by-products such as diacetyl, which affect the flavor. Rigid attention to cleansing the raw fruit and periodic interruption of processing for cleansing and sanitization delay bacterial deterioration (Hays and Riester, 1952).

Frozen Vegetables. In its early years the vegetable-processing industry used the same techniques and equipment as the canning industries. Although vegetables became heavily contaminated during post-blanch handling in processing for canning, contamination received little attention and apparently had little effect on the quality of the product.

By modern standards such concepts of plant sanitation were primitive. The equipment, not designed for easy cleaning, was conducive to severe contamination. Modern detergents were not developed until after 1945. The post-blanching traverse time to the freezers was prolonged, often taking several hours, in contrast with modern methods in which transit time between the blanchers and the freezers often is less than 30 minutes. The blanchers were poorly constructed and retained vegetable particles that became discolored, necessitating lengthy post-blanching inspection and sorting lines. The facilities for warehousing, transportation, and retail sales were crude and inefficient, shippers and retailers often being unaware that deep freezing temperatures must be maintained.

The very high counts of bacteria in specific frozen foods recorded during the early years do not reflect the conditions in the industry today. Smart (1939) stated that during the thirties the bacterial counts of frozen foods had been reduced from millions to hundreds of thousands per gram. Little apparent progress was made during World War II, for in 1947 the Frozen Food Institute advised its members to meet a standard of not more than 500,000 bacteria per gram (Fitzgerald, 1947). At present many frozen vegetables are produced with 20,000 bacteria per gram or less at the time of pack, that is, the same day as emergence from the freezer or the day after.

With few exceptions, vegetables to be frozen are blanched in boiling water or in steam to inactivate the enzymes catalase, peroxidase, and lipoxidase. Incidental benefits of blanching are the setting of the green color characteristic of many vegetables and the destruction of bacteria. Often only a small percent of particulate vegetables sampled at the blancher outlet yield spore-forming bacteria. Bacteria that do not form spores, frequently of the coliform type, survive in the insulated centers of conglomerates of greens (Weiser, 1951) that have not been separated by dispersers at the beginning of blanching.

Blanched vegetables promptly become recontaminated with a wide variety of airborne plant and soil bacteria that enter the processing areas (Hucker et al., 1952). *Achromobacter, Bacillus, Chromobacterium, Corynebacterium, Enterobacter, Flavobacterium, Lactobacillus, Micrococcus, Phytomonas, Pseudomonas, Sarcina* and *Serratia* are among the genera that have been identified. Microflora that become dominant during the processing of saccharine vegetables are the lactobacilli, *Streptococcus faecalis,* and *S. faecalis*-like strains; *S. faecium* and *S. faecium*-like strains; *S. faecium* var. *casseliflavus; S. lactis* and *S. lactis*-like bacteria; *Leuconostoc* spp.; *Aerococcus;* and *Pediococcus.* As the processing season progresses the lactic-acid-producing bacteria tend to stabilize on equipment (Splittstoesser and Gadjo, 1966). The bacteria thrive on the substances in the cell sap that are leached from the blanched vegetables by the cooling and fluming waters, or are expressed upon spongy compression as the vegetable particulates drop from one conveyor to another. In test-tube studies the soluble solids, measurable in micrograms per ml of water, have provided adequate food to support populations of millions of bacteria per ml of water (Mundt et al., 1966). This microflora produces adherent slimes that are difficult to remove. They grow in the cracks of older, rubberized conveyor belts. Bacterial counts of processed green beans may increase 10 fold during a 60 second passage on a 12 foot rubberized belt (Haynes and Mundt, 1948). The actively growing bacteria continue growth on the vegetable passing into the freezers until stopped by temperature reduction.

Improvements in several major areas have helped lessen bacterial counts. Copious use of cold waters reduces the temperature of the blanched product. Air conditioning reduces the ambient post-blanching temperature and also the temperature of the equipment, lengthening generation times. More rapid movement of produce after blanching means shorter retention time before freezing. In-plant chlorination of post-blanching waters exerts a continuing disinfecting action even during prolonged operating days. Improved equipment design enables more effective cleansing. Modern freezing equipment is able to cope with prepackaged foods without temporary fluctuations in freezing temperatures, and can attain lower freezing temperatures than were formerly possible. Particulate vegetables may be individually frozen prior to packaging, sometimes within seconds at the surface of the particle. Prepackaged foods are frozen upon contact with freezing plates, or in cross-current or countercurrent high-velocity blasts of cold air.

Precooked Frozen Foods. Most precooked frozen foods are pot pies or TV dinners. As in the vegetable freezing industry, improvements in handling and equipment design have lowered bacterial counts of products. Unlike that of blanched vegetables, the pH of many precooked items is near neutrality, and the component meats and gravies support rapid growth of bacteria if temperatures permit. In-plant chlorination cannot be used except when cleaning equipment. The industry makes extensive use of metalware and less use of flexible belts for moving moist or wet components. Because modern detergents are coupled with sanitizing chlorination with water containing 50 or more ppm available chlorine, microbiological cleanliness of equipment surfaces approaches sterility. From cookers through cooling to portioning, fluid components are transported in completely enclosed systems, and hand portioning has yielded to metered portioning. Raw product storage, cooking rooms, and packaging areas are physically separated. Post-cooking handling of meats is done in rooms with temperatures at 10°C or less. By reducing the potential for contamination and prolonging the generation times of bacteria, these precautions have improved quality. Precooked frozen foods prepared under modern conditions are not likely to possess the very high microbial counts noted in 1957 (Canale-Parola and Ordal) or the enzymic activity noted in meat pies in 1960 (Peterson and Gunderson).

Death and Survival during Freezing

Freezing causes the apparent death of 10 to 60% of the viable population, and gradually increases the percentage if frozen storage continues. In foods with highly mixed microbial populations, the decline may at first be rather rapid, as the more susceptible species die, then become more gradual for a protracted period of storage; long-continued storage produces little further observable loss. The highly resistant bacteria have been described as **freeze-resistant** (Hucker, 1954). They have been observed to persist through 10 years of frozen storage.

In general gram-negative bacteria are less resistant to freezing death than are gram-positive nonsporulating rods, and spherical bacteria are the most resistant. In frozen eggs the high-temperature mesophiles were observed to die more rapidly than the low-temperature mesophiles (Nielson and Garnatz, 1941). Bacterial spores, unaffected by freezing, are usually present in minimal numbers.

Most frozen foods bear a mixture of microflora. Which ones are most numerous at the time of analysis will depend on which were dominant at the time of freezing and on their relative rate of death or survival during continued frozen storage. Commonly encountered genera include *Pseudomonas, Achromobacter, Flavobacter,* micrococci, lactobacilli, *Corynebacterium*-like catalase-positive rods, enterococci, *Streptococcus lactis* and *S. lactis*-like streptococci, a wide variety of unspeciated plant streptococci, *Aerococcus,* and *Pediococcus.* Deliberate searches of frozen vegetables have yielded no staphylococci (Splittstoesser and Wettergreen, 1964; Splittstoesser et al., 1965) or salmonellae.

Slow freezing to a final temperature of −10°C is more lethal than rapid freezing to a final temperature of −20°C. During slow freezing ice crystals form to concentrate the soluble solids that affect the stability of the cellular proteins. During rapid freezing the temperature passes through the eutectic point to inactivate the solutes. Modern freezing to very low temperatures does not kill bacteria (Borgstrom, 1955). Rapid freezing was developed to induce formation of small ice crystals and to maximally reduce vapor pressure to prevent freezing dehydration. Ice-crystal formation is external and does not penetrate internally through the membrane barrier (Ling, 1967). During rapid freezing to −20°C and below, 90% of the cellular water is frozen and the remaining 10% is bound water (Masur, 1966).

During the freezing of liquids, bacteria and soluble solids become concentrated in the centers of cone-shaped areas as ice crystals form selectively at the periphery. The bacteria subsequently die more rapidly in the center than at the periphery (McFarlane, 1940). Some cells do not die for the reason that solutes are unevenly distributed in the intercrystalline film (Squires and Hartsell, 1940). High concentrations of sucrose protect both bacteria and yeasts, but bacteria survive better in concentrated solutions near neutrality, whereas yeasts survive better in more acidic solutions (McFarlane, 1941).

Cold Shock. Bacteria that undergo abrupt transition from 37 to 0°C in the exponential phase of growth lose viability. Alteration in the permeability of the cytoplasmic membrane causes leakage of amino acids and adenosine triphosphate. This effect is more pronounced with gram-negative mesophiles such as *Escherichia coli, Pseudomonas aeruginosa, Enterobacter aerogenes,* and *Salmonella typhimurium* grown at lower temperatures. Mesophiles maintained at low temperature contain more unsaturated fatty acid residues in the membrane lipids. The effect is not as noticeable among gram-positive bacteria (Farrell and Rose, 1967).

Freezing Injury. Many cells that are damaged during freezing do not die (Jansen and Busta, 1973). They are unable to initiate growth on minimally nutritive media or media containing inhibiting agents. But if given an energy source in the absence of inhibiting agents, the cells rapidly repair the damage and become able to grow in minimally nutritive media. The recovery period persists for one to six hours. Damaged cells initiate RNA synthesis about five hours after thawing (Sinskey and Silverman, 1970). Freezing alters the permeability of the cell membranes, leading to loss of 70S ribosomes and amino acids.

Thawed Foods. Frozen foods are not sterile and prolonged frozen storage does not sterilize them. Once frozen, foods should be maintained at a constant frozen temperature until they are to be prepared for consumption. Warming, even to below thawing temperatures, mobilizes water that, during subsequent freezing, recrystallizes and alters texture. Foods in which enzymes have been produced during the prefreezing period are subject to flavor changes (Peterson, 1962).

Upon thawing, the low-temperature mesophiles initiate growth that accelerates with time and rising temperature. Pathogens, if present, do not initiate growth until the temperature rises to approximately 10°C. Frozen vegetables acquire a creamy white slimy growth and an offensive odor and changed appearance. This markedly altered appearance warns the consumer that the food has spoiled before pathogens are able to multiply.

Loss in Refrigeration. Major loss in commercial refrigeration facilities usually occurs because equipment breaks down or wind and water damage the power supply. Temporary loss of power or minor equipment breakdown in commercial freezers seldom causes food deterioration. The large, well-insulated storage chambers offer a relatively low radiating surface area in proportion to volume for the dissipation of cold. If a breakdown continues long enough for the internal temperature of the products to rise above $-10°C$, partial thawing, as described above, could result. Home freezers are not as well insulated and offer a larger radiating surface in proportion to storage volume. Home freezer owners suffering power loss or equipment failure are advised not to open the freezer repeatedly, and, where possible, to maintain refrigeration with dry ice at the rate of 1 lb/ft^3 capacity per day. Unequivocally, foods warmed to $4.4°C$ or above should be discarded. If they have not risen above this temperature they may be refrozen, although the texture may be impaired.

References

Barber, M. 1955. J. Dairy Sci. *38*, 233.

Borgstrom, G. 1955. In: *Advances in Food Research*. (E. M. Mrak and G. F. Stewart, eds.) *9*, 163, Academic Press, New York.

Burton, M. A. 1949. Food Res. *14*, 434.

Buttiaux, R., and D. A. A. Mossel. 1961. J. Appl. Bacteriol. *34*, 353.

Canale-Parola, E., and Z. J. Ordal, 1957. Food Technol. *11*, 578.

Cooney, D. G., and R. Emerson. 1964. *Thermophilic Fungi*, W. H. Freeman & Co., San Francisco.

de Figueiredo, M. P. 1970. Food Technol. *24*, 157.

Dickerson, R. W., and R. B. Read, Jr. 1973. J. Milk Food Technol. *36*, 167.

Elliott, R. P., and H. D. Michener. 1961. Appl. Microbiol. *9*, 452.

Elliott, R. P., and H. D. Michener. 1965. USDA Tech. Bull. 1320.

Farrell, J., and A. H. Rose. 1967. In: *Thermobacteriology* (Rose, A. H. ed.), Academic Press, London.

Fitzgerald, G. A. 1947. Food Technol. *1*, 575.

Hales, K. C. 1963. In: *Advances in Food Research*. (E. M. Mrak and G. F. Stewart, eds.) 12, 147, Academic Press, New York.

Haynes, R., and J. O. Mundt. 1948. Food Inds. *20*, 977.

Hesseltine, C. W., R. R. Graves, R. Rogers, and H. R. Burmeister. 1969. Appl. Microbiol. *18*, 848.

Hucker, G. J. 1954. Food Technol. *8*, 79.

Hucker, G. J., R. F. Brooks, and A. J. Emery. 1952. Food Technol. *6*, 147.

Hunter, A. C. 1939. Food Res. *4*, 531.

Ingraham, J. L. 1958. J. Bacteriol. *76*, 75.

Ingraham, J. L. 1961. In: *The Bacteria* (I. C. Gunsalus and R. V. Stanier, eds.) *4*, 265, Academic Press, New York.

Ingraham, J. L. 1969, Cryobiology *6*, 188.

Ingraham, J. L., and J. L. Stokes. 1959. Bacteriol. Revs. *23*, 97.

Jackson, H. W., and M. E. Morgan. 1954. J. Dairy Sci. *37*, 1316.

Jansen, D. W., and F. Busta. 1973. J. Milk Food Technol. *36*, 520.

Leininger, H. V., L. R. Shelton, and K. H. Lewis. 1971. Food Technol. *25*, 224.

Ling, G. N. 1967. In: *Thermobiology* (A. H. Rose, ed.), Academic Press, London.

McFarlane, V. H. 1940. Food Res. *5*, 43.

McFarlane, V. H. 1941. Food Res. *6*, 481.

Mallmann, W. L. 1932. J. Agric. Res. *44*, 913.

Martin, J. D., and J. O. Mundt. 1972. Appl. Microbiol. *24*, 575.

Martinez, J., and M. D. Appleman. 1949. Food Technol. *3*, 392.

Masur, P. 1966. In: *Cryobiology* (H. T. Merryman, ed.), Academic Press, London.

Miller, A. T., H. M. El-Bisi, and F. M. Sawyer. 1964. Univ. Mass. Exp. Sta. Bull. 548.

Mundt, J. O. 1976. J. Milk Food Technol. *39*, 43.

Mundt, J. O., G. Shuey, and I. E. McCarty. 1954. J. Milk Food Technol. *17*, 12.

Mundt, J. O., S. A. Larsen, and I. E. McCarty. 1966. Appl. Microbiol. *14*, 115.

Mundt, J. O., and W. F. Graham. 1968. J. Bacteriol. *95*, 2005.

Nashif, S. A., and F. E. Nelson, 1953. *J. Appl. Microbiol. 1,* 47.

Overcast, W. W., and K. Atmaram. 1974. *J. Milk and Food Technol. 37,* 233.

Pedersen, C. F. 1959. Appl. Microbiol. *7*, 1, 7, 12.

Peterson, A. C. 1962. In: *Proceedings Low Temperature Microbiology Symposium—1961,* Campbell Soup Co., Camden, N.J.

Peterson, A. C., and M. F. Gunderson. 1960. Food Technol. *14*, 413.

Plateniu, H. 1939. J. Agric. Res. *59*, 41.

Rouf, M. A., and M. M. Rigney. 1971. Appl. Microbiol. *22*, 503.

Schelhorn, M. von. 1951. In: *Advances in Food Research.* (E. M. Mrak and G. F. Stewart, eds.) *3,* 431, Academic Press, New York.

Sinskey, T. J., and G. J. Silverman. 1970. J. Bacteriol. *101*, 429.

Smart, H. F. 1937. Food Res. *2*, 515.

Smart, H. F. 1939. Food Res. *4*, 293.

Splittstoesser, D. F., and W. P. Wettergreen. 1964. Food Technol. *18*, 134.

Splittstoesser, D. F., G. E. R. Hervy, II, and W. P. Wettergreen. 1965. J. Milk Food Technol. *28*, 149.

Squires, H., and S. Hartsell. 1940. J. Appl. Bacteriol. *3*, 40.

Stanier, R. Y., N. Palleroni, and M. Douderoff. 1966. J. Gen. Microbiol. *43*, 159.

Surkiewicz, B. F., R. J. Groomes, and A. P. Padron. 1967. Appl. Microbiol. *15*, 1324.

Surkiewicz, B. F., M. E. Harris, and R. W. Johnston. 1973. Appl. Microbiol. *26*, 574.

Tarrant, P. J. V., N. Jenkins, A. M. Pearson, and T. R. Dutson. 1973. Appl. Microbiol. *25*, 996.

Thatcher, F. S., and D. S. Clark. 1968. *Microorganisms in Foods: Their Significance and Methods of Enumeration,* Univ. of Toronto Press, Toronto.

Tompkin, R. B., and A. B. Shaparis. 1972. Appl. Microbiol. *24*, 1003.

Virgilio, R., C. Gonzalez, S. Mendoza, S. Avedano, and N. Munoz. 1970. J. Food Sci. *35*, 842.

Waskell, L., and D. A. Glaser. 1974. J. Bacteriol. *118*, 1027.

Weiser, H. H. 1951. Quick Frozen Foods *13*(7), 50.

3

Control of Moisture

Although microorganisms require available moisture for metabolism and growth, molds, yeasts, and bacteria differ widely in how much they need. Generally, molds require less moisture than do yeasts, which in turn require less than bacteria. However, even within one family, genus, or species of molds, yeasts, and bacteria, the minimal moisture requirements for growth can differ widely. Below the minimum water content required for growth, some molds, yeasts, and bacteria die rapidly, others exhibit prolonged endurance to drying. Thus, some molds, yeasts, and bacteria will persist in very dry environments, such as in stored grains, cereal products, starches, sugars, dried meats, and fish. Given sufficient time, some molds will grow in a nearly saturated solution of sugar, and fungal enzymes are known to remain functional in solutions of salt approaching saturation. At the other extreme, some species of molds, yeasts, and bacteria grow only in very humid, moist, or wet environments.

Organisms that persist in dry environments or in those with high percentages of salts or sugars are described by various terms. For example, organisms that thrive in very dry environments are called **xerophiles;** those that endure are **xeroduric.** Similarly, microorganisms that flourish in jams, jellies, honey, syrups, brines, and pickles or other foods high in sugar or salt content are **osmophilic;** those that can persist in these substrates are **osmoduric.** Organisms that demand salt concentrations exceeding those of physiological salt solutions receive a more restrictive term: **halophilic.**

Water Activity (a_w)

The cells of plant and animal tissue contain a large proportion of water. But not all of this water may be available to microorganisms. Consequently, foods with high proportions of fat may make less moisture available to organisms for growth than their "average" water content suggests. This is true of products made from meat, fish, poultry, and eggs, in which adipose tissue, fat deposits, or yolk make up a significant proportion of total composition. Nevertheless, the microenvironment in which the organism proliferates may contain much available moisture for biological and chemical reactions. The "free" moisture, or **water activity** (a_w), is determined by comparing the chemical potential of the available water to that of pure water at the same temperature. By definition $a_w = p/P_o = ERH/100$, where p represents the water vapor pressure of the solution (food), p_o the vapor pressure of the solvent (water), and ERH equilibrium relative humidity. For example, the value for fresh meat is $a_w = 0.99$ (see Figure 3-1). When the surrounding atmosphere is drier than this value, the surface of the meat begins to dry and loses its desirable moist, glistening appearance, or "bloom." The relative humidity of the atmosphere in equilibrium with the food is calculated by multiplying the $a_w \times 100$. The surface of meat given prolonged storage in an atmosphere of low relative humidity becomes so dehydrated that it becomes almost impervious to further loss of water. The lowest a_w at which *Pseudomonas* spp. grow is 0.97. Also, the limiting a_w value for *Escherichia coli* and *Clostridium botulinum* is 0.95; that of *Staphylococcus aureus* is 0.86. Thus, fresh meats held in cold storage for a few days at $a_w = < 0.95$ are free from danger of growth of *E. coli* and *C. botulinum* or slime formation by *P.*

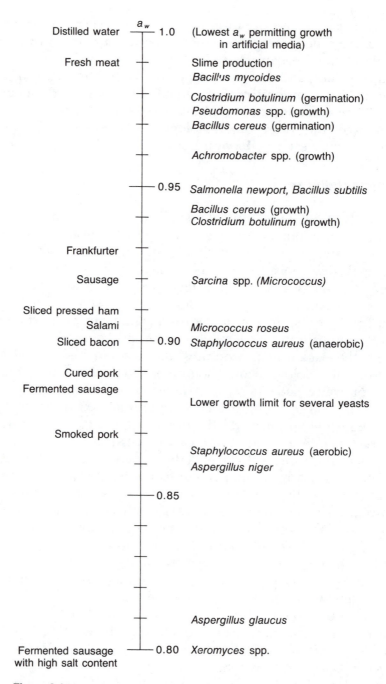

Figure 3-1
Comparison of water activity (a_w) values for various meats with those limiting the germination or growth of microorganisms.

geniculata or other pseudomonads. Fortunately, *S. aureus* grows poorly at refrigeration temperatures and is not a prominent contaminant of fresh meats.

On the other hand, when the a_w levels of certain types of cured meats are too high, staphylococci cause problems. With cured meats, a_w values needed to assure a desirable product vary with the fat content, the amount of smoke received, salt content, and the degree of microbial action to which the product is subjected.

Staphylococci are by no means competitive with xerophilic molds or yeasts upon relatively "dry" substrates. Grain at an a_w of 0.80 contains sufficient available water for the growth of *Aspergillus glaucus* but not of *S. aureus*. Scott (1961) reported that substrates exceeding an a_w of 0.65 to 0.75 contain sufficient moisture for *A. glaucus* but that *S. aureus* requires a_w values of 0.86 to 0.88.

Measuring a_w

Measuring the a_w of a particular food involves determining the relative humidity of the atmosphere in equilibrium with the food. A number of methods have been proposed. One is to use the hygroscopic characteristics of various salts of different known affinities for water: when a given salt placed in a chamber with a given food becomes wet, the relative humidity of the atmosphere in equilibrium with the food can be estimated. Hygrometers that depend on changes in electrical conductivity also can be used, as can be gas chromatography. Instruments designed for determining a_w are commercially available, and a review of methodology has been published (Labuza, 1975; Labuza et al., 1976).

Drying

Drying is the oldest and one of the most widely used methods of preserving many foods. As stated earlier, to metabolize and proliferate, microorganisms require a water bridge. The amount of moisture available determines which organisms are able to grow. Many agricultural commodities are artificially dried so that fungi cannot attack them during storage. For safe storage of peanuts, the amount of water must be reduced to about 8%, but wheat, rice, or corn with a moisture content of less than 12 to 14% can be stored for years without spoiling or becoming moldy. Owing to their high sugar content, certain fruits keep for months if dried to a moisture content of from 6 to 25% and not later exposed to humid atmospheric conditions.

Methods of Drying

From time immemorial sun-drying, using the sun's heat and warm, dry winds, has been an effective natural way of preserving grain, nuts, legumes, and other seeds. Early humans learned to use this process for drying strips of meat (pemmican or jerky) and fish, and certain sliced vegetables and fruits—especially potatoes, cassava, figs, dates, prunes, and grapes. Artificial heat was later applied in chambers, kilns, or tunnels, thus enhancing the human diet to include a number of cereal products, starchy foods, and additional fruits and vegetables. In more recent times, plant juices, syrups, milk, and other fluids have been dehydrated by evaporation. The drying industry has been revolutionized by recent technological developments—such as drum, spray, foam-mat, vacuum, and freeze drying, to dry food products quickly and thoroughly—and today many dehydrated foods are available as convenience items.

Dehydration can also be accomplished with dry salts or sugars or with brines or syrups. A heavy layer of salt (NaCl or NaCl plus other curing salts), applied to the surface or rubbed into the flesh of red meats or fish, withdraws some of the moisture from the tissue. Alternatively, when brines are used, as for meats, cabbage, or cucumbers, water and salts diffuse through cell walls by osmotic action, pickling the product—lowering the water content but raising the salt content.

Certain meats, fish, shellfish, and poultry are preserved by smoking. Smoking is effective largely because it dries the meat and introduces a film of phenolic, cresylic, and aldehyde compounds. High-temperature, low-moisture methods, such as smoking, can remove moisture from the meat more rapidly than it can diffuse to the surface, causing the outer layer to become horny or "case hardened." This condition is not restricted to smoked foods, however. Case hardening can occur during the drying of fruits, vegetables, or meats, whenever vaporization from the surface takes place so quickly that an impervious boundary is set up that impedes diffusion of water from the interior. When the initial rate of drying fruits is too rapid, cell saps form a glaze at the surface, making it difficult for internal moisture to escape. The product appears to be fully dehydrated during storage, but the moisture equilibrates and supports the growth of yeasts and/or molds.

Microbial Behavior on Dried Foods

Yeast, molds, and bacteria are abundant on dried foods. Sanitary measures diminish microbial growth (see Table 3-1), but unsanitary sorting procedures, trimming tables, and storage receptacles support large populations of microorganisms. Table 3-2 shows a study of some of these sources in a dried-fruit plant. Molds of the genera *Penicillium, Aspergillus, Cladosporium, Alternaria,* and *Scopulariopsis* are the most commonly encountered on dried foods. Among yeasts, *Zygosaccharomyces, Hanseniaspora, Saccharomyces, Hansenula, Pichia,* and *Torulopsis* are typically found.

As the moisture content of dried dates decreases, so does the rate of heat penetration (Fellers, 1930; Clague and Fellers, 1933). However, there is no exact relationship among the water content of the fruit, temperature, and the time-temperature

Table 3-1
Number of yeasts and bacteria on samples collected in dried fruit plant.

Sample	Number of microorganisms
Wash water taken from container at packing table. Water used to wash hands	184,000 bacteria per ml. 700 yeasts per ml
Damp cloths used at sorting table	400 million yeasts and bacteria per square inch
Damp cloths used at cutting and trimming table	32 million yeasts and bacteria per square inch
Water from buckets used for washing hands	104 million yeasts and bacteria per ml. About half were yeasts

Source: Mrak and Stadtman, 1946.

Table 3-2
Number of mold colonies obtained from field run Deglet Noor dates.[a]

Time of sampling	Number of mold spores		Mold types (genera)
	Per gram	Per date	
When delivered to the packing house	4600	53,300	Chiefly *Alternaria*
After 16 hours of fumigation with methyl bromide	2900	36,000	*Aspergillus* and *Penicillium;* few *Alternaria*
At the end of shaker	100	1400	*Aspergillus* and *Penicillium;* few *Alternaria*
At the end of the sorting belt	200	2400	*Aspergillus* and *Penicillium;* few *Alternaria*

[a] Moisture content 26.6%.

Source: Mrak and Stadtman, 1946.

value necessary to achieve sterilization. A limiting moisture level for dried vegetables stored at 20°C (Schelhorn, 1951) may be obtained by superimposing a relative air humidity of 72.5% on absorption isotherms constructed by Heiss (1949). (See Figure 3-2.)

For small particulate materials, such as grains, a new equilibrium is established within hours. Cereal grains to be stored for longer than six months must be dried to a 12–14% moisture content; an intermediate range of 14–16% is satisfactory for three to six months. Above 18%, extensive mold growth takes place in weeks. Damage includes lipolysis, with increase in free fatty acids, and enzymic digestion of starch to sugars; as moisture and temperature in the silo or enclosed storage area increase, respiration of the embryo becomes elevated. The process may become catalytic, and, if seed grains are involved, the death rate of embryos is proportionate to the elevated moisture content and the time in storage. Also, grains in storage undergo a "distillation" effect. As the sun warms the storage bin's southward-facing metal surfaces, vaporizing the internal moisture, water condenses on the grains stored on the unheated, and colder, northern exposure. During late winter and early

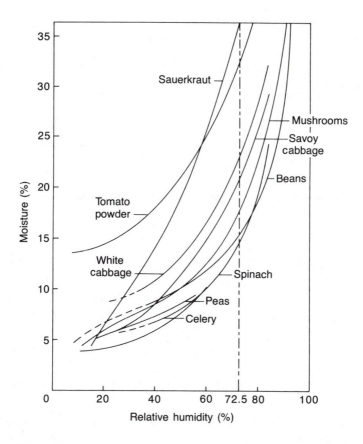

Figure 3-2
Sorption isotherms of dried vegetables at 20°C (Heiss, 1949).

spring, cells of *Apsergillus niger* may begin to grow in the top 2–3 feet of federal storage bins unless air has been circulated or heat driven through the bins.

With larger masses, such as large clumps of dried fruits, molds grow when surface moisture exceeds the minimal required a_w level, and growth cumulates as the relative humidity fluctuates above and below that supportive level (see Figure 3-3). Eventually, after days, weeks, or even months, the accumulation is visible to the naked eye.

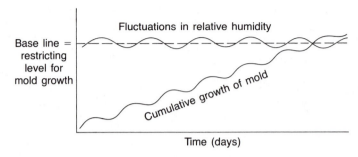

Figure 3-3
Effect of fluctuating relative humidity upon mold growth.

Many dried fruits and vegetables are hygroscopic: they collect water from the atmosphere unless stored in packaging materials that are good vapor barriers. Otherwise, as the vapor pressure of the microenvironment changes, the cells are alternately exposed to growth-limiting and growth-permitting conditions. Active dry-yeast preparations can be maintained only in a dry microatmosphere, that is, $a_w = 0.45$. If water gets into the package, respiration increases and the yeast cells quickly die.

The bacterial forms most associated with dried products are spore-forming aerobes and anaerobes, lactic acid bacteria (lactobacilli, *Leuconostoc* spp., streptococci, and the like), micrococci, and coliforms. Lactobacilli frequently have been associated with the souring of onions, and growth of *Aerobacter (Enterobacter)* spp. during the drying of potato flakes, producing souring. Staphylococci, *Achromobacter, Pseudomonas,* and *Microbacterium* spp. have been isolated from dried foods. These bacteria require a much wetter environment for long survival.

Microorganisms can survive for long periods in some dry foods. Higginbottom (1953) reported that when commercially dried whole-milk powder was stored at relative humidities of from 5 to 70%, from 30 to 70% of the bacteria initially present survived for more than two years (see Table 3-3). While milk powder dehydrated to a moisture content of 11% is microbiologically stable, it continues to undergo enzymatic de-

Table 3-3
Survival of bacteria in dried whole milk powders in prolonged storage at various relative humidities at 38°C.

Relative humidity (%)	Viable bacteria (%) after	
	72 weeks	103 weeks
70	51	—
60	55	—
50	57	—
40	59	—
30	69	60
20	70	—
10	68	69
5	67	30
0	0.1	0.02

Source: Adapted from Higginbottom, 1953.

terioration, and opened packages undergo rapid oxidative changes unless dried to 2%. Similarly, viable salmonellae can be recovered from pan-dried egg albumen (< 3% moisture) after two years of storage. When egg white is carefully spray-dried to a residual moisture level of 1.5, 3.0, 6.0, or 12.0% and is stored at 50, 60, or 70°C, the number of surviving salmonellae varies inversely with moisture level and storage temperature.

The removal of water by **sublimation,** that is, as a vapor from a substrate that remains frozen solid, is called freeze drying, or **lyophilization.** Although an expensive procedure, it is used extensively for food items such as meats, poultry, fish, shellfish, coffee, fruits, yeasts, and bacterial cultures. Freeze drying not only reduces the weight of foods but also preserves functional properties that other drying procedures destroy in many foods, and halts as well all microbial growth until the product is reconstituted.

Dried fruit is usually spoiled by a group of yeasts peculiarly tolerant to high concentrations of sugar. These organisms grow slowly, fermenting the sugars in the fruit and gradually changing the flavor, appearance, and structure of the tissue.

Concentrates

Concentrates are food products in which the soluble solids are in the form of sugars; fruit or vegetable juices or pulp; or milk solids. Water content is so reduced that spoilage is restricted to osmophilic and/or halophilic organisms.

Prunes, figs, dates, and raisins are natural food concentrates high in soluble solids. When their moisture content is too high these products are subject to spoilage by yeast and molds. Such spoilage may be detected by the presence of a white powdery or cottony formation, particularly on the surface of the fruit nearest the inner surface of the package, where the most moisture is likely to be absorbed. If moisture is absorbed and storage temperatures are favorable, molds grow readily on dried fruit. They develop rapidly on fruit held at 8°C when the humidity approaches 80–85% for several weeks; under these conditions some molding takes place even at 4.4°C. The pectins that cause jelling in jams and jellies are concentrated by boiling and evaporation, or the pectin content can be supplemented with apple pectin. Jams and jellies are acidic, with a pH of 4.0 or less and an average sugar content of 67%. The only spoilage agents are sugar-tolerant yeasts and filamentous molds. Since these are quickly destroyed during boiling, spoilage occurs when the molds and, more rarely, yeasts gain access, with oxygen, through improper seals. The molds most often seen are aspergilli and penicillia; phycomycetes and black molds (family Dematiaceae) are rarely observed on spoiled jellies and jams.

In areas of the United States where the climatic index (that is, the typical relative humidity) exceeds 65%, jams and jellies in opened containers undergo mild ethanolic fermentation unless they are promptly consumed. A thin film of surface moisture provides a site with reduced solids content upon which mold may proliferate and secrete enzymes that hydrolyze the substrate to usable metabolic substances.

The same phenomenon is considered responsible for the spoilage of honey by osmophilic yeasts. When the moisture content of honey exceeds 20%, the critical level, the surface of the honey is diluted by its hygroscopic property and contaminative yeasts quickly adapt and bring about spoilage.

In preparation of citrus juice concentrates, falling film evaporators concentrate the juice to 58% solids. The juice is then pumped to cold-wall tanks where pulpy single-strength cut-back juice reduces the solids—to 42% for orange juice, 39% for grapefruit juice—before it is slushfrozen and put into cans. The acidity, sugar concentration, and low storage temperature of these products cause them to be germicidal for many organisms, but enterococci, other lactic acid bacteria, yeasts, and coliforms persist. The most important index of processing sanitation is the lactic acid bacteria (lactobacilli, *Leuconostoc* spp., and streptococci), which can be indirectly measured by a colorimetric test for diacetyl (Murdock, 1967). *Streptococcus faecalis* and *S. liquefaciens* in frozen orange juice concentrate are important contaminants only if the juice is mishandled prior to evaporation. Organisms resembling *Escherichia coli*, isolated from orange juice, may possibly be members of the genus *Erwinia* that have lost their ability to decompose pectin. Also, colonies of certain lactose-fermenting yeasts present in frozen orange juice are identical in appearance to those of *E. coli* on eosin methylene blue agar (Martinez and Appleman, 1949).

Sweetened condensed milk is prepared by adding 18 to 20 pounds of sucrose or dextrose to whole milk and then evaporating sufficient water from this mixture to provide a condensed product containing 8.5% fat and 28% total milk solids. Sweetened condensed skim milk contains 20% total milk solids. After condensation, the products are retailed in small sealed cans or sold industrially in steel drums or bulk tanks. Adding sugar serves a twofold function: it (1) increases the osmotic pressure and (2) makes the water unavailable for microbial metabolism. Similarly, the evaporation of water increases the concentration of milk solids and, in turn, raises the osmotic pressure and binds water. While sweetened condensed milk is microbiologically stable even during prolonged storage on the grocer's shelves, it contains many viable microorganisms.

Ordinarily, sweetened condensed milk is sufficiently bacteriostatic that refrigeration is unnecessary. However, if it is recontaminated by molds during forewarming, "buttons" form in the product. Such buttons may appear as white-to-brownish globules composed of mycelial strands and coagulated casein. Also, faulty sanitation after forewarming occasionally results in gas formation, "bloat," or bulged or blown cans of sweetened condensed milk. Yeasts of the genus *Torulopsis* often are responsible.

Sugar solutions such as maple syrup, molasses, sorghum, and corn syrup readily spoil after the container has been opened. The water that these products absorb from the air provides sufficient dilution to allow growth of saccharolytic yeasts and aspergilli (*A. repens* and similar species). Osmophilic yeasts have been reported to grow in syrups with a_w values as low as 0.62 to 0.65. A minimal a_w value of less than 0.62 is necessary to halt all mold growth, but most food spoilage molds are inhibited at a_w values less than 0.70.

Intermediate-Moisture Foods

Intermediate-moisture foods are partially dehydrated but have sufficient dissolved solids to bind the remaining water so that the product does not require sterilization to prevent microbial attack. These products have enjoyed a considerable market as pet foods but not yet as human foods, because no known solute is both desirable in taste and effective in sufficiently lowering a_w values.

Food pieces can be infused with glycerol, sorbic acid, sorbitol, or sorbates by one of two methods. In one the food, either vacuum dried or freeze dried, is so infused that, after draining, it has both microbial stability and the proper moisture content for a stable shelf life. In the second method, food having a normal water content is soaked in an equilibrating solution until a desired moisture level has been attained so that it is safe from microbiological spoilage.

References

Clague, J. A., and C. R. Fellers. 1933. Arch. Mikrobiol. *4*, 419.
Fellers, C. R. 1930. Amer. Public Health *20*, 175.
Goldblith, S. A., M. A. Joslyn, and J. T. R. Nickerson. 1961. *Introduction to Thermal Processing of Foods*, AVI Publishing Co., Westport, Conn.
Hansen, N. W., and H. Riemann. 1962. Die Fleischwirtschaft *14*, 861.
Heiss, R. 1949. Z. Lebensm. Untersuch. u. Forsch *89*, 173.

Higginbottom. 1953. Ayrshire J. Dairy Sci. *20*, (1), 65. Cited in J. Dairy Sci. *36*, A72.

ICMSF, 1974. (International Commission on Microbiological Specifications for Foods, of the International Association of Microbiological Societies.) *Microorganisms in Foods.* 2. *Sampling of Microbiological Analysis: Principles and Specific Applications,* Univ. of Toronto Press, Toronto, Canada.

Labuza, T. P. 1975. In: *Water Relations of Foods* (R. B. Duckworth, ed.), Academic Press, New York.

Labuza, T. P., K. Acott, S. R. Tatini, and R. Y. Lee. 1976. J. Food Sci. *41*, 910.

Manson, J. E., and J. F. Cullen. 1974. J. Food Sci. *39*, 1084.

Martinez, J., and M. D. Appleman. 1949. Food Technol. *3*, 392.

Mrak, E. M., and T. Stadtman. 1946. Rept. of 23rd Annual Date Grower's Institute, Coachella, Calif., pp. 19–23.

Murdock, D. I. 1967. Food Technol. *22*, 90.

National Canners Association. 1966. *Processes for Low-acid Canned Foods in Metal Containers.* Bulletin 26-L.

National Canners Association. 1968. *Laboratory Manual for Food Canners and Processors.* AVI Publishing Company. Westport, Connecticut.

National Canners Association. 1971. *Processes for Low-acid Canned Foods in Glass Containers.* Bulletin 30-L.

Schelhorn, M. 1951. In: *Advances in Food Research* (E. M. Mrak and G. F. Stewart, eds.) *3*, 431, Academic Press, New York.

Schmidt, C. A. 1964. In: *Botulism. Proceedings of a Symposium* (K. H. Lewis and K. Cassel, Jr., eds.), USDHEW Public Health Svce., Cincinnati, Ohio.

Scott, W. J. 1957. In: *Advances in Food Research* (E. M. Mrak and G. F. Stewart, eds.) *7*, Academic Press Inc., New York, N.Y.

4
Canning

Canning is the process, used in the home or in industry, in which food is enclosed in hermetically sealed containers to which heat is applied so that the product can be stored without spoilage. A canning process that has become popular in recent years, particularly for fluid foods, is the high-temperature–short-time (HTST) process, which drastically reduces the processing time when used in combination with aseptic filling of presterilized containers. The end result of the conventional and HTST processes is the same. Nearly everyone is familiar with the old wheeze "We eat what we can and what we can't, we can." While the adage is no longer true, canning is still the most commonly used procedure for keeping many foods.

Each food has its own characteristics of fluidity, rate of heat transfer, particulate nature, and stratification. Consequently each has its own optimal size of container, temperature of fill, and acidity. Each of these in turn affects the total amount of heat that must be applied to the container or to the food, with respect to temperature and time of exposure to it. Moreover, hermetically sealed or sealable food containers can be made of tin-coated iron sheet, or of aluminum, glass, laminated paper-board, or may be laminated plastic pouches. The heat treatment must therefore be suitable to each food to be preserved and to each size and type of container.

Brief History of Canning

Early in the nineteenth century, following the disastrous experiences in the war with Russia, during which the French army lost large numbers of troops to inclement weather and to starvation, the French government offered a prize of 12,000 francs for the invention of a process of food preservation that would be useful to the military. In 1809 Nicholas Appert, a confectioner, claimed the prize. From his observations of the lactic acid souring of peas and other foods, Appert formulated a process that laid the foundations of our modern canning industry. Appert preserved a wide range of foods including meats, vegetables, fruits, and even milk by placing them in widemouth glass bottles that were then carefully corked and heated in boiling water. From his observations, Appert reasoned that "fire or heat applied to foods sealed in a container which was impervious to air" had the peculiar quality of preventing the food from spoiling. He then concluded, erroneously, that spoilage was caused by ferments and the presence of air. Apparently he did not realize that he was destroying microorganisms, and he had no knowledge of oxygen, which the chemist Antoine Lavoisier had discovered only a few years earlier. Appert used large glass containers closed with laminated corks in which the cement was fish glue. He paid meticulous attention to the quality of the foods to be canned and to the canning process. In 1810 he published his famous work *Le Livre de tous les Menages ou l'Art de Conserver pendant Plusieurs Annees Toutes les Substances Animales et Vegetales.* There he ascribed losses to "no other cause than that of bad application of the principle, or of forgetfulness of hygiene in the preparatory processes, according to the account of them that I have rendered." The following year Black, Perry, and Kingsbury published the first English translation of Appert's treatise, *The Art of Preserving All Kinds of Animal and Vegetable Substances for Several Years.*

Appert was an amazingly resourceful individual. Not only did he develop the art of canning and institute quality control, but he anticipated Pasteur by over half a century. He "pasteurized" day-old eggs at 60 and 90°C and preserved wines by heating. A true scientist, he also gave credit to the earlier work of Fabricus, who had studied the effect of heat on grape must; to Thenard, who had studied the effect of heat on currants, cherries, and other fruits; and to Vilaris and Cazales, who had studied the effect of heat on the putrefaction of meats.

A 1938 study of old cannisters of food conveys the effectiveness of Appert's primitive method of canning (Drummond et al., 1938). Cans of food dating back to nautical expeditions of the 1820s and foods packed for the Crimean War in 1855 were studied. Dehydrated tinned soups showed no apparent change. Perry's roast veal, packed about 1824, was in good chemical condition, with low iodine numbers and low content of free fatty acids; vitamin D was still active. Carrots and gravy were edible, although the sucrose in the carrots had inverted. The investigators recovered 6 cultures of thermophiles that apparently had remained dormant for more than 100 years.

In 1819 the Englishman Peter Durand conceived and patented the idea of using, instead of glass bottles, cannisters made of thin sheets of steel. However, plain steel has the major drawback of poor resistance to rust and corrosion. In the United States, in 1821, William Underwood of Boston began the commercial preservation of foods in cans. When tin-coated steel containers were developed in 1839 they soon replaced glass containers, and ultimately came to be known by the abbreviation "tin can." Such cans are now made by machine, but in the early nineteenth century they were made by hand. Before "hot fill" was known, filled cans were sealed, heated in boiling water, vented to relieve the pressure and to establish a vacuum when cooled, and heated again for several hours. In 1861, Isaac Solomon discovered that calcium chloride added to boiling water raised its temperature to 115.6°C or higher. However, the practice was soon abandoned because many cans exploded and others corroded. In Baltimore, in 1872, A. K. Shriver developed the closed steam-pressure retort. The pressure cooker, the autoclave, and the canning retort operate on the same principle as the retort developed by Shriver.

Although Louis Pasteur proved in 1860 that microorganisms were responsible for fermentations, the canning industry apparently did not associate bacteria with spoilage until 1895. In that year, H. L. Russell, of the University of Wisconsin, demonstrated that bacteria were responsible for the swelling, odors, and the gaseous spoilage of canned peas processed at 110°C for 10 to 12 minutes, and recommended that the processing time be increased to 116.7°C for 15 minutes (Bitting, 1937; Goldblith et al., 1961). In 1897 Prescott and Underwood of the Massachusetts Institute of Technology presented papers on the role of bacteria in the spoilage of canned clams and lobsters, and in 1898 traced spoilage of certain lots of sweet corn to imperfect sterilization. Russell and Prescott and Underwood all advocated the use of the retort in canning vegetables, as did the Canadian Andrew McPhail in canning lobsters, to prevent bacterial blackening, or smut, and putrefactive souring. More than 15 years later, in a classical study, Barlow (1913) described the spoilage of canned corn by thermophilic anaerobes, indicators of spoilage being a cheesy odor and hydrogen production.

The sanitary, or open-top, can was developed commercially by G. W. Cobb and introduced on a mass scale by American Can Company in 1908. It eliminated the use of solder in sealing, and the ingenious double seaming of top and bottom lids to the body of the can guaranteed a perfect closure. By 1938 paraffined paper milk containers were rapidly replacing the glass milk bottle, and by 1960 aluminum and plastic containers, unseamed except for the lids, were found to be commercially feasible. Also, flexible and semiflexible cans made from aluminum and paperboard laminates came into widespread use. Pouches of polyester-aluminum-nylon laminate can be used successfully at high retort temperatures if compensating air pressure is maintained during cooling to prevent rupture.

Today worldwide growth of the canned food industry is attributable to the development of the art of container manufacture (tin cans, glass jars, and bottles), refinements in canning processes, and improvements in canning machinery. The United States is the world's largest producer and consumer of canned foods. The modern tin can, which is about 98.5% sheet steel with a thin electroplating of tin, is the cheapest, most easily handled, serviceable, and convenient container for mass use. Foods can be processed by automatic machinery at rates exceeding 300 cans per minute. In a typical year, the U.S. canning industry uses about 30 billion tin cans and glass jars. The annual pack in the United States alone approximates 800 million cases (24 cans per case) of vegetables, fruits, juices, soft drinks, specialties, milk, meat, and fish.

Purpose and Processes in Canning

Quality of Product

While the primary objectives of the canning process are food preservation and consumer safety, the quality of the product must also be impaired as little as possible. Satisfying these opposing requirements necessitates careful attention to each of the following:

1. The Product. There is considerable range in the ability of various products to withstand heat. The consistency of the container contents is of great importance. When dry sugar rather than syrup is added, swollen cans often result. The sugar stratifies, becoming more concentrated at the bottom, and, unless agitated, does not dissolve readily. An organism has different resistances in concentrated sugar and in the natural product. Fruits are acidic and can be readily sterilized at the boiling point of water, whereas vegetables require "heavier cooks." Some products, such as meat and several varieties of fish, are injured by "high-short cooks" in which temperatures are maintained above 115.6°C.

2. Organisms Present. Although most microorganisms found on the raw material are ordinary soil forms, different species of microorganisms vary in resistance. A cooking period considered safe in one locality may not be in another region. For this reason, a margin of safety, based on cumulative experience, is given in accepted processing tables.

Actinomycetes are often responsible for the earthy odor of root crops. Molds are important spoilage agents in food before it is canned, whereas yeasts are of special significance in foods high in sugar and acid. These organisms are sensitive to heat; the optimum temperature for most ranges from 20 to 32°C.

As we saw in Chapter 1, bacteria are divided into two groups: those without spores and those with. The latter are of most concern because the bacterial spore has great resistance to heat. One important spore-former, *Clostridium botulinum,* produces a potent toxin. The thermophiles are even more resistant to heat. The spores of some strains can withstand 17 hours at the boiling point and 22 minutes at 121°C. Sterilizing foods at temperatures that kill large numbers of these renders the product unfit to eat.

3. *Natural Acidity.* The more acid the product, the more readily it can be sterilized. Some berries, such as cranberries, which are high in benzoic acid content, can be submerged in water and kept there easily for weeks without heating or sealing.

4. *Initial Temperature.* Only if all of the raw product to be processed is uniform in temperature can the proper processing time and temperature be accurately determined. Moreover, differences in the efficiency with which the product is handled during and after blanching and exhausting can result in differences of as much as 20 to 40°C in the temperature of cans entering the retort. Spores are destroyed in 78 minutes of heating if the product's initial temperature is 82.2°C, but require 83 minutes of heating if the initial temperature is 71.1°C. This is of great importance in viscous products that have poor conductivity and in foodstuffs processed in still retorts; initial temperature in agitating cookers is less important.

5. *Heat Penetration.* The rate at which heat penetrates a food is influenced by the size of the can: the larger the container, the longer the heating required to be lethal in the center. And because the contents of a can neither heat nor cool instantane-

ously or uniformly, both "come-up" time (the time required for the product to reach retort temperature) and cooling period are part of the process. Another critical factor is whether heat is transferred to the center of the contents by **convection,** which is rapid if brine within the can circulates freely, or by **conduction,** in which heat is transferred by molecular excitation. How quickly the commodity reaches retort temperature is also influenced by particle size; for example, large pieces of beets reach retort temperature at the center more slowly than do the smaller peas. Other factors\that influence heat penetration are starch gelation, as in cream style corn, and stratification, as in spinach and other leafy greens. Sucrose and fats exert a protective action, but salt has no effect until a concentration of 4% is used, at which point an enhanced lethal effect of heating is observed.

Steps in Canning

There are eight basic steps in canning: (1) preparation, (2) blanching, (3) filling, (4) exhausting, (5) sealing, (6) processing, (7) cooling, and (8) storage.

Preparation. Foods selected for canning are sorted, sized, graded, and otherwise prepared in the field or in sizing and grading sheds, receiving rooms, storage cellars, refrigerated warehouses and on husking, peeling, shelling, slicing, dicing, sorting ing, or trimming tables. These operations provide much opportunity for spreading contamination and permitting microbial growth to accelerate. Handling should be kept to a minimum, and where possible gloves, not bare hands, should touch the food. Storage, when necessary, should be at as low a temperature and as brief as possible to prevent or halt the growth of microorganisms already present.

Blanching. Blanching is the preliminary treatment of raw products with hot water or steam. Many foods—and particularly vegetables—are blanched for one or more of the following purposes: to reduce volume, to exclude air or reduce oxidation, to remove gums and waxes, to inactivate enzymes (e.g., peroxidases and oxidases), to reduce microbial populations, to fix color, to aid peeling, to improve flavor, to aid cleansing, and to comply with the desired or legal weight of the product in its final container.

Filling. Good technique must be used so that food placed in containers does not slop over the rim, remain on the lip, or leave large air pockets. Leafy products, such as spinach, may undergo slack fill unless the leaves are wilted beforehand. Some meat products may require prior heating to compensate for shrinkage. Other products—particularly those that heat by conduction—may expand and cause overfilling unless compensation is provided for such expansion. Materials entering the can are measured or metered. Before the can is sealed it passes under a tamper that compresses the solids and forces out excess fluid to ensure the proper headspace so that after the can is processed and cooled it is under a low partial pressure (i.e., high vacuum).

Exhausting. In this operation, the food product, before it is sealed, is placed for a short period in a hot water or steam bath, called an **exhaust** box. This operation is essential for several reasons. First, exhaust before sealing prevents excessive strain on large cans during processing. Otherwise, the expansion that takes place at increased temperature and pressure might cause paneling and buckling of the can. A standard #2 can withstands up to 30 psi without buckling, but a #10 (or institution size) can can withstand only 10 psi. Second, exhausting helps drive out dissolved gases, replacing much of the air with steam and thus halting oxidative deterioration. Third, exhausting ensures more uniform temperature of product entering the retort and permits the canner to adhere to standard processing schedules. The product's increased temperature and reduced level of dissolved gases at the time of sealing ensure a low partial pressure. In addition, exhausting prevents overfilling, while with soft products such as berries it may permit greater fill. Exhausting may reduce numbers of heat-sensitive microorganisms, but this role is of little consequence because the thermal process quickly destroys these organisms.

Sealing. Placing lids loosely on top of the can body and injecting live steam to sweep free air from the headspace may be part of exhausting or it may be a separate operation. To permit a good seal, the rim of the can to be sealed must be free of food tissue and all other debris. Where the lid is in contact with the can it usually has a thin layer of adhesive. A closing machine hooks the lid and can body together by means of a fairly tight but unwrinkled double seam. Although the operation seems quite simple, it is highly exacting and must be constantly examined to prevent seaming that is too loose or too tight, undercaps, underlaps, and seam failures (British Food Mfg. Inds. Res. Assn., 1941).

Processing. Thermal processing is the exposure of a product at a given temperature for a specific time to inhibit microbial and destabilizing enzymatic activity; the result is termed "commercial sterilization." Since overcooking harms certain organoleptic and nutritive properties of foods, care is employed to use only enough heat to accomplish the above-stated purposes while preserving maximum palatability and nutritive properties.

Commercially sterilized canned foods may not be completely sterile in the microbiological sense, but the surviving spores or contaminating cells have no capacity to initiate growth or to grow. In addition, commercial sterility implies that the product will have acceptable shelf life when stored at ordinary temperature. Products to be shipped to a hot climate require commercial sterilization times different from those going to a more temperate region. And because different food products have different temperature requirements for adequate processing, foods are usually grouped by their pH. Acid foods, those with pH from 4.5 to 3.2, include tomatoes, figs, pears, apricots, pineapple, plums, cherries, berries, citrus fruits, pickles, sauerkraut, etc. (see Figure 4-1). Several texts treat the mathematical principles of thermal processing.

Spoilage organisms in the most highly acidic foods such as citrus, plums, cherries, pickles, sauerkraut, etc.—i.e., those with pH 3.7 or less—consist principally of yeasts, molds, lactobacilli, and certain other non-spore-forming bacteria. These are quickly destroyed by heating to a temperature of 76.5°C for

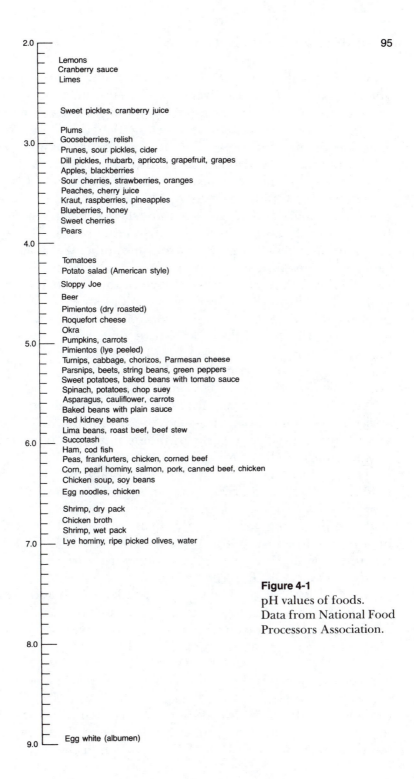

Figure 4-1
pH values of foods.
Data from National Food
Processors Association.

only a few minutes; surviving bacterial spores are ignored. Some nonacid foods are given a similar low-heat treatment if some factor of the food prevents bacterial growth—for example, the high sugar content of sweetened condensed milk prevents germination of spores. Other acidic foods that are high in solids content as well—such as jams, jellies, preserves, marmalades, syrups, purees, catsup, and other concentrates—are easily preserved by heating to 76.5°C. Even the most resistant species of molds and yeasts, the important spoilage agents for these products, cannot survive at this temperature. It is often forgotten that these organisms are osmophilic: if viable spores remain on nonsterile lids or are trapped in the headspace above the preserves, they can cause spoilage of the product after several weeks of storage. A germinating mold spore is capable of growing in catsup with a solids content of 34% and 1.5% of acid, or in the thin film of moisture that condenses on the surface of jams, jellies, and syrups (Townsend and Esty, 1939). After capping, such containers should be inverted while still hot, to sterilize the headspace and lid.

The less acidic of the acid foods, such as tomatoes, pears, ripe apricots, figs, pineapple, etc.—i.e., those with pH 3.7 to 4.5—are subject to spoilage not only by molds and yeasts but also by a group of mesophilic spore-forming bacteria that have comparatively high heat resistance. The latter organisms, which grow readily at pH 4.6 and slowly down to pH 3.7, are referred to as the "butyrics": they produce considerable amounts of butyric acid and hydrogen gas, but only small amounts of CO_2. Their optimum temperature for growth is 21 to 33°C and their maximum is 35°C. Spores of *Clostridium pasteurianum, Bacillus subtilis, B. coagulans, B. polymyxa,* and *B. macerans* germinate and grow in this pH range. The bacteria produce acid, and because they often are unable to sporulate under increased acidity, the food undergoes autosterilization with the passage of time. These foods are heated to an internal temperature of 95 to 100°C.

Low-acid, or neutral, foods such as meats, poultry, eggs, fish, shellfish, bananas, melons, most vegetables, and milk—i.e., those with a pH above 4.6—require severe heat processes. Boiling in water is inadequate; these foods are usually processed at temperatures of from 110 to 121°C. Although these foods can be spoiled by many types of mesophilic and thermophilic, aerobic and anaerobic spore-forming bacteria, the thermal processes used are calculated primarily from the heat resistance of *Bacillus stearothermophilus* (see Chapter 13). Nearly all vegetables have a pH of 5.0 or more, and peas and corn are usually 6.1 or more. Because bacterial spores can germinate freely when these foods have been canned, temperatures well above 100°C are employed to ensure sterility.

All modern thermal processing is based on two concepts advanced in 1920. The **thermal death point** is defined as the temperature required to completely destroy, within a stated time, a specified concentration of spores in a medium of known hydrogen ion concentration (Bigelow and Esty, 1920a). The **thermal death time** (TDT) is the time required at a given temperature to accomplish the same objective. An organism's heat resistance is expressed in terms of its TDT; that is, in minutes required to inactivate its population at a given temperature. The TDT decreases logarithmically as temperature increases (see Figure 4-2). This relationship can also be expressed as **lethality** (L). Since 10 minutes at 116°C is equivalent to 2.78 minutes at 121.6°C, 1 minute of 116°C and 0.278 minute at 121.6°C are similar, and each represents 0.1 of lethality. From these and other temperature increments a lethality curve can be plotted when the temperature and time at the point of slowest heating are known (Figure 4-3).

Bigelow and Esty demonstrated that strains of thermophilic bacteria differ in resistance to destruction by heat. Where Strain #1595 was destroyed in not more than 450 minutes (7.5 hours) at 100°C, the spores of Strain #1503 required exposure for 1020 minutes (17 hours) to be destroyed. They also demonstrated that as the temperature increased, the time required for destruction of the spores decreased. As a result of these studies Strain #1518 of *Bacillus stearothermophilus* (*stearo-*, fat; *thermo-*, heat; *-philus,* loving) was adopted as the standard test organism for studies in heat processing of low-acid foods. Subsequently, there developed the concept of the F value, defined as the time in minutes required to kill a standard suspension of spores in the food under test. While the temperature of 250°F (121.6°C) is frequently employed (designated as F_{250}), other temperatures are used, and designated in the subscript. Figure 4-2 shows that

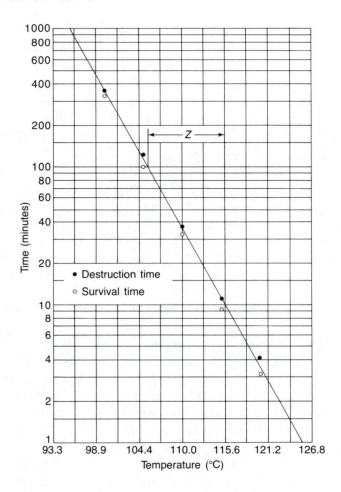

Figure 4-2
Thermal death time curve (for *Clostridium botulinum*)
illustrating a *Z* value of 10°C or degrees traversed in one log
cycle of time (Esty and Meyer, 1921. The University of
Chicago Press © 1921).

a process of 2.78 minutes at 250°F is equivalent to a process of
10 minutes at 240°F (116°C) or of 36 minutes at 230°F
(110.4°C). For convenience, the *slope* of the TDT is expressed as
the number of degrees Fahrenheit required for the curve to
traverse one log cycle on the time scale. The symbol *Z* (Figure
4.3) is used to represent the slope of the TDT curve (Bigelow et
al., 1920). Also, Figure 4-3 shows, if the slope of the TDT curve

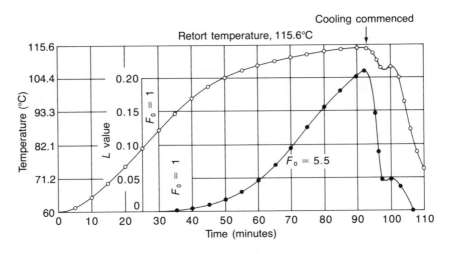

Figure 4-3

Heat penetration curve of a 303 × 406 can of solid food and the corresponding L value vs. time curve. Areas representing F_0 values of 1 are also shown (Board, 1965).

is 18°F, or $Z = 18$, the special condition for describing the number of minutes necessary to effect microbial destruction at 250°F is referred to as an F_0 value (for example, $F_0 = 2.78$). Using the F and Z values, the processor can select the time and temperature relationship required for specific commodities, since few foods are processed at 250°F (Manson and Cullen, 1974).

Refined sugar was once a major vehicle for introducing thermophiles into the canning plant (Cameron and Williams, 1928) (see also Chapter 13). The spores of the thermophiles lodge in warm or hot areas of the flow lines in canning operations. There they germinate and multiply to the point where the numbers of spores entering the container exceed the normal load expectancy for the thermal process. While thermophiles do not often accompany the raw products, they may contaminate wooden tanks, blanchers, cookers, conveyors, and conduits (Cameron and Loomis, 1938). They are often associated with soy products. Today most thermophilic spoilage occurs in pet foods (Segner, 1978).

To confirm theoretical processing schedules processors use experimental packs. Time–temperature relationships for a specific pack are determined with experimental packs of 100 cans at each level of heating at 5 minute increments. The temperature range is selected so that all or nearly all cans heated to the lowest temperature spoil, while no spoilage occurs among cans heated to the highest level. That heating level selected for the actual commercial pack is the one at which no spoilage occurs upon subsequent incubation. A level of 2% spoilage would not be tolerated by the industry.

Cooling. A minimum requirement for water used to cool cans after processing is that it be potable. Hot cans placed in the cooling canal are swollen, and cooling of the contents creates a partial pressure within the can. The hot cement in the liquid is still in the plastic state, and if the seal is not perfect it yields to the differential in partial pressure. Under such strain, water may be pulled inside the can if the seams are not completely tight. Thus, any viable organisms in the water "inoculate" the contents and bring about spoilage.

Storage. Canned food should be stored in areas that remain cool and dry. Cans should not be stacked near steam pipes, radiators, or furnaces, or immediately adjacent to uninsulated metal walls or roofs. Conversely, canned foods should not be permitted to freeze. It may be necessary to heat and/or ventilate the storage area and to maintain the cans at the same temperature as the surrounding atmosphere. While high temperatures adversely affect color, flavor, and texture, freezing is primarily detrimental to texture. Sudden temperature changes may cause the cans to "sweat" and later to rust. In coastal areas dry storage conditions are particularly crucial, because salt is especially corrosive to metals.

Safety

Clostridium botulinum. Types A and B of *Clostridium botulinum* are of concern in canned foods because their spores are resistant to heat. Types C and D are seldom if ever encountered in foods as human pathogens, and the spores of fish and marine types are readily destroyed by heat. Botulism is rarely caused by consumption of commercially canned foods; in the 75 years from 1899 through 1973, only 62 of the 688 recorded outbreaks were attributed to commercially canned foods. The chance of being killed by lightning is 100 times greater than that of being poisoned by commercially canned foods (Riester, 1974). Nevertheless, botulism is among the diseases most feared by consumers.

In what appears to be one of the earliest studies of heat resistance of the spores of *Clostridium botulinum*, it was reported that a suspension of 900 million spores were completely destroyed only after heating for 330 minutes at 100°C, but heating to 120°C reduced the time to 4 minutes (Esty and Meyer, 1921). Some strains sporulated irregularly and poorly, and the heat resistance of the spores of 109 strains was noted to vary widely. Subcultures of heated spore suspensions germinated as late as 378 days after heating. Dickson (1928) also noted dormancy among heated spore suspensions, with 14% of 490 such suspensions germinating at the end of 36 months. It should be noted that the earlier investigators often employed litmus milk containing reduced iron in the subculture of heated spore suspensions. With the culture media used today, germination of spores occurs rapidly, and dormancy is no longer prolonged.

As is true of other sporulating bacteria, the spores of *Clostridium botulinum*, when heated to the same temperature, are killed more quickly in acidic foods than in foods with low acidity. A month-old spore suspension can be destroyed in 50 minutes in a medium poised at pH 3.85 or less, in 60 to 90 minutes in a moderately acidic food with pH 4.2 to 5.1, 90 to 120 minutes at pH between 5.1 and 5.35, and 150 to 180 minutes for foods in the range of pH 7.0 (Weiss, 1921). However, with refined laboratory techniques, in which come-up time and other factors are easily controlled, the maximum recorded thermal death time of a concentration of 6×10^{10} spores of *C. botulinum* in phosphate buffer at pH 7.0 is 10 minutes (Lewis and Cassel, 1964; Schmidt, 1964). Early workers (Townsend et al., 1954) considered pH 4.7 to be the limiting level for germination and growth of *C. botulinum*. Ten years later Schmidt (1964) indicated that the lowest pH at which growth is initiated is pH 4.8—the limit that the Food and Drug Administration recognizes. Adding a safety factor of 0.2 units, the FDA has set an upper limit of pH 4.6 for acid foods not subject to thermal processing.

The safety of canned foods is assured through laboratory and pilot plant studies that are conducted independently of the studies determining the sterilizing process. Foods are subjected to the "botulinal cook," a thermal process in which a concentration of 1×10^{12} spores is reduced to 1×10^{0} (the $12\,D$ concept). Laboratory results are confirmed in canning practice: foods are inoculated with spores of *Putrefactive Anaerobe* NCA #3679 (an anaerobe similar to, if not identical with, *Clostridium sporogenes*), to prevent spores of *C. botulinum* from being carried into the canning plant. The concentration of spores of PA 3679 is selected to have a heat resistance equivalent to the resistance of the botulinal spores used in the laboratory.

All thermal death times are determined in accordance with the characteristics of the food being canned. When changes in formulation could affect rate of heat penetration, or when altered mechanical operations result in heavier fill of product or change some other product attribute, new laboratory studies must be conducted. Processors neglect precautions at their peril. For example, had the thermal characteristics of a widely distributed canned soup been checked by the laboratory when a more viscous formulation was adopted, *Clostridium botulinum* could have been detected before the product went into commercial production. Such action could have spared the company considerable embarrassment, anxiety, and expense. During 1972, in several separate incidents with another product, new mechanical fillers packed mushrooms more tightly than did hand filling and altered the heat penetration rate so much that the conventional process was no longer adequate to destroy *C. botulinum*.

Spoilage

Sources of Spoilage

A number of factors contribute to the spoilage of canned foods. Containers are understerilized if not subjected to the proper temperature for a long enough time. Putrefactive anaerobes may cause as much as 83% of spoilage of foods receiving insufficient heat treatment (Segner, 1978). The following factors are responsible for faulty processing:

improperly vented retorts

excessive condensate in retorts

failure to rehydrate dry products

very slow product heating characteristics

change in physical nature of product

inadequate fill, excessive headspace

faulty thermal process

Because understerilization and oversterilization alike were frequent occurrences when manual controls were in use, the Food and Drug Administration requires that thermometers and recording charts be used to determine the adequacy of heat processes. Nearly all time and temperature applications are now instrument controlled. In addition, retort operators are required to attend school to study the principles of retort operations.

If spore loads in the product are excessive, a small number of heat-resistant spores may survive. Very large spore loads usually follow the failure of plant sanitizing personnel to eliminate scums and slimes at critical points in the flow lines. Critical points are those with temperatures conducive to rapid multiplication of the bacteria.

If the temperature of the container contents entering the retort changes—and especially if it is reduced from the expected temperature—the contents at the center of the container take longer to reach retort temperature. In modern canning practice it is quite easy to heat or preheat foods to a specified temperature at the time of fill. But at one time equipment limitations prevented preheating of items prepared by hand labor between the cook and fill; such items included pumpkin, squash, greens, and sweet potatoes. Delay between filling and retorting, or a decrease in the ambient temperature, can also significantly lower product temperature.

Reaction between Tin and Food Acid. When the metal of the can and the acid of the food react, hydrogen is liberated and a swollen can results. Although foods in such cans are not spoiled, they are not acceptable to the consumer. Cans receiving highly acidic foods are now protected with a lacquer coating.

Mechanical Damage. Dents, rust, and punctures are external conditions that normally occur after the can has left the warehouse. Denting per se, if not severe, does no damage to the contents, since containers have some amount of headspace. It may, however, weaken some portion of the side or end seam, breaking the hermetic seal. Rust occurs when cans are stored in a damp atmosphere or under leaky roofs or undergo flooding. Puncture occurs through accidental contact with a sharp object.

Defective Containers. Figure 4-4 shows the steps in manufacturing the sanitary can. Defects may occur at any step during manufacture, but they occur most often in the alignment of the flange (step d); in poor compression, caused by worn-out rollers, on the flange or on the end plate and the lid (steps g and i); in the formation of the curl of the end plate (step h); and in imperfect deposit of the rubber sealing compound on the edge of the lid. Any of the defects result in a leaky container, which usually is detected by the can manufacturer or in the canning plant. Rarely do foods in defective containers reach the market. "Leakers" may also be caused by faulty sealing. Such cans often become contaminated when placed in the cooling canal. Even without such contamination, portions of the contents ooze out and foul the surfaces of adjoining cans in the stack or case. It is estimated that 60% of all spoilage is caused by leakers (Segner, 1978).

Nature of Spoilage

Spoilage of canned foods may be either flat sour or gassy. In **flat sour spoilage,** acid without gas is produced, and although outwardly the can seems normal, the food has a tart, metallic, medicinal taste. Except for evaporated milk, flat sour spoilage cannot be detected until the container is opened. In evaporated milk, however, a curd is formed that imparts a characteristic

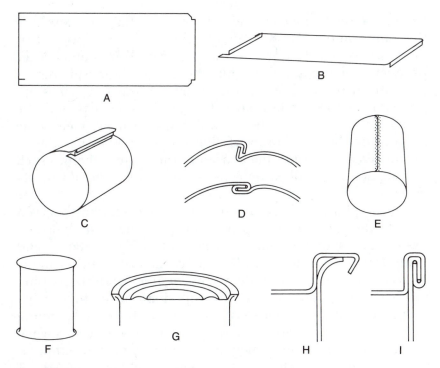

Figure 4-4
The tin can. Steps A–G represent manufacture of the can; steps H–I the
sealing process: A. formation and cutting the metal sheet; B. bending the
notched ends; C. rolling the body; D. interlocking and compression of the
notched ends; E. the finished body after treatment with flux and solder; F. the
curled end or flange, prepared to receive the bottom and the lid; G. the
bottom in place before cutting; H. placement of the lid after filling the can; I.
the end seam showing the sealed lid on the can. The cement is between the
rolled edge of the can body and the lid.

thumping sound when the container is shaken. Bacteria producing flat sour spoilage include *Bacillus stearothermophilus, B.
coagulans,* and some species of *Lactobacillus.*

Gassy spoilage is observed as the bulge of either one or both
ends of the metal can, or the bulge of the lid of the glass containers, or gas bubbles in retortable pouches. Several terms describe the extent of gas production: the term *flipper* is used if
the lid springs inward and outward when compressed; *springer,*
if the opposite end springs out when the bulged end is com-

pressed; and *soft* and *hard swells,* if both ends bulge outward. As the names imply, the lid of a *soft swell* can be compressed by moderate pressure, whereas that of a *hard swell* cannot. Gas-producing microorganisms tolerate several atmospheres of pressure in the can. When the internal pressure reaches approximately five atmospheres, the side seam parts and the liquid contents may be spewed a distance of many feet. Some flippers are the result of overfilling and are not attributable to microbial activity. Since the consumer cannot tell whether such a container is overfilled or contains spoilage, it should be rejected.

Thermophilic spoilage of canned foods is seldom observed outside the cannery. The processing program is designed to destroy any spores present at a normal load expectancy, and many canneries systematically sample, incubate, and examine foods taken directly from the processing line. Underprocessing can have any of several causes: composition of the food being canned may change; solids may increase, with a corresponding decrease in brine content; or spore loads may exceed normal expectancy. Precise instrumental control of time and temperature ensures that such microbial spoilage seldom occurs.

Some spoilage agents are characteristically associated with a particular food. Yeasts, for example, produce spoilage most commonly in highly acid foods such as citrus juices and pickles, although they may also produce ethanolic spoilage in other saccharine canned foods. Heterofermentative lactobacilli cause gassy spoilage of sauerkraut, tomatoes, and tomato juice. Clostridia are generally found in spoiled meats and meat products and animal foods, but *Bacillus* spp. are capable of reducing nitrate to produce gas in canned, nitrate-cured meat items. Growth of fermentative yeasts in sweetened condensed milk usually is halted by the combination of high sucrose and the small amount of ethanol, but fermentation may proceed to the point where the side seam bursts, with a slow oozing of the viscous contents. Members of the *Bacillus macerans polymyxa* group cause a gaseous deterioration of canned vegetables (Vaughan et al., 1952). Spores of these are highly heat resistant and tolerate a low pH.

Nonsporulating bacteria may be recovered from foods recently canned, but viable cells are seldom recovered from canned goods procured in the marketplace. Even though the organisms may be observed in stained preparations, the combination of high acid production and age apparently results in autosterilization.

Streptococci are often seen on stained vegetable brine preparations, but may have been resident on the plant before harvesting. Although *Streptococcus faecium* has been recovered in culture from European meats, especially hams, there is no evidence that it causes spoilage.

Molds are rarely seen in canned foods. An exception is *Byssochlamys fulvus*, cells of which survive the thermal process given fruits such as peaches, and grape juice. Ascospores of the mold tend to remain in the ascus, and seemingly are protected. The mold is controlled by stringent sanitization of containers used in harvesting, by the use of metal baskets, and by plant sanitation to minimize the spore load.

Tomatoes and tomato juice are an exceptional class of foods, because the ionizing acidity (pH) lies between that of fruits and vegetables, that is, between pH 4.1 and 4.5. Spoilage agents include *Clostridium pasteurianum, Bacillus coagulans* (also termed *B. thermoacidurans*), and *Lactobacillus brevis* (also known in the tomato canning industry as *L. lykopersicki*). Spoilage by *C. pasteurianum,* which may occur in tomatoes and juice with pH values above 4.3, is controlled by acidification with citric acid to pH 4.1 and heating the contents of the container to 93.8°C (Bowen et al., 1954). *B. coagulans* may accompany the tomato into the processing plant as a soil contaminant, and may become established in portions of flow lines that offer suitable temperature for growth. Effective washing practices and sanitation control the organism (American Can Company, 1948).

Home Canning

The glass canning jar, or "fruit jar," became available for home use during the last century. It was a welcome addition to the methods of food preservation then in use, but it antedated the pressure canner by several decades. Foods were preserved by open-kettle canning or processing in a boiling water bath, or by heating in the oven, where the highest temperature attainable was that of boiling water or of the dilute salt brine.

Because consumption of home-canned foods occasioned relatively frequent outbreaks of botulism during the early part of the century, the U.S. Bureau of Home Economics and its state counterparts embarked on an education program promoting the use of the pressure canner. The pressure canner was not universally adopted, however, and even today some home canners process vegetables in boiling water, a practice prompting the statement that "people who could probably never be reached by any method of warning have been victims of their own ignorance" (Meyer, 1931).

Also during the earlier years of home canning there developed the unscrupulous practice of selling the home canner so-called canning compounds to be added to the container before processing. Some tablets consisted merely of NaCl; these at least were innocuous. Others, however, are reputed to have been composed of sodium bicarbonate, aspirin compounds, or salicylates. $NaHCO_3$ has an alkalinizing effect. Presumptive evidence indicates that home canners were using salicylates as recently as 1976. Salicylate added to tomato juice causes a darkening in color, and when added to excess turns the serum nearly black.

Studies conducted in the mid-1940s revealed that approximately 2% of home-canned foods spoil, most because of under-sterilization, and many because of improper seal (Esselen and Tischer, 1945). Tests with bacteria recovered during the study revealed that 5.5 to 12 hours of processing in boiling water was inadequate to kill the test organisms.

As early as 1917, Dickson (1917) associated outbreaks of botulism with the cold pack method of canning. Between 1899 and 1973, 495 recorded outbreaks of botulism involved home-canned foods, while 62 incriminated commercially canned foods (U.S. Dept. of Health, Education, and Welfare, 1974). Many outbreaks were traced to the consumption of green beans, corn, olives, spinach, asparagus, and beets. *Clostridium botulinum* does not germinate at a pH of less than 4.8, yet intoxication has been associated with such acidic foods as apricots, pears, chili peppers, catsup, tomato relish, and green tomatoes. *C. botulinum* has sometimes been shown to grow as the result of a synergistic action with alkalinizing microorganisms (Meyer and Gunnison, 1929), but synergism in acid foods has not always been demonstrated. It is possible that the synergist and *C. botulinum* have grown in a specific particle of food.

It is incorrect to assume—as is often done—that foods in which *Clostridium botulinum* has grown will show obvious spoilage. Foods may be unchanged, or may be so little changed that less discerning consumers accept them. Some consumers also fail to understand that spores introduced from the soil firmly adhere to the vegetable, and no amount of washing dislodges them. Therefore, the only means to ensure freedom from intoxication remain the dual recommendations that

1. home-canned foods are not to be tasted until they have been heated; and

2. foods must be boiled for 10 minutes to ensure destruction of the toxin.

Having developed a tomato with a higher sugar content that masks the tomato's acid flavor, plant breeders introduced the term "low-acid tomato." The term was mistakenly thought to mean a true mild tomato—and thus one capable of breeding botulism. However, none of the 55 varieties of tomatoes recently tested has a pH high enough to support the growth of *C. botulinum*. On the other hand, the pH of very overripe tomatoes can reach a low acid range and may enable growth of the bacterium.

References

Many of the classical papers about canning are not readily available in libraries. The American Can Company (1948), Bitting (1937), and Goldblith, Joslyn, and Nickerson (1961) have compiled a number of papers in a bulletin or in texts. The volume by Bitting reproduces fascinating articles published during the early part of the nineteenth century that indicate a keen insight into heat preservation by food processors who had no knowledge of microorganisms and their role in the spoilage processes.

American Can Company, 1948. *Control of Flat Sour Spoilage in Tomato Juice.* American Can Company, Maywood, Ill.

Ball, C. O. 1923. *Thermal Process Time for Canned Food.* Natl. Academy of Sciences, Washington, D.C.

Barlow, B. 1913. A Spoilage of Canned Corn Due to Thermophilic Bacteria. Thesis. Univ. of Ill., Urbana, Ill.

Bigelow, W. D., G. S. Bohart, A. C. Richardson, and C. O. Ball. 1920. Bull. 16-L., National Canners Assn.

Bigelow, W. D. and J. R. Esty. 1920a. J. Infect. Dis. *27,* 602.

Bigelow, W. D. and J. R. Esty. 1920b. Bull. 16-L., Natl. Canners Assn.

Bigelow, W. D., H. R. Smith, C. A. Greenleaf. C. W. Bohrer and J. M. Reed. 1950. *Tomato Products: Pulp, Paste, Catsup,* and *Chili Sauce.* Bull. Natl. Canners Assn., Washington, D.C.

Bitting, A. W. 1937. *Appertizing or the Art of Canning: Its History and Development.* Trade Pressroom, San Francisco, Calif.

Board, P. W. 1965. Determination of Thermal Processes for Canned Foods. CSIRO, Div. Food Preserv. Circ. 7-P, Melbourne, Austr.

Bowen, J. F., G. C. Strackan, and A. W. Moyles. 1954. Food Technol. *14,* 471.

British Food Mfg. Inds. Res. Assn., 1941. Can Seaming Efficiency. Food Res. Rept. #41.

Cameron, E. J., and H. M. Loomis. Food Res. *3,* 91.

Cameron, E. J., and C. C. Williams. 1928. Zentralbl f. Bakteriol. Parasitenk u Infektionskr. II Abt. *76,* 28.

Dickson, E. C. 1917. J. Amer. Medical Assn. *99,* 966.

Dickson, E. C. 1928. Proc. Exper. Biol. and Med. *25,* 426.

Drummond, J. C., W. R. Lewis, T. Macara, G. S. Wilson, and H. L. Shipp. 1938. Food Indus. *57,* 808.

Esselen, W. B., and R. G. Tischer. 1945. Food Res. *10,* 197.

Esty, J. R., and K. F. Meyer. 1921. J. Infect. Dis. *31,* 650.

Goldblith, S. A., M. A. Joslyn, and J. T. R. Nickerson. 1961. *Introduction to Thermal Processing of Foods.* Avi Publ. Co., Inc., Westport, Conn.

International Commission on Microbiological Specifications for Foods (ICMSF) of the Intl. Assn. of Microbiological Societies. 1974. *Microorganisms in Foods.* 2. *Sampling for Microbiological Analysis: Principles and Specific Applications,* Univ. of Toronto Press, Toronto, Canada.

Lewis, K. H., and R. Cassel, Jr., eds., 1964. *Botulism Proceedings of a Symposium.* U.S. Dept. of Health, Education, and Welfare, Public Health Service, Cincinnati, Ohio.

Manson, J. E., and J. F. Cullen. 1974. J. Food Sci. *39,* 1084.

Meyer, K. F. 1931. Amer. J. Public Health *21,* 762.

Meyer, K. F., and J. B. Gunnison. 1929. J. Infect. Dis. *45,* 135.

National Canners Association. 1966. *Processes for Low-acid Canned Foods in Metal Containers.* Bulletin 26-L.

National Canners Association. 1968. *Laboratory Manual for Food Canners and Processors.* AVI Publishing Company, Westport, Connecticut.

National Canners Association. 1971. *Processes for Low-acid Canned Foods in Glass Containers.* Bulletin 30-L.

Riester, D. 1974. FDA Consumer *8*(6), 10.

Schmidt, C. A. 1964. Spores of *C. botulinum:* Formation, resistance, germination. In *Botulism: Proceedings of a Symposium,* K. H. Lewis and K. Cassel, Jr., eds. U.S. Dept. of Health, Education, and Welfare, Public Health Service, Cincinnati, Ohio.

Segner, W. 1978. Update on Food Spoilage Organisms. Paper presented for Food Microbiology Division Program, 38th Annual Meeting, Institute of Food Technologists, Dallas, June 4–7.

Stumbo, C. R. 1965. *Thermobacteriology in Food Processing.* Academic Press, New York, N.Y.

Townsend, C. T. and J. R. Esty. 1939. Western Canner and Packer (3) (4) and (5).

Townsend, C. T., L. Yee, and W. A. Mercer. 1954. Food Res. *19,* 536.

U.S. Dept. of Health, Education, and Welfare, 1974. *Botulism,* Publ. No (CDC) 74-8279.

Vaughn, R. H., I. H. Kreulewith, and W. A. Mercer. 1952. Food Res. *17,* 560.

Weiss, H. 1921. J. Infect. Dis. *29,* 362.

5

Radiations and Electromagnetic Waves

Wilhelm Roentgen's discovery of X rays, in 1895, began an era that, in the decades since World War II, has culminated in the experimental application of various types of particle and electromagnetic radiations in the foods industries. Several technological developments gave much of the impetus to research: nuclear reactors, which have made radioisotopes inexpensively available; electron accelerators such as the Van de Graaf accelerator, which discharges electrons with a high velocity in the steady state; and the linear accelerator and the resonant transformer, which discharge electrons in pulses. Such technology has present and potential application in the foods industries in sterilization, pasteurization, disinfection, disinfestation, sprout inhibition, and heating.

Types of Radiations

Naturally occurring ionizing radiations and artificially produced electron particles elevate electrons to energy states that release them from atoms, producing ions and free radicals. Table 5-1 lists types of ionizing radiations and some of their

Table 5-1
Ionizing Radiations

Type of particle	Description
Alpha	Occurs through decay of radioactive elements; positive charge; extremely short travel distance; no penetration
Beta	Occurs through decay of radioactive elements; negative charge; deep penetration
Proton	No penetration; loses energy rapidly
Meson	Very high energy particle with no present application
Positron	Insufficient energy to penetrate
Neutron	No electric charge; deep penetration; radioactive atoms produced
Electron	Accelerated to extremely high velocities

Table 5-2
Electromagnetic radiations

Type	Wavelength (meters)	Source
Gamma	1×10^{-15} to 10^{-13}	Decay of radioactive particles
X ray	1×10^{-9} to 10^{-10}	Bombardment of metal target
Ultraviolet	1×10^{-8} to 3×10^{-6}	Selective emission through cathode tube
Infrared	7.5×10^{-3} to 4×10	Resonance
Microwave	0.3 to 3.0×10^{2}	Oscillatory electric current produced in magnetron generator
Ultrasonic	2.5 to 4.5×10^{4}	Resonance

sources. Among this group only the electron particles have present or potential applications to foods. Table 5-2 lists types of electromagnetic radiations. Of these, gamma rays and X rays have sufficient energy to cause ionization in the target. Gamma rays and ultraviolet rays have germicidal uses. Applicability of X-rays is limited by the expense of producing and delivering a small quantity of energy to the target. The remaining electromagnetic radiations have uses in which lethality is not the primary objective.

Alpha and beta particles originate from decay of radioactive elements. Alpha particles have a very short travel distance and do not penetrate. The negatively charged beta particles penetrate deeply, but have limited intensity. Protons and positrons do not possess sufficient energy to penetrate surfaces. Mesons are very high energy particles with no present applications. Neutrons penetrate surfaces, but they also produce radioactive atoms in the target.

Electron particles, also termed cathode rays, are streams of beta particles that have been accelerated to extremely high velocities in a Van de Graaf generator (Figure 5-1), linear accelerator, or resonant transformer. The emitted beam of electron particles can be directed at a target. Velocities of the accelerated particles have reached as high as 12 meV (million electron volts). The energy absorbed by the target is measured

A   B

Figure 5-1
Van de Graaf generator: A, Accelerator and, B, electromagnet for capturing particles.

in **Roentgen equivalent particles** (reps) and rads. The rep is defined as the amount of energy that releases one electrostatic unit of charge in passage through 1 cm^3 of air under standard conditions. The **rad** is the quantity of ionizing radiation required to absorb 100 ergs per gram of irradiated material. Rad values may be expressed as the krad, or 1000 rads; and the mrad, one million rads. Sterilizing treatments are termed radappertization, or radsterilization, and pasteurizing treatments are termed radpasteurization, or radurization.

Accelerated electron particles produce no measurable radioactivity in the target (Meinke, 1954), although they increase temperature by approximately 2°C per 2 meV. Foods intentionally radiated with cathode rays are considered to be adulterated, and food radiations must be performed in accordance with regulations of the Food and Drug Administration.

During passage into the target, accelerated electron particles and gamma rays produce ions and free radicals, as the following equations illustrate (Proctor and Goldblith, 1952):

$$H_2O \longrightarrow H_2O^+ + e^- \quad \text{(liberation of electron)}$$

$$H_2O + e^- \longrightarrow H_2O^- \quad \text{(negative water ion produced)}$$

$$H_2O^+ \longrightarrow H^+ + OH \quad \text{(free OH radical produced)}$$

$$H_2O^- \longrightarrow H + OH^- \quad \text{(free H radical produced)}$$

Recombination of the H^+ and the OH^- ions produces water. The OH radical is a strong oxidizing agent that is responsible, at least in part, for the lethal effect on microorganisms through because the two radicals combine to form peroxide. Catalase in the irradiated medium exerts a protective action on cells (Williams and Kempe, 1957). It is estimated that 99% of the free radicals recombine with each other (Miller et al., 1954).

When logarithms of surviving numbers of irradiated cells were plotted against dosage, cell deaths used to appear to be of first-order magnitude, that is, linear. However, such relations appear only in studies in which survivor curves were generally established on the plating of irradiated substances immediately after exposure. It now appears that many cells are damaged rather than killed, and accumulating evidence reviewed by Desrosier and Rosenstock (1960) suggests that cells cultured in media with the proper nutrients undergo repair and are able to resume their normal activities. Because of the mixture of sensitive and resistant cells in a culture, the dosage required to kill 100% of the cells may be 10 times the dosage required to kill 99% (O'Meara, 1952). O'Meara states that the energy required to kill by radiation is 2% that of steam energy, but heat energy is uniformly distributed whereas radiation energy is transmitted by ionizing the occasional molecule. Thus, complete destruction in fact requires a high dosage if the "scatter effect" is to work. The effectiveness in killing of microorganisms is sometimes conceived as the "one hit" theory, or the effect of a single particle upon a cell. The theory has been questioned by Wheaton and Pratt (1962), who estimate that as many as 13 impacts are required to inactivate bacterial spores.

Susceptibility to Radiations

Organisms are resistant to radiations in the following ascending order: insects, gram-negative bacteria, asporulating gram-positive bacteria, molds and yeasts, sporulating bacteria, viruses and enzymes. Wide ranges of variation have been reported within each group, the variations occurring among genera, species, and even strains of microorganisms being radiated. Variable factors affecting resistance are the age of the cells, young cells being more susceptible than mature cells; the temperature of the object being irradiated; the characteristics of the substrate; the physical state, whether fluid or solid; and the presence or absence of oxygen.

The presence of insects in grain increases a_w, and microbial deterioration follows. When bulk grain has been experimentally irradiated, by flowing at a controlled rate over the radiation source, the insects have been destroyed. The radiation affects the insect's midgut. However, there is no commercial grain radiating installation in the United States (Tilton and Brower, 1973).

The pseudomonads are markedly less resistant to the lethal effects of radiation than are cells of *Moraxella* and *Acinetobacter* spp. Tiwari and Maxcy (1971 and 1972) observed that the percent of these latter bacteria encountered on platings of ground beef was eight % prior to radiation of the beef and as high as 52% after. Among pure cultures of the species of *Moraxella* and *Acinetobacter*, the *D* values ranged between 16 and 112, and some strains were more radiation-resistant than *Micrococcus radiodurans* (Welch and Maxcy, 1975). Cells of the latter bacterium, which resembles *M. roseus*, were six to eight times as resistant as cells of *Staphylococcus aureus* (Anderson et al., 1956) and greater than that of spore-forming bacteria tested at that time. Spores of *Clostridium botulinum* Type E exhibit 45 to 55% the resistance of spores of Types A and B (Schmidt et al., 1962); however, it should be kept in mind that the spores of Type E differ from those of Types A and B in other respects, notably in having a lower resistance to heat. Yeasts are generally more resistant than are the gram-negative bacteria. The resistance of mold spores varies, a dosage of 250,000 reps being required to destroy spores of *Aspergillus niger*, more than one million reps to destroy spores of *Mucor* sp. (Dunn et al., 1948).

Radiation dosages required for sterilization vary directly with the numbers of cells to be irradiated. A suspension of 46,000 spores of *Clostridium botulinum* may be destroyed with a dosage of 180 krads, but 82 million spores suspended in phosphate buffer require 3.8 mrads (Denny et al., 1959). To destroy spores frozen in phosphate buffer requires 40% more energy than is required to destroy comparable numbers in liquid suspension. Edwards et al. (1954) reported similar observations during studies with *Bacillus subtilis*.

Sterilization

In food canning, bacterial spores able to cause spoilage are more likely to survive than the less heat-resistant spores of *Clostridium botulinum*. The spores of spoilage mechanisms have not been observed to survive radiation dosages lethal to spores of *C. botulinum*. Many radiosterilization experiments have been directed to ascertaining a dosage at which no spores of *C. botulinum* survive. That dosage is an ionizing radiation of 4.8 mrads (Kraybill, 1965). Ross outlines the experimental procedures for applying the $12D$ concept, or the probability that no more than one spore among an inoculum of 1×10^{12} will survive the radiating process. To show how this recommended dosage was arrived at, several of the many reports in the literature are summarized here.

When containers received inocula of 10,000 spores, toxic swells were noted in canned peas radiated with 2.5 mrads but not with 3 mrads, and in chicken radiated with 3.5 mrads but not with 3.8 mrads (Pratt et al., 1959). Nontoxic swells were observed at radiation levels of 3.7 mrads but none at 4 mrads. Canned meats inoculated with putrefactive anaerobes were sterilized by radiation with 2.5 mrads, but meats inoculated with *Clostridium botulinum* required 3.5 mrads (Kempe et al., 1954). Anellis et al. (1969) reported the calculated theoretical 12-log reduction in numbers of spores of *C. botulinum* in bacon to be 2.65 to 2.87 mrads, with observed minimum sterilizing dosages to be 2.0 mrads; below this level of radiation viable spores can exist for at least eight months without producing visible or toxic spoilage in bacon stored at 30°C. Later studies indicated that the minimum radiation dose required for the 12 *D* equivalent to be in excess of 4 mrads (Anellis et al., 1969).

Radiation of foods prior to heat sterilization reduced the heat required in proportion to the level of radiation, but no effect was observed if sublethal heat was applied prior to radiation (Kempe, 1955). Radiation of canned hams, which may contain large numbers of lactic acid bacteria, appears to have no advantage over the combination of heat treatment and radiation at a level of 1 mrad (Drake et al., 1960).

Authors in the proceedings of several symposia (Kraybill, 1965; Morgan and Siu, 1957; Raymon and Byrne, 1957) review studies of numerous foods including meats, milk and milk products, poultry, sea foods, cereals, baked products, and vegetables. Radiation at levels far below sterilizing values produced changes in flavor; the flavors are variously described as burnt, metallic, bitter, cured, cheesy, goaty, and in beef, wet dog. Flour made from irradiated wheat yielded smaller cakes than did the same amount of nonirradiated flour. Vegetables irradiated with 2 to 3 mrads lost bouquet and texture and became darkened. Radiated fruits show similar effects.

Flavor defects in radiated meats have been an object of serious criticism; in order of decreasing sensitivity, beef, pork, ham, bacon, and poultry are affected (Josephson et al., 1968). Radiating in the frozen state at low temperatures (−30 to −80°C) may reduce flavor defects; however, because bacteria have greater resistance at lower temperature, and because impedance to penetration is greater in the frozen than in the liquid state, higher dosages are needed and the cost is correspondingly higher. Foods subjected to flavor changes may be protected by irradiation in vacuum (Proctor et al., 1956) or by the addition of reducing substances such as ascorbic acid and its sodium salt (Proctor and Goldblith, 1952).

Bacterial radsterilizing dosages do not inactivate enzymes in foods. Milk may be pasteurized with 200 krads and sterilized with 750 mrads, but phosphatase is detectable after a treatment level of 10 krads, and 25% of the diastase in barley malt was active after an exposure to 15 mrads (O'Meara, 1952). To inactivate enzymes, then, foods to be radsterilized are given an initial heat treatment.

Radpasteurization or Radurization

Radpasteurization is done to prolong the shelf life of perishable commodities, by reducing the number of viable microorganisms, and to free foods from objectionable microorganisms such as salmonellae. Low-intensity radiations are also used to delay sprouting of such vegetables as potatoes and onions (Kraybill, 1965) and the ripening of such fruits as bananas. Radpasteurization increases the storage life of perishable foods, but only by a finite amount, because eventually fungal growth on fruits and vegetables and bacterial growth on meats reoccur, as does senescence through the continuing activity of naturally present enzymes. Radpasteurization of fruits and vegetables is not at present considered practical.

It has been proposed that shell eggs be treated with radpasteurization of 600 mrads (Desrosier et al., 1955). When whole egg and yolk solids and frozen egg yolk were radiated with 400 to 650 mrads, the number of salmonellae was reduced by 10^7—equivalent to the possible survival of one cell in 2.2 million pounds of egg containing an inoculum of 100 cells of *Salmonella typhimurium* and *S. senftenburg* per gram of egg (Brogle et al., 1957). If ground beef is radiated in the range of 34 to 68 mrads, most of the gram-positive and gram-negative asporulating bacteria are destroyed, but spoilage occurs during refrigerated storage because psychroduric bacteria renew growth

(Tiwari and Maxcy, 1971). Radiation of cut beef surfaces destroys cells of *Pseudomonas geniculata,* the bacterium commonly associated with the darkening of beef; ultimately, however, resistant bacteria, chiefly cells of *Microbacterium thermosphactum,* bring about darkening and spoilage (Wolin et al., 1957). Horse meat, imported into Great Britain in large quantities generally is contaminated with salmonellae, but can be rendered safe with as little as 0.65 mrad radiation.

If fresh ocean fish are irradiated within three days after being caught the shelf life can be extended to as many as 20 days if they are refrigerated at or below 4°C, making lengthy inland shipments of fresh fish possible. Spoilage bacteria are reduced in number by as much as 95%, but yeasts of several genera survive (Dassow and Miyauchi, 1965). Radpasteurization in the range of 150 to 250 mrads prolongs the life of fillets and 450 mrads that of clam meats.

Electromagnetic Radiations

Microwaves

Microwaves are produced from direct current by radiofrequency power tubes. Water molecules exposed to alterations of current undergo rapid molecular reorientation, producing heat through friction. It is not the microwaves but this resultant heat that induces lethality (Carroll and Lopez, 1969) and inactivation of enzymes (Lopez and Baganis, 1971).

Microwave ovens are used widely in homes and restaurants because foods can be rapidly heated and cooked. Such applications have been criticized because the ovens allow small numbers of asporulating pathogens (Baldwin et al., 1971; Woodburn et al., 1962) and the spores of *Clostridium perfringens* (Craven and Lillard, 1974) to survive. Bacteria survive because microwave ovens could not induce an internal temperature of 55°C in every portion of the food being heated.

Microwave energy is used to temper large blocks of frozen foods, the temperature being elevated from the deeply frozen to near thaw in minutes, thereby precluding the growth of psychroduric bacteria during thawing (Bezanson, 1976). Foods in plastic pouches can now be thermally processed (Kenyon et al., 1971).

Infrared Rays

Infrared rays induce molecules to vibrate; the energy from vibration is translated into heat on surfaces. Infrared is used to surface dry foods and to keep food hot in food service establishments. Molin and Oestlund (1975) have achieved rapidly attained, high, and well-defined surface temperatures that destroy spores of *Bacillus subtilis.* They observed mixtures of heat-sensitive and heat-resistant fractions in the spore suspensions and obtained a two-phase survival surve, the resistant group being over seven times more resistant than the sensitive group.

Ultraviolet Rays

Ultraviolet (UV) rays elevate the energy in the target to alter the relationship of electrons within the atoms. Although UV rays extend over a broad range in wavelength, the specific wavelength of 25.6×10^{-7} meters, or 2560 Å gives the most effective germicidal action. UV rays are used to retard the growth of molds during the high-temperature aging of beef; over packing lines to assure aseptic packaging; and to sterilize beverage waters when chlorine, if used, would react with substances in the water to produce an objectionable flavor. Mack et al. (1959) reported that counts of microorganisms in radiated apple juice were reduced to as low as 0.003%. Maunder (1977) has reviewed the recent development of a high-intensity ultraviolet lamp (capable of a range of 0.1 to 1.0 watts/cm^2, whereas conventional lamps emit in the range of 1 to 10 microwatts) and its possible application to sterilizing plastic pouches for aseptic packaging.

Miscellaneous Forms of Electric Energy

Cavitation, or the foaming of protoplasm, occurs when ultrasonic waves bombard surfaces at a very high frequency to dislodge soil. Sound waves at a frequency of 280 to 300 kilocycles have been investigated for application in processing tomato juice (Havens et al., 1952). Direct ohmic heating, in which the food is part of the electrical circuit, may be useful for rapidly heating large masses of food (Decareau et al., 1976). Dielectric heat, generated by contact between electrodes, may be useful to heat treat fabrics. It has also been proposed as a way of destroying mold spores in Boston brown bread (Bartholomew et al., 1948).

References

Anderson, A. W., H. C. Jordan, R. F. Cain, G. Parrish, and D. Duggan. 1956. Food Technol. *10*, 575.

Anellis, A., D. Berkowitz, C. Jarboe, and H. M. El-Bisi. 1969. Appl. Microbiol. *18*, 604.

Anellis, A., N. Grecz, D. A. Huber, D. Berkowitz, M. D. Schneider, and M. Simon. 1965. Appl. Microbiol. *13*, 37.

Baldwin, R. E., M. Cloninger, and M. L. Fields. 1968. Appl. Microbiol. *16*, 1929.

Baldwin, R. E., M. D. Fields, W. C. Poon, and B. Korschgen. 1971. J. Milk Food Technol. *34*, 467.

Bartholomew, J. W., R. G. Harris, and F. Sussex. 1948. Food Technol. *2*, 91–94.

Bezanson, A. F. 1976. Food Technol. *30*(12), 34.

Brogle, R. C., J. T. Nickerson, B. E. Proctor, A. Pyne, C. Campbell, S. Charm, and H. Lineweaver. 1957. Food Res. *22*, 572.

Carroll, D. E., and A. Lopez. 1969. J. Food Sci. *34*, 320.

Craven, S. E., and H. S. Lillard. 1974. J. Food Sci. *39*, 221.

Dassow, J. A., and D. T. Miyauchi. 1965. Publication 1273, S. A. Goldblith, ed. National Academy of Sciences–National Research Council, Washington.

Decareau, J. K., J. M. Prifti, and D. E. Westcott. 1976. Food Technol. *30* (12), 36.

Denny, C. B., C. W. Bohrer, W. E. Perkins, and C. T. Townsend. 1959. Food Res. *24*, 44.

Desrosier, N. W., F. J. McArdle, H. A. Hollender, and W. W. Marion. 1955. Food Engr. *27*(5), 78.

Desrosier, N. W., and H. M. Rosenstock. 1960. *Radiation Technology in Food Agriculture and Biology.* AVI Publishing Company, Westport, Conn.

Drake, S. D., J. B. Evans, and C. F. Niven, Jr. 1960. Food Res. *25*, 270.

Dunn, C. C., W. C. Campbell, H. Fram, and A. Hutchins. 1948. J. Appl. Physics *19*, 605.

Edwards, R. B., L. Peterson, and D. G. Cummings. 1954. Food Technol. *8*, 284.

Havens, R. A., B. E. Orr, and P. E. Pitts, Jr. 1952. Food Engr. *24*(8), 184.

International Atomic Energy Agency. 1968. *Preservation of fruit and vegetables by radiation. Proceedings of a Panel.* Vienna.

International Atomic Energy Agency. 1973. *Radiation Preservation of Food.* Proceedings of a Symposium. Vienna.

Josephson, E. S., Brynjolfsson, and E. Wierbicki. 1968. Trans. N.Y. Acad. Sci. *30*, 600 (Ser. II).

Kempe, L. L. 1955. Appl. Microbiol. *3:*346.

Kempe, L. L., J. T. Graikoski, and R. A. Gillies. 1954. Appl. Microbiol. *2*, 330.

Kenyon, E. M., D. E. Westcott, P. A. La Casse, and W. J. Gould. 1971. J. Food Sci. *36*, 289.

Kraybill, H. F. 1965. In: *Radiation Preservation of Foods.* National Academy of Sciences–National Research Council, Publication 1273.

Lopez, A. and N. A. Baganis. 1971. J. Food Sci. *36*, 911.

Mack, S. D., J. E. Albrecht, J. H. Litchfield, and M. E. Parker. 1959. Food Res. *24*, 383.
Maunder, D. T. 1977. Food Technol. *31*(4), 36.
Meinke, W. S. 1954. Nucleonics *12*(10), 37.
Miller, W. C., B. E. Proctor, and S. A. Goldblith. 1954. J. Milk Food Technol. *17*, 159.
Molin, G., and K. Oestlund. 1975. Ant. v. Leeuw. *41*, 329.
Morgan, B. H., and K. G. H. Siu. 1957. In: *Radiation Preservation of Foods*. Bailey, S. M. et al., eds. U. S. Government Printing Office, Washington, D.C.
O'Meara, J. 1952. Nucleonics *10*(2), 19.
Pratt, G. B., E. Wheaton, C. W. Bohrer, and C. B. Denny. 1959. Food Res. *24*, 51.
Proctor, B. E., and S. A. Goldblith. 1952. Nucleonics *10*(4), 64.
Proctor, B. E., J. T. Nickerson, and J. J. Licciardello. 1956. Food Res. *21*, 11.
Raymon, M. M., and A. F. Byrne. 1957. In: *Radiation Preservation of Foods*. Bailey, S. M. et al., eds. U.S. Government Printing Office, Washington, D.C.
Ross, E. W., Jr. 1974. J. Food Sci. *39*, 800.
Schmidt, C. F., and W. K. Nank. 1960. Food Res. *25*, 321.
Schmidt, C. F., W. K. Nank, and R. V. Lechowich. 1962. Food Res. *21*, 77.
Tilton, W. E., and J. H. Brower. 1973. In: *Radiation Preservation of Foods*. International Atomic Energy Agency, Vienna.
Tiwari, N. P., and R. B. Maxcy. 1971. J. Food Sci. *36*, 833.
Tiwari, N. P., and R. B. Maxcy. 1972. J. Food Sci. *37*, 901.
U.S. Dept. of Commerce. 1965. *Radiation Preservation of Food*. Washington, D.C.
Welch, A. G., and R. B. Maxcy. 1975. Appl. Microbiol. *30*, 242.
Wheaton, E., and G. B. Pratt. 1962. Food Res. *27*, 327.
Williams, N. J., and L. L. Kempe. 1957. Appl. Microbiol. *5*, 366.
Wolin, E. F., J. B. Evans, and C. F. Niven, Jr. 1957. Food Res. *22*, 682.
Woodburn, M., M. Bennion, and G. E. Vail, 1962. Food Technol. *16*, 98.

6

Chemicals in Foods

Adding chemicals to foods is another modern practice with ancient roots. For example, Hebrew writings do not make clear whether lye, obtained by leaching wood ashes, was given uses other than medical by the Hebrews, but it is clear that the early Chinese used lye to preserve eggs, and burned kerosene to ripen bananas and peas. (The Chinese were not aware that such combustion produced the ripening gases ethylene and propylene.) Moreover, if such common items as salt and vinegar are thought of as chemicals, then ancient peoples intentionally used additives as preservatives and seasonings, as Old Testament and ancient Persian writings attest. Spices and condiments were so eagerly sought that they inspired many an exploration.

Although used primarily to enhance the taste and appearance of already desirable dishes, additives were often employed also as adulterants, to make inferior food seem acceptable. The number of ancient decrees forbidding various techniques of defrauding the public is evidence that adulteration was an early widespread practice. In the nineteenth century the list of known chemical substances grew; at the same time, through the pioneering work of Appert, Pasteur, Prescott, and their contemporaries throughout the world, it was being learned how harmful microorganisms could be to food. It was natural, then, that the effects of chemicals on microorganisms would become a subject of study. In the generations since, a wide array of chemicals has been used or proposed for use in all facets of the foods industry.

To assure adequate harvests despite some of the vagaries of nature, farmers employ chemicals as fertilizers, as insecticides, herbicides, fungicides, and rodenticides, and as feed additives and plant growth regulators to increase yields and stabilize growth. To convert the raw commodities into desirable products, one or more of the following additives are then used: coloring and flavoring materials, vitamins, essential amino acids, preservative agents, antioxidants, stabilizers and thickeners, emulsifiers, firming agents, coatings, humectants, buffers, and neutralizing agents. Other end products are derived from processing changes induced by heat, freezing, dehydration, fermentation, radiation, and/or enzymatic action. And many food products reach the consumer packaged in a rigid or flexible container, or have been smoked or otherwise treated.

General Considerations

The natural substances in food products are chemicals; the remainder must be considered intentional additives to retard spoilage, decomposition, fermentation, oxidation, rancidity, instability, flaccidness, staling, acidification, or other undesirable changes, or to improve nutritional properties and/or aesthetic quality such as color, flavor, and texture (Larkin, 1976). Tolerances have been established for many of the chemicals used to preserve foods. If an excessive amount of the chemical is found, such foods are considered adulterated and subject to seizure.

The need for food additives has been increased by the need to transport foods great distances and to store them for a long time, and to satisfy the consumer's desire for products that are uniform, attractive, convenient to use, and available throughout the year. However, government agencies and industry representatives alike stipulate that the use of such additives is warranted only if they are (General Foods Corp., 1976):

1. inhibitory or bactericidal for microorganisms;

2. capable of retarding or preventing deteriorative changes;

3. nonirritating;

4. noninjurious to the health of the consumer;

5. readily available or conveniently prepared;

6. detectable without excessive difficulty;

7. not too costly for ready usage.

Although any additive should be safe at its intended level of use, no additive should be used indiscriminately, such as in foods that can be prepared satisfactorily without it. Nor should an additive make possible the use of unfit raw material, serve as a substitute for cleanliness, or permit poor manufacturing practices (Sanders, 1966). Under no circumstances should an additive be metabolized or otherwise converted to a substance of greater toxicity than the parent material. Some additives are used to improve foods nutritionally, or to make them more appealing to the senses, such as by clarifying wines, fruit juices, etc., or to accelerate aging (as of flour). Cleansers, disinfectants, and sanitizers must be easily and completely removable from surfaces with which foods come into contact.

Relatively few of the thousands of known inorganic and organic chemicals are approved for use in the foods industry. The vast majority have been prohibited because of toxicity for the consumer; because they produce undesirable flavor, color, odor, or side reactions; or because they have no apparent use. Those used are under rigid scrutiny of such agencies as the Food and Drug Administration, which must give final approval for the use of any chemical with or in foods, and can do so only after the additive has been subjected to rigorous and extensive toxicological testing. Even the chemicals on a prepared list of those "generally recognized as safe" (GRAS) are subject to review. Foods cannot be imported into this country if they contain unapproved chemicals—even if those chemicals have been approved for use by other countries. As a result, the American consumer is protected by the most stringent food laws in the world.

There is no known single, universally applicable preservative agent. Rather, an intelligent selection must be made from a wide variety of those available, according to such considerations as:

1. the site of application;
2. the nature of the food to be protected;
3. the specific microbial agent to be controlled;
4. the possible adverse effect upon the flavor, color, odor, or such nutritive qualities as preservation of vitamins; and
5. potential corrosion of the equipment or the packaging materials that will receive the food.

Inorganic Compounds

Several inorganic compounds have specific uses in the food industry. The sulfites and sulfurous acid are extremely useful in preventing enzymatic browning and in protecting the flavor of sliced potatoes, apples, and other fruits and vegetables. Alkalies and alkaline salts are employed principally as cleansing agents or detergents. Phosphoric acid is used as an acidifying agent in many soft drinks. Until it was outlawed over 65 years ago for use in food products in this country, boric acid was added to milk to prevent spoilage, and also was sold to housewives as a "canning compound." Nitrites are used as "curing agents" for meats, poultry, and fish.

Sulfur Dioxide

Sulfur dioxide (SO_2 is usable as an additive in any of several forms: as a gas; in solution in water as sulfurous acid; or as a constituent of the sodium and potassium salts sulfite and metabisulfite. SO_2 is lethal to microorganisms in concentrations ranging from 50 to 1000 parts per million (ppm). It is most effective as a germicide in acidic solutions and relatively ineffective near neutrality. There have been several postulates as to how the substance works. One is that the sulfur dioxide dissolves within the cell to form sulfurous acid, which then reacts with essential components of the cell or with metabolic intermediates to produce unionizable and unavailable complexes. Sulfurous acid forms hydroxysulfonic acid by reacting with carbonyl compounds but does not generally react with ketone-carbon compounds. During the yeast fermentation of sugars, sulfurous acid reacts with the acetaldehyde to prevent the reoxidation of reduced nicotinic acid diphosphate, the hydrogen-transporting agent in the Meyerhof-Embden pathway of glycolysis. The yeasts then produce glycerol, using dihydroxy-acetone-phosphate as the hydrogen acceptor. This process was used extensively during wartime in Europe to produce glycerol.

Sulfurous acid reacts with amines to yield amine bisulfites, and it reduces the oxygen tension below the point at which aerobes will grow. Yeasts are less sensitive to SO_2 than are bacteria, and fermentative yeasts are less sensitive than are strongly oxidative yeasts. For this reason, grape must (unfermented juice) is treated with SO_2 to eliminate undesirable bacteria and yeasts.

In the United States sulfurous acid is used only in wineries and as an agent for bleaching dried fruits that otherwise would be discolored an unsightly tan or brown. SO_2 has been used for preparatory bleaching of the fruit that is made into maraschino cherries. It reacts very rapidly with anthocyanin pigments, and cherries and strawberries become pale or white. In wineries it is used as a sanitizing agent, being sprayed on walls, floors, and equipment, and into soil around the winery and pomace piles (Joslyn and Braverman, 1954).

In other countries sulfurous acid is added to all forms of dried fruits, to preserve juices in warm climates, as an antioxidant, to preserve fish, and in sugar production. In sugar manufacture SO_2 is both a bleaching agent and a neutralizer of alkalies. In open containers, the preserving effect is temporary, for the sulfur dioxide escapes as a gas or is spontaneously oxidized to the sulfate ion.

Carbon Dioxide

Of the several fugitive gases employed for preserving foods, carbon dioxide (CO_2) is the most widely used. At levels of 10 to

15%, CO_2 in the atmosphere delays maturation of the fresh fruits. When under pressure in a variety of carbonated beverages it acts *synergistically* with the acidulants commonly used to kill contaminating microorganisms. Nevertheless, some molds, such as those of the *Alternaria* spp., and a variety of yeasts of the genera *Saccharomyces* and *Torulopsis* may not only survive but may grow. CO_2 is used in controlled concentrations with un-baked biscuits and bacon. In these applications carbon dioxide is an effective preservative. In small containers CO_2 appears to function by depriving the obligately aerobic microorganisms of the oxygen they require. There is ample evidence of an inverse relation between carbon dioxide content and reproduction of bacterial cells (see Figure 6-1).

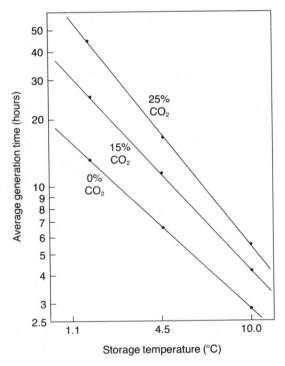

Figure 6-1

Relations of temperature to microbial generation time at various carbon dioxide concentrations (Ogilvy and Ayres, 1951. Reprinted from Food Technology, Vol. 5, pp. 97–120, 1950. Copyright © by Institute of Food Technologists).

Ozone

Ozone is a form of oxygen highly reactive to both living and nonliving organic matter. Ozone is used to free shellfish of coliform bacteria, to treat water used in the manufacture of beverages, and to terminally disinfect effluents. It is nonselective: it kills all microbial forms. It is an excellent topical germicide and is effective at quite low concentrations; for example, at 0.6 mg per cubic meter of air, or 3 to 5 ppm. However, several undesirable effects limit its uses: it oxidizes fats, producing rancid flavor; discolors or bleaches carotene pigments in items such as eggs and butter; and alters flavor in other foodstuffs. Continued inhalation, even at very low concentrations, temporarily depresses the protective lysozyme in the lungs, breaks the tryptophan ring, fragments double-bonded molecules, and depresses alkaline phosphatase levels.

Halogens

In food processing plants the halogens (Cl_2, Br_2, I_2, F_2) and their compounds are used to sterilize water and, carried in the water, to chemically disinfect products and surfaces. Chlorine —as *available chlorine*, not as the ion—is undoubtedly one of the world's most widely used disinfectants. It is readily available in many relatively inexpensive forms. Chlorine gas added directly to water reacts to produce the hypochlorous ion,

$$Cl_2 + H_2O \longrightarrow HOCl + H^+ + Cl^-$$

$$HOCl \rightleftharpoons H^+ + OCl^-$$

The hypochlorous ion is also yielded by the sodium (NaOCl) and calcium ($Ca(OCl)_2$) hypochlorites. The hypochlorite salts are available in solution, such as in the common household bleaches (which at 5.25% concentration contain 49,500 ppm available chlorine); and in concentrated form as powders. Organic forms of chlorine are sodium-*p*-toluene sulfochloramide (chloramine-T), *p*-toluene sulfodichloramide (dichloramine-T), 1,3-dichloro-5,5-dimethyl hydantoin (dichloromethyl hydantoin), *p*-sulfondichloramidobenzoic acid (Halazone tablets), succinochlorimide, chlorinated cyanuric acid derivatives, chlorazodin (or azochloramide) and trichloropelamine. Each compound has rather specific release properties, and specific needs often dictate the proper chlorine compound.

As OCl^-, chlorine reacts very rapidly with all forms of microbial and inert organic matter; in solutions the ion reacts with itself to form molecular O_2 and Cl_2. The entire foods industries encompass a broad array of needs, and different chlorine compounds meet different needs: some compounds react instantaneously and with only evanescent disinfecting action; on the other hand, some organic carriers impart a prolonged disinfecting action by releasing chlorine at a fairly constant rate over a period of time. The latter compounds are useful when disinfection is accomplished in tanks or sinks following the washing of disassembled equipment and of dishes in restaurants, bars, and roadside shops.

A number of theories have been advanced to describe how chlorine kills microorganisms. The old theory, once widely maintained, that chlorine results in the formation of nascent or highly active oxygen, appears untenable. Chlorine may combine with proteins of the cell membrane to form compounds of the chloramine type; these compounds interfere with cellular metabolism, disrupting the cell membrane and causing the cell's contents to leak out. Or chlorine may oxidize the –SH group of essential enzymes and destroy essential enzymes. Unlike that of some disinfectants, the killing action is instantaneous and irreversible. At concentrations of 50 ppm or less, vegetative cells of bacteria, yeasts, and molds are killed within seconds; at a concentration of 200 ppm spores are destroyed within two minutes.

Gaseous chlorine, widely used in the final treatment of drinking water and of effluents from sewage treatment plants, is also introduced into waters entering food processing plants to deliver levels of approximately 5 ppm at the point where the water is used. The process is termed **in-plant chlorination.** At this rather low concentration, the chlorine is extremely effective in preventing the growth of microorganisms that produce slime and odor on conveyor belts and on supporting equipment, and the process reduces contamination of foods to a very low level. Animal carcasses are given a terminal wash of chlorinated water at 50 to 200 ppm to reduce the load of contaminating microorganisms introduced during slaughtering. Also at 50 ppm, chlorine is used as a sanitizing agent. (The terms **sanitization** and **sanitizing agents** have peculiar meanings: they imply, not sanitation, which involves cleansing, but rather the use of disinfecting agents to further reduce microbial contamination of surfaces that have already been cleansed. Because no sanitizing agents, with the possible exception of the iodophors, have cleansing ability or the ability to penetrate debris, surfaces to be sanitized must first be cleaned thoroughly.)

Hypochlorites are useful when it is not practical to use gaseous chlorine. Hypochlorites are effective bactericides when there is little organic matter present, but animal or vegetable tissue quickly neutralizes them. While hypochlorites are used primarily for water treatment, restaurants and food service organizations, food plants, and dairy farms also make extensive use of them. To be accepted as a sanitizing agent, a hypochlorite must meet the Chambers test: "in recommended concentration, the compound must produce a 99.999% kill of 75 to 125 million cells of *Escherichia coli* ATCC 11229 and 75 to 125 million cells of *Staphylococcus pyogenes* var. *aureus* ATCC 6539 within 30 seconds at 70 to 75 F." Not all hypochlorites react this rapidly, and an alkaline hypochlorite, which releases available chlorine more slowly, as in tank sanitization, is sometimes more useful. All hypochlorites are unstable when exposed to the atmosphere because they rapidly absorb moisture, and subsequent spontaneous chemical reactions liberate molecular chlorine. Therefore, all containers should be kept tightly closed when not in use.

Chlorine dioxide is an exceedingly reactive compound produced by the reaction of gaseous chlorine with sodium hypochlorite, or by the action of sulfuric acid on potassium perchlorate. It is used primarily as a bleaching agent for cellulose, paper pulp, flour, and fats and oils. It is also used to control the taste and odor of water.

Iodine is used exclusively as an iodophore, in which the iodine is loosely bound to an organic carrier. Because the carrier frequently has detergent properties, the compound is extremely useful for washing the udders and teats of cows before milking. As the active iodine is dissipated, the carrier releases more available iodine. Iodophores are quite effective under acidic conditions, yet are less corrosive to equipment and less damaging to skin than are the chlorine compounds. They have the added advantage that the disappearance of iodine color indicates the loss of sanitizing ability. And on stainless steel equipment the iodine reacts with residual film, indicating incomplete cleansing.

Owing to the high toxicity of fluorine, city and state health authorities carefully regulate its use, which is restricted to the fluoridation of water supplies in extremely dilute solutions (1 to 2 ppm).

Sodium Chloride (NaCl)

Sodium chloride is without question the most widely used chemical in foods. It may be applied directly to foods or may be used in brines to retard certain kinds of microbial growth and spoilage without destroying acid-producing organisms. However, this common chemical also may be toxic if improperly used. Its uses include enhancing flavor; withdrawing cell saps from materials subjected to fermentation; providing an additive or complementary or inhibiting effect to preservatives such as acetic and sorbic acids; and acting as a preserving agent in its own right. Most bacteria and yeasts are relatively sensitive to low concentrations of NaCl, being inhibited at concentrations between 3 to 5% in fluids. As a preserving agent it is commonly used in concentrations of 15% and more. It is often applied to fish as a dry salt, in a ratio of, say, 3 parts salt to 10 parts fish. Meats being cured for unrefrigerated storage are salted at a typical ratio of 8 pounds salt applied to the surface of 100 pounds of meat. Salt curing of meat achieves a dual effect: first, salt penetrates the meat, withdrawing water from its interior with an approximate 10% loss in weight; second, the ultimate osmotic pressure prevents the growth of the clostridia, the major internal spoilage agents of cured meats.

Halotolerant and halophilic anaerobic bacteria have been reported in foods, but these are inconsequential in the spoilage of brined and salted foods. Salted and brined foods exposed to air ultimately develop a luxuriant growth of a variety of fungi, usually of the classes Ascomycetes and the Fungi Imperfecti, and of bacteria that frequently are pigmented and resemble (but are not identical with) cells of *Serratia* spp. For this reason, brining for preservation is most effective when combined with anaerobiosis or with temperature reduction. To protect salted, unrefrigerated fish, it has been advocated that 0.2% sorbic acid be added to the salt mix.

Spoilage agents termed **haloduric** tolerate and grow slowly in extremely high concentrations of NaCl. The minimum concentrations that prevent growth differ among microorganisms, environments, and substrates. Haloduric bacteria and molds are common in the terrestrial environment, and if precautions are taken to prevent dehydration, may be obtained during routine culture of soils and water upon salt-containing media with prolonged incubation (three to five weeks).

Salt originates from two sources. *Mined salt* (see Figure 6-2) is sterile *in situ*. Any microorganisms present are contaminants incidental to mining and handling of the salt. *Solar salt* is obtained by evaporation of seawater, and is produced extensively along the coasts of Africa and the Mideast. These salts contain appreciable numbers of haloduric and halophilic bacteria.

Figure 6-2
Underground salt mining: an electric-powered shovel loads rock salt into a primary crusher. (Courtesy of International Salt Company, Clarks Summit, Pa. A part of Akzona, Inc.)

Because salt is so common a preserving agent and is sometimes used as an inhibitor of groups of bacteria, an extensive literature on the bacteriology of salt has accumulated in which the absence of a fixed terminology has occasioned confusion, and in which bacteria have been assigned to the wrong genera. Larsen's excellent detailed review (1962) should be consulted. He divides the bacteria into several groups according to their degree of tolerance or preference for salt. The *extreme* **halophiles** require 15% and grow best at 25 to 30% NaCl. These are gram-negative rods, typified by members of the family Halobacteriaceae, which produce a vivid red pigment and had been mistaken for members of *Serratia*—a physiologically different bacterium. Cells of the *Sarcina-Micrococcus* group also produce red pigments, are obligately aerobic, nonmotile, and non-spore-forming. The moderate halophiles grow best in solutions ranging from 5 to 20% NaCl; Larsen names 11 genera, including several autotrophs. The *slight* halophiles, which grow in solutions containing 1.5 to 5% NaCl, include members of the genera *Pseudomonas*, *Vibrio*, *Achromobacter*, and *Flavobacterium*. The recently constituted genus *Beneckia* belongs to this group. The **nonhalophiles** are those that thrive in the presence of minimal NaCl or none at all, although some grow more rapidly or to higher populations in media containing 0.5% NaCl than they do in NaCl-free media. Many of the nonhalophiles are quite tolerant of NaCl. Some of these—for example, some clostridia and staphylococci and enterococci—grow quite well at levels of 5 and even 10% NaCl.

Quaternaries

The quaternary ammonium compounds, often termed "quats" or QAC, are molecules of pentavalent nitrogen, which bears four organic moieties and an anion, the latter often a halogen. Quaternaries have a sanitizing action only if at least one of the organic moieties is quite complex; if the moiety is a long-chain fatty acid, it is most effective with hydrocarbon chains of 12 to 14 carbon atoms. At high dilutions quaternaries are lethal to fungi and vegetative bacterial cells, and are used in dilutions ranging from 1:500 to 1:2500. In normal concentrations they are colorless, odorless, tasteless, and noncorrosive to metal and to hands, but at very high concentrations may make foods taste bitter. They are very stable in storage. They affect the bacterial cell by interfering with the passage of molecules through the cell membrane and by promoting leakage of soluble nitrogen and phosphorous compounds from the cell.

The quaternary ammonium compounds are compatible with alkalies such as trisodium phosphate, carbonates, and borates, but not with anionic detergents, iodine, or soaps. They are quite effective in neutral to slightly alkaline solution. They are widely used in the direct sanitization of equipment and utensils at the site of use in wash tanks, and as sprays on walls and floors, where they leave a residual microbiostatic action. They are used in cleansing eggs of soil and surface bacteria before cracking, but they cannot penetrate the shell; they are used in washing tomatoes and cucumbers prior to packing. When used as sanitizing agents for the udders of milking cattle, they do not irritate the udders or teats.

Hydrogen Peroxide

Hydrogen peroxide (H_2O_2), a potent oxidizing agent, is used at much higher concentrations than are chlorine and iodine compounds. It is highly corrosive to metal equipment and damaging to skin. In the United States the use of hydrogen peroxide is permitted with milk to be used in making Colby, Cheddar, Swiss, and other washed-curd types of cheese, to reduce the numbers of undesired bacteria. It is not permitted in the treatment of fluid milk because it fails to destroy *Mycobacterium tuberculosis*. It is used in the treatment of milk in some countries where fuel for heat pasteurization is expensive. There hydrogen peroxide is added to milk heated to 49°C. Both in domestic cheesemaking and foreign fluid milk treatment, residual peroxide is destroyed by the addition of catalase.

Nitrates and Nitrites

Nitrates and nitrites have been used for centuries as "curing agents" for flesh foods. Some waters and certain vegetables, such as spinach, beets, celery, collards, eggplant, lettuce, radishes, and turnip greens, have high nitrate content, and several microorganisms are able to reduce nitrates to nitrites. Since nitrites have bacteriostatic effect in acid solution, they have been recommended for preserving fish. Nitrites reduce the redox potential, thus reacting with the heme pigment of heated pork muscle to produce the desired pink color of ham.

Recently there has been great concern that, at acid pH, secondary amines react with nitrous acid, nitrite, or oxides of nitrogen to produce N-nitrosamines. In 1954 Barnes and Magee (1954) established a causative relationship between the occurrence of liver disease in two industrial workers and the introduction of N-nitrosodimethylamine as a plant solvent. Between 1963 and 1965 interest in the nitrosamines became more widespread when liver toxicity was observed in sheep and mink fed fishmeal containing N-nitrosodimethylamine that had been produced by adding large quantities of nitrite as a preservative prior to high-temperature drying (see Wolff, 1972). Nitrosamines are present in cured meats and fish and other food products, but the authenticity of some of these findings is questionable (Wolff and Wasserman, 1972). The functions of nitrite in meat processing and curing are discussed in Chapter 17.

Metallic Salts

Metals and metalloid elements have recognized toxic properties and marked bacteriostatic or bactericidal properties. These substances include antimony, arsenic, barium, bismuth, boron, cadmium, chromium, cobalt, copper, gold, iron, lead, manganese, mercury, nickel, phosphorus, selenium, silver, tin, and zinc. Some of the metallic cations and their salts such as copper (Cu), gold (Au), and mercury (Hg) are used extensively as bactericides, fungicides, or algacides.

Copper salts have relatively little effect on bacteria but destroy algal growth in water supplies and prevent fungal infection of plants. Since gold salts and complex organic gold compounds are expensive, they are used only in the therapy of tuberculosis and leprosy. Silver compounds—especially those in the form of nitrate, citrate, lactate, and preparations of protein silver or colloidal silver—are useful antiseptics. None of the metals or their salts is now used in foods. Silver, however, is believed to be responsible for the purification of water stored in vessels made of clay contaminated with silver.

Simple mercury salts such as mercuric chloride, mercury oxycyanide, and potassium mercuric iodide are often used as disinfectants, but their lack of specificity makes them hazardous to humans and domestic animals. Fildes (1940) demonstrated that mercury combines with sulfhydryl (R–SH) groups in the cell and deprives the cell of those –SH groups essential for metabolism. In mid-1970 unexpected levels of mercury were recovered from canned tuna and swordfish, and later the association of this contaminant with industrially polluted water caused widespread concern that certain kinds of seafoods might not be wholesome. Bacteria living in water convert mercury to the extremely toxic methyl mercury in the cell. Accumulated in the tissue of successive predators in the food chain, methyl mercury concentrations may exceed 0.5 ppm in ultimate predators such as the swordfish. Scandinavian research has shown that 90% of the mercury in freshwater fish is methyl mercury. However, not all mercury is in this form in all tissues.

Sources of mercury in the environment include seed-treating agents, slimicides used in the pulp and paper industry, turf fungicides, and mercury paints. In 1970 a farmer fatally misused seed grain treated with a mercurial fungicide when he fed the grain to his hogs and later butchered them for consumption by his family.

Inorganic arsenic occurs naturally in fish and shellfish. Arsenicals are quite toxic to protozoan parasites, spirochaetes, yeasts, and other microorganisms but much less poisonous to humans and higher animals. Organic arsenic compounds have not only helped to free poultry of parasites but when administered at sufficiently high dilution have stimulated appetite and growth. Because arsenicals are used extensively as insecticides, the FAO/WHO Expert Committee on Food Additives has established a maximum acceptable load of 0.05 mg/kg body weight—permitting, that is, on average 1 ppm in all foods, whereas estimated average human consumption is now 0.05 ppm.

Organic Compounds

Antimicrobial organic compounds are classified as being acids, antibiotics, oxidizing agents, and metabolic inhibitors. Some, especially the acids, may occur naturally in foods. Generally, the antimicrobial organic compounds act directly to inhibit the microbial cell. Sugars, not inhibitory per se, become inhibitory through bacterial fermentation to lactic acid, which occurs, for example, when sugars are added to sausage meats.

Organic Acids

Organic acids and their salts are common food additives. Several of these, such as acetic, ascorbic, citric, lactic, malic, and tartaric, are present initially in various plant juices or develop as a result of microbial fermentation. In addition, benzoic, oxalic, salicylic, and sorbic acids often are found in plant tissues. Benzoic acid occurs naturally in cranberries and is an effective yeast inhibitor in oleomargarine, fruit juices, and several other acid foods. Oxalic acid is found in rhubarb, spinach, and sorrel, and small amounts of salicylic acid occur in most fruits. Propionic acid occurs in Swiss cheese and is what gives it its flavor. Sorbic acid occurs naturally in the berries of the mountain ash.

In recent years, dehydroacetic acid, which has a preservative effect, has been allowed limited use as a pesticide. Peracetic acid is composed of a molecule of acetic acid with an atom of oxygen loosely bonded to the carboxyl carbon atom. In contact with organic matter it decomposes spontaneously to active oxygen and acetic acid. Although now used primarily in the textile industries, it has been proposed as a treatment for harvested tomatoes, to reduce the growth of harvest mold and to discourage the fruit fly from laying eggs in damaged tomatoes.

By increasing the hydrogen ion concentration in foods, organic acids exert an antiseptic effect; in addition, the concentration of the undissociated acid controls microbial growth. Merely acidulating the substrates will preserve many foods or, in some instances, prevent the growth of pathogenic or toxigenic microorganisms. Many major groups, particularly some species of bacteria, fail to grow if concentrations of hydrogen ions are below the equivalent of a certain pH. A few of the important groups of bacteria and their limiting pH values for growth will now be cited (others are found elsewhere in the text). *Bacillus subtilis* var. *panis* is a spore-forming bacterium whose spores are found in flours and survive baking; its cells grow in bread, converting the gluten to an unpalatable, unsightly, sticky tan-brown mass. *Bacillus subtilis* var. *aterrimus,* a soil organism, also survives the baking process and grows as blue-black colonies on slices of bread (Galloway and Rymer, 1967). Both bacteria are readily controlled if sufficient sodium diacetate or propionic acid is added to bring the pH below 6.0. Members of *Pseudomonas* spp. grow poorly if at all below pH 5.5, and micrococci and staphylococci fail to grow below pH 4.5. These are important values in the preparation for marketing of a variety of perishable foods. Cells of *Bacillus coagulans,* a preferential thermophile, begin growing in tomatoes unless citric acid is added to lower the pH to 4.2 or less. Although very few microorganisms can initiate growth in foods containing 1% acetic acid, the fungus *Monilia acetoabuteus* (now identified as *Geotrichum candidum*) proliferates in concentrations in excess of 5% acetic acid, and within several months reduces the acid level to less than 1% (Dakin and Stolk, 1968). Use of acids alone in microbial control renders some foods unpalatable. A few microorganisms, notably lactobacilli, some yeasts, and specific members of *Leuconostoc, Xanthomonas,* and *Achromobacter,* are exceedingly acid tolerant. In such instances the additive use of several agents is effective. Salt (NaCl) forces ions to reassociate with the acetic acid molecule. Sugar, usually sucrose, also has a binding effect on water molecules: by limiting the number of free water molecules it helps force the reassociation of the acid molecule. The acids thus affected are chiefly acetic, lactic, and citric. The phenomenon of inhibition is termed the *whole molecule* effect, but the principle extends to several other acids, including benzoic, propionic, and sorbic. Foods in which several agents have an additive effect as preservatives include pickles, relishes, catsup, salad dressings, and beverages.

Benzoic, sorbic, and propionic acids, at concentrations of 0.1% in acidic foods, are effective fungistatic agents. They are most effective at pH 5.5 or less, where the undissociated form of the molecule predominates (Bell et al., 1959). Fungistatic agents prevent growth without necessarily killing fungi during prolonged exposure. Yet because there is gradual death, a fungistat often becomes a fungicide with time. Benzoic acid damages the cell membranes; so do oxalic and salicylic acids, but these two are not permitted as food additives.

Used in excess, any of the organic acids damage cell membranes, but benzoic, oxalic, sorbic, and salicylic acids also interfere with enzymatic processes. A tolerance level of 0.1% for sodium benzoate has been established for fruit juices; also, ice used for preserving fish can contain benzoic acid. This acid or its salts has no adverse effect on humans if less than 0.5 gram per day is consumed. Occasionally there are reports of acute oxalic acid poisoning, caused by eating raw rhubarb stalks or leaves. Some European countries permit the use of salicylic acid and salicylates as food preservatives, but the United States does not. Ingesting overdoses of aspirin results in over 200 deaths a year mostly among infants, young children, and those with suicidal intent.

Propionic acid, CH_3CH_2COOH, a fermentation product of the propionic bacteria, is the major flavor constituent in Swiss cheese. Propionic acid interferes with the synthesis of proteins by yeast and mold cells. This acid has been used to acidify bread dough and to retard the growth of mold in bread and bakery products. It can also be incorporated into wrapping materials that come into contact with surfaces of foods, such as butter and cheese, that are subject to mold growth.

Sorbic acid, $CH_3CH=CH-CH=CH-COOH$, while readily absorbed in the body, is a polyunsaturated organic acid that is an effective yeast and mold retardant because of the alpha-beta unsaturation of the molecule. The gamma-delta unsaturation has no effect. Compounds with alpha-beta unsaturation act aerobically to depress oxidative phosphorylation by an uncoupling mechanism and anaerobically through reaction with cysteine to inactivate essential cellular enzymes. In acidic foods, if sorbic acid is used at approximately 0.1% concentration, only lactobacilli, streptococci, leuconostocs, pediococci, and clostridia grow. At 0.075%, sorbic acid is fungistatic for a variety of molds isolated from strawberries and tomatoes (Beneke and Fabian, 1955). NaCl enhances the inhibitory action. It has been claimed that sorbic acid inhibits microorganisms that produce catalase and that have the respiratory enzymes of the cytochrome system, but such claims are not wholly true: staphylococci, which are strong catalase producers, grow in mildly acidic conditions.

Sorbic acid is readily metabolized with no untoward effects. Most other alpha-beta unsaturated compounds that are soluble impart undesired flavors and odors reminiscent of garlic, onion, and horseradish.

Antibiotics

More than 30 years ago a special group of organic antimicrobial agents came into use: the antibiotics. These very complex compounds, produced in small quantities by bacteria, streptomycetes, and fungi, are detrimental to other forms of microbial life. Their use in food protection remains an intriguing possibility. But in the United States none is permitted in foods, with the exception of nisin, a specific anticlostridial agent produced naturally during milk fermentation by many strains of *Streptococcus lactis*. There are several reasons for prohibition, a few of which are:

consumers might object to drinking milk from diseased cows or milk containing, say, penicillin;

some consumers might be made sensitive to the antibiotic;

consumers already sensitized might have physiological reactions;

microorganisms might develop resistance to the antibiotics;

antibiotics might be used as substitutes for good sanitation and temperature control;

antibiotics might interfere with starters in milk cultures and in cheesemaking.

In some areas of the world chlortetracyclenes are perfused into beef immediately before slaughter, to protect it during unrefrigerated transportation and storage. In the United States, excellent modern refrigeration renders the perfusion of beef unnecessary.

The polypeptide antibiotics subtilin and tylosin are heat-stable agents that damage the cell membranes of germinating spores, and so were proposed for use in canned foods. Subtilin prevents outgrowth of the bacteria after they germinate, but does not halt growth of important food spoilage organisms such as *Clostridium sporogenes* (NCA PA Strain #3679). The more heat-resistant macrolide, tylosin, also has the disadvantage that it inhibits cell growth rather than inhibiting spore germination.

In the mid-1950s the tetracyclines chlortetracycline (CTC) and oxtetracycline (OTC) were used commercially to double the storage life of meats, fish, and poultry, and in 1955 and 1956 the Food and Drug Administration approved the use of CTC and OTC, respectively, in ice or cooling water, at a residual level not exceeding 7 ppm, to preserve dressed poultry. Unfortunately, yeasts still grow on antibiotically treated flesh, and shelf life was terminated by fermented, rather than putrefactive odors.

Oxidizing Agents

Three organic compounds, formaldehyde, ethylene oxide, and propylene oxide, are of some use as preservatives in the gaseous state. Formaldehyde gas penetrates minute cracks and crevices, but its use is limited to fumigating chambers, and is prohibited if its residues might come into contact with foods. Formaldehyde and several phenolic and cresylic compounds occur naturally in cured or smoked meat, fish, poultry, and cheese as a minor constituent of smoke. These agents combine with proteins in cells to form protoplasmic poisons and insoluble proteinates.

The epoxides, ethylene oxide and propylene oxide, have been used as fumigants for preserving a wide range of particulate and pulverized materials not readily sterilized by other means. These include starches, gums, nuts, dried fruits, glacé fruits, dates, and spices. Lethality was formerly thought to be due to the alkylation of such active radicals as carboxyls, carbonyls, and amino and sulfhydryl groups by the alkene moiety. But lethality is in fact brought about by the alkylation of the guanosine triphosphate within the microbial cell. Only three-membered rings are effective germicides. Strangely, the epoxides are most effective when relative humidity is between 35 and 40 percent, and become less effective as relative humidity rises. The rate of kill is slow: several hours are required to kill vegetative cells, and 8 to 24 hours to destroy bacterial spores. Because the gases are extremely explosive in pure form they are generally marketed as "nitroxide," a mixture of gases containing 10% epoxide in 90% nitrogen. The epoxides combine with inorganic chlorides to form very toxic chlorhydrins, and present federal regulations forbid the use of the epoxides with all foods except nuts and spices.

Metabolic Inhibitors

A number of compounds interfere with the metabolism of microorganisms, usually fungi. All metabolic inhibitors are prohibited from use in foods because they are toxic when ingested, and might be ingested if, say, they were applied to citrus fruit skins to prevent spoilage such as stem end rot. Inhibitors in this group include thioacetamide, thiourea, and copper-9-hydroxyquinolate. At one time it was proposed that sodium pentachlorophenol be incorporated in the paper liners and fillers of egg crates, where continuing, very low vapor tension would retard the mold spoilage of eggs in prolonged storage.

Esters

Alkylesters of parahydroxybenzoic and vanillic acids are fungistatic in both neutral and slightly acidic foods. Because certain esters impart mild but distinct flavors, they are used only in foods having flavors with which the ester flavors blend.

Water-insoluble aliphatic acids with 7 to 12 carbon atoms in the chain exert some fungistatic activity as their sodium salts in alkaline foods. Three of these acids, caproic, capric, and caprylic, are believed responsible for the selection of fungi used in curing Roquefort and Camembert cheeses, which are made with milks in which these acids are prominent in esterified form.

Other Compounds

In addition to those agents already discussed, many other chemicals have been proposed for use as antimicrobial agents —among them the sodium salt of undecylenic acid, thymol, tetrachlorbenzoquinone, dichloranaphthoquinone, crotonic-acid, and trichlorophenyl acetate. Trichloracetic acid is a strongly ionizing acid the chlorine ions of which impart anti-microbial properties.

Limonene, a fungistatic constituent of orange peel oil, is reportedly more effective near neutrality than at more acidic pH. The diethyl ester of pyrocarbonic acid has been proposed as an aqueous treatment of small fruits such as strawberries. Japanese researchers report that 125 ppm of ethyl-*b*-benzoyl acrylate controls lactobacilli in ethanolic fermentations.

Ammonium persulfate, at levels of 1 percent in a highly acidic solution, has been advocated as a wash treatment of yeast slurries to destroy troublesome bacteria in breweries.

Chemicals as food additives are the subject of much distrust, but it must be understood that removing all chemicals from foods—in all stages, from farming through processing and packaging to final preparation in homes and restaurants—will not improve our foods. Indeed they will be improved only by continuing research and judicious application of new knowledge. Meanwhile, the American consumer, protected by the world's most stringent food laws, is probably provided with the best and most nutritive foods that modern science and technology can produce.

References

Barnes, J. M., and P. N. Magee. 1954. Brit. J. Ind. Med. *11*, 167.

Bell, T. A., J. L. Etchells, and A. F. Borg. 1959. J. Bacteriol. 77, 573.

Beneke, E. S., and F. W. Fabian. 1955. Food Technol. *9*, 486.

Chambers, C. W. 1956. J. Milk Food Technol. *19*, 183.

Dakin, J. C., and C. A. Stolk. 1968. J. Food Technol. *3*, 49.

Fildes, P. 1940. Brit. J. Exptl. Pathol. *21*, 67.

Galloway, L. D., and T. E. Rymer. 1967. J. Food Technol. 2, 95.

General Foods Corp. 1976. *Today's Food and Additives*. White Plains, N.Y.

Joslyn, M. A., and J. B. S. Braverman. 1954. In: *Advances in Food Research. V*, 97. E. M. Mrak and G. F. Stewart, eds. Academic Press, New York.

Larkin, T. 1976. FDA Consumer. HEW Publ. No (FDA) 76-2020.

Larsen, H. 1962. Halophilism. In: *The Bacteria*. Academic Press, New York.

Ogilvy, W. S., and J. C. Ayres. 1951. Food Technol. *5*, 97.

Sanders, H. J. 1966. Chem. and Eng. News 44, Oct. 10, p. 100; ibid. Oct. 17, p. 108.

Sebranek, J. B., and R. G. Cassens. 1973. J. Milk Food Technol. *36*, 76.

Wolff, I. A., and A. E. Wasserman. 1972. Science *177*, 15.

Part II
Fermentations

7

Alcoholic Yeast Fermentations

In a variety of natural foods, indigenous flora direct fermentation. The nature of the fermentation is determined by several factors, among them the temperature and the composition of the food. For example, high-carbohydrate, acid foods such as fruit encourage yeast growth, making alcoholic fermentation likely. Foods such as milk and raw silage, rich in vitamins, nitrogenous constituents, and carbohydrates, favor lactic acid bacteria and consequently lactic fermentation. Although the nature of fermentation came to be understood in relatively recent times, humans have taken advantage of these and other fermentations for thousands of years, to preserve foods and add variety to the diet. For millennia bread has been leavened, alcoholic beverages have been brewed, milk has been transformed into cheese, meat has been aged, and sauerkraut has been made from cabbage, pickles from cucumbers, olives from olive plums, and soy sauces from beans.

According to Aibe et al. (1965), Thenard proposed, in 1803, that yeasts used by French winemakers are alive and responsible for the formation of alcohol. Thus, an ancient art began to yield to the science of fermentation.

Prior to 1855 influence of the German chemist von Liebig was such that fermentation was believed to be an alteration in the state of nitrogenous matter caused by oxygen. Belief that oxygen is involved was founded on experiments by Gay-Lussac. Studying Appert's method of using heat in food preservations, Gay-Lussac observed that heated grape juice would remain unchanged for many days until exposed to oxygen, which, he concluded, was the causative agent for fermentation. This belief was proven incorrect by Bechamp, who observed, in 1855, that molds appeared in pure cane sugar solutions that had undergone inversion when exposed to air; a full account of his experiments was published two years later (Bechamp, 1857).

Although Louis Pasteur's *Memoire sur la fermentation appelée lactique,* published in 1857, did not demonstrate why lactic acid appeared in fermenting natural products, it presented concepts that remain fundamental to our present understanding of fermentation. Discovery that fermentation is caused by living airborne organisms, long credited to Pasteur, now appears to have been made by Bechamp (Hume, 1947).

Yeast Fermentation

Various end products of food fermentation are important to the texture, flavor, and keeping quality of the foods. Before examining how they are produced, some definitions must be given, and energy must be discussed, since fermentation, like respiration, is an energy-yielding process.

Respiration may be defined as energy-yielding reactions in which molecular oxygen serves as the hydrogen acceptor; **fermentation** as energy-yielding reactions in which organic compounds serve as hydrogen acceptors. Both processes are biological oxidations. Respiration yields more energy; fermentation can occur in the absence of oxygen. During these processes, loss of hydrogen atoms and electrons brings compounds in a reduced state to a less reduced or more oxidized state. The electrons are carried by coenzymes made up of vitamins—

explaining in part why vitamins are essential in the nutrition of humans and microorganisms. Cells are classified as aerobic, anaerobic, or microaerophilic according to their respective capacities to use nutrients in the presence, absence, or near-absence of oxygen.

In all types of microorganisms the energy released by oxidation, except that lost as heat, is ultimately transformed into adenosine triphosphate (ATP), which drives the various energy-requiring processes. ATP can change its energy level, enzymatically transferring its phosphate group to phosphate-accepting molecules, which thereby gain energy. In each reaction during fermentation the free energy change is determined by the oxidation-reduction potential of the substances involved. This change is measurable with the standard hydrogen electrode, as a value of kilocalories per mole (Kcal/mol). Hydrolysis of various phosphorylated compounds yields free energy that is also measurable, and compounds such as ATP, which have values at pH 7.0 more negative than 7 Kcal/mol are classified as high-energy compounds. A variety of biosynthetic reactions can be driven by a number of other energy-rich compounds that occur in cells—for example, phosphoenolpyruvate, acetyl phosphate, aminoacyl adenylates, and thiol esters such as coenzyme A and lipoic acid. Energy-containing phosphate compounds with more negative free energies of hydrolysis lose their phosphate groups to those with less negative values.

In the absence or near-absence of oxygen, various species of bacteria important in fermenting foods use glucose by three pathways (see Figure 7-1):

1. the glycolytic, also called the hexose diphosphate and Emden-Meyerhof-Parnas pathways and the alcoholic or homolactic fermentation;

2. the heterolactic, also called the hexose monophosphate, pentose, and 3-2 split pathways; and

3. the mixed acid pathway.

Although these pathways share some features in common, each has distinctive end products.

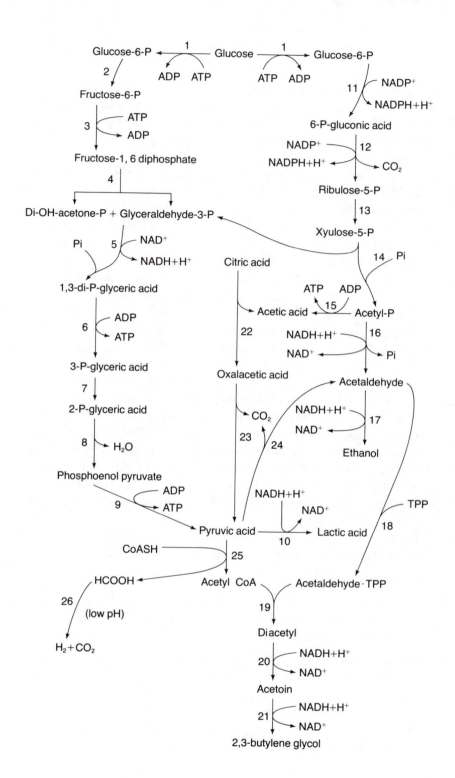

Figure 7-1

The glycolytic, heterolactic, and mixed acid pathways. The reactions and enzymes are numbered as follows: 1, hexokinase; 2, phosphoglucose isomerase; 3, phosphofructokinase; 4, aldolase; 5, glyceraldehyde-3-phosphate dehydrogenase; 6, 3-phosphoglycerate kinase; 7, phosphoglyceromutase; 8, enolase; 9, pyruvate kinase; 10, lactic dehydrogenase; 11, glucosec-6-phosphate dehydrogenase; 12, 6-phosphogluconate dehydrogenase; 13, ribulosephosphate-3-epimerase; 14, phosphoketolase; 15, acetokinase; 16, aldehyde dehydrogenase; 17, alcohol dehydrogenase; 18, see reaction 24; 19, diacetyl synthetase; 20, diacetyl reductase; 21, acetoin reductase; 22, citritase; 23, oxaloacetic acid decarboxylase; 24, pyruvate decarboxylase system; 25, thioclastic system; 26, formic hydrogenlyase.

If oxygen is ample, some organisms further oxidize pyruvic acid by means of the citric acid cycle (see Figure 7-2), which begins when acetyl coenzyme A reacts with oxaloacetic acid to form citric acid. Citrate then undergoes a series of reactions that, with yeasts, yields 38 moles of ATP per mole of glucose used. Under anaerobic conditions, yeasts or homofermentative lactic acid bacteria are restricted to glycolysis and produce only 2 moles of ATP per mole of hexose. Certain intermediate substances of the citric acid cycle—such as α-ketoglutaric, fumaric, and oxaloacetic acids—are involved in amino acid biosynthesis.

Yeast-Fermentation Food Products

Whereas some fermented foods are produced by joint action of yeasts and bacteria, others—including bread, beer, wine, distilled liquors, and some vitamins and yeast food—are produced primarily or exclusively by yeast action.

Beer

The first fermented beverages may have been consumed by humans as early as the first breads, if not earlier, and must have been mead, wine, or beer, depending on whether the available raw material was honey, fruit, or grain. Like bread, beer is

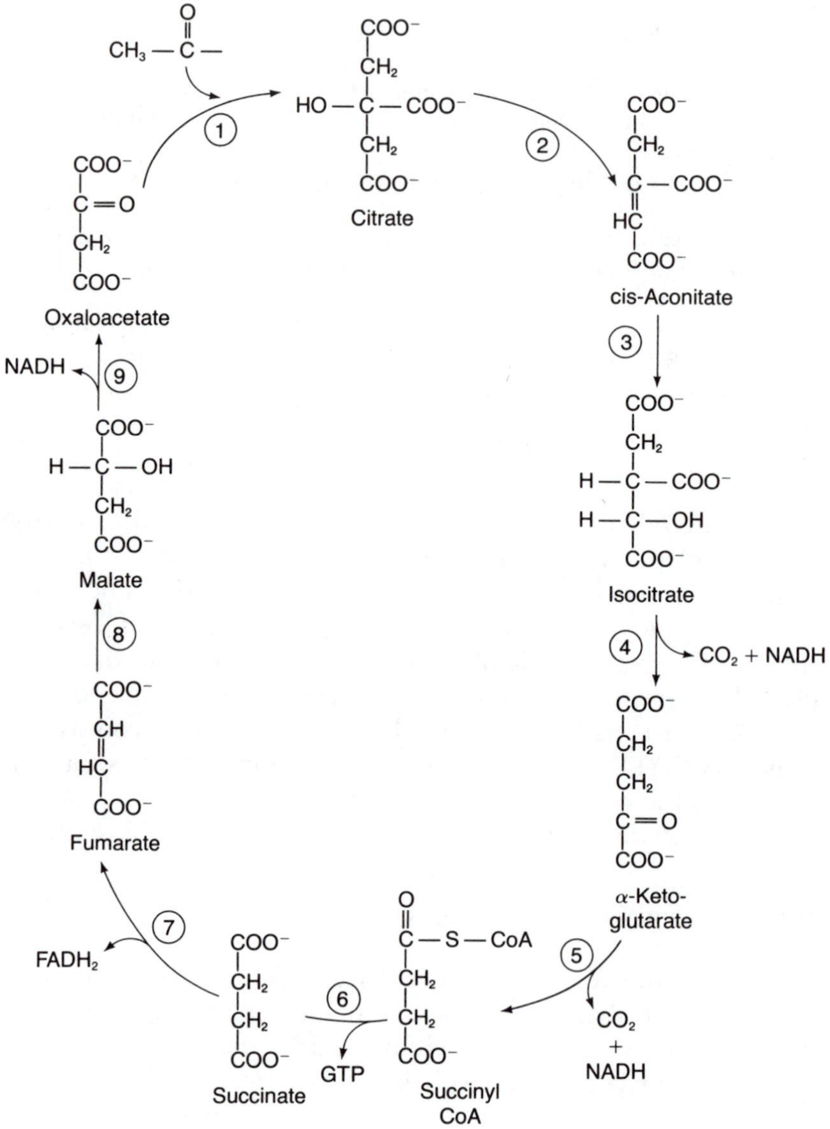

Figure 7-2
The citric acid cycle. (From *Biochemistry* by Lubert Stryer. W. H. Freeman and Company. Copyright © 1975.)

made today basically as it was in ancient times, but modern technological developments have greatly improved product uniformity, quality, and stability.

According to archeological evidence, Egyptians were practicing the art of brewing over 6000 years ago along the Nile River (see Figure 7-3); and also in the distant past people in the Orient were making fermented rice beverages such as sake. Christopher Columbus found Indians in the New World who were producing an alcoholic beverage—in 1502 the Spanish explorer was presented with "a sort of wine made from maize resembling English beer." In 1643 Charles I of England, recognizing beer as a source of tax revenue, imposed the first beer duty. In the nineteenth century, Germany was already famous for producing fine beer, which was shipped across the Rhine River into France; perturbed by this, Louis Pasteur began, in

Figure 7-3
A relief on the tomb of Achet-hetep-her, showing steps in brewing as practiced in ancient Egypt, during the Fifth Dynasty. (Courtesy of Dr. A. Klasens, Director of the Rijksmuseum, Leiden, The Netherlands.)

1871, investigations into improving brewing techniques. For his 1876 book *Studies on Beer* he was honored by the French brewers in the Congress of 1889. In Denmark, in 1883, Hanson prepared pure cultures of various species of *Saccharomyces* and showed that different pure strains yielded different fermentation reactions. He also showed that some groups of the wild yeasts cause detrimental changes in beer while others are harmless.

Brewing Processes

Beer manufacturing methods vary somewhat from plant to plant, but in all large breweries scientists and technicians work on quality control and product improvement. Still, brewmasters differ on how to produce the best product. Figure 7-4 presents a general outline of a typical batch brewing process. A recently developed technique, the continuous brewing process, is in operation in New Zealand and Great Britain but not at present in the United States (Hough and Button, 1972). Variations in procedures, raw products, and yeast types result in products with different characteristic flavors and colors. For example, top- or bottom-fermenting yeasts may be used, the former rising to the top of the tank and the latter settling to the bottom during active fermentation. *Lager,* or aged (one to two months), beer is made with bottom-fermenting yeasts. *Bock* beer, a dark, sweet, heavy-bodied product traditionally marketed in the spring for festival celebrations, is made with a

---

Figure 7-4

A flow diagram of the typical steps in beer manufacture. A. Barleycorns are ground and placed in the malt hopper. B. From thence they go to the mash tub. C. Corn, or other adjuncts, are placed in the adjunct hopper. D. They are boiled in the adjunct cooker and added to the mash. E. The malt wort is then filtered in the lauter tub. F. The clear wort goes to the brew kettle, where hops are added. G. The hops are removed by the hop strainer. H. Other substances are removed by a settling tank. I. The wort is then cooled and sent to, J, open and, K, closed fermenting tanks. In the closed tank carbon dioxide is collected and later used to carbonate the beer. L, N, P. After fermentation the green beer goes to lagering tanks. M, O. As it passes from one tank to the next it is filtered. From "Beer," by Anthony H. Rose. Copyright © 1959 by Scientific American, Inc. All rights reserved.

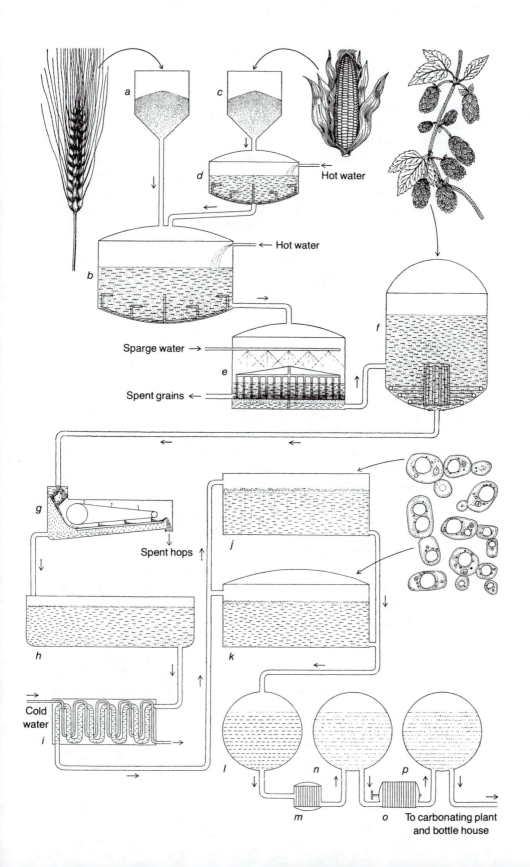

Hot water

Sparge water →

Spent grains ←

Spent hops

Cold water

To carbonating plant
and bottle house

higher percentage of dry malt. Carmelized malt or other sugars are added before or after fermentation. *Porter* and *Stout* are also dark, thick beverages made by top-fermenting heavy worts without such malt adjuncts as rice; they are traditional English beers, often drunk mixed with ale.

Germinated barley, called *malt,* is of prime importance in beer production, and is produced by *malting:* cleaned barley is soaked in water at 15 to 20°C and then transferred to aerated chambers where temperature and moisture are controlled to encourage maximum and uniform germination of the seeds. Once germination is at a satisfactory stage, the malt is dried (by *kilning*) for future use. The malt provides not only starch-hydrolyzing enzymes (amylases) but also starch, protein for flavor, yeast nutrients, and proteolytic enzymes.

Brewing proper begins with the *mashing* process: crushed malt and other starchy materials *(adjuncts)* are mixed with water and allowed to stand at controlled temperatures. During mashing the brewmaster varies the temperature to obtain the final sugar content desired for the type of beer being produced. The malt is mixed with water at about 40°C and held for 30 minutes while, in a separate tank, the cereal adjuncts (usually corn grits, rice, or wheat, and in the United States, sucrose) are mixed with water and brought to a boil. When they thicken, and while hot, the adjuncts are added to the malt mash, resulting in a temperature near 70°C which is held for another 30 minutes at pH 5.1 to 5.2. During the first holding period, the proteolytic enzymes produce peptides and amino acids that later serve as nutrients for yeast growth. During the second holding period, at the higher temperature and lower pH, the starch-digesting amylases are active. Starch consists of amylose, straight chains of glucose from which maltose units are liberated by action of β-amylase; and amylopectin, branched chains of glucose from which straight chains are liberated by the enzyme α-amylase, thus providing more substrate for β-amylase. Because the optimum temperature for β-amylase is about 60°C and for α-amylase is about 70°C, mashing at the higher temperature yields less fermentable sugar and more dextrins in the final beer. The result is more body and a stable foam. The clear amber liquid, called *wort,* is then filtered through a filter press, a lauter tub with a perforated false bottom and a pad of barley husks. The husks are sparged, or leached, with hot water for further extraction and the spent grains and settled proteins are harvested from the lauter tub for use as cattle feed.

Usually, hops are added at the rate of about 0.50 to 0.75 pounds per barrel of wort and the mixture boiled for 30 to 60 minutes in large copper or stainless steel brew kettles. Hops, dried *strobiles*, or flowers, of the female vine *Humulus lupulis*, assist in albumin coagulation, inhibit most gram-positive bacteria, aid in foam formation, and contribute flavor. Hops and substances coagulated by boiling are then strained as the wort is pumped into the hop-wort tank (Figure 7-5 A). The wort is gradually cooled to yeast *pitching* temperature (8 to 11°C) and pumped into the starting, or primary fermentation, tank. Liquid yeast, prepared as described below, is added at the rate of 0.75 to 1.5 pounds per barrel (31 gallons; 280 pounds) of wort and the inoculated wort incubated for 12 to 24 hours. During this time, dead or weak yeast cells will settle out along with precipitated proteins and hop resins. The wort is then pumped into the fermenter, where it is held at temperatures of 3 to 15°C for from 8 to 12 days, depending on plant conditions and the brewmaster's preferences. During fermentation the pH falls from around 5.5 to 4.0, but yeast autolysis then increases it to about 4.2. The final alcohol content, controlled by the amount of starch conversion allowed during mashing, varies from 3.6 to 5.2%.

After fermentation the beer is held for several weeks in *ruh* (storage) tanks (Figure 7-5 B). Here the temperature is maintained near 0°C to encourage settling of coagulated nitrogenous materials, phosphates, and yeast. The beer is pumped from the aging to the finishing tank through a diatomaceous earth filter bed and then is held for a week or so. During finishing the CO_2 content of the beer is adjusted to between 2.5 and 3.0 volumes. It then receives a final *polishing* filtration and is transferred to the final holding tank to await packaging (Figure 7-5 C). Beer placed in kegs is not pasteurized and has limited storage life; beer placed in bottles or cans has a long shelf life, since after packaging the containers are passed through a pasteurizer at 55 to 60°C for 15 to 30 minutes. Certain breweries have sterilized beer by a combination of filtration and aseptic packaging.

158

A

B

Figure 7-5
A. Modern brewing kettles
where wort is boiled and hops
are added. B. Ruh storage
tanks where beer is held near
0°C for several weeks. C.
Long filter press where beer
receives its final "polish"
before packaging.
(Reproduced with permission
of Adolph Coors Company,
Golden, Colorado.)

C

Most beer of the lager type consumed in the United States and Europe is made with a bottom-fermenting yeast strain of *Saccharomyces cerevisiae* classified separately as S. *carlsbergensis* (after the famous Carlsberg brewery, in Denmark). Sometimes the organism used in beer manufacture is classified as S. *uvarum;* these strains have stronger fermentative and weaker respiratory abilities than S. *cerevisiae.*

One of the brewery microbiologist's most important duties is to maintain an active, healthy yeast and to build up the pure culture for use in fermentation. Over the years a number of quality-control procedures have been developed (Haas, 1960); among the variables that can be determined are degree of attenuation, number of dead cells, fermentation power, generation time, flocculation, growth requirements, morphology, and flavor consistency in final product.

Measuring the alteration in solids content of the wort during fermentation yields the degree of attenuation *(A)*. A sudden change in this property of the yeast may indicate that the yeast is contaminated or is an undesirable variant. Attenuation can be determined by mixing known quantities of wort and yeast and fermenting, with occasional shaking, for 48 hours at 25°C. The level of dissolved solids (called extract) is measured before and after fermentation and is used to calculate the degree of attenuation:

$$A = [(E_0 - E_t)/E_0] \times 100$$

where E_0 = the extract at the beginning of fermentation and E_t = the extract at the end of the fermentation.

Autolysis of cells or other degenerative changes can be recognized by microscopic examination of the yeast. Staining with methylene blue will reveal the ratio of dead cells to live cells. The latter are refractory to the dye, while the dead cells readily take up the stain. Cultures with more than 2–3% stained cells are undesirable.

The fermenting power of the yeast may be quantified: a refrigerated Warburg respirometer measures the volume of CO_2 evolved per hour per gram of moist yeast from a standard substrate at, for example, 20°C.

The generation time g of the yeast may be determined from:

$$g = \frac{T \times \log 2}{\log b - \log a}$$

where T is the time the cells were allowed to grow, a the number of cells at time 0, and b the number of cells at time T. Generation times may vary from 20 hours at 4°C to 6.5 hours at 23°C.

Flocculation—agglomeration, or clumping, of yeast cells—is what causes the yeast to settle to the bottom of the tank. Premature flocculation causes poor attenuation; insufficient flocculation can cause off-flavors, as suspended cells degenerate, and failure of the beer to clarify well in storage. Various yeasts will show different flocculation rates in the same medium. Usually a yeast has a characteristic rate, but this may be altered by such environmental factors as wort composition (see Haas, 1960). In the most widely used test for yeast flocculation, liquid yeast is mixed with an equal volume of water (say 100 ml of each) and dried by filtering through a Buchner funnel. The yeast (5.0 g) is mixed with 100 ml of water and an aliquot (10.0 ml) suspended in a 15 ml conical centrifuge tube with 1.0 ml of pH 4.6 acetate buffer. After being vigorously shaken by hand, the suspension is allowed to settle for 10 minutes and the volume of settled yeast is measured. Highly flocculent yeasts yield 1.0 ml of settled cells, weak ones yield 0.5 ml or less.

The Yeasts in Brewing

A yeast's growth requirements may be determined by measuring its growth rate in synthetic media lacking various nutrients. A particular yeast strain's nutritional biotype may thus be determined; if a yeast being used in a brewery shows a change in its biotype there is cause for concern. While all strains of *Saccharomyces cerevisiae* and *S. carlsbergensis* require biotin for growth, their requirements for inosine, pantothenic acid, and other vitamins vary.

Most breweries maintain stock cultures of pure yeast strains on wort agar; however, it is advisable to grow cultures alternately in broths and agars so that no type is selectively developed that flourishes in culture but grows poorly during actual fermentation in the wort. When a new or revitalized strain is introduced from the laboratory into the brewery it is necessary that the desired brewery fermentation temperature be reached only gradually by decreases through several batch fermentations. These batches would also progressively increase in size to obtain the desired volume of yeast to introduce. Examples of successive volumes and temperatures might be 1 liter at 14 to 18°C, 50 liters at 12 to 14°C, 500 liters at 10 to 12°C, and finally 1500 liters at 8 to 10°C.

Contaminants in Beer

Growth of contaminating bacteria tends to be discouraged by brewing conditions and ingredients, such as:

1. boiling the wort;
2. the low fermentation temperature;
3. the low pH;
4. microbial inhibitors present in hops;
5. anaerobiosis prevailing during the vigorous yeast growth;
6. pasteurization of the beer; and
7. high sanitation standards in the processing plant.

Although a variety of types of microorganisms may contaminate wort or beer, few grow and cause beer spoilage. The most frequent contaminants of beer are lactobacilli that, being widely distributed on green plant material, are carried into breweries on the barley or malt. The organism usually encountered, *Lactobacillus pastorianus,* is a gram-positive, non-spore-forming, narrow rod-shaped species. Because it is heterofermentative, producing lactic and acetic acids, ethanol, and CO_2, it changes the taste and acidity of the beer. Cells of *Pediococcus cerevisiae,* a homofermentative bacterium, may also grow in beer and increase the acidity. Cells of *P. cerevisiae* are highly fastidious and gram-positive, grow in tetrads (in a neutral medium they may occur as diplococci), and are often difficult to cultivate on laboratory media, especially in the absence of CO_2. Organisms of the genus *Acetobacter* may grow in beer, using the ethanol and, if oxygen is present, oxidizing it to acetic acid. These are aerobic, gram-negative, acid-tolerant rods able to cause acidity, turbidity, and sometimes ropiness in beer. Also contaminating and growing in beer are other bacteria, of questionable taxonomic status—a species of *Leuconostoc* (Szilvinyi, 1958); *Flavobacterium proteus* (Shimwell, 1940); *Zymomonas anaerobia (Achromobacter anaerobium)* (Shimwell, 1937). The latter, a gram-negative non-spore-forming rod, produces dense turbidity and off-flavor in beer with added glucose-containing syrups; it has not been reported a problem in the United States. Cells of *F. proteus,* the principal contaminant in fermenting wort, may impart a parsnip odor to the beer. Certain anaerobic streptococci of unknown taxonomic relation to other streptococci may also contaminate and grow in beer, producing acid and diacetyl off-flavors. Coliform bacteria, such as cells of *Enterobacter aerogenes, E. cloacae,* and *Escherichia coli,* also can grow in wort and produce objectionable flavors and odors.

Quality Control. During different stages of manufacture, brewery laboratories determine the microbial content of beer as a quality control test, especially to detect spoilage bacteria and undesirable yeasts. This may be done by plating on universal beer agar (Kozulis and Page, 1968), which will enumerate the total number of microorganisms present. This medium contains tomato juice broth, 25.0 g; dextrose, 10.0 g; peptonized milk, 15.0 g; agar, 12.0 g; tap water, 750 ml, and beer, 250 ml. The beer is added after the other ingredients have been dissolved by heating. After autoclaving for 10 minutes at 15 pounds pressure, the pH should be 6.3 and the alcohol content about 0.7%. Lactic acid bacteria in beer may be determined by plating on a medium in which the following are added to 1000 ml of distilled water: yeast extract, 5.0 g; trypticase, 20.0 g; liver concentrate 202-3 (Sigma), 1.0 g; maltose, 10.0 g; fructose, 10.0 g; betaine, 2.0 g; di-ammonium citrate, 2.0 g; potassium aspartate, 2.5 g; potassium glutamate, 2.5 g; $MgSO_4 \cdot 7H_2O$, 2.0 g; $MnSO_4 \cdot H_2O$, 0.5 g; K_2HPO_4, 2.0 g; N-acetyl glucosamine, 0.5 g; Tween 80, 10.0 ml; agar, 20.0 g (Saha et al., 1974). The pH after autoclaving should be 5.4. Common beer spoilage bacteria can be differentiated by a medium containing 2.0% tomato juice solids, 2.0% peptonized milk, 1.0% yeast extract, 1.0% glucose, 2.0% calcium pantothenate, 1.5% agar, 0.5% $CaCO_3$, 0.11% citric acid monohydrate, K_2HPO_4, KH_2PO_4, $MgSO_4$, $MnSO_4$, $FeSO_4$, NaCl, Tween 80, bromocresol green, and actidione (Lee et al., 1975).

Beer can also be contaminated by many different species of wild yeasts, most of which sporulate more readily than *Saccharomyces carlsbergensis*. Therefore, yeast suspected of being contaminated can be placed under conditions favoring sporulation (on filter paper saturated with 0.5% sodium acetate) and then examined microscopically. Wild yeast spores also frequently differ from spores of *S. carlsbergensis*. If appreciable sporulation occurs in 40 hours or less it is likely that wild yeasts are present. In addition to *S. cerevisiae* or wild strains of *S. carlsbergensis*, other species that can cause beer infections are *S. pastorianus, S. ellipsoideus, S. exiguus, S. turbidans, S. williamus, Candida cerevisiae, C. krusei, Torula utilis, Hansenuela anomala,* and *Kloeckera apiculata*.

Wild yeasts may be enumerated by plating on a differential medium (Lin, 1974) in which the following are added to 1000 ml of distilled water: yeast extract, 4.0 g; malt extract, 2.0 g; peptone, 2.0 g; dextrose, 10.0 g; K_2HPO_4, 1.0 g; NH_4Cl, 0.5 g; crystal violet, 0.0004 g (0.4 ppm); fuchsin-sulfite mix (4.0 g basic fuchsin plus 25.0 g anhydrous sodium sulfite plus 1.0 g dextrin mixed in a mortar, desiccated three days and stored in a tight brown bottle), 1.0 g; agar 20.0 g. Also in use is a fluorescent antibody technique for identifying wild yeast contaminants (Richards, 1968 and 1969).

Diacetyl Off-Flavor. To make their product appealing to more people, especially to women, U.S. brewers have made milder beers by reducing the hops content. Flavor defects are more noticeable under these conditions. One abnormality in particular, diacetyl off-flavor, has become a problem of some magnitude. Diacetyl is a metabolic product of many organisms; it can be caused in beer by growth of contaminating lactic acid bacteria, or can be produced by yeast cells during fermentation.

Normally, this metabolic product is reduced to acetoin (acetyl-methylcarbinol) so that only small amounts (less than 0.1 μg per ml) remain in the final product. Figure 7-6 shows how the amount of diacetyl produced by *Saccharomyces cerevisiae* changes with time in a wort medium. Milder beer requires more extensive lagering to achieve diacetyl levels below 0.2 μg/ml (Rice et al., 1973). Thus, large breweries need greater storage capacity.

The literature contains much controversy about how yeast and bacteria form diacetyl. According to some investigators (Speckman and Collins, 1968; Chuang and Collins, 1968), both these microorganisms synthesize the compound by condensing active acetaldehyde and acetyl CoA. However, another mechanism may also function in yeast: the nonenzymatic oxidative decarboxylation of α-acetolactic acid formed as part of the pathway for valine biosynthesis: α-acetolactate apparently leaks from the cell into the wort and is then converted to diacetyl (Owades et al., 1959; Inoue et al., 1968; Lewis, 1968; Collins, 1972). Figure 7-7 presents these mechanisms diagrammatically. α-acetolactic acid is nonenzymatically converted to diacetyl under acidic conditions but is not excreted into the medium during growth of *Saccharomyces cerevisiae* (Collins and Speckman, 1974).

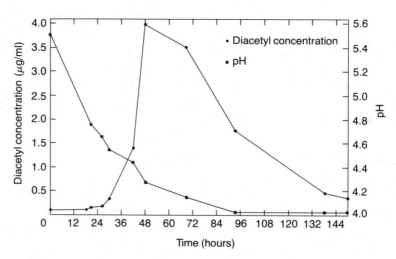

Figure 7-6
Amounts of diacetyl found at different times in a wort medium fermented by *Saccharomyces cerevisiae*. (From Tolls et al., 1970.)

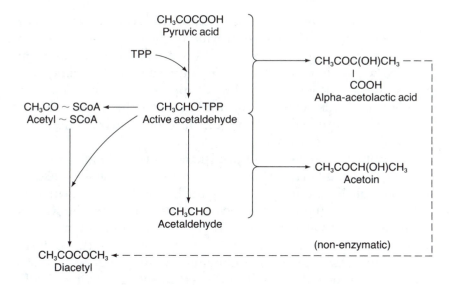

Figure 7-7
The mechanisms of diacetyl formation in yeast and beer.

The enzyme diacetyl reductase, widely distributed in yeast and bacteria, catalyzes the irreversible reduction of diacetyl to acetoin (Seitz et al., 1963; Tolls et al., 1970); as the diacetyl in beer is caused to decline, the diacetyl off-flavor is eliminated (Thompson et al., 1970). A patent has been issued for a microencapsulation technique to allow reuse of the enzyme extracted from *Enterobacter aerogenes* in removing diacetyl from beer and wine (Shovers and Sandine, 1973). Encapsulated whole yeast cells may also be used.

Wine

Bread, cheese, and wine have made many a meal for royalty, peasants, warring armies, and for people of today. These foods typify the artful use of microorganisms to make raw perishables more stable. Like beer, wine has been a staple food beverage for millennia. Wines are alcoholic beverages made by yeast fermentation of fruit and fruit juices. Since the juice from grapes best lends itself to conversion into a liquid essence that improves with age, most wine is produced from this fruit. In fact, some people restrict the definition of wine to the beverage produced by grape juice fermentation, calling all other fermentations fruit wines. Other fruits used include cherries, blackberries, elderberries, gooseberries, plums, apples, pears, raisins, and peaches. Wine also can be produced from less likely plant materials, for example, the dandelion or other herbs. These are generally fortified with raisins.

The critical factor in wine production is the climate in which the grapes are grown. Certain regions of the world have excellent soil and climate for producing the best grapes for fermentation: the Bordeau and Burgundy regions of France, the Rhineland district of Germany, the Tokay region of Hungary, and parts of Spain, Switzerland, and Italy are famous in this regard. France and Italy have ideal climates for growing grapes and together produce about half of the world supply of wine. Probably 90% of our domestic wines are produced in California, although there are sizable industries in New York, Washington, Oregon, Ohio, Arkansas, Virginia, and North Carolina.

The native American grape *Vitis labrusca* is used to some extent, especially in the East, but California wines are produced from different varieties of *Vitis vinifera*. Of European origin, this species was grafted to about 9 different natural wild vines, now consisting of about 5000 varieties. The muscadine grape of southern Europe, *Vitis rotundifolia,* also is used, but less so commercially than the other two.

Between 1865 and 1875 the European wine industry, particularly in France, was devastated by widespread loss of grape vineyards to infection by the root louse aphid *Phylloxera vastatrix.* Vineyards in this country also were affected, but many grape varieties flourished in the desert valleys of California after *Vitis vinifera* were grafted to the state's native wild grape vines, which were resistant to *P. vastatrix.*

Winemaking

Grapes selected for winemaking are crushed without stemming and then treated with some form of sulfur dioxide (for example, sodium metabisulfite or sodium bisulfite) to prevent excessive contamination by wild yeasts, molds, and bacteria. Crushed grapes are called *must* and consist of 85 to 95% juice, 5 to 12%

skins, and 0 to 4% seeds. The seeds are rich in tannins and oil and also contain resins. The skin, covered with a layer of cutin, and the cell layers immediately below contain the coloring, flavoring, and aromatic constituents. During ripening on the vine the pH increases from about 2.8 to 3.1 and free malic and tartaric acids decrease while fructose increases. At or near maturity crystalline deposits of potassium acid tartrate also will form just under the skin. Anthocyanin pigments occur in the skin cells of red grapes.

Table wine is wine produced by fermentation of must from red or white grapes with no additives; its alcohol content ranges from 9 to 14% by volume. There are several categories of table wine. Preferred table wines, called vineyard wines, are sold under their vineyard names; their flavor generally improves within up to five years or more. Good table wines, called regional wines, are sold under the names of the districts or towns of origin; these usually are drunk within two or three years of the vintage year, the year in which their grapes were grown. Other table wines, generally called standard, represent about 90% of the volume of production. These wines usually contain about 12% alcohol. Well-known American standard table white wines are called *chablis, Rhine wine* (Moselle, Riesling) and *sauternes;* pink wines may be called *grenache rosé,* being named after the grape. Standard red table wines are *Burgundy, claret, chianti,* and *zinfandel,* the last also being named after the grape. Familiar American standard wines also are named after the famous wine-producing districts of Europe.

Table wines in which CO_2 is produced by a secondary fermentation are called sparkling wines; champagnes are examples. The addition of herbs will produce aromatic wines, such as vermouth, with alcohol content being as much as 21%. The addition of brandy produces *fortified* wines like sherry and port; these range from 14 to 21% alcohol.

The Yeast Fermentation

A number of yeasts of several genera occur naturally on grapes and are therefore present during fermenting. Among those occasionally found are species of *Candida, Hansenula, Kloeckera, Pichia, Saccharomycodes (Schizosaccharomyces)* and *Torulopsis* (Reed and Peppler, 1973; Galzy, 1956). Winemakers, however, ferment musts by using strains of *Saccharomyces cerevisiae* var. *ellipsoideus,* cells of which are larger and more oval than beer yeasts. Musts are purposely inoculated with yeasts of the genus *Saccharomyces* because they are the strongest fermentors, producing 18 to 20% ethanol, and will thus dominate the fermentation.

Because must fermentation is begun by adding two or three gallons of yeast starter culture to each 100 gallons of juice, the volume of starter must be built up, sometimes from a pure culture from a test tube slant. Compressed and active dry wine yeasts also are commercially available with stabilities of 3 to 4 weeks and 6 to 12 months, respectively. *Saccharomyces cerevisiae* var. *ellipsoideus, S. beticus,* and *S. boyanus* are now available in both compressed bulk quantity and the active dry state.

A number of factors can influence the fermentation of must by wine yeasts (Amerine, 1967; Reed and Peppler, 1973).

Carbon Source. Of the two principal sugars in must, most wine yeasts ferment glucose more readily than they do fructose; however, *S. elegans,* a yeast used in sauterne wine manufacture, uses fructose more rapidly than glucose. Normally, fermentation of the carbohydrate naturally present in the must readily yields 11 to 12% alcohol. In cold climates, especially in the Eastern United States, where *Vitis labrusca* grapes are grown for winemaking, the wine may be fortified with sugar (a practice called *amelioration*). Also, if weather conditions allow the grapes to mature too quickly, water may be added to produce a wine of normal composition.

Alcohol. Ethanol produced by the yeasts will inhibit the cells, the degree of inhibition increasing as temperature rises. The lower the temperature, the higher is the yield of alcohol, because fermentation is more complete and less alcohol is lost by evaporation and by entrainment in escaping CO_2.

Carbon Dioxide. CO_2 pressures of about 7.2 atmospheres halt yeast growth, and 30 atmospheres completely prevent production of alcohol. The effect of CO_2 pressure is important in bottle- or tank-fermented wine or if the fermentation rate is regulated by pressure. In table wines about 0.1 to 0.5 g of CO_2 is dissolved per liter; 12.0 g of CO_2 per liter produce pressures of 4.0, 4.8, 5.8, 6.6, and 7.5 atmospheres at temperatures of 0, 5, 10, 15, and 30°C. In sparkling wines six to eight atmospheres of CO_2 pressure are desired in the final product.

Temperature. Most wine yeasts grow best at temperatures of 27 to 30°C, but many grow at lower temperatures and can ferment must at 7°C or below. It has been noted that among wine yeast strains, 23 days at 0°C, 8 days at 6°C, and 3.5 days at 12°C are average times needed to produce 5 g of ethanol from 1 liter of grape juice. Long slow fermentation at low temperatures produces wines with flavors more bitter than rapid fermentation at high temperatures produces. Common temperatures for white wine production are 7 to 18°C; for sparkling wine, 12 to 18°C. In red wines, because tannins and pigments must be released from the skin, higher fermentation temperatures of 21 to 27°C are used. Fermentation tanks are cooled to maintain the desired temperature despite the heat produced by fermentation.

Sulfur Dioxide. Since ancient times sulfur dioxide has been used to preserve foodstuffs (Joslyn and Braverman, 1954). SO_2 has been used for over 40 years in California to control botrytis rot of fresh grapes after harvest. It is also used extensively in musts to inhibit natural grape yeasts and bacteria, especially lactobacilli. Starter yeast can be adapted to grow in more than 200 ppm of SO_2; in wineries, from 50 to 100 ppm are added to musts (by adding 3 to 6 ounces of potassium metabisulfite per ton of grapes). Such treatment will kill more than 99.9% of natural must organisms. Because free SO_2 has antiseptic properties, its protective power is reduced if the must has a relatively high concentration of aldehydes, ketones, and sugars with which the SO_2 will form bisulfite addition products. The antimicrobial efficiency of SO_2 is also affected by temperature, pH, and the type of microorganisms present. SO_2 liquified under pressure is now readily available and can be metered into the must to the desired level.

Tannins. Tannins are widely distributed polyphenolic vegetable products of variable composition. When dissolved in water, they form solutions of astringent taste; when combined with ferric ions they give blue-black or green colors. They are precipitated by proteins. The chief sources are oak and hemlock bark and nut galls. About 3 to 6% of the skin of red grapes are tannins and they help stabilize wine color. Tannin compounds vary widely in their antiseptic properties, and natural yeasts may be quite sensitive to them, although wine starter yeasts are resistant to them.

Available Nutrients. Satisfactory fermentation of grape musts requires sufficient nitrogen, minerals, and other nutrients. Yeasts do not require exogenously supplied amino acids for growth and, except in unusual circumstances, the nitrogen level of must is adequate. To be made into wine, musts of fruits other than grapes may need to be supplemented with a salt such as ammonium phosphate. Grapes contain adequate supplies of minerals, but excessive amounts of some, such as copper or iron, may be inhibitory. If any of a number of vitamins are added in some media, yeasts increase growth; all yeasts require an exogenous supply of biotin. White grapes generally contain less biotin than red types, but still contain sufficient amounts of this vitamin and others to allow a normal fermentation rate. The fermentation rate of the juice is affected by the biotin and total nitrogen content of the grapes when both are present, but is not affected when biotin, or any one of several B vitamins, is present alone (Ough and Kunkee, 1968).

As among breweries, fermentation practices vary considerably among wineries according to the type of wine being produced and the volume, the maturity of the grapes, and the available equipment. Whatever the variations, however, the basic steps are followed in every winery: harvesting and crushing the grapes; separating the stems; transferring pulp to fermentation vats; treating with SO_2; adding starter yeast; fermenting; separating the wine from the pomace (skins, seeds, and stem fragments); completing fermentation; racking, or drawing off the wine from the *lees,* or sediment; storing and aging; clarifying; and bottling. The steps after the first fermentation are generally called cellar practices. Figure 7-8 shows flow diagrams for wine manufacture as practiced in the Eastern and Western United States.

Defects in Wine

Microbial organisms can cause several defects in wine. Unless oxygen is excluded from the must or bottled wine, acetic acid bacteria will grow and produce vinegar. A number of lactic acid bacteria, especially the heterofermentative lactobacilli, may grow in wine and produce an acid defect. Cells of *Leuconostoc mesenteroides* and *Pediococcus cerevisiae* may produce off-flavors; the former will produce ropiness if sucrose is present to allow dextran production.

Malo-lactic fermentation in wine has recently received attention (Kunkee, 1967). As much as 50% of the natural acidity of grapes is due to malic acid, and, during a secondary fermentation of the wine, many lactic acid bacteria, including lactobacilli, cells of *Leuconostoc* spp., and pediococci, are able to convert malic to lactic acid. A "soured" flavor is thus produced that is undesirable in low-acid wines. Other flavor changes may also occur at this time.

Seasonal conditions sometimes cause a grape crop with higher than normal malic acidity, making deacidification by fermentation desirable. Cells of an acid-tolerant wine bacterium, *Leuconostoc oenos,* may be used to stimulate fermentation (Stamer and Stoyola, 1970). If acid is excessive, conversion may take place too slowly; hence it is suggested that *Schizosaccharomyces pombe* be added to reduce the amount of malic acid (Yang, 1973).

Yeasts such as *Candida sp.* and *Pichia membranefaciens* may form surface film on musts or in wine, and osmophilic types such as *Saccharomyces rouxii* may cause spoilage of sweet wines. *Saccharomyces bailii,* another osmophilic yeast, sometimes causes sandy to flocculent deposits in wine, especially in white and rose table wines. This yeast is highly resistant to SO_2 (Rankine and Pilone, 1973).

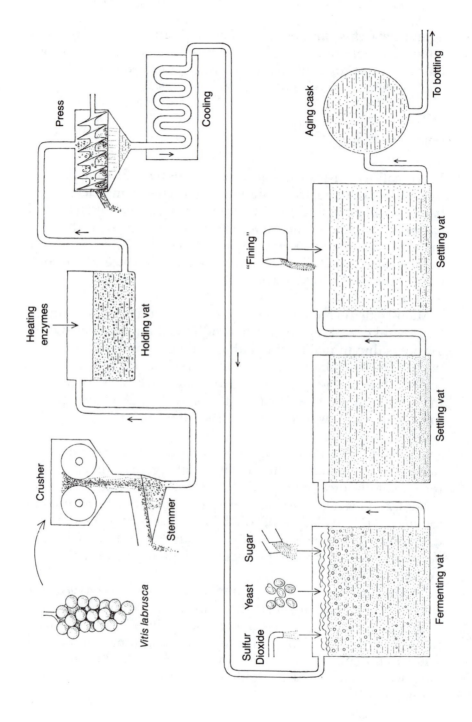

Crusher

Press

Heating enzymes

Holding vat

Cooling

Vitis labrusca

Stemmer

Sulfur Dioxide

Yeast

Sugar

"Fining"

Fermenting vat

Settling vat

Settling vat

Aging cask

To bottling

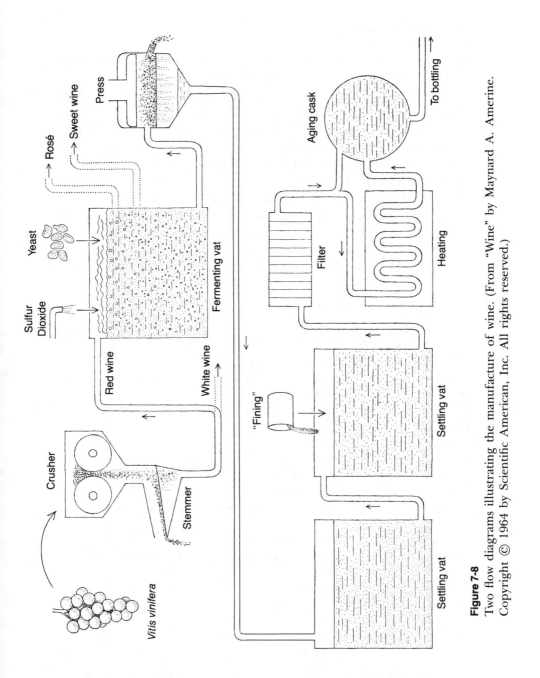

Figure 7-8

Two flow diagrams illustrating the manufacture of wine. (From "Wine" by Maynard A. Amerine. Copyright © 1964 by Scientific American, Inc. All rights reserved.)

Distilled Liquors

A number of grains, molasses, sugar cane products and other plant materials may be fermented by yeasts to produce alcohol that is then distilled to produce **spirits** such as brandy, whiskey, rum, gin, liqueurs, tequila, aquavit, or vodka. The fermentation of the grains to produce distilled liquors is similar to that already described for beer. For example, the steps in manufacture of Scotch whiskey are malting, mashing, fermenting, distilling, maturing and blending. Table 7-1 lists the best-known spiritous liquors and the material usually fermented in their production. **Liqueurs** and **cordials,** including such concoctions as Benedictine, Cherry Heering, Drambuie, anisette, crème de cacao, sloe gin, and bitters, are exotic alcoholic beverages made from rectified alcohol, refined sugar cane, and flavoring and

Table 7-1
Raw materials fermented to produce distilled spiritous liquors

Liquor	Material fermented
Scotch whiskey	Barley
Irish whiskey	Malt, unmalted barley, wheat, rye, and oats
Bourbon	Corn
Canadian whiskey	Rye
Gin	Grain mash, flavored with juniper berries, anise, etc.
Rum	Sugar cane
Aquavit	Grain or potatoes, flavored with caraway seeds
Vodka	Grain or potatoes, not flavored
Tequila (Mescal)	Juice from the hearts of the cactus Agave tequilana
Brandy	Various fruits
Kirschwasser	Cherry juice
Applejack	Apple juice
Cognac	White grapes produced in the Cognac region of France

Source: Compiled from data in *Chemistry and Technology of Wine and Liquors,* 2nd ed., 1948, by K. M. Herstein and M. B. Jacobs. D. Van Nostrand Company, Inc., New York.

aromatic substances extracted from fruits, herbs, seeds, and roots. They are high in sugar content and are served in small quantity as appetizers or as after-dinner drinks. Bitters are special liquors used for their tonic properties and in small portions to flavor other beverages.

Hard Cider

In the United States the unfermented juice of the apple (see Chapter 14) is generally called cider; the fermented product containing from 0.5 to 8.0% alcohol is called hard cider or applejack (in England the fermented juice is called cider; see Beech, 1972). Fermented apple juice with greater than 8.5% alcohol generally is called apple wine. Improved harvesting, processing, and transportation have increased the availability of a variety of fruits, making more surplus apples available for cider production. In fact, California now makes a considerable amount of apple wine. The best apples for cider production are those high in sugar and tannin—these are generally true cider apples or cull dessert and culinary apples. They contain about 10% fructose, 2% sucrose and 1.5% glucose; the malic acid content varies from 0.2 to 0.9% and the tannins from 0.1 to 0.5%.

Figure 7-9 shows a flow diagram of cider manufacturer. After the fruit is harvested, essential steps include washing; milling; pressing to extract juice; inoculating the juice with desired yeast, which settles to become the lees; and racking, or siphoning of the liquor into a clean vat. The cider then is filtered and packaged; before packaging it may also be blended and carbonated (Carr and Whiting, 1971).

Before inoculation and fermentation, the juice is usually treated to eliminate undesirable microorganisms. It may be pasteurized by flash heating to 85°C for 30 seconds, or, sometimes, given filter sterilization; or given centrifugation in a bactofuge, which removes 99.8% of the microbial counts. However, the juice microflora are most often controlled by the addition of sulfur dioxide. The amount required depends on the pH of the juice, its oxygen content, and the amount of sulfite-binding compounds present.

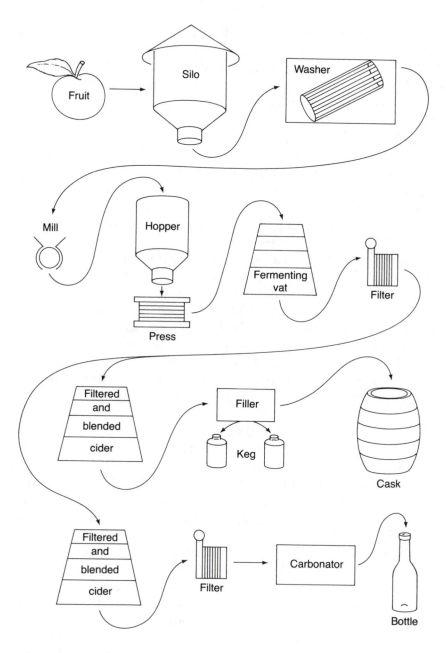

Figure 7-9
Flow diagram showing steps in the manufacture of cider. (Reproduced from *Modern Methods of Cider Making,* with the permission of the National Association of Cider Makers, Dorchester, England.)

Following SO_2 treatment the juice sets overnight. It should then contain at least 30 ppm SO_2 at pH 3.5 or correspondingly greater amounts at the higher pH levels. The juice is inoculated with strains of *Saccharomyces uvarum*, selected because they have desirable properties and no undesirable properties. A suitable yeast should be a pure culture, stable in fermentation and flocculation characters, unable to produce high levels of undesirable metabolites such as H_2S and diacetyl, resistant to SO_2, and so on. The yeast used must also be able to degrade de-esterified pectin to galacturonic acid; otherwise, the juice pectin remains unchanged throughout the fermentation. In addition to CO_2 and ethyl alcohol (up to 8.5%), the yeasts also produce aromatic compounds, organic acids, and higher alcohols, including butyl, hexyl, and propyl, totaling from 200 to 250 ppm on an average.

The fermentation has three phases—lag, intermediate, and stationary—with a total duration, usually, of several weeks at temperatures around 15 to 25°C. The fermentation ceases when the specific gravity, initially ranging from 1.037 to 1.082, reaches about 1.002. During or after the yeast fermentation, lactic acid bacteria carry out a second fermentation: the malolactic fermentation mentioned above in connection with wine. During the second fermentation the heterofermentative bacteria *Lactobacillus pastorianus* var. *quinicus* and *Leuconostoc mesenteroides* are usually involved in producing lactic acid and CO_2 from malic acid. Homofermentative types occur less often, but occasionally *Lactobacillus plantarum*, *L. mali* (Carr and Davies, 1970), and *Pediococcus cerevisiae* are involved.

During the malo-lactic fermentation the dicarboxylic malic acid is converted to monocarboxylic lactic acid, reducing acidity of the cider by 50%. In addition, lactic acid bacteria also catabolize other acids present in cider such as citric, quinic, shikimic, caffeic, and chlorogenic acids. From the sugars present lactic acid bacteria may also produce lactic and acetic acids. Thus, the second fermentation can produce somewhat varying results in different batches of cider, according to what flora are present. Reduced acidity is desirable because the cider then has a mellower, smoother flavor. Synthetic DL-malic acid can be added to cider made from low-acid varieties of apples.

Because its alcoholic content is somewhat lower, cider is more subject to disorders than is wine. One of the most common disorders is acetification: growth of acetic acid bacteria in the presence of oxygen converts ethanol to acetic acid. However, with large storage vessels (e.g., 40,000 gallons), the ratio of surface area to volume is so small that the amount of acetate produced has negligible effect on the flavor of the final mixed and packaged product. Another defect of cider is caused by cells of *Zymomonas anaerobia,* which produce acetaldehyde; reacting with tannins, this substance causes a milky white haze and a flavor like banana skin. Oiliness or ropiness may also be caused by lactobacilli such as *Lactobacillus pastorianus* var. *brownii* (Millis et al., 1958), *Leuconostoc, Streptococcus damnosus* var. *viscosus, Acetobacter visciosium,* and *A. capsulatum.* These defects are controlled in commercial cider manufacture by such measures as (1) treating juice to eliminate undesirable microorganisms, (2) fermenting to dryness, (3) minimizing contact with air, (4) carrying over SO_2, (5) keeping the acidity to at least 0.45%, and (6) using sterile filtration or flash pasteurization under pressure.

References

Aibe, S., A. E. Humphrey, and N. F. Millis. 1965. Biochem. Eng., Academic Press, New York.

Amerine, M. A. 1964. Wine. Scientific American. August.

Amerine, M. A. 1967. *Technology of Wine Making,* Avi. Publishing Co. Westport, Conn.

Bechamp, P. J. A. 1857. Annales de Chimie et de Physique, 3e serie *54,* 28.

Beech, F. W. 1972. Progress in Industrial Microbiol. *11,* 133.

Carr, J. G., and P. A. Davies. 1970. J. Appl. Bacteriol. *33,* 768.

Carr, J. G., and G. C. Whiting. 1971. J. Appl. Bacteriol. *34,* 81.

Chuang, L. F., and E. B. Collins. 1968. J. Bacteriol. *95,* 2083.

Collins, E. B. 1972. J. Dairy Sci. *55,* 1022.

Collins, E. B., and R. A. Speckman. 1974. Can. J. Microbiol. *20,* 805.

Galzy, P. 1956. Ann. Technol. Agr. *5,* 473.

Haas, G. J. 1960. Appl. Microbiol. *2,* 113.

Hough, J. S., and A. H. Button. 1972. *Progress in Industrial Microbiology* 11, 89, Churchill Livingstone, London.

Hume, E. O. 1947. *Bechamp or Pasteur.* 3rd ed. C. W. Daniel Co., Ltd., London.

Inoue, T., K. Masuyama, Y. Yamamoto, K. Okada, and Y. Kuroiwa. 1968. Proc. Am. Soc. Brewing Chem. *158.*

Joslyn, M. A., and J. B. S. Braverman. 1954. Adv. Food Res. *5,* 97.

Kozulis, J. A., and H. E. Page. 1968. Proc. Am. Soc. Brewing Chem. *52.*

Kunkee, R. E. 1967. Adv. Appl. Microbiol. *9,* 235.

Lee, S. Y., N. O. Jangaard, and J. H. Coors. 1975. Proc. Am. Soc. Brewing Chem. *33*(1), 18.

Lewis, M. J. 1968. Brewers Digest *43*, 75.

Lin, Y. 1974. Proc. Am. Soc. Brewing Chem. *32*(2), 69.

Millis, N. F., I. Husain, A. N. Hall, and T. K. Walker. 1958. J. Appl. Bact. *21*, 299.

Ough, C. S., and R. E. Kunkee. 1968. Appl. Microbiol. *16*, 572.

Owades, J. L., L. Maresca, and G. Rubin. 1959. Proc. Am. Soc. Brewing Chem. *22*.

Rankine, B. C., and D. A. Pilone. 1973. Am. J. Enol. and Viticult. *24*, 55.

Reed, G. and Peppler, H. J. 1973. *Yeast Technology*, Avi Publishing Co., Westport, Conn.

Rice, J. F., M. Y. Pack, and J. R. Helbert. 1973. Proc. Am. Soc. Brewing Chem. *31*.

Richards, M. 1968. J. Inst. Brewing *74*, 433.

Richards, M. 1969. J. Inst. Brewing *75*, 476.

Rose, A. H. 1959. Scientific American, June.

Saha, R. B., R. J. Sondag, and J. E. Middlekauff. 1974. Proc. Am. Soc. Brewing Chem. *5*(1), 9.

Seitz, E. W., W. E. Sandine, P. R. Elliker, and E. A. Day. 1963. J. Dairy Sci. *46*, 186.

Shimwell, J. L. 1937. J. Inst. Brewing *43*, 507.

Shimwell, J. L. 1940. J. Inst. Brewing *46*, 207.

Shovers, J., and W. E. Sandine. 1973. U.S. Patent #3,733,205.

Speckman, R. A., and E. B. Collins. 1968. J. Bacteriol. *95*, 2083.

Stamer, J. R., and B. O. Stoyola. 1970. Appl. Microbiol. *20*, 672.

Szilvinyi, A. 1958. Am. Brewer *91*, 72.

Thompson, J. W., J. Shovers, W. E. Sandine, and P. R. Elliker. 1970. Appl. Microbiol. *19*, 833.

Tolls, T. N., J. Shovers, W. E. Sandine, and P. R. Elliker. 1970. Appl. Microbiol. *19*, 649.

Yang, H. Y. 1973. Am. J. Enol. Viticult. *24*, 1.

8

Other Yeast Fermentations

Bread

Bread is one of the most basic human foods, and is one of the earliest to be processed. Although most bread is now baked in bakeries instead of in the home as in past centuries, the basic principles of bread making have remained substantially the same for thousands of years. In the British Museum are examples of Egyptian bread, dating from 2100 B.C., the texture of which makes clear that leavening was practiced then. There is no record as to when dough was first fermented as a regular procedure, but yeasts are naturally present on grain made into flour, and would be introduced into any flour that was mixed with water in the vicinity of wine making. In fact, early Egyptian bakers obtained yeasts from the settlings in beer vats as Greeks and Romans later did from wine vats. And in fifteenth-century France and later in England brewer's yeast, or barm, was the leavening agent for bread.

Bread making involves grinding (milling) grain into flour, which is then mixed with water, salt, and sometimes other ingredients to make the dough. Yeast is added and the dough ferments for two to five hours. The carbon dioxide produced during this time provides the light, porous texture. After leavening, the bulk dough is cut into pieces of uniform size and weight and placed in pans for a final fermentation, or proofing. Proofing may last from 15 minutes to an hour under careful temperature control (28 to 32°C). The bread is then baked at 100 to 250°C for periods ranging from a few minutes to several hours, depending on the type of bread being made. To make sourdough bread, fermented dough from a previous batch is usually used as the leavening and souring agents, since both yeasts and lactic acid bacteria are involved.

In the United States most bread flour is made from wheat, although other cereals such as barley, maize, oats, and rye are also used. Different classes, subclasses, and grades of wheat are recognized (USDA, 1964) and knowledge of these is important in selecting the type to be used in bread manufacture. Also, a particular wheat can be milled in different ways to produce different types of flour, each with its own properties.

The wheat kernel has three parts: the germ, or embryo (2% by weight), the bran (13%), and the endosperm (85%). The embryo is where the new plant originates; the bran give protective covering; and the endosperm, a starchy substance, provides food for the new plant. White flour is milled for the purpose of separating the endosperm from the bran and germ, as a powdered fraction suitable for use in baking. The milled flour will contain approximately 11% protein, 1% fat, 0.4% ash, 14% moisture, and 73% starch.

Cells of *Saccharomyces cerevisiae* (Figure 8-1) are usually grown on molasses, a by-product of sugar refining, supplemented with biotin, thiamine, and ammonium salts (Reed and Peppler, 1973). Among other growth conditions, aeration is provided; by limiting sugar fermentation, aeration encourages yeast growth by respiration and also minimizes alcohol production, to less than 0.1%. Cells are harvested by centrifugation, washed with water, and further concentrated by pressing or filtration. The cake yeast is then blended in mixers; emulsifiers such as lecithin and soy bean oil are added to aid in extrusion and cutting of the yeast. The mixture is extruded through tapered nozzles and cut into one-pound cakes for commercial use or one- to two-ounce cakes for home use. The cakes are wrapped

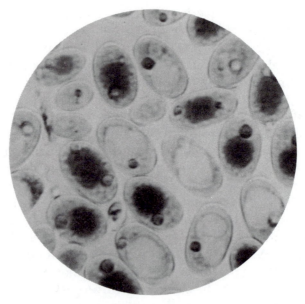

Figure 8-1
Ultraviolet photomicrograph of *Saccharomyces cerevisiae* cells. The dark-appearing cells have vacuoles with S-adenosylmethionine, because they were grown in a medium containing 0.5 mg of methionine per 100 ml. (Courtesy of K. D. Nakamura and G. Svihla, Argonne National Laboratory.)

in waxed paper or foil, stored, and shipped in refrigerated trucks at 4°C. In this form the baker's yeast is stable for a week to 10 days; after 3 or 4 weeks, molds develop on its surface, making it unusable in bread making. Recent technology has produced an active dry yeast, which has replaced compressed yeast cakes for home baking; however, commercial bakers continue to use the one-pound cakes. Dry yeast for home use is put in polyethylene-lined envelopes of 7 grams capacity. The envelopes are flushed with nitrogen gas after filling and heat sealed. Such yeast can be stored at room temperature for as long as 18 months and still provide suitable leavening action in fermenting dough. The moisture content is reduced to an a_w of 0.45 or a water content of 8.0% (Crane et al., 1952).

Bread leavened by microbial fermentation—usually by action of yeasts—may be prepared by mixing all of the ingredients together either in a single operation, the *straight dough process*, or in two steps, called the *sponge and dough process* because the yeast, yeast food, and a portion of the flour and water are mixed together to form a stiff *sponge*. These ingredients undergo a preliminary fermentation before the remaining flour, water, salt, and other ingredients are blended with the sponge, and before the final mixture undergoes a second fermentation.

Incubation temperature is usually 25 to 28°C for a straight dough, 22 to 25°C for a sponge and dough. The straight dough process is quicker and easier, and gives higher yield and better flavor than the sponge and dough process, but requires a larger inoculum of yeast and produces bread of inferior volume and texture (Pyler, 1967). Also, scheduling is more exacting. Doughs prepared by the straight dough process must be baked when ripe, without delay. Adjustments to correct defects or alter ripening time are difficult to make. However, the flour used for a straight dough need not be as strong as that required for the sponge and dough system.

Fermentation

The principal ingredients of dough are wheat flour, water, yeast, and salt. Other ingredients may be added, such as sugar, milk or milk products, shortening, malt flour or fungal amylase, yeast food, and dough conditioners. Flour used for bread making should have good elasticity, strength, stability, amylase activity, color, and moisture content. The major proteins in the flour, *glutenin* and *gliadin*, combine with some of the water to form an elastic gluten complex. The gluten must be able to stretch sufficiently to permit the dough to rise adequately, and yet be strong enough to prevent the carbon dioxide from escaping too readily. Flour prepared from a strong (hard) wheat requires more water than that from a weak (soft) wheat. During hydration, water penetrates the protein structure, causing the protein to form a continuous network or film that contains the starch granules in the flour.

Alpha (α) and beta (β) amylases in the flour hydrolyze the starch to maltose. Natural flours contain ample β-amylase but may not contain enough α amylase. For this reason, malt flour or fungal amylase may be used to supplement the α amylase activity of wheat flour.

The yeast, *Saccharomyces cerevisiae*, is a special baker's strain selected for its ability to produce the hydrolytic enzymes maltase, invertase, and the zymase complex that allow maltose and sucrose to be efficiently converted to simple sugars, carbon dioxide, and alcohol. The quantity of yeast needed to complete the dough fermentation depends on time and temperature. Low dough temperatures or short times require a large inoculum.

Salt aids the development of flavor, toughens the gluten so that the dough is less sticky, and retards fermentation. Ammonium sulfate and the phosphate salts of magnesium and potassium are added to enhance the yeast's enzymatic activity so that adequate quantities of carbon dioxide are produced quickly—leavening and expanding the dough. Dough improvers are often added to flours to modify the physical properties of gluten, for example, by making the dough more elastic.

Mixed sponge or dough is placed in large troughs in temperature- and humidity-regulated rooms, where it rises. When the dough is ready to be made into loaves, the trough may be sent to the make-up room, either via a chute or by being wheeled. In the make-up room the dough is fed into a machine called a "divider."

At this stage dough has many of the characteristics of a liquid, and definite quantities flowing through the chute are cut into exact size by weight, rounded, and rolled into a ball. The formed pieces of dough may then be held briefly in individual pans, to regain lightness. They then go into a specially constructed compartment called a *proofing* room, in which the temperature and atmospheric moisture are carefully controlled to promote maximum, yet uniform, rise. After proofing the loaves travel to the baking oven. These operations may be by batch processes or continuous as in some of the newer and larger bakeries. Ordinarily, the ovens operate continuously for long periods of time, the loaves of dough entering at one end and the completely baked bread being automatically discharged from the bake pans at the other. If baking times are very short—as for rolls—the yeast may survive.

The baked loaves are cooled before being sliced and wrapped. Unless sufficient time is allowed for this step, the warm, moist loaf fosters accelerated microbial growth. The spore-forming bacteria are the only organisms that survive baking, but, if adequate precautions are not taken, the air in the cooling area may be laden with mold spores, and the loaf may become contaminated while cooling. It may also be contaminated by contact with unsanitary bread racks, flour dust, slicing knives and other equipment, handling, wrapping paper and wrapping machines, returned bread, and bread storage boxes. While proper sealing of the wrapper prevents further contamination, the atmosphere and substrate within the sealed package are ideal for germination and growth of mold spores.

Moldy Bread. A variety of molds grow in cracks and breaks on the crust surfaces, on the surfaces of slices, and throughout the crumb. When bread is stored under fairly humid conditions, cells of the well-known bread mold *Rhizopus nigricans* grow luxuriantly. *Monilia (Neurospora) sitophila,* the red or pink bread mold, is of special concern in bakeries: once established on the premises, spores of this pernicious mold are virtually impossible to eliminate. *Penicillium* spp. (Figure 8-2) and *Aspergillus niger* spp. are also common contaminants, but any of a score or more of other species and genera may be present. Figure 8-2 B illustrates bread spoilage by mold that results when bread containing no preservatives is cut with a knife contaminated with spores of *Penicillium* and held three days at room temperature (25°C).

To protect commercially prepared bread from mold development sodium or calcium propionate, acetates, or diacetates may be added to the flour at a level of 0.1% (Ponte and Tseu, 1978). The air of the bakery may be filtered, walls and floors may be periodically washed, and wrapping and slicing areas may be irradiated with ultraviolet light.

Figure 8-2
Top. Fresh bread. Bottom. Bread that became spoiled after being cut by a knife contaminated with *Penicillium* spores and then being held in the wrapper for three days at 25°C. The bread contained no preservatives.

Ropy Bread. Although ropy bread is a rarity today, bread stored for some time in certain conditions may acquire a discolored, soft, and sticky interior. When the crumb or slices of such bread are separated, long, stringlike threads are evident (giving rise to the term *ropy bread*). Capsular materials produced by spore-forming aerobes such as *Bacillus subtilis* and *B. mesentericus,* together with hydrolyzed proteins and starches, provide the materials considered essential for the development of rope. Whole-grain and cracked-grain flours or flours that are not highly refined are more apt to be sources of rope organisms than are patent or highly purified flours. Bacilli capable of producing rope are common in the soil, the air, and on grains, and consequently are more likely to be retained in large numbers by flours having a high proportion of bran or residual crease dirt. While flour is considered a major source of the spore-forming organisms, so also are yeast, malt, milk, and sugar. When baking temperatures do not exceed 100°C for a significant time, surviving spores germinate after the bread has baked—especially if the bread is stored where it remains warm and moist.

Slicing knives and other equipment occasionally become contaminated by spore-forming bacteria and "seed" these organisms in bread sponges, doughs, and directly into the interior of the loaf. This defect occurs much more commonly in home-baked breads than in commercially baked breads.

Sourdough Bread

A number of sourdough breads are made commercially in Europe and the United States, among them the internationally known San Francisco sourdough bread (Figure 8-3) and a sourdough bread made in Italy. Both breads have essentially the same characteristics and microflora (Galli and Ottagalli, 1973). Sourdough bread has been made in the San Francisco area for more than a hundred years, no doubt because prospectors arrived who carried fermenting dough in their saddlebags for making biscuits. (Hence, prospectors were affectionately called "old sourdoughs.") Until recently, however, little was known about the microbiology of the commercially made bread (Kline et al., 1970; Sugihara et al., 1970; Sugihara et al., 1971). The important yeasts identified are *Saccharomyces exiguus (Torulopsis holmii)*, an acid-tolerant yeast capable of initiating growth at pH 4.5, and *S. inusitatus*. No types of bakers' yeast are found.

The sourdough bacterium has been tentatively identified as *Lactobacillus sanfrancisco* (Kline and Sugihara, 1971; Sriranganathan et al., 1973). (See Figure 8-4.) It is unusual in that it ferments maltose but not glucose, galactose, sucrose, xylose,

Figure 8-3
San Francisco sourdough bread. (Courtesy of R. Floyd, Oregon
Agricultural Experiment Station Publications.)

arabinose, rhamnose, or raffinose. The yeasts use other carbohydrates, primarily glucose, but do not use maltose. The
flour used in San Francisco sourdough bread provides mostly
maltose, formed in the dough by amylase action on the starch
in the flour.

To make bread of the San Francisco type 100 parts of a
special high-gluten flour and 100 parts of dough from a previously fermented batch are mixed with 50 parts of water. During
incubation, at 27°C for 8 hours, the pH falls from about 5.2 to
3.8 and the populations of yeasts and bacteria reach about 10^7
and 10^9 per gram of dough, respectively. The activity of the
starter sponge is maintained by transfer every eight hours to
sustain good leavening and souring activity in the organisms.
The dough is prepared by mixing 20 parts of starter, 100 parts
of flour, 60 parts of water, and 2 parts of salt. The dough is then
weighed, molded, allowed to ferment (in what is called the
proofing stage) for seven to eight hours at 30°C, unpanned on
canvas dusted with cornmeal or rice flour, and baked directly

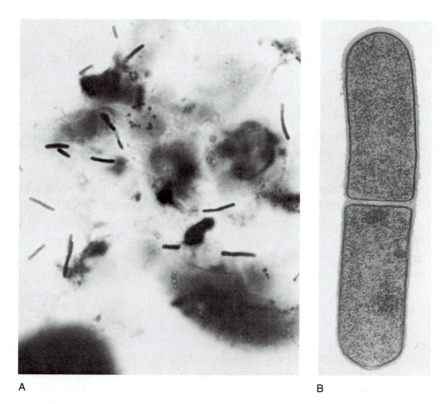

Figure 8-4
A. Sourdough "mother" sponge showing cells of *L. sanfrancisco* (× 1000). B.
Ultrathin section of the strain (× 107,000). (Courtesy of Kline and Sugihara,
1971.)

on a hearth for 45 to 50 minutes at 190 to 200°C. The crust is
browned by injecting steam into the ovens until a light tan color
develops. The bread is packaged in open nonplastic bags.

Surprisingly few bacteria other than lactic types are present
in sourdough starters, no doubt because they are inhibited by
lactic and acetic acids produced by the lactic acid bacteria.
When considerable air is available, *L. sanfrancisco*, which is
heterofermentative, produces as much as 20 to 30% of the total
acidity expressed as acetic acid in fermenting dough. Because
heterofermentative lactics anaerobically produce significant
amounts of ethanol, only 3.0% of the total acidity in broth

culture under static conditions is acetic acid (Ng, 1972). Under acidic conditions, acetate is primarily in the undissociated form, which inhibits most bacteria. Starters for sourdough bread may also come from other fermented products, such as buttermilk, yogurt, silage, and unpasteurized beer.

Sweet Breads, Cakes, Pastries, and Pies. Sweet breads are made from yeast-leavened doughs and cakes and pastries from chemically leavened doughs, and pie crusts are usually unleavened. Several antimicrobial agents prevent or retard mold spoilage. Propionates, sorbic acid, or potassium sorbate may be added at levels of 0.1 to 0.4% weight of the batter or dough (Brachfield, 1969). Methyl and propyl parabens may be added to chemically leavened doughs, pastries, and pie crusts.

Crackers

The soda cracker has one of the simplest compositions of all foods. Flour, shortening, salt, sodium carbonate, yeast, water, and flour bacteria are essential components. The fermentation occurs in two steps: the sponge stage process, which requires 18 hours, and the dough stage process, which requires four hours. The sponge is stiff when set and has a pH of about 6.0, but after fermentation at 27°C it is much more acid. The inoculum of yeast, 2 pounds of compressed yeast per 1000 pounds of flour, is one-tenth that used in bread making (Micka, 1948). The lactic acid bacteria increase the acidity in the sponge to a pH of 4.2 to 5.0 by the time it is ready for remix. After the rest of the ingredients—flour, salt, and sodium bicarbonate—have been added and the dough stage has fermented for about four hours, pH is approximately 6.8. The lactic acid bacteria may be introduced by adding 0.2 to 0.5% of old sponge to the remix, or may come from thin films of dough remaining in the trough from a previous fermentation. These organisms "seed" the dough and contribute to the acidity, pH, and flavor of the finished product. When insufficient acid is produced during fermentation, neutralization by sodium carbonate results in a product with a pH as high as 8.5. Inaction by lactic acid bacteria results in soapy, dull, or easily scorched crackers. The dominant bacterium is *Lactobacillus plantarum,* but *L. delbrueckii* and *L. leichmanni* are present in substantial numbers and *L. brevis, L. casei,* and *L. fermenti* in lesser numbers. If a liquid sponge is prepared with the first three bacteria, the stiff sponge and dough stages may be reduced to four and two hours, respectively (Sugihara, 1978).

Vitamins

Fermentation wastes from breweries and distilleries can have significant vitamin content. Because antipollution pressures prohibit dispelling these wastes into streams, they are dried and used as animal feed supplements. The dry residue from a distillery producing grain alcohol may contain the following approximate amounts of vitamins, in μg per g: thiamine, 7; riboflavin, 18; pantothenic acid, 31; niacin, 132; pyridoxine, 9; biotin, 0.5; and choline, 6500.

Yeasts also produce a number of sterols, the principal one being ergosterol, the commercial source of vitamin D. The vitamin is produced by ultraviolet irradiation of the sterol. In culture, ergosterol is best produced when aeration is vigorous and carbon is readily available. *Saccharomyces uvarum* is the organism most widely used to produce ergosterol for vitamin D manufacture.

Some yeasts also can be grown under conditions that allow production of 50 to 100 μg of riboflavin per g of dry cell weight. When *Candida* spp. are used, practically all the riboflavin is excreted into the medium, but commercial use of these species has been hampered because the flavin-synthesizing system is extremely sensitive to traces of iron in the fermentation medium. At present, industrial production of this vitamin relies almost exclusively on the higher fungi *Ashbya gossypii* and *Eremothecium ashbyii,* plant pathogens that parasitize coffee, citrus, tomato, and cotton plants (Goodwin, 1959).

Single-Cell Protein

As the world food supply has worsened in recent years, attention has focused on mass cultivation of microorganisms as sources of food. Microbial cells grown on wastes and meant to be used as food are called **single-cell proteins.** Studies are being made (Kihlberg, 1972) to assess the feasibility of producing crops of algal, fungal, yeast, lichen, and bacterial cells on various waste products. These wastes include cellulosic grain straws, cheese whey, sulfite waste liquors, molasses, fruit-processing residues, starchy wastes, and petroleum by-products. (Since nitrogen is 7 to 10% of the dry weight of most microorganisms, they, too, are rich sources of protein.) At present most single-cell protein in the human diet comes from cultivated mushrooms and from yeasts. Nearly all mushrooms are consumed fresh or canned, whereas dietary food and feed yeasts are dried. Several hundred thousand tons of the mushroom *Agaricus campestris* are grown annually in the United States on composted horse manure. Mushrooms of the *Morchella* sp. are adaptable to cultivation but are not now produced (Litchfield, 1967). Over 100,000 tons of food and feed yeasts are produced annually in the United States from the yeasts *Saccharomyces cerevisiae, Kluyveromyces fragilis,* and *Candida utilis.*

Food Yeasts

Of all the microorganisms studied, yeasts are the most promising source of single-cell protein. In pure culture they can be easily produced in large tonnages, worldwide, on readily available carbohydrates. Nearly 100% of the carbohydrates, supplemented with the necessary minerals and ammonia, is converted into yeast cell substance. The yeasts can be grown on molasses, juices of fruits indigenous to an area, hydrolyzed starches obtained from a variety of cereals, lactose in whey, methanol, petroleum fractions, stickwater and fish-processing wastes, and fruit- and vegetable-processing wastes and effluents.

Saccharomyces cerevisiae is grown primarily on molasses and grain hydrolysates; *Kluyveromyces fragilis,* which can use lactose efficiently, is grown on whey. In England and Russia, *Candida lipolytica* is grown on alkane petroleum fractions. *Geotrichum candidum* has a variety of applications because it can use not only the carbohydrate in food-processing wastes but many other molecules as well.

Although yeasts do not grow as rapidly as some bacteria, nevertheless 1000 pounds of yeast cells would produce 100,000 pounds of protein per day, whereas the same weight of soybeans or cattle produces 100 pounds and one pound of protein per day, respectively (Ogur, 1966).

Since live yeasts can deplete B-vitamins in the intestines and cause avitaminosis, the cells are heat-killed and dried. And their relatively high purine bases (adenine and guanine) cause an increase of their metabolic end product (uric acid) in the plasma and urine. Uric acid is only slightly soluble, and quantities present in the plasma may precipitate in the tissues and joints, causing a painful gout condition; stones may also form in the kidneys and bladder. A number of heating and extracting processes reduce the nucleic acid content of harvested yeast cells. These will no doubt find future commercial use.

In the United States *Candida utilis* is used to produce over 50,000 tons of yeast per year by fermenting sulfite liquor wastes from pulp mills. The liquor is treated with sulfurous acid to hydrolyze hemicelluloses to hexose and pentose sugars and acetic, galacturonic, and formic acids. The waste liquor is then steam stripped to reduce the sulfite content and neutralized for fermentation at from 32 to 38°C. *C. utilis* ferments both the hexoses and pentoses present, but ammonia nitrogen, phosphate, and potassium must be added. Fermentations are maintained as continuous operations with the pH kept between 4.5 and 5.5. Fermentor liquor is removed at the same rate as stripped liquor is added, usually at dilution rates of from 0.25 to 0.40. Yeasts are collected by centrifugation, washed several times with water, pasteurized in solution at 15 percent yeast solids, and dried on steam-heated drums. Figure 8-5 shows a flow diagram of the process.

Another so-called waste fermented to produce food yeast is cheese whey, which generally contains about 93% water, 0.3% fat, 0.9% protein, 4.0% lactose, 0.6% ash, and 0.2% lactic acid. Whey is sweet or acid, according to the type of source cheese (see Figure 8-6). Sweet whey (pH 5.9 to 5.1) is produced from cheddar and Swiss cheese; acid whey (pH 4.0 to 4.6) from cottage and cream cheese. Nearly all dried whey marketed is sweet, from cheddar cheese. While dried whey is quite nutri-

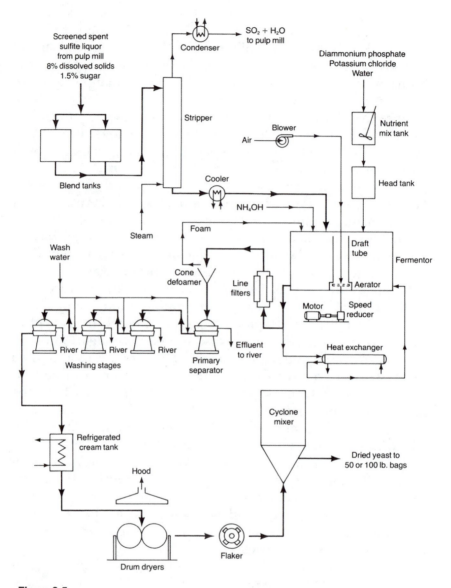

Figure 8-5

Flow diagram showing production of *Candida utilis* food yeast from paper pulp sulfite liquor waste. (From Inskeep et al., 1951. Reproduced with permission from Ind. Eng. Chem. Copyright by the American Chemical Society.)

A

B

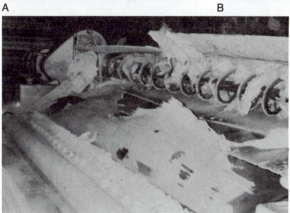
C

D

Figure 8-6
Equipment used to produce Wheast, a yeast food product made by cheese whey fermentation. A. The fermentor. B. Separating the yeast cells. C. The dried yeast (Wheast) coming from the roller dryer. D. Packaging. (Courtesy of Bromley Mayer, Knudsen Corporation, Los Angeles, California.)

tious, it is expensive to dry. Most cottage cheese is produced in small, scattered plants, and the volumes of the highly acid whey are too small to make recovery of by-products economical. For these manufacturers the ultimate solution appears to be to oxidize the components through the combined actions of several yeasts and yeastlike microorganisms, specifically those of *Geotrichum candidum* and *Candida lipolytica,* to reduce the **biochemical oxygen demand** (BOD).

Other Microbial Sources of Single-Cell Protein. The alga *Spirulina,* which has been used as a food for centuries, grows naturally and in pure culture in the highly alkaline lakes of Mexico and the Lake Chad area of Africa. The *Spirulina* is harvested, sun-dried, washed to remove sand, and incorporated into meal cakes. Algae have been cultivated to a limited extent in China and as a delicacy in Japan. In Indonesia they are grown on bamboo meshes in brackish waters. Algae growing in waters off the coasts of the Channel Islands, in the English Channel, are used as thickening agents in cookery and as feed for grazing sheep.

At least a dozen species of algae have been investigated for their adaptability to algal culture. *Scenedesmus* and *Chlorella* spp. are commonly cultivated in sewage effluents in the western United States, Central America, Israel, and the Far East. The effluents provide complete nutrition for the algae, but cultivation is limited to those parts of the world below 35° latitude, where the days are long year-round. They are generally cultivated in shallow vessels to accommodate limited light penetration. Algal culture presents certain problems, including: difficulty in maintaining pure culture and keeping it free of algal pathogens and protozoan scavengers, and the potential presence of viral pathogens, making human feeding possibly hazardous.

In Taiwan and India *Pseudomonas* spp. are used to produce single-cell protein from petroleum fractions. In India the bacteria are grown on the otherwise useless waxy residues of crude petroleum. *Bacillus megaterium* and *Hydrogenomonas eutrophica* have been produced for experimental studies in human and animal feeding; they lead to intestinal discomfort and distress in humans when 8.6 to 17.2 g is ingested per feeding, but are tolerated well by rats and chicks (Waslien et al., 1970). In areas of concentrated cheese production, wheys are pasteurized, cooled, and inoculated with *Kluyveromyces fragilis,* one of the few yeasts that use lactose.

References

Brachfield, B. A. 1969. Bakers Dig. *43*(5), 60.

Crane, J. C., H. K. Steele, and S. Redferre. 1952. Food Technol. *6,* 220.

Eoff, J. R., W. V. Linder, and G. F. Beyer. 1919. Ind. Eng. Chem. *11,* 842.

Food Processing. 1964. *22,* 117.

Galli, A., and G. Ottagalli. 1973. Ann. Microbiol. *23,* 39.

Goodwin, T. W. 1959. Progress in Ind. Microbiol. *1,* 137.

Inskeep, G. C., A. J. Wiley, J. M. Holderby, and L. P. Hughes. 1951. Ind. Eng. Chem. *43,* 1702.

Kent, N. L. 1966. *Technology of Cereals.* Pergamon Press, New York, N.Y.

Kihlberg, R. 1972. Ann. Rev. Microbiol. *26,* 427.

Kline, L., T. F. Sugihara, and L. B. McCready. 1970. Bakers Digest, *44*(2), 48.

Kline, L., and T. F. Sugihara. 1971. Appl. Microbiol. *21,* 459.

Litchfield, J. C. 1967. Food Technol. 21:159.

Marshall, P. G., W. Dunkley, and E. Lowe. 1968. Food Technol. *22,* 969.

Micka, J. 1948. Food Inds. *20*(1), 34.

Ng, H. 1972. Appl. Microbiol. *23,* 1153.

Ogur, M. 1966. Develop. Ind. Microbiol. *7,* 216.

Pasteur, L. 1858. Compt. Rend. *46,* 857.

Ponte, J. G., and C. C. Tseu. 1978. In: *Food and Beverage Mycology,* L. R. Beuchat, ed. Avi Publishing Co., Westport, Conn.

Pyler, E. J. 1967. Bakers Digest *41*(6), 55.

Reed, G., and H. J. Peppler. 1973. Yeast Technology, Avi Publishing Co., Westport, Conn.

Sriranganathan, N., R. J. Seidler, W. E. Sandine, and P. R. Elliker. 1973. Appl. Microbiol. *25,* 461.

Sugihara, T. F. 1978. J. Food Protect. *41,* 977, 980.

Sugihara, T. F., L. Kline, and L. B. McCreary. 1970. Bakers Digest *44*(2), 51.

Sugihara, T. F., L. Kline, and M. W. Miller. 1971. Appl. Microbiol. *21,* 456.

USDA. 1964. Official Grain Standards of the United States, Agric. Mktg. Services, Washington, D.C.

Waslien, C. I., D. M. Calloway, S. Margen, and F. Costa. 1970. J. Food Sci. 35:294.

9

Lactic Fermentations

Brine Preservation of Vegetables

Foods are preserved by brining and by lactic acid fermentations throughout the world's Temperate Zones. Both practices are far more than two millennia old. A Roman writer of the first century A.D., Lucius Junius Columella, in his *De re rustica,* described Roman brining practices. (He makes no mention of fermentative preservation of vegetables—apparently the Romans used fermentation only of grapes, in winemaking.) The Romans used "hard brine," obtained by letting saturated solutions of solar salt stand until impurities had settled out; such brine had a NaCl concentration of 26.4%. They also preserved foods in mixtures of hard brine, vinegar, and "must." The must was prepared by boiling concentrations of fruit juices, usually of grapes, to half volume. Containers of brined or pickled foods were sealed with pitch.

As early as the building of the Great Wall, in the third century B.C., the Chinese were preserving vegetables by fermentation. To seal out oxygen the mouths of their earthenware crocks were (and still are) circled by moats to receive flanged covers and water (Chao, 1949). In Oriental practice several vegetables might be fermented together in a mixture, containing, for example, cabbages, sweet turnips, carrots, peppers, radishes, cooked soybeans, and mung bean sprouts (Orillo et al., 1969). In American and European practice single vegetables are fermented. Also in the Orient mixtures of fish and rice and shrimp and rice are fermented (Orillo and Pederson, 1968). Such mixtures, known in Indonesia and the Philippines as *burong dalag* and *burong hiphon,* are colored a deep red to purple with *angkak,* or rice grains upon which cells of the ascomycete *Monascus purpureus* have grown (Figure 9-1). *Monascus* spp. are used in the Orient to make such foods as red rice wine, red Shao-hsing wine, and red soybean cheese (Lin, 1973).

At one time in the United States beets, green beans, green tomatoes, carrots, brussels sprouts, and some greens were commonly fermented (LeFevre, 1927). Many vegetables were brined successfully during World War II (Etchells and Jones, 1943; Fabian and Blum, 1943). Brined vegetables must be freshened in one or more changes of water to render them palatable. Most vegetables contain 1.8% or more of fermentable sugar; however, because green tomatoes have a low sugar content, fermentation is accelerated by addition of sugar to the vat (Fabian and Erickson, 1940).

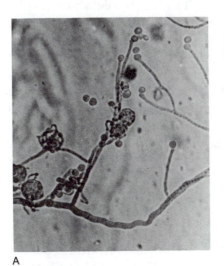
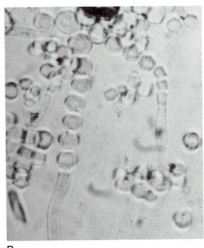

A   B

Figure 9-1
Monascus purpureus. A. Conidia and young cleistothecia (× 400).
B. Conidiosphore and conidia (× 950).

Preservation with Salt

Sodium chloride (NaCl) is the salt of choice in all vegetable lactic acid fermentations. The amount of salt present (expressed as percent of weight) affects the fermentation rate. Fermentation is fairly rapid at NaCl concentrations of 6 to 10% or less; is slower at slightly higher concentrations; and does not occur at concentrations of 15% or more. Other salts, such as $CaCl_2$, $MgSO_4$, etc., are bitter or toxic or cathartic. Sometimes cucumbers are treated with slaked lime during home fermentations, presumably to harden them. Used for the same purpose, various forms of alum may actually reduce the firmness of fresh-pack pasteurized dill pickles (Etchells et al., 1972). All NaCl, whether solar or mined, rock or granulated, is equally effective in cucumber fermentation (Borg et al., 1972). Salt reclaimed from spent cucumber brines has a high content of potassium that was leached from the cucumbers, but it can be reused for preparing brine (Durkee et al., 1974).

In dry brine processes 2.25 to 3.5% of dry salt is added to shredded, chopped, or highly compact vegetables. The salt withdraws fluids from the vegetable tissues. Larger vegetables are immersed in brine solutions. The vegetable absorbs 40 to 60% of the salt, depending on the ratio of fresh vegetable to brine, and equilibrium is established within several days.

Salt helps withdraw from vegetable tissues several substances, among them fermentable sugars and other nutrients needed by the fastidious lactic acid bacteria and substances inhibitory to most gram-negative bacteria. It also either completely inhibits or greatly retards the growth of most bacteria residing in the soil or on the plant surfaces. Salt contributes to the flavor and the texture of the fermented vegetable.

Temperature

All vegetable fermentations proceed optimally between 23.9 and 26.7°C. The fermentation rate decreases in direct proportion to decrease in temperature. Below 12.8°C cabbage fermentations proceed only through the first, or *Leuconostoc*, stage. In late fall the fermentation may cease, resuming the following spring without impairing the quality of the sauerkraut. At high temperatures the salt-tolerant aerobic and anaerobic bacteria and the yeasts outgrow the lactic acid bacteria, producing putrid, foul, discolored, and softened products. The temperature of cucumber fruits may be adjusted by heating or chilling brines (Etchells et al., 1973).

Anaerobiosis

The surfaces of all vegetables bear large numbers of resident and secondary microflora, many of them obligately aerobic. Washing in water reduces their numbers but does not eliminate them entirely. Insofar as the available oxygen permits, the remainder will grow in the fermentation crocks and vats, producing undesirable changes in flavor, odor, and texture. For this reason, all dry-salted vegetables should be firmly packed to prevent air pockets. Oxygen that remains entrapped during packing is consumed during the respiration of the vegetable tissue and is removed mechanically by formation of the brine and by the sweeping action of CO_2 produced by *Leuconostoc*.

Abnormal Color Changes

Most vegetables contain tannin or tanninlike compounds that react with iron to produce dark brown, gray, and black discolorations. For this reason, vegetables are scrupulously kept from touching rusty metal surfaces, and cooperage is not nailed. At a concentration of 5 to 10 ppm, copper replaces the magnesium in chlorophyll, noticeably changing the vegetable's color. At one time the copper-green color was deemed desirable and copper salts were added to fermented and pickled products; the practice became illegal in 1912. Green discolorations occur when copperware is used in home pickling. Copper discolorations were known to the Romans; Columella recommended that when vinegar was used vegetables should be boiled in pewter or lead vessels—a dubious practice in light of our present knowledge of lead poisoning.

Fermenting and fermented ciders also become discolored by iron and copper. Iron reacting with tannin imparts a "mousy" taste, termed cider sickness or *La Tourne,* when the fermentations occur during warm weather, and copper dissolved from copper fittings brings about a greening discoloration (Arengo-Jones, 1941).

Species of *Desulfovibrio* and *Desulfotomaculum* (including the former species *Clostridium nigrificans,* which, however, is not likely to grow in pickle brines) reduce sulfate to H_2S, which reacts with iron to form the dark FeS. The rapid formation of lactic acid usually prevents sulfate reduction. Another darkening discoloration is attributed to the microbial, enzymatic, or spontaneous formation of melanin or melaninlike pigments, usually at the surface. Such discolorations occur during fermentation and also after packing. Glass-packed sauerkraut and horseradish darken when oxygen remains in the head space at the time of sealing.

Preservation of Fermented Products

All proper fermentations yield a final pH of approximately 3.6 and leave no fermentable sugar unused. If anaerobiosis is maintained in the depth of the liquid, the lactic acid or lactic and acetic acids will preserve the fermented product indefinitely. At the surface, however, film-forming, acid-oxidizing yeasts grow freely, lowering the acidity or raising the pH; and many yeasts bring about softening. Members of the genera *Debaromyces, Candida, Endomycopsis, Hansenula, Rhodotorula,* and *Zygopichia* elaborate both polygalacturonase (pectin polygalacturonase) and pectin methylesterase (Bell and Etchells, 1956). Fermented vegetables are preserved through anaerobiosis, refrigeration, or canning. Natural (cellar) refrigeration can be used in the northern United States if vegetables are fermented in the late fall and if they are consumed before the film yeasts have grown to significant numbers.

The Lactic-Acid-Producing Bacteria

The lactic-acid-producing bacteria are fastidious gram-positive, mostly catalase-negative, non-spore-forming spheres and rods. Aged cultures of some species stain gram-negatively. Species of *Pediococcus* grown on media with very low sugar content may exhibit slight catalase activity. *Streptococcus faecalis* produces catalase if the heme precursor is incorporated into the medium.

All members of the group require carbohydrates for energy, and are unable to synthesize many of the amino acids and growth factors essential for reproduction. Further, they function under microaerophilic to anaerobic conditions—although they can be cultivated on agar surfaces on suitable media. The rapid production of lactic acid inhibits the growth of most detrimental, acid-sensitive bacteria of the soil and of plant surfaces.

Fermentation usually is initiated by the spherical bacteria, most commonly *Leuconostoc mesenteroides* and *Pediococcus* spp. although streptococci related to *Streptococcus faecalis* and *Aerococcus* may either replace them or occur together with them. *A. viridans* has been isolated only during slow, spontaneous fermentations of lima beans and Southern peas. It dies rapidly when the pH is reduced to approximately 5.2.

The rod-shaped bacteria *Lactobacillus plantarum*, of the group Streptobacterium, and *L. brevis*, of the subgroup Betabacterium, normally succeed the spherical bacteria and lower the pH to the desired value of approximately 3.6. However, if some malfermentations occur, such as by rapid destruction of lactic acid or rapid pectinolysis, these bacteria may not be effective acid producers.

Other genera and species of the lactic acid bacteria occur seldom in fermenting vegetables, although some that are culturally similar, such as *Streptococcus lactis*, *S. faecium*, and streptococci, are very common on plants (Mundt et al., 1967; Mundt and Hammer, 1968). The species active in fermentations occur commonly on parts of succulent plants, such as the surfaces of cucumbers and soft-textured squashes and less frequently and less populously on dry and hard-textured surfaces such as leaves and stems. Cells of *Leuconostoc mesenteroides*, by far the most cosmopolitan species, may occur in numbers approximating 10^5 per g of plant tissue. The lactobacilli often are present at less than 10 per g of plant tissue, and are detected and enumerated in selective liquid media that inhibit *Leuconostoc* (Mundt and Hammer, 1966).

Cells of *Leuconostoc mesenteroides* are spherical to slightly elongated and occur normally in pairs and in short chains. They ferment hexose sugars to 50% lactic acid, the remainder of the hexose molecule being converted into CO_2 and a mixture of ethanol and acetic acid. In the presence of sucrose the fructose moiety is fermented and the glucose is converted into a translucent, highly branched, viscous dextran through a nonphosphorylating, transglycosidic process. It produces copious slime. (Long before the nature of the bacterium was known, the slime masses forming in fluids in sugar refineries were known as "frog spawn.") During the lactobacillary stages of fermentations the slime disappears. The bacterium attains a titratable acidity of about 0.67% calculated as lactic acid, and a pH 4.4 to 4.6. A portion of the glucose is reduced to the bitter sugar alcohol mannitol. The acetic acid, ethanol, and that portion of the mannitol not subsequently destroyed by the lactobacilli contribute to the flavor of the fermented foods.

A salt content higher than 5% or a temperature higher than 29.4°C suppresses *Leuconostoc mesenteroides*, giving a fermented food a flat, soured taste. The brines for some fermentations are prepared with more than 5% salt, but equalization, or absorption of salt by the food to be fermented, reduces the salt concentration. The bacterium is unique in being the only lactic acid bacterium that is a leavening agent. Idli, which is prepared in India (Muckerjee et al., 1965) from ground and moistened blackgram mungo and rice flours, is steamed after overnight fermentation (see Chapter 10).

Pediococcus pentosaceus and *P. acidilactici,* usually found as plants residents, are so similar in cultural characteristics that some taxonomists consider them to be one species. In nature they exist in low numbers on the surfaces of succulent plants, and they grow on the surfaces of harvesting equipment where cell juices, soil, and debris accumulate. During growth a single sphere divides into a pair of cells that then divides on the long axis, becoming a tetrad that then separates into single spheres. In some culture media, the species tend to form short chains and clusters rather than well-defined tetrads. The plant pediococci grow well between 23.9 and 37.8°C in plant juices, in weak to moderately strong salt brines, and in culture media. During the homolactic fermentation they lower the pH to 4.2 or less.

Cells of the homofermentative *Streptococcus faecalis* may initiate fermentations, or they may accompany or follow the pediococci. They do not normally play a significant role in vegetable fermentations. The lower limits for acid production are pH 4.0 to 4.5. Two parallel groups of the species are known. One group, a common resident of human and animal intestinal tracts, rapidly acidifies in litmus milk and sometimes produces the acid-proteolytic digestion. The second group, widespread on plants, generally is rennin-proteolytic or soft-curd producing in litmus milk. Some strains lack the ability to ferment lactose and others to digest casein (Mundt, 1973). The bacteria apparently reside in insects and are distributed to plants during the growing season.

The lactobacilli most often isolated from fermenting vegetables are the well-defined species *Lactobacillus plantarum* and *L. brevis*. Both species grow well between 23.9 and 35°C, and do not grow below 15°C. *L. plantarum* is homofermentative, fermenting hexoses to lactic acid, while *L. brevis* converts sugars to lactic and acetic acids, ethanol, and CO_2. Many strains of *L. plantarum* ferment pentoses, are motile, and reduce nitrates; some strains produce a light- to deep-yellow pigment.

Sauerkraut

Cabbage was first fermented to produce sauerkraut in the sixteenth century, in Northern Europe. An old practice now abandoned was to add fruits, particularly apples, and spices to the cabbage. A Yugoslavian variation is to ferment whole heads of white or red cabbage, which are either immersed in brine or packed with shredded cabbage. Europeans enjoy sauerkraut juice as a delicious beverage before and with meals, but it has only a limited market in the United States.

Sauerkraut is defined as the product of characteristic acid flavor, obtained by the full fermentation, chiefly lactic, of properly prepared and shredded cabbage in the presence of not less than 2% nor more than 3% salt. When fermentation is completed, sauerkraut contains no less than 1.5% acid expressed as lactic acid. Sauerkraut is graded by color, cut, absence of defects, crispness, and flavor. Of these, flavor is given 45 of the 100 possible points on the grading scale. Although the definition specifies shredded cabbage, chopped cabbage is also used. Good sauerkraut may have as much as 1.7% acid calculated as lactic acid, a pH of 3.6, from 2.0 to 2.5% NaCl, not over 0.13%

ethanol, and 0.25% volatile or acetic acid. A higher ethanol content indicates the growth of fermenting yeasts; a lower volatile acid content indicates a depressed heterolactic fermentation (Pederson and Albury, 1969).

Cabbage contains 1.2 to 2.2% sugar, those with the higher sugar content producing finer sauerkraut, as do the late-maturing varieties. The loose-headed, crinkled-leaf varieties commonly grown during the early summer generally produce poorer sauerkraut. Since all varieties of cabbage used for making sauerkraut have enough sugar for complete fermentation, adding fermentable sugars is ineffectual.

Preparatory to making sauerkraut cabbage heads are wilted for two or more days to improve shredding, because fresh, crisp heads fracture easily. The heads are trimmed of the coarse outer leaves and washed to reduce the soil- and plant-resident bacteria; the core is either removed or broken with a chisel drill, and the head is shredded to strips 1/32 inch wide. The shreds are mixed with 2.0 to 2.5% salt and, to eliminate air pockets, packed firmly but without crushing.

Cabbage is fermented in wood-stave or plastic-lined concrete vats as large as 12 to 14 feet in diameter and 8 feet deep. The vats are covered with plastic sheets that are weighted with water to ensure anaerobic conditions. Before plastic was used the mouth of the vat was exposed to the atmosphere, and a variable quantity of cabbage, as much as one foot in depth or one ton in weight, was routinely lost through the combined activities of acid-consuming film yeasts, pectinolytic bacteria, and enzymic oxidative darkening.

Sauerkraut is marketed as either the raw or the canned product. Raw sauerkraut, packed either in barrels or in plastic pouches, is highly perishable and must be kept under continuous refrigeration. Sauerkraut canned in metal or in glass is heated to 74°C to destroy the lactic acid-producing bacteria and yeasts. Canned sauerkraut rarely spoils. Gassy cans are caused by underpasteurization and the fermentation of the residual sugar by cells of heat-resistant *Lactobacillus brevis*.

The acidity rises in three stages, first to approximately 0.67%, then to 1.20–1.25%, and finally to 1.5–1.7%, corresponding to the successive populations of *Leuconostoc mesenteroides, Lactobacillus plantarum,* and *L. brevis.* The rapid fall in pH brought about by *L. mesenteroides* discourages the growth of extraneous bacteria. The acetic acid, ethanol, and mannitol contribute to the flavor of the sauerkraut. Some of the mannitol is subsequently fermented by the lactobacilli. If it is not fermented, the product will be bitter. The CO_2 produced by *L. mesenteroides* helps establish anaerobic conditions by sweeping through the vat to remove residual air. During the first several days of fermentation the vats appear frothy.

The optimum temperature of fermentation lies between 18.3 and 21.1°C (Martin et al., 1939). Above 26.7°C pediococci and cells of *Streptococcus faecalis* initiate a homolactic fermentation and the product is sound, or edible, but the flavor is raw and sour.

The cells of homofermentative *Lactobacillus plantarum* succeed those of *Leuconostoc mesenteroides* and continue to produce lactic acid. At the peak of its activity, and population, there may be more than 4×10^8 colony-forming units per ml of brine. If, as sometimes happens, the temperature of the fermenting cabbage drops below 12.8°C, the lactobacilli do not develop until the following spring. The cabbage is unharmed, and once fermentation is complete the product is still good. Cells of the heterofermentative *Lactobacillus brevis* complete the fermentation, often so slowly that no production of CO_2 is apparent. In rare instances, strains of this bacterium have produced a water-soluble, heat-stable red pigment in cabbage juice and sauerkraut (Stamer et al., 1973). The pigment has been characterized as a saturated ester, aldehyde, or diketone, or its structure may contain a five-membered ketone ring.

Because the cell wall is mucilaginous, washing the cabbage does not remove the gram-negative, resident bacteria, most of which are in the genera *Pseudomonas, Flavobacterium,* and *Achromobacter.* However, when most varieties of cabbage are shredded a thermolabile, bactericidal substance is released (Pederson and Fisher, 1944) that drastically reduces the bacterial numbers. Most of those bacteria not so affected are normally suppressed by the salt, the anaerobic conditions, or rapid lactic acid fermentation. If present in large numbers cells of *Aerobacter (Enterobacter) cloacae* and *Flavobacterium rhenans* prevent the normal formation of acid, cause darkening, and produce a radishlike flavor (Fulde and Fabian, 1953). Clostridia, whose growth is favored by high temperature or delayed acid production, produce butyric, isobutyric, n-valerianic, propionic, and caproic acids to impart a cheeselike and putrid flavor (Vorbeck et al., 1961). Good sauerkraut should have less than one ppm butyric acid.

Normally the fermentative yeasts are easily controlled; washing the cabbage reduces them to low initial numbers, and the rapid onset of fermentation and intense competition with the bacteria for the available sugar keep them low in number. If salt is unevenly distributed during packing, the oversalted areas support growth of the unsightly pink yeast *Rhodotorula,* and the undersalted portions of cabbage support growth of pectinolytic bacteria. Although the salt is in solution in the brine, packing the cabbage too densely prevents the normal diffusion needed for equal distribution, and the same vat may contain rotted cabbage with oversalted areas in close proximity to rotted cabbage with undersalted areas. Also, if the temperature in the vat is above the optimum for lactic acid fermentations, or if the shredded cabbage is a variety that does not furnish the proper nitrogen compounds needed for growth of the lactic-acid-producing bacteria, cells of *Rhodotorula* spp. may grow (Pederson and Kelly, 1938).

When the sauerkraut has been removed, the brine remaining in the vat has a high content of organic matter, expressed as approximately 24,000 BOD (biochemical oxygen demand) and 2% lactic acid. Having a very limited market, this brine has been discarded as a waste product. All the acid and much of the organic matter can be oxidized by *Geotrichum candidum* during a four-day growth cycle. The yeast is recovered for use as cattle feed (Hang et al., 1974).

Cucumber Pickles

In 1972, 23 million bushels, or 571,000 tons, of pickling cucumbers were grown on more than 128,000 acres of land (Pickle Packers International, Inc., 1972). The crop is the fourth largest in acreage of national truck crops. About half the acreage is in Michigan and Wisconsin; about one-fifth is in the South Atlantic states.

Cucumbers are harvested at the desired state of maturity, usually when small, green, and firm. Upon delivery to the buying or brining stations, the defective cucumbers are removed and the sound cucumbers are washed and graded for size (according to how many are required to fill a 45-gallon cask). Cucumbers have a flesh acidity of pH 5.1 to 5.8 and a sugar content of 1.2 to 2.2%. A small amount of the harvest is converted directly into a variety of fresh pickle items. Some cucumbers are fermented with dill to produce fresh or genuine dill pickles, and the rest is converted into salt stock. By salt-stock fermentation, large volumes of cucumbers can be preserved with a minimum of effort and investment in processing facilities, to cope with the relatively short harvest season.

A pickle is defined as an immature cucumber, properly prepared without taking up any metallic compound other than salt (NaCl), and preserved in any kind of vinegar, with or without spices and sugar. However, the word pickle is also applied colloquially to pickled ripe cucumbers and a variety of fruits and vegetables. Fabian and Switzer (1941) list 11 types of fermented and unfermented dill pickles; six types of sour pickles, including relishes; and 18 types of sweet pickles ranging from whole cucumbers to comminuted relishes.

Even thorough washing does not remove the variety of heterogeneous bacteria, yeasts, and molds present on cucumber surfaces at harvest. Because these organisms complicate the lactic acid fermentations, much research in the past four decades has been devoted to finding mechanisms to suppress them and to encourage the desired lactic-acid-producing bacteria. Many of the references in this section are to recent publications of the several aspects of the studies.

Fresh-Fermented or Genuine Dill Pickles

To make genuine dill pickles, medium to large (four inches or more in length) cucumbers are packed in a salt solution of 5% or more and dill weed, seed, or oil is added. During the approximate 20 day fermentation the cucumbers absorb about half the salt. The fermentation begins slowly, reaches a peak in microbial activity within several days, and gradually subsides to the end of the fermentation period. Often fresh-fermented dill pickles are produced in headed barrels which remain unsealed during the early stage of fermentation to permit the froth and the CO_2 to escape.

Cells of *Leuconostoc mesenteroides* initiate a vigorous fermentation in low-salt brines; those of *Pediococcus* spp. and *Streptococcus faecalis* may be active in high-salt brines. The lactobacilli complete the fermentation. Properly fermented dill pickles contain 1.1 to 1.2% titratable acidity expressed as lactic acid, although acetic acid may also be present. During fermentation free fatty acid and neutral fat contents increase and caproic and caprylic acids, absent in the fresh cucumber, appear. There is a decrease in phospholipids, accompanied by the formation of acetyl- and lactyl-choline by *Lactobacillus plantarum* (Pederson et al., 1964).

Salt-Stock Fermentations

With salt-stock fermentations cucumbers can be preserved in brine indefinitely, sometimes for several years. In principle, the initial salt concentration is adjusted to reduce malfermentations and yet permit the lactic acid bacteria to ferment the sugars withdrawn from the cucumbers. Fermentation is in outdoor vats, some large enough to hold 1200 bushels of cucumbers. The ratio of cucumbers to brine in the vat is slightly less than 2 to 1 by volume.

The initial brine concentration ranges between 6.0 and 7.5% salt in the Northern states, to ensure that fermentation is completed before the onset of cold weather, and between 8 to 10% in the Southern states, where fermentations may take several months. Toward the end of the fermentation the brine concentration is raised by small weekly additions of salt to a final concentration of 15%. This value reads as 60° on a *salometer,* a hydrometer, with a scale from 0 to 100, that measures salt concentration of a brine as a percent of saturation. A saturated solution contains 26.4% NaCl. Completed fermentations contain between 0.45 and 1.0% titratable acidity, the higher acidity being associated with low-salt brines. Certain lactic acid bacteria—*Pediococcus* spp., *Lactobacillus plantarum,* and *L. brevis*—operate in succession to initiate fermentation and bring it to completion (Etchells et al., 1964). No other species of lactic acid bacteria tolerate the salt concentrations common in cucumber brine fermentations. They produce normal growth and acidity in brines up to 8% NaCl content. Other species of lactobacilli tested—*Lactobacillus thermophilus, L. lactis, L. helveticus, L. fermentans,* and *L. delbrueckii*—will tolerate no more than 2.5 to 4% NaCl.

The brined cucumber may have poor color and texture, and may be shriveled and flat and have an unclean odor and taste (Etchells and Moore, 1971). Among the factors responsible are unfavorable climatic conditions during the growing season; temperatures during fermentation that are above or below the optimum for the lactic acid bacteria; too rapid salting and oversalting; and a habit of brining by trial and error. Cucumbers grown in good rainfall are meaty, solid, and have few well-developed seeds; they absorb salt rapidly and undergo rapid fermentation. In contrast, cucumbers grown in dry weather have well-developed seeds, absorb salt slowly, and undergo slow fermentation (Pederson and Albury, 1961).

Abnormal Fermentations

If cells of *Enterobacter* spp. and fermentative yeasts grow to excessive numbers, malfermentation, the conversion of sugars to products other than lactic acid, may occur. Cells of *Enterobacter* spp. grow rapidly in low salt brines of 20° salometer, or 5% salt, reaching maximum population within two days and vigorously producing CO_2 and H_2. The cells are severely repressed in 40°, or 10% brine, reaching a maximum population of about 1% that achieved in the lower salt brine. In 60° brine (15% salt) they grow only after a period of adaptation, achieving a maximum population in 12 to 14 days. In brine of this strength the lactic acid bacteria do not function. In low-salt brines, when a surplus of sugar is present, yeasts are active, but during later stages of the fermentation they compete poorly with the lactic acid bacteria for the sugar. Figure 9-2 shows the interplay among the cells of *Enterobacter* spp., the lactic acid bacteria, and the yeasts. Figure 9-3 illustrates the effect of lactic acid bacteria and the persistence of cells of *Enterobacter* spp. When a cucumber is immersed in brine, the permeability of its tissue changes, resulting in the rapid extraction of the sugars and

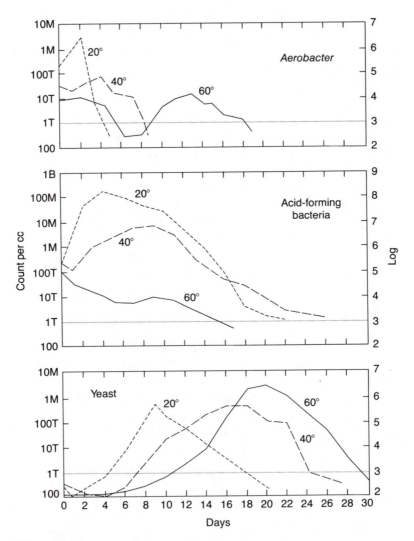

Figure 9-2
Growth of predominating microorganisms in cucumber fermentations
at initial brine concentrations of 20, 40, and 60° on a salometer. The 20°
brine was increased 10° per week to 60°, and the 40° brine was increased
5° per week to 60°. (Etchells and Jones, 1943. *Fruit Products Journal,* Avi
Publishing Company, Westport, Connecticut.)

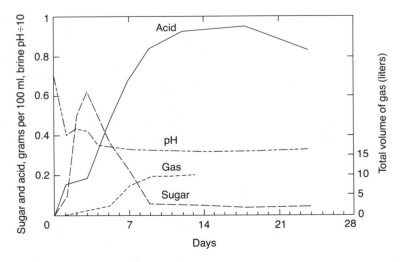

Figure 9-3
The relationship among changes in titratable brine acidity, brine sugar concentration, and gas evolution from brine surface for a typical dill fermentation. (Jones and Etchells, 1943.)

other soluble constituents that the microorganisms require for growth.

The gases produced during heterofermentations accumulate within the cucumber, forming either small, lens-shaped gas pockets or, in extreme instances, one large cavity in which the carpels are separated and the entire seed portion is pressed toward the skin. The gas is chiefly CO_2, but it has been postulated that when *Enterobacter* members are active they migrate through the site of detachment from the vine to ferment internally. Later the soluble CO_2 passes out into the brine, leaving ignitable concentrations of H_2 in the cucumber. Gas formation has been attributed not only to the heterofermentative lactic acid bacteria but also to *Lactobacillus plantarum* (Fleming et al., 1973). Because gassy cucumbers tend to float they are termed floaters. Such cucumbers can be used chiefly to produce comminuted products such as relishes.

To control undesirable microflora it has been suggested that the surface of the cucumber be blanched and that pure culture inocula then be used. When only the homolactic bacteria *Pediococcus* spp. and *Lactobacillus plantarum* are used, final titratable acidity is high, final pH is low, and residual sugars are not fermented. In consequence, the vats develop a yeast population. If the heterofermentative bacteria, such as those of *Leuconostoc* spp. and *Lactobacillus brevis,* are included, they convert some of the sugars to ethanol and the slightly ionized acetic acid. A higher pH is established, assuring that fermentation by *L. brevis* continues until the sugars are exhausted.

After many years of studies a procedure has been suggested by which pickling cucumbers can be commercially brined in low-salt brines under controlled fermentation that is complete in as few as 12 days (Etchells et al., 1973). In this procedure brine water is chlorinated to 80 ppm to destroy the unwanted microflora; the brine is acidified with glacial acetic acid or vinegar, to establish an environment that favors the lactic acid bacteria; sodium acetate is added as a buffer; water temperature is kept between 25.5 and 29.4°C; inocula of the homofermentative pediococci and *Lactobacillus plantarum* are applied; and the vats are intermittently purged with gaseous nitrogen, to sweep out the metabolically produced CO_2.

All fermentations are affected drastically by the lactic-acid-oxidizing, film-forming yeasts, which, before yeast classification became widely adopted, were commonly named *Mycoderma.* Among the film-forming and fermentative yeasts identified in cucumber brines are *Brettanomyces versatalis, Candida krusei, Debaromyces membranefaciens, Endomycopsis* and *Hanseniospora* spp., *Hansenula anomala, H. subpelliculosa, Kloeckera* spp., *Pichia alcoholophila, Torulopsis caroliniana, T. holmii, Torulospora rosea, Zygosaccharomyces* (now *Saccharomyces*) *globiformis,* and *Z. halomembranefaciens* (Etchells and Bell, 1950; Etchells et al., 1953a, 1953b, 1961).

Some yeasts are exceedingly halotolerant and grow equally well in high-salt and salt-free media. To keep yeast films from growing, uncovered fermentation vats have been placed in open fields to take advantage of the lethal effects of sunlight. However, this technique is not entirely practical: the fermentative yeasts are not kept from growing in the depth of the brine, and the vats are exposed to dilution by rain and snow and to a variety of airborne debris. Expensive and relatively impractical

Figure 9-4
Vats of fermenting cucumbers covered with plastic and water to
prevent atmospheric oxidation. (Courtesy of J. L. Etchells, North
Carolina Experimental Station.)

attempts to control the yeasts under roof have included install-
ing batteries of ultraviolet lights, covering the vats with mineral
oil, and using sorbic acid. As Figure 9-4 shows, the most effective control is to deprive the yeasts of oxygen by covering the
vats with plastic sheets holding several inches of water.

Pectinolysis and Cellulolytic Softening

During fermentation during later brine storage, pectinolysis
and cellulolytic softening may occur. In softening, the cucumber
or pickle becomes slippery and soft; but this condition may not
become apparent until some months after the brine stock has
been converted into finished pickles. For many years the cause
was thought to be the pectinolytic enzymes produced by cells of
the halotolerant *Bacillus subtilis* and *B. pumilis*. But the bacterial
enzymes are active under alkaline conditions, with pH optima
at 8.5 and 9.4, and are rapidly inactivated in moderately acidic
solutions (Nortje and Vaughn, 1953). Further studies have revealed that the enzymes originate in the filamentous molds that
grow in the large cucurbital flower (Etchells et al., 1958 a and b;
Bell et al., 1958). The enzymes are leached from the adherent

fragments of the flowers during brining and are adsorbed on the surface of the cucumber. How long softening takes to appear depends on the quantity of flower remnant accompanying the cucumber into the vat. If the flower remnants are removed before brining, no softening occurs. The remnants on the larger cucumbers are usually dry and brittle and are likely to drop from the cucumber during harvest, grading, and washing. The blossoms of the smaller cucumbers are wilted and strongly attached, and can be removed only with expensive hand labor. Vats containing small cucumbers sometimes are drained of their brines about two days after brining, when the enzymes are in brine solution.

Five genera and 31 species of the filamentous molds make up more than 80% of all identified isolations from cucumber flowers. The 10 most frequently isolated species have been *Penicillium oxalicum*, *P. janthinellum*, *Aschyta cucumis*, *Fusarium oxysporium*, *F. solani*, *Cladosporium* sp., *Alternaria tenuis*, *Trichoderma viride*, *Fusarium roseum*, and *Mucor silvaticus*. All except the last two species produce enzymes and, in laboratory studies, a great deal of cucumber softening. The enzymes are inhibited by tannins or tanninlike compounds in grape leaves, which is probably why the rural American housewife lines dill pickle crocks with these leaves (Porter et al., 1961). Other sources of enzymes that produce softening of finished products include garlic molded by *Penicillium canescens* and onion on which *Alternaria humicola*, *Fusarium oxysporium*, and *Botrytis cinerea* have grown (Misekow and Fabian, 1953; Meloche and Fabian, 1955).

Preserving Pickles by Pasteurization

Most pickles are spoiled by yeasts and lactobacilli. The pickling fluid may be clear, spoilage being indicated only by a creamy to gray deposit on the pickles touching the shoulders of the jar or as a sediment on the bottom of the container. Correct pasteurization of the pickles prevents spoilage. The containers should be heated slowly to an internal temperature of 74°C and held there for 15 minutes. The pickling fluid should have 0.6% acid, and pasteurization should occur immediately after preparation, while pH is low and before the acid can equalize with the cucumber (Bell et al., 1972). The proper ratio of quantity of pickle to brine is important. With less than 66% product, temperature rises higher than desired and the pickle becomes overcooked and soft; with more than 66%, the desired internal temperature is not attained throughout the container.

When homemade pickles spoil, gas forms and the lid bulges. Usually vegetables other than cucumbers have been pickled—often being a mixture of several vegetables. Viable microorganisms are rarely recovered from such containers, but microscopic examination reveals gram-positive rods resembling those of *Lactobacillus brevis*. The pH of such pickles tends to be high: pH 4.5 or above. Most such spoilage occurs because recipes specify too little vinegar, or vinegar-water mixtures are carelessly prepared, or vinegar with low acid content is used. Molded cucumbers used in pickling impart earthy or woody flavor (Etchells and Bell, 1942). Earthy flavor may also have resulted from spraying with insecticides immediately before harvest.

Preserving Pickles without Pasteurization

Unpasteurized pickles can be preserved by inhibitory concentrations of one ingredient or the additive effects of two or more preserving substances. For example, the proper ratio of acetic acid and sugar preserves cucumber pickles. If two components are additive, decreasing one necessitates increasing the other. Yeasts closely related to *Zygosaccharomyces globiformis* are the most common spoilage agent. Figure 9-5, a **preservation prediction chart,** shows the exact concentrations of acetic acid and sucrose or glucose that assure preservation. Selecting the concentration of one ingredient allows the concentration of the other to be calculated by formula (Bell and Etchells, 1952):

Sugar used	Sugar needed, %	Acid needed, %
Sucrose	80 − (20 × acid)	$\dfrac{80 - S_{sucrose}}{20}$
Glucose	70 − (20 × acid)	$\dfrac{70 - S_{glucose}}{20}$

Thus, if 2% acetic acid is specified, the pickling solution must contain 40% sucrose or 30% glucose. If 30% of either sugar is

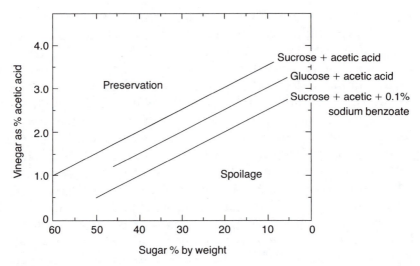

Figure 9-5
Preservation prediction chart. The area above each line represents
preservation for that particular combination of acid, sugar, and/or sodium
benzoate; the area below each line represents the spoilage zone. (Bell and
Etchells, 1952. Reprinted from Food Technology, Vol. 6, p. 468, 1952.
Copyright © by the Institute of Food Technologists.)

selected, there must be 2.5% acetic acid in sucrose solution or
2.0% acid in glucose solution. If acetic acid, sucrose, and benzoic acid are used, the formula becomes

$$A \ (\% \ \text{acetic acid}) + B \ (\% \ \text{benzoic acid}) = \frac{60 - S_{\text{sucrose}}}{20}$$

Other Vegetables

Many other vegetables besides cucumbers are successfully preserved by fermentation or by brining. Among them are beets,
broccoli, brussels sprouts, cauliflower, carrots; chard, kale, and
other greens; corn, green beans, lima beans, green tomatoes,
okra, English peas, green and red peppers, and turnips. Small
vegetables from which fluid can be extracted are packed with
2.5% dry salt. Dry vegetables, such as lima beans, and large
vegetables are immersed in salt brines. As in other vegetable
fermentations, several lactic acid bacteria dominate in succes-

sion; the usual sequence is *Leuconostoc mesentericus, Lactobacillus plantarum,* and *L. brevis.* Because these bacteria are not always present on the vegetable to be fermented, the outer leaves of a cabbage should be added to supply them. Fermentation causes some vegetables, such as corn and English peas, to become unpalatable and sour. Brining these in 12 to 15% salt prevents lactic acid fermentations. Greens and green tomatoes are deficient in fermentable sugars, but adding 1 to 2% sugar assures a proper fermentation. Ripe tomatoes and some varieties of summer squash contain pectinolytic enzymes that cause softening and spoilage. The pectinolytic enzymes of the green tomato are locked into very small, thick-walled cells. During maturation the cell walls enlarge and become thinner, and the enzyme is released.

All the difficulties of the fermentation of cabbage and cucumbers affect fermentations of other vegetables. Sound fermentations require initial cleanliness, thorough washing, proper temperature during fermentation, and the control of film-forming yeasts. Plastic sheets draped over the container and weighted with water effectively maintain anaerobiosis.

Olives

For millennia olives have been used as a source of oil and as an aperitif as well. In the Mediterranean area olives are preserved in high-salt brines or packed in dry salt. The use of lye to hydrolyze the bitter glycoside, oleuropein, appears to be a modern innovation. Columella described the Roman method of preservation: the olives were placed "in a solution composed of one-third brine, one-third vinegar, and one-third boiled must," then placed in vessels sealed with pitch. The olives were not very pleasant to the taste, according to Columella, who recommended that they be chopped and mixed with rue, parsley, and mint.

A small quantity of the American olive crop is marketed directly as fresh ripe olives and as olives packed with salt. The remainder are harvested as green olives when they show a cherry-red blush. A portion of the green olive crop is ripened by oxidation and the remainder are fermented in brine to produce the green Spanish and Sicilian types of olive. Most tree-ripened olives are not considered acceptable for eating by most

---

Box 9-1

Five varieties of olive are grown in California. The *Mission* variety, with 20% or more oil, and the *Ascolano* variety, with less than 15% oil, are most often used to produce ripe olives; the *Seveillano* variety, with 15% oil, is used for both ripe and green olives; and the *Manzanillo variety, with 16 to 18%* oil, is an all-purpose olive. The *Barouni* variety is sold on the fresh market.

Olives are graded according to size. Grades differ by 1/16 inch increments, starting with the small or standard size, which are 7/16 inch in diameter and average 135 olives per pound, and ending with the super-colossal size with 32 olives per pound. Grading prior to lye treatment ensures more uniform curing.

American consumers, but may be pressed for their oil. Because olives bruise easily they are handled during harvest and processing in ways that minimize bruising.

Lye Treatment

Lye (NaOH) concentrations ranging between 0.5 and 2.0% hydrolyze the bitter oleuropein in the olive. The concentration of the lye and the number of lye treatments vary with the type and size of olive and the processor's preferences. Lye is allowed to penetrate to the pit when ripe olives are desired, and about two-thirds of the distance toward the pit of green olives. The time of exposure to lye varies from 4 to 7 hours, depending upon the size of the olive and the temperature. The unhydrolyzed oleuropein remaining near the pit gradually disappears during fermentation and storage of green olives. Too strong a lye solution and too high a temperature (above 27°C) causes blistering by dissolving pectins. Blisters are prevented by using 1.6 to 1.8% lye containing 3 to 6% NaCl and by reducing the temperature in the pickling sheds (Samish, 1955).

California olives are harvested in the fall, and the cool temperature sometimes delays fermentation. After the lye treatment, the olives are washed in several changes of water to leach the lye from them. To prevent softening of the flesh during sterilization, the flesh aklalinity must be reduced below pH 8.0.

If there is too much residual lye, the final pH at the end of the fermentation will be too high. The depth of penetration of the lye and its subsequent removal are monitored with phenol-phthalein.

Sugars

Olives contain glucose, fructose, sucrose, and mannitol in concentrations between 3.7 and 7.5%. During lye treatment and leaching, as much as 65% of the sugars may be removed and lost. Because the small remainder may limit lactic acid fermentation, glucose or sucrose may be added to assure an adequate fermentation.

Ripe Olives

As was said, the common dark-purple-to-black ripe olive is not tree-ripened but is harvested instead when it is green but shows a cherry-red blush. The olives are made dark purple by the oxidation of tannins or tanninlike compounds. The oxidation is encouraged by alternately treating the olives with lye solution and water and then aerating them by stirring with wooden paddles or injecting air into the vat. The color may be intensified with ferrous gluconate. After lye treatment, oxidation, and leaching, the olives are cured in dilute brine to absorb salt and prevent shrivelling during canning. The canned olives are processed at 116°C for 60 minutes. Cruess (1931) states that with respect to spoilage and the possibility of botulism, olives probably are the safest canned food on the market.

Green-Fermented Olives

Spanish-type olives are treated with lye, Sicilian-type olives are not. After lye treatment and leaching, Spanish-type olives are fermented in redwood barrels or concrete vats. The generally high salt brines, with 5 to 8% or more salt, and the moderate late fall temperatures prolong the lactic acid fermentation. Some fermentations are completed in heated tanks or heated rooms. The active lactic acid bacteria are *Leuconostoc mesenteroides* and *Lactobacillus plantarum*. The titratable acidity ranges from 0.4 to 0.6%, and the final pH is about 3.6. During the fermentation the sugar is exhausted and the olives become the characteristic "olive green." Olives exposed to too much air during changes in the lye solutions and during leaching, become gray. Cells of *L. plantarum* may cause blisters or yeast spots to develop in the stomatal openings of the olives (Vaughn et al., 1953). Sluggish fermentations of olives in Israel have been associated with poor agricultural practices leading to low concentrations of essential nutrients in the olive (Samish, 1955). In olives there is a thermolabile, nonhydrolyzable compound that infrared spectra have shown to contain phenolic groups and unsaturated and conjugated carbonyl groups (Fleming et al., 1969). *L. mesenteroides* is markedly sensitive to the compound.

Olive brines contain ethanol in various amounts, reflecting the activity of yeasts and heterofermentative bacteria, and contain also 2-butanol, methyl sulfide, and acetaldehyde (Fleming et al., 1969). Growths of species of *Fusarium, Penicillium, Aspergillus,* and *Paecilomyces* taken from olive brines exhibit pectinolytic activity (Balatsouras and Vaughn, 1957). If the salt content of the brines is reduced because of ecological factors (such as government regulations), gas pockets have resulted, as have softening of the olives by *Hansenula anomala* and pectinolysis by *Saccharomyces kluyveri* and *S. oleaginosus* (Vaughn et al., 1972). Growths of *Rhodotorula* spp. cause stem end softening. The fermentative anaerobes *Clostridium beijerinckii, C. multifermentans, C. fallax, C. butyricum,* and *C. acetobutylicum* produce gas pockets, gassy spoilage, odoriferous compounds, and rancid and bitter flavor. Cheesy, sagey, foul, and fecal odors occur in brines that have not become acidified to pH 4.5 (Kawatomari and Vaughn, 1956), while cells of species of *Propionibacterium* produce a cheeselike odor. The "Zapatera" spoilage, a malodorous fermentation, may be produced by propionic acid bacteria (Plastourgos and Vaughn, 1957).

Fermented Milks

Once cattle were domesticated it was learned that milk left undisturbed for a day or so would sour. The temperature and microorganisms present determined the soured milk's flavor and texture, which were not unpleasant.

Box 9-2

Virtually every human culture in which there are domesticated milk-bearing animals has maintained to this day a tradition of making some kind of fermented milk. In Western nations several of these products are made commercially. They vary in nature considerably: some are mildly acidic, others strongly acidic; some also contain fruit; some contain alcohol. Among low-acid products (<1% lactic acid) are cultured buttermilk and sour cream, made extensively in the United States, and filmjolk or filia made in Finland using capsule-producing (ropy) strains of *Srreptococcus lactis* var. *hollandicus.* Medium- to high-acid products (2 to 3% lactic acid) include yogurt, made in Europe, Africa, Asia, and the United States (and known as leben in Egypt; matzoon or mazum in Armenia; naja in Bulgaria; and dahi in India). Other medium-acid fermented milks are tarho (Hungary), gioddu (Sardinia), kos (Albania), fru-fru (Switzerland), grusavin (Chile), skorup (Balkan countries), kaimac (Yugoslavia), Bulgarian buttermilk (Bulgaria), and acidophilus milk (United States). The alcoholic (1 to 3%) sour milks include koumiss, kefir, and araka (all, the Soviet Union), fuli and puma (Finland), taette (Norway), and lang (Sweden).

The most popular fermented milk produced in the United States is cultured buttermilk. Originally the term buttermilk meant the serum containing fat (butter granules) removed from churns during butter making. Because such cream for butter was gravity-separated, it was sour by the time it had risen. When the centrifugal separator was introduced, around 1900, sweet-cream buttermilk became common. However, because most people preferred sour buttermilk, the milk was allowed to sour, either naturally or by adding culture from a previously soured milk. As demand for buttermilk increased, dairy processors began culturing skim milk with a lactic streptococcal culture and selling the product as buttermilk, either with or without blending in sweet or sour buttermilk from the churn. Today, the butter-making by-product sweet-cream buttermilk is spray-dried and sold to the baking industry; some is also commercially available for home use. Cultured buttermilk, on the other hand, is now a separate product having nothing to do with butter making.

Buttermilk

The United States Federal Food and Drug Act of 1906 defines cultured buttermilk as "the product obtained by souring pasteurized skimmed or partially skimmed milk by means of a suitable culture of lactic bacteria. It contains not less than 8.5 percent of milk solids non-fat." A satisfactory method of manufacture includes pasteurizing freshly separated skim or low-fat (up to 2%) milk at 85°C for 30 minutes, homogenizing (2500 pounds the first stage, 500 pounds the second stage), cooling to 22°C, and inoculating with a lactic culture containing acid (*Streptococcus cremoris*) and aroma-producing bacteria (*Streptococcus diacetylactis* and *Leuconostoc cremoris*). If the milk is deficient in citric acid, 0.1% sodium citrate is added before pasteurization. Also, the milk may be fortified with up to 1% of nonfat dry milk solids. Inoculation may be made with 1% by volume of a fresh liquid culture or a frozen concentrated buttermilk culture. The product is incubated for 12 to 16 hours, then cooled to 4.5°C and packaged. The marketed product should have pleasant acid flavor, buttery aroma, slight effervescence from the CO_2 produced, and a smooth, slightly viscous texture in the mouth. In the absence of significant postpasteurization contamination from psychrotropic bacteria or temperature abuse, the product should maintain these characteristics for 10 days to 2 weeks. Milk intended for use in cultured buttermilk may also be pasteurized by the HTST method, but the milk tends to be less firm than desirable, because the heat treatment fails to denature the whey proteins.

Some strains of both *Streptococcus diacetylactis* and *Leuconostoc cremoris* are common flavor organisms in buttermilk. The former produce diacetyl and acetaldehyde from the milk citrate; the latter have an active alcohol dehydrogenase that drains off the acetaldehyde and prevents harsh, green flavors. Since sufficient numbers (10^8/ml) of *L. cremoris* are difficult to maintain in buttermilk cultures, a frozen concentrate inoculum of this organism is added at the same time that the lactic culture is added. While certain combinations of *Streptococcus cremoris* and *L. cremoris* yield fine-flavored buttermilk low in acetaldehyde (Cogan, 1975), equal proportions of these two species are difficult to maintain. For this reason, frozen concentrates of the two are used, either separately grown and blended or added as two separate culture inoculations.

Cultured buttermilk can suffer loss of flavor when contaminated by psychrotropic and other bacteria during packaging. Such contaminants (members of *Pseudomonas, Alcaligenes, Es-*

cherichia, and *Enterobacter* species) contain diacetyl reductase (Seitz et al., 1963), which reduce the diacetyl to acetoin, a flavorless, odorless product. Some strains of *Streptococcus diacetylactis* should not be used as flavor organisms in buttermilk because they actively reduce diacetyl to acetoin, contributing to the green flavor defect.

Sour Cream

Cultured sour cream, a smooth viscous product, is used extensively as a dressing for salads, potatoes, soups, and various meat, fish, and vegetable dishes, and as an ingredient of desserts such as cakes and cookies. It is usually made by pasteurizing, homogenizing, and culturing cream containing 18 to 20% milk fat. The same species of bacteria recommended for making cultured buttermilk are also used to make cultured sour cream.

Commercially cultured sour cream is prepared by one of three incubation methods:

1. inoculation, packaging, and incubation in cartons;
2. inoculation, incubation, and cooling in cans, followed by stirring or suctioning to remove the product for packaging; and
3. inoculation and incubation in vats.

The second method yields a satisfactory product if the cooled sour cream is not overstirred when the cans are emptied. Overstirring yields a product with a grainy texture. In the vat method the sour cream is forced slowly through orifices constricted by a 60 mesh wire screen or a perforated metal disc in the line feeding the packaging machine, or through an inverted homogenizer valve by air pressure.

If fresh, sweet cream is used and if contamination from equipment is minimal, cultured sour cream maintains desirable flavor and body characteristics for two to three weeks at 4.4°C. However, temperature abuse frequently occurs in retail markets (Bodyfelt and Davidson, 1975), and the low pH may catalyze flavor deterioration of the fat (oxidation).

Yogurt

Yogurt is a cultured fluid or frozen milk to which fruits or fruit flavors may be added. Cells of *Streptococcus thermophilus* and *Lactobacillus bulgaricus* in approximately equal numbers bring about the fermentation (Davis et al., 1971). If the cultures are not in balance, *S. thermophilus* overgrows the *L. bulgaricus*, making the yogurt harsh and sour. Traces of acetaldehyde, acetone, and diacetyl give the yogurt its flavor. Compatibility between the yogurt strains may be due to a synergistic action in which *L. bulgaricus* liberates valine, histidine, and glycine, which are essential for the growth of *S. thermophilus*, while the streptococcus produces the formate needed by the lactobacillus (Bautista et al., 1966; Veringa et al., 1968).

Yogurt is made with nonfat or low-fat milk to which stabilizers such as gelatin or vegetable gums and sugar are added. The milk is pasteurized, homogenized, cooled to incubation temperature, inoculated, and incubated (Mann, 1974; Yaeger, 1975; Kroger, 1976). Fruits or fruit flavors are pasteurized separately at 60°C for 10 minutes. The inoculated milk is incubated for different durations at different temperatures, ranging from 25 minutes to 5 hours at 42 to 46°C, 8 to 10 hours at 32°C, or 14 to 16 hours at 29 to 30°C. It is then promptly cooled to 4°C to prevent further development of acid. Yogurt has a shelf life of approximately six weeks.

Pasteurizing yogurt destroys lactase, to the detriment of individuals sensitive to lactose in the diet (Speck, 1977). In unpasteurized yogurt the enzyme is released by bacteria in the gut to hydrolyze the lactose.

Koumiss

Originally made from mare's milk by nomadic horsemen of Asiatic Russia, this fermented milk has become famous throughout much of the Soviet Union, where in some places it is now also made from cow's milk. It is reportedly consumed in that country as a treatment for pulmonary tuberculosis (Kosikowski, 1977). The major acid-producing bacterium is *Lactobacillus bulgaricus*; the important yeasts are species of *Torula* and *Mycoderma* (Makhanta, 1962).

Bulk starter for koumiss is prepared in a mixture of cow's and mare's milk; for each of three days increasing volumes of the latter are added to maintain the yeast-lactic starter mixture at an acidity of 0.65 to 0.70%. When the acidity reaches 1.4%, on the fourth day, it is ready for use as an inoculum. Mare's milk is warmed to 28°C and 30% of the bulk starter is added with strong agitation to incorporate air. After about two hours, the product is bottled warm, incubated for about two more hours, cooled to 4.4°C, and marketed (Berlin, 1962).

Kefir

Produced primarily in mountainous regions of Russia, Bulgaria, and Yugoslavia, this fermented milk is made by adding dried kefir granules to fresh whole milk, which ferments for 24 hours or longer at the prevailing indoor or outdoor temperature. The granules contain about 10.3% dry matter; the solids consist of 3.5% fat, 32.6% protein, 6% ash, and 56.2% carbohydrate. Microorganisms present vary somewhat from region to region, but have been shown to include *Streptococcus lactis, Lactobacillus brevis, L. acidophilus, Leuconostoc kefir, Saccharomyces lactis, Saccharomyces carlsbergensis, Candida tenuis, Candida pseudotropicalis, Acetobacter rancens,* and *Bacillus subtilis.* The organisms are imbedded in a glucose-galactose containing polysaccharide elaborated by *L. acidophilus* (Ottogalli et al., 1973); this polysaccharide gives the granules their elasticity. The granules develop during the fermentation, float to the top, and are removed and dried.

Originally the fermentation (Elliker, 1949) was carried out in leather skins; hung in the shade in summer or indoors in winter and shaken frequently to mix. During the fermentation considerable gas and as much as 1.0% each of alcohol and lactic acid are formed.

Acidophilus Milk

To prepare acidophilus milk, skim milk that has been heated at 93.3°C for an hour and cooled to 37.8°C receives 2 to 5% of an active culture of *L. acidophilus.* The fermentation is allowed to proceed until a titratable acidity of about 0.7% is reached. Then the product is rapidly cooled to 4.4°C and is ready for consumption. Acidities up to 2.0% may develop, and the cells die off rapidly if the product is not kept cold. Acidophilus milk is consumed primarily for its possible therapeutic value, but its sour taste makes it quite unpalatable. Now widely available in this country (Speck, 1978) is a sweet-tasting acidophilus milk prepared by adding a frozen concentrated culture of *L. acidophilus* to pasteurized milk, yielding a product which has 10^6 to 10^7 viable cells per ml (Duggan et al., 1959).

A number of closely related lactic acid bacteria have been shown to play an important role in human health, especially in neonates. Vaginal and intestinal normalcy also appear to depend on the presence of lactobacilli, including *L. acidophilus* (Sandine et al., 1972; Speck, 1976).

References

Alvarez, R. S. 1926. J. Bacteriol. *12,* 359.

Arengo-Jones, A. W. 1941. Fruit Prod. J. *20,* 338.

Balatsouras, J. D. and Vaughn, R. H. 1957. Food Res. *23,* 235.

Bautista, E. S., R. S. Dahiya, and M. L. Speck. 1966. J. Dairy Res. *33,* 299.

Bell, T. A. 1951. Botan. Gazette *113,* 216.

Bell, T. A. and J. L. Etchells. 1952. Food Technol. *6,* 468.

Bell, T. A. and J. L. Etchells. 1956. Appl. Microbiol. *4,* 196.

Bell, T. A., J. L. Etchells, and R. N. Costilow. 1958. J. Food Sci. *23,* 198.

Bell, T. A., L. J. Turney, and J. L. Etchells. 1972. J. Food Sci. *37,* 446.

Berlin, P. 1962. Int. Dairy Fed. Ann. Bull. IV(A), 4.

Bodyfelt, F. W., and W. D. Davidson. 1975. J. Milk Food Technol. *38,* 734.

Borg, A. F., J. L. Etchells, and T. A. Bell. 1972. Pickle Pak Sci. *2,* 11.

Chao, H. H. 1949. Food Res. *14,* 405.

Cogan, T. M. 1975. J. Dairy Res. *72,* 139.

Columella, J. M. *De res rustica* (circa A.D. 75) (E. E. Forster and E. H. Heffner, trans.). Harvard Univ. Press, Cambridge, Mass.

Cruess, W. V. 1931. Calif. Agric. Exp. Sta. Bull. 498.

Davis, J. G., T. R. Ashton, and M. C. McCaskill. 1971. Dairy Ind. *36,* 569.

De Ley, J., and J. Frateur. 1974. In: *Bergey's Manual of Determinative Bacteriology,* 8th ed. R. E. Buchanan and N. E. Gibbons, eds. William & Wilkins, Baltimore, Md.

Duggan, D. E., A. W. Anderson, and P. R. Elliker. 1959. Food Technol. *13,* 465.

Durkee, E. L., E. Lowe, and E. A. Toochek. 1974. J. Food Sci. *39,* 1032.

Elliker, P. R. 1949. Practical Dairy Bacteriology. McGraw-Hill, New York, N.Y.

Etchells, J. L., and T. A. Bell. 1942. Fruit Prod. J. *21,* 330.

Etchells, J. L., and T. A. Bell. 1950. Food Technol. *4,* 77.

Etchells, J. L., T. A. Bell, H. P. Fleming, R. E. Kelling, and R. L. Thompson. 1973. Pickle Pak Sci. *3,* 4.

Etchells, J. L., T. A. Bell, and I. D. Jones. 1953a. Farlowia *4*, 265.

Etchells, J. L., T. A. Bell, R. J. Monroe, P. M. Mosley, and A. L. Demain. 1958b. Food Technol. *12*, 198.

Etchells, J. L., T. A. Bell, and L. J. Turney. 1972. J. Food Sci. *37*, 442.

Etchells, J. L., T. A. Bell, and C. F. Williams. 1958. Food Technol. *12*, 204.

Etchells, J. L., A. F. Borg, and T. A. Bell. 1961. Appl. Microbiol. *9*, 139.

Etchells, J. L., A. F. Borg, and T. A. Bell. 1968. Appl. Microbiol. *16*, 1029.

Etchells, J. L., R. N. Costilow, T. E. Anderson, and T. A. Bell. 1964. U.S. Dept. Agric. Yrbk. of Agric. p. 229.

Etchells, J. L., R. N. Costilow, and T. A. Bell. 1953b. Farlowia *4*, 249.

Etchells, J. L., H. P. Fleming, R. E. Kelling, and R. L. Thompson. 1970. Pickle Pak Sci. *1*, 4.

Etchells, J. L., and I. D. Jones. 1942. Fruit Prod. J. *21*, 330.

Etchells, J. L., and I. D. Jones. 1943. Fruit Prod. J. *22*, 242.

Etchells, J. L., and W. R. Moore, Jr. 1971. Pickle Pak Sci. *1*, 1.

Fabian, F. W., and H. B. Blum. 1943. Fruit Prod. J. *22*, 228.

Fabian, F. W., and J. F. Erickson. 1940. Fruit Prod. J. *19*, 263.

Fabian, F. W., and R. G. Switzer. 1941. Fruit Prod. J. *20*, 136.

Federal Register 1974. *39*, No. 89 (May 7).

Fleming, H. P., J. L. Etchells, and T. A. Bell. 1969. J. Food Sci. *34*, 319.

Fleming, H. P., J. L. Etchells, and T. A. Bell. 1975. J. Food Sci. *40*, 1304.

Fleming, H. P., R. L. Thompson, J. L. Etchells, R. E. Kelling, and T. A. Bell. 1973. J. Food Sci. *38*, 499.

Fleming, H. P., W. M. Walter, Jr., and J. L. Etchells. 1969. Appl. Microbiol. *18*, 856.

Fulde, R. C., and F. W. Fabian. 1953. Food Technol. 7, 486.

Gilliland, J. R., and R. H. Vaughn. 1943. J. Bacteriol. *4*, 315.

Hang, I. D., D. F. Splittstoesser, and R. L. Landshoot. 1974. Appl. Microbiol. *27*, 807.

Jones, I. D., and J. L. Etchells. 1943. Food Inds. *15*(1), 62.

Kawatomari, T., and R. H. Vaughn. 1956. Food Res. *21*, 481.

Kosikowski, F. V. 1977. *Cheese and Fermented Milk Foods*. Edwards Bros. Inc., Ann Arbor, Mich.

Kroger, M. 1976. J. Dairy Sci. *59*, 344.

Lee, S. Y., E. R. Vedomuther, C. J. Washam, and G. W. Reinbold. 1974. J. Milk Food Technol. *37*, 272.

LeFevre, E. 1927. U.S. Dept. Agric. Farmer's Bull. 1438.

Lin, C. F. 1973. J. Ferment. Technol. *51*, 407.

Makhanta, K. C. 1962. Dairy Sci. Abstr. *24*, 257.

Mann, E. J. 1974. Cultured Dairy Products J. *9*(3), 13.

Martin, E. A., W. E. Peterson, and E. B. Fred. 1939. J. Agric. Res. *39*, 285.

Meloche, H. P., and F. W. Fabian. 1955. J. Food Res. *20*, 1.

Misekow, R. W., and F. W. Fabian. 1953. J. Food Res. *18*, 1.

Monroe, R. J., J. L. Etchells, J. C. Pacillo, A. F. Borg, D. H. Wallace, M. P. Rogers, L. J. Turney, and Z. S. Schoene. 1969. Food Technol. *23*, 71.

Mrak, E. M., and L. Bomar. 1939. Zentralbl. f. Bakteriol. u. Parasitenk II Abt. *100*, 289.

Muckerjee, S. K., M. N. Albury, C. S. Pederson, A. G. van Veen, and K. H. Steinkraus. 1965. Appl. Microbiol. *13*, 227.

Mundt, J. O. 1973. J. Milk Food Technol. *36*, 364.

Mundt, J. O., W. F. Graham, and I. E. McCarty. 1967. Appl. Microbiol. *15*, 1303.

Mundt, J. O., and J. L. Hammer. 1966. Appl. Microbiol. *14*, 1044.

Mundt, J. O., and J. L. Hammer. 1968. Appl. Microbiol. *16*, 1326.

Nortje, B. K., and R. H. Vaughn. 1953. Food Res. *18*, 57.

Orillo, C. A., and C. S. Pederson. 1968. Appl. Microbiol. *16*, 1671.

Orillo, C. A., A. C. Sison, M. Luis, and C. S. Pederson. 1969. Appl. Microbiol. *17*, 10.

Ottogalli, G., A. Galli, P. Resmini, and J. Volonterio. 1973. Ann. di Microbiol. ed Enzimol *23*, 109.

Pederson, C. S. 1930. N.Y. State Agric. Exp. Sta. Tech. Bull. 168.

Pederson, C. S. 1974. Food Inds. *19*, 778.

Pederson, C. S., and M. N. Albury. 1961. Food Technol. *15*, 351.

Pederson, C. S., and M. N. Albury. 1969. N.Y. State Agric. Exp. Sta. Bull. 824.

Pederson, C. S., and P. Fisher. 1944. N.Y. State Agric. Exp. Sta. Tech. Bull. 273.

Pederson, C. S., and C. D. Kelly. 1938. Food Res. *3*, 583.

Pederson, C. S., L. R. Mattick, F. A. Lee, and R. M. Butts. 1964. Appl. Microbiol. *12*, 513.

Pickle Packers International, Inc. 1972. Printed leaflet. St. Charles, Ill.

Plastourgos, S., and R. H. Vaughn. 1957. Appl. Microbiol. *5*, 267.

Porter, W. L., J. H. Schwartz, T. Z. Bell, and J. L. Etchells. 1961. J. Food Sci. *26*, 600.

Robinson, R. K., and A. Y. Tanime. 1975. J. Soc. Dairy Technol. *28* (3), 149.

Samish, Z. 1955. Food Technol. *9*, 173.

Sandine, W. E., K. S. Muralidhara, P. R. Elliker, and D. E. England. 1972. J. Milk Food Technol. *35*, 691.

Seitz, D. W., W. E. Sandine, P. R. Elliker, and E. A. Day. 1963. J. Dairy Sci. *46*, 186.

Speck, M. L. 1976. J. Dairy Sci. *59*, 338.

Speck, M. L. 1977. J. Food Protection, *40*, 863.

Speck, M. L. 1978. Dairy & Ice Cream Field, *161* (3) 70A.

Stamer, J. R., G. Hrazdina, and B. O. Stoyola. 1973. Appl. Microbiol. 26, 161.

Vaughn, R. H., K. E. Stevenson, B. Dave, and H. C. Park. 1972. Appl. Microbiol. *23*, 316.

Vaughn, R. H., W. D. Wen, F. B. Spencer, D. Pappagianis, I. O. Foda, and P. H. Krumperman. 1953. Appl. Microbiol. *1*, 82.

Vaughn, R. H. 1954. In: *Industrial Fermentations,* L. A. Underkofler and R. J. Hickey, eds. Chemical Rubber Co., New York, N.Y.

Veldhuis, M. K., and J. L. Etchells. 1939. Food Res. *4*, 621.

Veringa, H. A., T. E. Galesloot, and H. Davelaar. 1968. Neth. Milk Dairy J. *22*, 114.

Vorbeck, M. C., L. R. Mattick, E. A. Lee, and C. S. Pederson. 1961. J. Food Sci. *26*, 569.

West, H. S., J. R. Gilliland, and R. H. Vaughn. 1941. J. Bacteriol. *41*, 341.

Yaeger, C. 1975. Cultured Dairy Products J. *10* (4), 10.

10

Acetic and Other Fermentations

Vinegar

References to the use of vinegar in cookery and in medicine are found in ancient writings. Because neither the skin bag nor the clay vessel used for storing wines provided anaerobic conditions, it is likely that vinegar became known when wine was first fermented. The Romans were making vinegar at least as early as the first century A.D. In his book *De res rustica,* Columella advised the Romans to "mix water with the figs and from time to time add very ripe fresh figs and allow them to dissolve in the liquid until the flavor of sufficiently sharp vinegar results." They were then to let the liquid percolate through small rush baskets or sacking of witches broom, and next were to boil the clarified vinegar until it was free of scum. Finally they added a little grilled salt to prevent worms and other animals—by which Columella presumably meant the larvae and adults of the fruit fly *Drosophila,* the vinegar mite, and the vinegar eel *Anguillula aceti.*

All vinegars contain acetic acid that has been produced from ethanol. They can be made from fermented fruits and juices and from grain hydrolysates and fermentation distillates. Probably every variety of fruit has at one time or another been made into vinegar, either commercially or avocationally. The vinegars most popular in an area usually are made from resources common to the area. For example, continental European vinegars are derived from grape wines; much British vinegar is made from malt, or grain hydrolysate fermented as in beer making, but with the hops omitted (Figure 10-1); pineapple vinegar is made in quantity in Hawaii and in the Far East; and much American vinegar is made from apples. White vinegar is prepared from the distillate of grain fermentations and of the mother liquors after yeast crops have been harvested.

To be legal, vinegars sold in the United States must contain a minimum of 4.0% acetic acid (this value equals a strength of 40 grains; a grain is the equivalent of 0.1% acid). Many vinegars marketed have greater acetic acid concentrations than the minimum required.

Vinegars made by slow processes have more aroma, or bouquet, and flavor than those made by quick processes. Vinegars may be steeped in herbs, or herbs may be added. In addition to acetic acid, vinegars contain small quantities of glycerol, organic phosphates, pyruvic, formic, propionic, and butyric acids, and other products of microbial respiration, including diacetyl, acetoin, isobutyraldehyde, isovaleraldehyde, α-methyl valeraldehyde, methylisobutyl ketone, and esters of methyl, ethyl, isobutyl and isoamyl alcohols. Free propyl, secondary butyl, isobutyl, and amyl alcohols have also been identified in vinegars (Aurand et al., 1966). A vinegar derives its distinctive flavor from its particular combination of these substances, not from the acetic acid.

The Microbiological Processes

Sugar is converted to vinegar in two phases, anaerobically and aerobically:

$$C_6H_{12}O_6 \xrightarrow{\text{yeasts}} 2\,C_2H_5OH + 2\,CO_2$$
$$(180) \qquad\qquad\qquad (92) \qquad (88)$$

aerobically,

$$2\,C_2H_5OH + O_2 \xrightarrow{\text{vinegar bacteria}} 2\,CH_3CHO + H_2O$$

$$2\,CH_3CHO + O_2 \longrightarrow 2\,CH_3COOH + H_2O$$
$$(120)$$

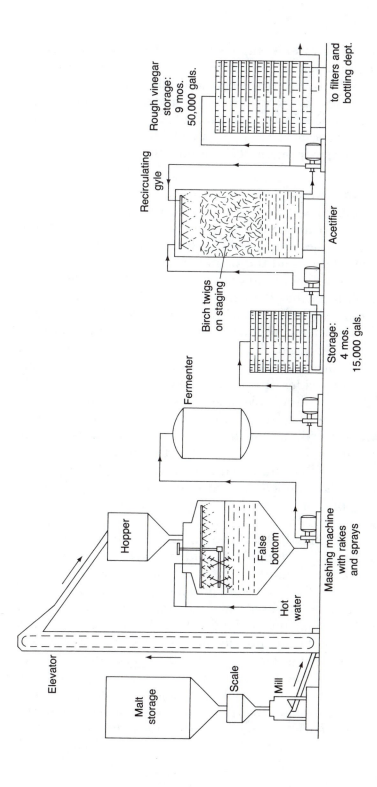

Figure 10-1

The process used in England for making vinegar from barley malt. (Food Industries, vol. 19, p. 1218, Sept. 1947.)

The theoretical yields suggested by the gram molecular weights are not obtained in practice, because appreciable quantities of the substrates, both sugar and ethanol, are converted into other products of metabolism, and some ethanol, aldehyde, and acetic acid are lost through volatilization. The actual yield of acetic acid varies between 38.4 to 40.1% of the weight of the sugar fermented (Vaughn, 1954).

Methods of Making Vinegar. Ethanol is converted to acetic acid by any one of four processes, two slow and two rapid. The oldest, and one still used in the home, is the slow process of natural acetification of wines and ciders. Such vinegar varies in quality, for the fermentative and the oxidative steps are influenced by the quality of the fruit, temperature, sanitation, and the strain of vinegar bacteria that happens to be in the container. Occasionally lactobacilli either are active or dominate the fermentation. Yeasts, known as the flowers of wine, grow on the surface to destroy the acetic acid.

The *Orleans* process, also a slow process, takes place in oak kegs laid on their sides. To permit the exchange of gases with the atmosphere a bung hole is left open (but is covered with cloth or ringed with turpentine to keep out *Drosophila*). Finished vinegar is diluted with wine stock (in various ratios but usually 1:1) to half fill the keg. The film-forming *Acetobacter xylinum* grows as a thick film on the surface. When the oxidation of the ethanol is completed, a portion of the vinegar is withdrawn and replaced in the keg with new wine stock. Each cycle requires several weeks.

In 1732, in Holland, Boerhave developed the quick vinegar process, or vinegar generator, and in 1823 Schutzenbach adapted it to the commercial production of vinegar in Germany. The generator is a cylinder, made of wood staves or, in home generators, glazed tiles, and ranging from several inches to many feet in diameter. As Figure 10-2 shows, the cylinder is filled with pieces of coke, noncalcareous stone, corn cobs, wood shavings (in the Frings generator), or other particulate material upon which vinegar bacteria become established. As the ethanolic solution or wine trickles over the surfaces, the bacteria oxidize the ethanol to acetic acid. Commercial generators use beechwood curls, or shavings. Packing materials must be

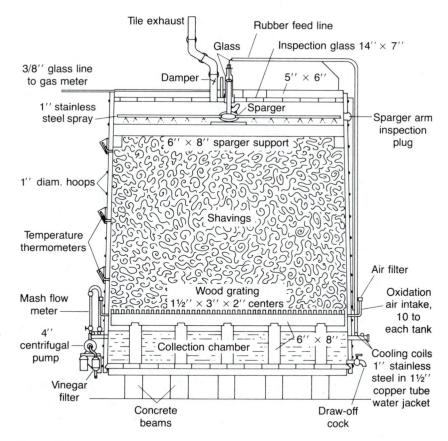

Tile exhaust

Rubber feed line

Glass

Inspection glass 14'' × 7''

3/8'' glass line
to gas meter

Damper

5'' × 6''

1'' stainless
steel spray

Sparger

Sparger arm
inspection
plug

6'' × 8'' sparger support

1'' diam. hoops

Shavings

Temperature
thermometers

Air filter

Oxidation
air intake,
10 to
each tank

Wood grating
1½'' × 3'' × 2'' centers

Mash flow
meter

4''
centrifugal
pump

Collection chamber

6'' × 8''

Cooling coils
1'' stainless
steel in 1½''
copper tube
water jacket

Vinegar
filter

Concrete
beams

Draw-off
cock

Figure 10-2
Cross section of the Frings generator. (Courtesy of A. E. Hansen, 1935.)

insoluble and indestructible and must not impart color, odor, or flavor to the vinegar. Commercial generators are equipped with dispersing mechanisms to prevent the wine from channeling through the generator, aerators, and catch basins. The effluent is divided; one portion is diverted to finished vinegar; the other dilutes fresh wine stock—or the wine-vinegar stock may be passed through successive generators. Cooling coils keep the internal temperature between 25 and 30°C.

During World War II the submerged culture fermentation method was developed for producing penicillin. It has since been adapted for making vinegar. In this process, tiny bubbles of air are sparged through the solution of ethanol, which is agitated by propellers, dispersing cells of the *Acetobacter* spp. throughout the liquid, which rapidly convert the ethanol to acetic acid (Hromatka and Ebner, 1949). Finished vinegars are filtered to render them clear and sparkling and then are pasteurized at 74°C.

The Vinegar Bacteria

Terminology used for identifying vinegar bacteria has been most confusing. Even in recent years the extreme variability among these bacteria has rendered classification almost impossible (Shimwell, 1959). De Ley and Frateur (1974) and Asai (1968) give taxonomic keys and descriptions of the genera.

De Ley and Frateur identify the two genera of vinegar bacteria as *Acetobacter* and *Gluconobacter* spp. Members of both genera are widespread on flowers and fruits. Whenever fermentation has produced ethanol, these bacteria, if present, oxidize the ethanol to acetic acid, reducing the pH to 3.6 or less and aldonic carbohydrates to their respective acids. These microorganisms are obligately aerobic and nonfermentative, and oxidize alcohols to keto compounds. Strains are used industrially to produce dihydroxyacetone from glycerol and sorbose from sorbitol in annual quantities of several hundred tons of ascorbic acid.

Cells of *Acetobacter* spp. are peritrichously flagellate or nonmotile and are termed "overoxidizers" because they oxidize acetic and lactic acids to CO_2 and water. They possess the mechanism of the Krebs citric acid cycle. Cells of *Gluconobacter oxydans*, the sole species in the genus, are polarly flagellate, do not oxidize organic acids, and do not oxidize glucose beyond gluconic acid. The species is used commercially to produce gluconic acid from glucose.

De Ley and Frateur (1974) divide the genus *Acetobacter* into three species and nine subspecies: *A. aceti*, with four subspecies; *A. pasteurianus*, with five subspecies; and *A. peroxidans*. Five of the nine subspecies need only mineral supplements and ethanol to support growth, which makes them indispensable if distilled ethanol is to be converted into acetic acid. The other five subspecies require nutrients supplied by fermented fruits and grains.

Strains of both *Acetobacter* and *Gluconobacter* produce 4.5 to 10.9% acetic acid and 2-, 5-, and 2.5-ketogluconates from glucose. These organisms also oxidize higher alcohols such as propanol, butanol, and pentanol. Some strains produce pink-red, reddish-brown, and brown pigments. The cells of *Acetobacter xylinum*, known as "mother of vinegar," form a thick, slimy layer of cellulose in which they are embedded.

The acetic acid bacteria can acidify and spoil beers, ales, wines, and ciders, and cause acetic odors in decaying fruits. Growths of *Acetobacter alcoholiphilus* produce fruity odors (Kesulis and Parsons, 1958); those of *A. pastioranus* produce a viscid, starchlike polysaccharide in beers.

Vinegar eels, mites, and flies are an aesthetic nuisance in vinegars. However, swarms of the vinegar eel *Anguilula aceti* (Figure 10-3) form a vibrating, turbid film at the surface of most unpasteurized vinegars. They can be of use in vinegar

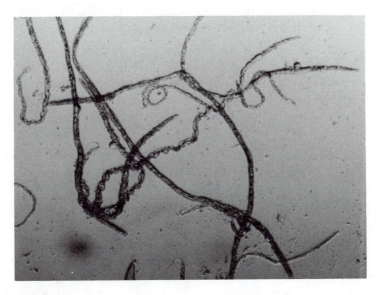

Figure 10-3
Anguilula aceti ($\times$ 100) obtained from cider vinegar and preserved in formalin.

generators, where they keep surfaces free of accumulated dead bacteria or convert substances into nutrients for the vinegar bacteria (Zalkan and Fabian, 1953). The mite *Iphidulus* and the fruit fly *Drosophila* abound as unsightly nuisances.

Other Microbial Food Fermentations

In the Far East, Southeast Asia, and islands of the Pacific many foods are prepared with the help of bacterial, mold, and/or yeast fermentations. There are too many to name here, but a few of the best known examples are: soy sauce, miso, sake, tamari sauce, ang-quac, sufu, ontjom, tempeh, natto, bongkrek, idli, poi, and pidan. These foods not only make use of novel procedures for preserving beans, wheat, rice, coconuts, and eggs, but also are palatable and nutritious.

Soy Sauce

Soy sauce, used as a seasoning of many foods, probably originated in China centuries ago and was gradually adopted by nearby countries. Soy sauce is prepared in three stages:

1. preparation of the inoculum, or **koji;**
2. rapid mycelial growth throughout the soybean-roasted wheat; and
3. slow enzymatic changes in heavy brine.

To prepare a satisfactory koji, a suitable culture of *Aspergillus oryzae* or *A. soyae* (Hesseltine and Wang, 1967) is selected and grown on wheat bran. Lactic acid organisms, e.g., those of *Lactobacillus delbrueckii,* are also present in the koji. After the koji has developed to the desired stage, it is added to trays containing a thick layer (3 to 5 inches) of cooked soybean and roasted crushed wheat. The ratio of wheat to soybean is one to four or five, by volume. The soybeans have been prepared by being soaked, raw, overnight in running water and, after draining, cooking under atmospheric pressure for 1 to 2 hours or at 15 psi for 15 minutes. The wheat has been roasted to dextrinize the starches in the berry and has then been crushed to allow proper spacing for mycelial growth and for aeration. The inoculated soybean-wheat mixture is placed in open koji boxes or trays, which are then placed in a warm room at 24 to 29.6°C. The mold spores germinate and grow; the mycelium develops rapidly, forming a thick cottony mass that completely encloses the mixture. Vegetative growth reaches its peak in three to five days. During that time cells of *Aspergillus oryzae* and of various lactic acid organisms such as *Lactobacillus delbrueckii* are rapidly producing enzymes.

When enzyme production has reached its peak but spore production has not become excessive, the bean-wheat cake (formed by penetration of ramifying mold mycelia) is crumbled and covered in a vessel with two volumes of 20 to 22% NaCl brine. The slurry, or **shoyu**, is stirred periodically to resuspend the solids to accelerate enzymatic action. In older methods of production a brown, tangy sauce is produced during a three to six month period. The use of selected cultures and the application of modern technology have reduced the time required for production of the sauce to several weeks.

Submersion in the brine greatly inhibits mold and bacterial growth, but proteolytic and amylolytic enzymes continue to hydrolyze the proteins and carbohydrates in the mash. Initially, lactic acid bacteria acidulate the young sauce, preventing bacterial spoilage and putrefaction. The aspergilli contribute proteinases that convert the soy and wheat proteins to glutamic acid and its salts, as well as to numerous other flavorful peptides and amino acids, and amylases that hydrolyze the starches and dextrins to sugars. Meanwhile, the carbohydrates are slowly converted to alcohol by the fermentative yeasts, which also mitigate the proteolytic activity of the aspergilli and themselves contribute to the characteristic tangy flavor, color, translucence of the final sauce. It is likely that during this complex fermentation other organisms also solubilize starch, making the aging less turbid and still more flavorful (Yokotsuka, 1960).

Miso

As with soy sauce, the primary fermenting organism of miso, a bean paste, is *Aspergillus oryzae* or *A. soyae*. The koji for this product is prepared by sprinkling the mold on steamed, polished rice or barley spread on trays. The inoculated trays of cereal are incubated at 29.6 to 35°C for about three or four days. By that time a creamy white mycelial mat has completely overgrown the cereal, but no spores are yet evident.

This preparation is mixed with finely chopped, cooked soybeans and the mixture is reincubated anaerobically for another two to three months. The final product is ground into a paste. Miso production makes use of the same organisms instrumental in soy sauce production.

Sake

Sake, or rice wine, is in fact Japanese rice beer with an alcoholic content of 14 to 16%. As with miso, the koji is made by growing cells of *Aspergillus oryzae* on steamed rice to provide optimal production of amylolytic and proteolytic enzymes. The koji and an inoculum of selected species of *Saccharomyces* are pitched into a slurry of steamed rice. As the amylases of the mold convert the starches to sugar, the yeasts in turn reduce these carbohydrates to alcohol. After 1½ to 2 weeks the alcoholic content reaches its maximum, and the sake is filtered and bottled.

Tamari Sauce

Tamari sauce, popular in Japan, involves a much shorter, simpler fermentation than do soy sauce and miso. In this preparation, rice, not wheat, provides discontinuity of the soybeans and additional carbohydrate material. The organism, *Aspergillus tamari,* is also a member of the Aspergillus flavus-oryzae group, but its cells have olive-brown fruiting heads rather than the yellow-green characteristic of *A. oryzae.*

Ang-quac

When cultured on steamed rice, cells of the organism *Monascus purpureus* (Figure 10-4) produce an intense reddish purple pigment that is much prized by the Chinese for coloring rice and giving it a friable texture. The "red rice" is used in turn to color a variety of food products, including wines, fish, and

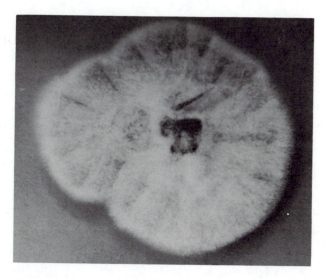

Figure 10-4
Colony of *Monascus purpureus* (× 10). This organism is
used to produce Ang-quac.

sauces (Skinner et al., 1947). This fungus has been found in
silage where it may cause the formation of large red masses or
"balls" up to a foot in diameter (Buchanan, 1910).

Tofu and Sufu

Tofu is a soft soybean curd consumed either as a fresh cheese or
incorporated into cooked fish, meat, and vegetable dishes. Sufu
(Chinese cheese) is made from tofu. Tofu is prepared from
washed and soaked soybeans that are ground to produce soybean milk. The milk is boiled to inactivate the enzymes and the
curd is precipitated with calcium or magnesium sulfate. After
separation from the whey, the curd is pressed lightly to form a
soft cheese with 83% water content. To make sufu the curd is
cut into small cubes, dipped in a solution of 2 to 6% salt and
2.5% citric acid, and heated in a hot air oven to pasteurize the
surface. The cubes are then inoculated with *Actinimucor elegans*
or one of several species of *Mucor* and incubated for several

days at room temperature to become coated with a white mycelium. They are then aged for six or more weeks in a fluid of rice wine containing 2 to 5% salt. The final product is soft, light yellow in color, and pleasant to the taste. The fungi secrete proteolytic and lipolytic enzymes that digest the proteins and the fats. Variations in the procedure for making sufu produce cheeses with different flavors, alcohol content, and color (Hesseltine and Wang, 1967; Wang and Hesseltine, 1970).

Ontjom

The Indonesian fermented food ontjom is unique in that it makes use of the orange-pink spored fungus *Neurospora sitophila*. Peanut press cakes from which peanut oil has been extracted are broken up and soaked in water for a day. After the remaining oil has floated free, the peanut mass is rewashed, steamed, placed in small, shallow (3 cm) rectangular molds, and covered with banana leaves. A pink starter (**ragi**) from a previous fermentation is sprinkled over the cakes, which are then incubated in shade until they are overgrown by the fungus. Ontjom fermentation requires good aeration. Ontjom is served fried; some report that it tastes like mincemeat (Hesseltine and Wang, 1967), others that it tastes like almonds.

Tempeh

To prepare tempeh, soybeans are soaked overnight in running water and then boiled for 20 to 30 minutes. A small amount of vinegar may be added to aid the removal of the seed coats, or testa, after which the bean cotyledons are placed on mats to dry or are blotted dry. Once cool, they are inoculated with a starter of spores of *Rhizopus oligosporus, R. oryzae,* or of other select strains of *Rhizopus,* and are rolled in banana leaves and placed in a tube of hollow bamboo or plastic and are incubated 20 to 30 hours at a temperature of 29.6 to 35°C. When fungal mycelia have completely covered the soybeans but black sporangia have not yet appeared, the product is ready for use. It is removed from the tube, sliced, and fried in coconut oil or other vegetable fat. Tempeh is a staple food in Indonesia.

Bongkrek

Bongkrek, an inexpensive and tasty fermented food consumed by large numbers of people in the central parts of the Indonesian island of Java, is kindred to tempeh. It is prepared from coconut press cake or grated coconut seeded with cells of the fungus *Rhizopus oryzae* and wrapped in banana leaves. The press cake is converted into a thick white mycelial mat within a day or two. Ordinarily, the product is used immediately. Occasionally, however, poorly prepared cakes or those kept too long are responsible for "bongkrek poisoning" (van Veen, 1973), attributed to substances produced by *Pseudomonas cocovenenans*. Bongkrek acid, a poison stable in the residual fat of the press cake, remains after heating. Below is the structural formula for bongkrek acid.

Bongkrek acid

Bongkrek acid has properties of an unsaturated fatty acid and prevents cells of *Rhizopus* spp. from growing properly, thereby upsetting oxidative phosphorylation (Welling et al., 1960). (Growths of *Pseudomonas cocovenenans*, however, do not produce bongkrek acid in soybeans or in peanut press cake.) Bongkrek acid interferes with glycogen metabolism, causing an hypoglycemic condition—and occasionally death. Despite the hazards of serious illness, a large segment of the population continues to consume bongkrek. Van Veen (1973) believes that health hazards attributed to *P. cocovenenans* could be reduced by adding oxalis leaves to reduce the pH to 5.5 during bongkrek preparation.

Natto

Unlike the aforementioned fermentations, bacteria rather than molds ferment natto. Soybeans are soaked in water until double their original weight and are steam cooked until soft; they are inoculated with *Bacillus natto,* wrapped in rice straw, and incubated in a hot room 40 to 43°C. Inoculum is seldom necessary. Cells of *Bacillus natto,* probably *B. subtilis,* are naturally present and grow quickly; in 18 to 20 hours, when the beans are considered properly fermented, they have an ammoniacal sour odor and feel sticky or slimy to the touch.

Idli

Washed and soaked rice is ground and mixed with an equal or larger amount of black gram mung, or urad, similarly treated. To this mixture water is added to form a batter, which is allowed to ferment overnight. Microbial activity causes the batter to rise. *Leuconostoc mesenteroides* is considered to be the primary leavening agent; *Pediococcus cerevisiae* contributes acidity. Small cakes (3 inches in diameter and 1 inch thick) of the leavened batter are then steam cooked for 15 to 20 minutes and eaten immediately.

Poi

Poi is prepared from taro, a plant of the arum family grown throughout the tropics for its edible starchy tuberous rootstalk. Taro is steamed, peeled, and finely ground. The ground taro is mixed with water and incubated in barrels or other containers for one to five days at room temperature. During the early stages of the fermentation coliforms and pseudomonads predominate, but acid-forming bacteria such as those of *Lactobacillus pastorianus, L. delbrueckii, L. brevis, Streptococcus lactis* and *Leuconostoc kefir* are primarily responsible for the characteristic CO_2, lactic and acetic acid development. Yeasts may contribute fruity flavors and alcohols.

Pidan and Shin-dan

Pidan is also known as houeidan, dsaoudan, the "thousand year-old," "century," or "Chinese" egg. First, a slurry of soda, slaked lime, burnt straw ash, and salt is dissolved and suspended in boiling water to form a paste. Duck eggs are then coated with the slurry, are covered with rice husks to prevent them from sticking to each other, and are stored in sealed clay jars for a month or more. The bacteria that penetrate the shell and the enzymes of the egg coagulate and partially digest the contents, and absorb salts from the slurry. The albumen becomes a translucent dark brown, the yolk a greenish gray. The egg has a sharp, salty, caustic flavor and ammoniacal odor. The bacteria are chiefly *Bacillus* spp. and coliform bacteria. Occasionally growth of species of *Escherichia, Serratia,* and *Proteus* causes the egg to become dark and to have a fecal or putrefactive odor. Shin-dan is made in Taiwan. Duck eggs are boiled in strong salt brine and then stored for a month, formerly in salt brine, now in a coating of clay. The albumen acquires a salty taste, while the flavor of the yolk remains unchanged. Shin-dan is now produced in the United States with hen's eggs.

References

Aurand, L. W., J. A. Singleton, T. A. Bell, and J. L. Etchells. 1966. J. Food Sci. *31,* 172.

Asai, T. 1968. *Acetic Acid Bacteria.* University of Tokyo Press, Tokyo.

Buchanan, R. E. 1910. Mycologia. *3,* 99.

Cruess, W. V. 1948. Commercial Fruit and Vegetable Products. McGraw-Hill Book Company, Inc.

De Ley, J., and J. Frateur. 1974. In: *Bergey's Manual of Determinative Bacteriology.* R. E. Buchanan and N. E. Gibbons, eds. 8th ed. Williams & Wilkins, Baltimore, Md., pp. 251–253, 276–279.

Hansen, A. E. 1935. Food Inds. 1, 277.

Hesseltine, C. W., and H. L. Wang. 1967. Biotech. and Bioeng. *9,* 275.

Hromatka, O., and H. Ebner. 1949. Enzymol. *13,* 369.

Kesulis, J. A., and R. H. Parsons. 1958. J. Inst. Brew. *64,* 47.

Pederson, C. S. 1971. *Microbiology of Food Fermentations.* Avi Publishing Company, Westport, Conn.

Prescott, S. C., and C. G. Dunn. 1959. *Industrial Microbiology.* 3rd ed., McGraw-Hill Book Co., New York.

Shimwell, J. L. 1959. Antonie v. Leeuw. *25*, 49.

Skinner, R. E., C. E. Emmons, and H. M. Tsichiya. 1947. *Henrici's Molds, Yeasts and Actinomycetes*, 2nd ed. John Wiley & Sons, New York.

U.S. Department of Agriculture, 1936. Making vinegar in the home and on the farm. U.S. Dept. Agr. Bull. 1424 (rev.).

van Veen, A. G. 1973. In: *Toxicants Occurring Naturally in Foods*. Committee on Food Protection. FNB Natl. Res. Coun., Natl. Acad. Sci., Washington, D.C.

Vaughn, R. H. 1954. In: *Industrial Fermentation*. L. A. Underkofler and R. J. Hickey, eds. Chemical Rubber Company, New York.

Wang, H. L., and C. W. Hesseltine. 1970. J. Agric. Food Chem. *18*, 572.

Welling, W. J., J. A. Cohen, and W. Berends. 1960. Biochem. Pharmacol. *3*, 122.

Yokotsuka, T. 1960. In: *Advances in Food Research*. C. O. Chichester, E. M. Mrak, and G. F. Stewart, eds. Academic Press, New York, N.Y.

Zalkan, R. C., and F. W. Fabian. 1953. Food Technol. *7*, 453.

Part III
Specific Food Products

11

Spices and Condiments

Spices

Spices are any of a variety of plant parts with pungent fragrance and flavor used in minor to ultraminute quantities to enrich, alter, or mask the flavors of foods. They include the true spices such as allspice, cinnamon, cloves, ginger, nutmeg, mace, pepper, and the capsicums (paprika and chili peppers); spice seeds such as those of anise, caraway, celery, coriander, cumin, dill, fennel, mustard, poppy and sesame; the herbs or the leafy spices such as bay, coriander, dill, fenugreek, mint, basil, marjoram, oregano, parsley, rosemary, sage, savory, and thyme; extracts such as peppermint; blends or mixtures such as chili powders and poultry seasonings; dehydrated powders of onion and garlic; and spice–nonspice mixtures such as cinnamon sugar, garlic salt, and onion salt.

Spices were discovered to be valuable condiments in prehistoric times, and no food or food component since has had a history as fascinating or played a role as profound in shaping the history of nations. The earliest known medical record, the Ebers Papyrus, verifies that spices were used by the Egyptians as early as 1550 B.C. not only for food preservation but for embalming as well.

Box 11-1

Spices from virtually all parts of plants—including the aril, bark, bulb, flowering part, green or ripe fruit, rhizome, root, and seed—although rarely do all parts of any one kind of plant yield spices. Saffron, for example, a pungent aromatic spice now used chiefly for its color, is produced by drying and grinding the stigmas of the herbaceous perennial *Crocus sativus.* Sarsaparilla, a woody plant of the genus *Smilax,* has been used to flavor carbonated beverages and as a spring tea. Flavors and aromas derive from mixtures of aldehydes, alcohols, ketones, carboxylic structures, or terpenes. All spices except garlic, horseradish, mustards, and onions derive their flavors and aromas from volatile or fixed oils the molecules of which include either simple benzene rings only, or complex inner ring structures with attached sidechains. The bacteriostatic or microbicidal properties present in some spices are associated with the side chains and specifically those possessing the *alpha-beta* unsaturation. This unsaturation inhibits enzymes containing sulfhydryl groups (Whitaker, 1959).

No spice in which the essential component bears an aromatic nucleus in fact owes its flavor or aroma to any one component. Eugenol, commonly considered to be the essential ingredient in cloves, is also the major component in allspice, and is present in laurel or bay leaves, sweet marjoram, and some nutmegs. Although *D*-limonene, commonly associated with citrus oils, is also found in caraway, celery, dill, peppermint, spearmint, and star anise, it is not considered a major ingredient of any spice or herb. That fraction of the oil of cardomom in which the components have been identified contains 17 components.

As Figure 11-1 illustrates, spices are produced throughout the world, in the tropics and the temperate zones. Many people not only use the spices sold in markets but also harvest locally grown plants. The Southern Appalachian mountaineers, for example, make use of the fruit of the spice bush or the Carolina allspice *(Lindera benzoin),* bird's pepper, butter cress, calamint and other mints, cancer jalap, tansy, wood sorrel, sumac, and bergamot. The bulbous plants garlic *(Allium sativum),* leek *(A.*

Figure 11-1
Major sources of the world's spices. (Courtesy of the American Spice Trade Assn.)

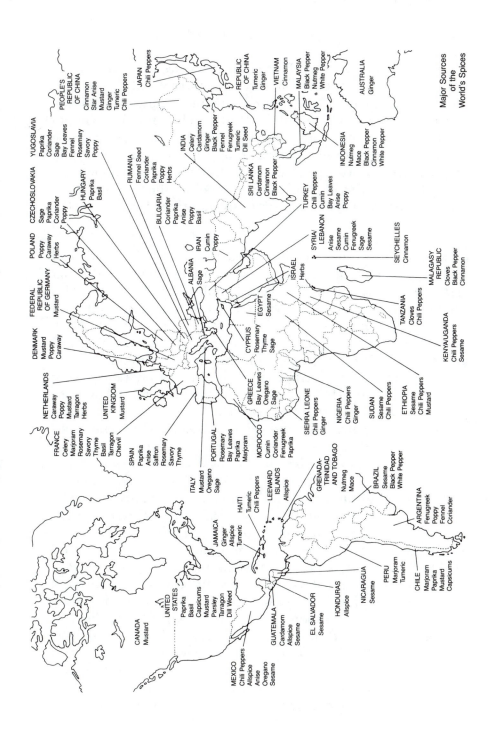

Major Sources
of the
World's Spices

porrum), onion *(A. cepa),* and shallot *(A. ascalonicum)* grow worldwide.

Because spices were early used as food preservatives, many ignorant people continue to believe that virtually all spices have antimicrobial properties. But for many spices quite the opposite is true. Anton van Leeuwenhoek, the Dutch lensmaker who, in the late seventeenth century, constructed the first microscope powerful enough to see bacteria, found millions of these "little animalcules" (microorganisms) in an infusion of whole pepper-corns. However, it took 250 more years before the supposed anti-septic property of spices was seriously questioned.

Bacteria. Early studies demonstrated that imported spices had high populations of bacteria (Yesair and Williams, 1942) that remained viable during marketing (Fabian et al., 1939). This situation seems to have prevailed to the present (Julseth and Deibel, 1974; Powers et al., 1975). Bacterial counts of black pepper, domestic and imported paprika, ginger, and mustard and celery seed often range from nearly one million to more than 100 million per gram (Karlson and Gunderson, 1965; Christensen et al., 1967). Total counts in excess of 500,000 per gram have been reported for bay leaves, curry powder, paprika, pepper, thyme, turmeric, and celery flakes (Karlson and Gun-derson, 1965). (See Table 11-1.) Spore counts on many of these spices range from a few hundred to over a quarter million per gram, but most spore counts of anaerobes are well under a thousand per gram. Anaerobic spores that are recovered seem to be randomly distributed on spice samplings. *Escherichia coli, Citrobacter fruendii,* and species of *Serratia, Klebsiella, Bacillus, Staphylococcus,* and *Streptococcus* are among the bacteria found on black and red pepper (Christensen et al., 1967). *Bacillus stearothermophilus* and *B. coagulans* are usually present in black pepper, ginger, and imported paprika. Proteolytic and amy-lolytic spores are numerous. *B. cereus,* which sometimes is pathogenic, is found in a high percentage of spices. Shigellae and salmonellae have not been reported, and coagulase-positive *Staphylococcus aureus* have been reported only occasion-ally (Powers et al., 1975). Cassia, cinnamon, cloves, oregano, allspice, onion, and garlic are toxic to the salmonellae.

Table 11-1
Numbers of bacteria in spices.

Type of spice		Total aerobes	Coliforms	Yeasts & molds	Aerobic spores	Anaerobic spores
		\multicolumn{5}{c}{Number of microorganisms per gram of dry sample}				
Bay leaves		520,000	0	3300	9200	<2
Cloves		3000	0	5	<2	<2
Curry	1	>7,500,000	0	70	>240,000	>24,0000
	2	400,000	30	70	>110,000	430
Marjoram	1	370,000	0	18,000	54,000	>24,000
	2	160,000	0	5000	9	150
Paprika	1	>5,500,000	600	2300	>240,000	620
	2	6,500,000	80	30,000	2900	43
Pepper		>2,000,000	0	15	>240,000	>24,000
Sage		6800	0	10	7	1700
Thyme	1	1,900,000	0	11,000	160,000	>24,000
	2	1,300,000	200	8300	150	...
Turmeric	1	>6,700,000	0	0	<2	<2
	2	1,300,000	50	70	>110,000	430

Source: Karlson and Gunderson, 1965. Reprinted from Food Technology vol. 19, p. 86, 1965. Copyright © by Institute of Food Technologists.

Spices, however, are not thought to threaten consumer health (Julseth and Deibel, 1974). Bacteria are often said to be contaminants of spices, but most such microbial populations were probably commensal residents on the plant that survived drying and storage. Moreover, most spices are collected in tropical areas by primitive methods and are exposed to many contaminants before they are dry enough to halt microbial growth, and are stored in conditions that make them easy prey to insects, rodents, and other vermin and dust. After storage, most spices are given only mechanical screening and grinding before being added to foods; they are not washed or disinfected. For these reasons spices should be carefully examined when they are received to be certain that they are free of insect eggs or fragments, soil and rodent excreta, and other filth. Often visual inspection is sufficient, but if such defects are suspected and not readily apparent, a flotation procedure for insect fragments should be used, such as that proposed by the Association of Official Analytical Chemists (1975).

Molds. Spices are subject not only to rodent and insect infestation but also to post-harvest deterioration by fungi. While consuming the centers of peppercorns, for example, insects can introduce fungi (Christensen et al., 1967; Christensen, 1972). Often the mycological quality of some spices on the market—especially of peppers—is quite poor, bearing many genera and species of fungi. Most fungi are of the storage type, which develop after harvest if relative humidity is not controlled during storage. Samples of whole or ground black pepper from various sources yield numerous colonies of several species of *Aspergillus.* Among the aspergilli isolated and the a_w at which they grow are: *Aspergillus glaucus,* 0.73–0.75; *A. restrictus,* 0.68–0.70; *A. ochraceous,* 0.78–0.80; and *A. flavus,* 0.83–0.85. In laboratory trials both *A. flavus* and *A. ochraceus* produced mycotoxins. Other species of molds are also isolated from spices, including those of *Penicillium, Spicaria, Scopulariopsis,* and *Sporendonema. Penicillia* and *Scopulariopsis* are abundant in nature, especially upon vegetation in the later stages of decay. The presence of molds can be readily demonstrated by scattering a thin film of ground spices on appropriate agar media and allowing the plates to set for two to four days to grow out. Mold contamination can be more accurately determined by appropriately diluting the material and plating aliquots in malt extract agar or potato dextrose agar.

Microbial Implications. The bacteria and molds in spices probably have little effect upon home-prepared foods consumed soon after preparation, but they may markedly shorten the shelf life and the overall quality of some commercially prepared foods. Contamination of commercially prepared foods can be measured by the "adding machine" approach (Karlson and Gunderson, 1965): separate plate counts are made of each component of unheated or unsterilized spiced foods to determine its individual microbial load. Spores that germinate in moist foods bring about changes. For example, *Bacillus* spp. introduced with spices (coriander and white pepper) in cured, canned hams reduce nitrates and produce CO_2 (Jensen et al., 1934), as was demonstrated when large numbers of these organisms in a canned meat product resulted in an outbreak of swollen containers. Pork sausage prepared with unsterilized spices may become unpalatable and rancid within 10 days (Pappas and Hall, 1952).

Processors should establish rigid specifications for the maximum permissible numbers of molds, coliforms, and total organisms, and insist that suppliers adhere to these limitations regardless of cost. In addition, spices received at processing plants must be kept clean, cool, and dry. A quality control program should be set up to assure that only fresh spices are used. This is feasible only if a regular, systematic inventory of the spice room and storage shelves is rigorously followed.

Sterilization of Spices. Not all methods of sterilization can be used on spices; some are ineffective, others are effective but destroy the spice. Ultraviolet light, for example, does not penetrate the surface of spices. And if sterilized by dry heat, spices become charred; if sterilized by steam, they lose volatile substances. For example, steam treatment of pepper for 5 minutes at 5 to 15 pounds steam pressure decreases the potency of its flavor by 10% (Yesair and Williams, 1942). Bacterial content, however, can be reduced to less than 0.1% of the original count by treatment with ethylene oxide (Pappas and Hall, 1952): the spices are placed in specially constructed chambers that can be evacuated. Ethylene oxide (ETO) is admitted at the rate of 1.6 pounds per 35 cubic feet and is held 2.5 hours, after which the residual ETO is withdrawn by evacuation. In typical observations, the bacterial count of ginger was reduced from 920,000 to 6000 bacteria per gram, black pepper from 4.2 million to 18,000, and sage from 11,000 to 500 viable bacteria per gram. However, because propylene and ethylene oxides both react with chlorine ions in foods to form the carcinogenic chlorhydrin, both sterilants are in disreptue. Another effective sterilant, gamma irradiation, is not currently permitted by the Food and Drug Administration.

Germicidal Action. Spices are fabled to have germicidal and medicinal properties, but only **cinnamic aldehyde,** found in cassia, cassia buds, and cinnamon; **eugonol,** a flavoring oil found in clove; and the mixtures of allyl sulfides and isothiocyanates of mustard, horseradish, garlic, and onion are markedly bacteriostatic in relatively low concentrations.

In 1887 Chamberland demonstrated that cinnamon oil is lethal to anthrax spores and that several other spice oils show some bacteriostatic action. An infusion of ground cloves was recommended (Prasad and Joshi, 1929) as a practical method for preserving native fruits. In early work Cochran and Perkins (1914) compared the preservative value of several essential oils in starch syrups. And tomato sauce was successfully preserved by aqueous and alcoholic extracts of ground cinnamon and cloves (Grove, 1918). Use of spices and their essential oils has also been advocated (Blum and Fabian, 1943) to prevent microbial scum formation on pickles, vinegars, and bottled wines. Eugenol is used commonly as a sterilant in dentistry. Mustard plasters, on the other hand, act as counterirritants and rubefacients without germicidal value.

Bacterial spores are quite resistant to the germicidal activity of spices; mold spores are generally quite susceptible, and the vegetative cells of bacteria vary in resistance. In test tube trials, 10% concentrations of bay leaves, ground mustard, paprika, ginger, nutmeg, and black and red peppers inhibit yeasts. Mustard oil may be germicidal at a 1% concentration, while anise, allspice, wintergreen, and onion are germicidal at concentrations above 5%. All these concentrations, however, are far above those common in food preparation. Peppers, on the other hand, appear to stimulate microbial growth (Webb and Tanner, 1945).

Growth of yeasts is markedly delayed by a 0.1% concentration cinnamon, cloves, and allspice (Webb and Tanner, 1945), and is inhibited in petri dish culture by oils of spices, suggesting that growth-inhibiting substances are volatilized. Inhibition can be demonstrated by the agar cup technique, either by placing the oil in the cup or by placing the oil or the spice in an inverted, inoculated petri plate. Fabian (1940) advocated that pickles be preserved by adding mustard oil to the liquor or to the paraffin sealant, and that cider vinegar be preserved by impregnating the paper liner with the oil. The oils of mustard, dill, and celery seed were found to be effective in preserving dill style pickles (Anderson et al., 1973). In the past two decades, however, advances in thermal and aseptic packaging have virtually eliminated the need for these spices and their oils as preservatives. The oil of lemon grass is ineffective against acid-fast bacteria (Bose et al., 1951). Because, to our knowledge, this is the only study of the effect of spices upon the mycobacteria, we cannot support the popular belief that the relatively high consumption of garlic and onion by Mediterranean peoples accounts for their low incidence of respiratory diseases. Also, because little is known about the viricidal action of spices, there seems to be no scientific basis for folk tales that consuming garlic and onion will ward off colds.

Onion and Garlic

Whitaker (1976) lists 28 sulfur compounds in the intact onion. Many are precursors to the substances responsible for the flavor, odor, and antibacterial activities. The enzyme alliinase, which is locked into cells that do not contain the alkyl and alkene compounds, frees the germicidal substances allicin, and propene disulfide.

Fresh onion and garlic juices are selectively germicidal: *Bacillus subtilis*, *Escherichia coli*, and *Saccharomyces cerevisiae* are inhibited; lactobacilli, *Enterobacter aerogenes*, and *Citrobacter freundii* are unaffected (Vaughn, 1951). When onion and garlic powders are being made, these bacteria and enterococci of the plant type (Mundt, 1976) persist and, during drying, grow on cut onion surfaces, causing souring. The number of microorganisms present in dried powders varies from day to day, and counts of viable organisms are often so high that no specifications can be conveyed to purchasers (Vaughn, 1970).

Owing to the bacteriostatic properties of fresh and dried onion and garlic, plate counts of them are unreliable. A 0.5% concentration of potassium bisulfite in diluents overcomes the inhibition (Wei et al., 1967). The same principle has now been advocated in performing pre-enrichment procedures when allspice, cinnamon, clove, and oregano are cultured for the presence of salmonellae (Wilson and Andrews, 1976).

Aspergillus allium and several species of *Penicillium* and *Leuconostoc mesenteroides* grow on stored bulbs. The penicillia produce pectinolytic enzymes. When bulbs used in pickling support a mold growth that is not yet apparent, the pickled product becomes softened.

Condiments

Catsup and Related Sauces

Catsup, chili sauce, Worcestershire sauce, and Tabasco sauce are highly acidic preparations commonly used to flavor meats, meat preparations, and other dishes. Catsup and chili sauce are made with tomatoes, but catsup may also be made with apples, cranberries, elderberries, grapes, currants, pineapple, and plums. Worcestershire sauce is a blend of vinegar and soy sauce; Tabasco sauce is prepared with peppers. Catsup and chili sauce are boiled with spices and reduced to the desired concentration before being bottled while hot into hot or cold sterilized containers and sealed. They may or may not receive further heat treatment.

Today catsup and chili sauce rarely spoil. In earlier years wooden equipment, primitive sanitation, and the atmosphere were sources of contamination by aerogenic and nonaerogenic lactobacilli (Pederson and Breed, 1929) and yeasts. Often unsterilized corks were responsible. Bacterial growth altered the products giving them flavors described as flat and tasteless. Sometimes yeasts produced gases, and when the cork was removed from the bottle its contents spewed out. Fresh tomatoes awaiting harvest readily become molded. (The quality of tomatoes used in catsup is monitored by the use of the Howard mold count procedure, described in Chapter 15.)

After containers are opened, molds and—more rarely—yeasts grow on the sauces if they are not consumed within a few days. Acetic acid, salt, and sugar, when combined in proper concentrations, will control the growth of yeasts in these products (Pederson and Breed, 1926). If 0.3% acetic acid is used, the inhibiting concentrations of sugar and salt are 15 and 3.0% respectively; and at 0.4% acetic acid, must be 15 and 2.9% or 10 and 3.5%, respectively.

Mayonnaise and Salad Dressings

Mayonnaise and salad dressings are emulsions of oil and other emulsifiable ingredients such as egg in water and vinegar. The ingredients are usually not heated during preparation but are preserved by the high concentration of acetic acid coupled with the additive effect of salt and sugar. The initial microbial load is that of the ingredients. While the vegetative cells of the bacteria soon die, both aerobic and anaerobic spores can be isolated. The acidifying agent in all dressings is the acetic acid in vinegar, which is more inhibitory to yeasts and molds at a given pH than are other organic acids (Nunheimer and Fabian, 1940). Fabian and Wethington (1950) found the pH of 110 samples of mayonnaise, salad and French dressings, and tartar sauces to range between 2.9 and 4.1.

During the early years of marketing salad dressings, spoilage was thought to be due to spore-forming bacteria. While the high acidity of the dressings makes this appear inconceivable today, it must be remembered that manufacturers were ignorant of the microbiological principles involved and used arbitrary formulas, inaccurate measurements, and often poor sanitation. Today spoilage would appear to be limited to highly acid-tolerant bacteria such as *Lactobacillus fructivorans* and the osmophilic yeast *Saccharomyces bailii* (Kurtzman et al., 1971). Because *S. bailii* exhibits delayed fermentation of sucrose, spoilage of the salad dressing may not become apparent until long after it has left the manufacturer. Like other highly acidic foods, salad dressing becomes susceptible to mold spoilage after the container has been opened. Mold, however, grows quite slowly and becomes evident only after prolonged storage at improper temperature.

Because most salad dressings are added to foods at the time of serving, they do not present a public health problem. Dressings of the mayonnaise type, however, often are used in sandwiches and are combined with neutral ingredients to prepare meat, poultry, and potato salads well before they are consumed. While the dressings per se do not support the growth of pathogenic bacteria, and in fact are slowly lethal to salmonellae and staphylococci, they introduce moisture that permits pathogens to grow. Conversely, if the salad dressing is added to poultry and potato salads immediately after they are prepared, growth of staphylococci is retarded, and is suppressed if the pH is maintained below 4.6 (Longree et al., 1959).

References

Anderson, E. E., W. B. Esselen, and A. R. Handelman. 1973. Food Res. *18*, 40.

Association of Official Analytical Chemists. 1975. *Official Methods of Analysis.* 12th ed.

Blum, H. P., and F. W. Fabian. 1943. Fruit Product Journal *22*, 326.

Bose, S. M., C. N. Bhima Rao, and V. Subrahmanyan. 1951. J. Sci. Indus. Res. *8*, 157.

Christensen, C. M. 1972. Pure spices—how pure? ASM News *38*, 165.

Christensen, C. M., H. A. Fanse, C. H. Nelson, F. Bates, and C. J. Mirocha. 1967. Appl. Microbiol. *15*, 622.

Cochran, C., and J. Perkins. 1914. J. Ind. Eng. Chem. *6*, 304.

Fabian, F. W. 1940. New method of preventing scum-forming yeast growth. Food Indus. *12*, 99.

Fabian, F. W., C. F. Krehl, and N. W. Little. 1939. Food Res. *4*, 269.

Fabian, F. W., and M. C. Wethington. 1950. Food Res. *15*, 138.

Grove, O. 1918. Ann. Rept. Agr. and Hort. Res. Sta., Long Ashton, Bristol, England, *29*.

Jensen, L. B., I. H. Wood, and C. E. Jansen. 1934. Indus. and Engin. Chem. *25*, 1128.

Julseth, R. M., and R. H. Deibel. 1974. J. of Milk Food Technol. *37*, 414.

Karlson, K. E., and M. F. Gunderson. 1965. Food Technol. *19*, 86.

Kurtzman, C. P., C. W. Hesseltine, and R. Rogers. 1971. Appl. Microbiol. *21*, 870.

Longree, K., J. C. White, K. Cutlar, and A. R. Willman. 1959. J. Amer. Dietetic Assoc. *35*, 38.

Mundt, J. O. 1976. J. Milk Food Technol. *39*, 413.

Nunheimer, T. D., and F. W. Fabian. 1940. Amer. J. of Public Health *30*, 1040.

Pappas, H. G., and L. A. Hall. 1952. Food Technol. *6*, 456.

Parry, J. W. 1969. *Spices.* Chemical Publishing Company.

Pederson, C. S., and R. S. Breed. 1929. *New York Agricultural Experiment Station Bulletin* No. 570.

Powers, E. M., R. Lawyer, and Y. Masuoka. 1975. J. Milk Food Technol. *38*, 683.

Prasad, H. H. and N. V. Joshi. 1929. Agric. J. of India *24*, 402.

Vaughn, R. H. 1951. Food Res. *16*, 429.

Vaughn, R. H. 1970. Food Technol. *24*, 189.

Webb, A. H., and F. W. Tanner. 1945. Food Res. *10*, 273.

Wei, L. S., J. A. Siregar, and M. P. Steinberg. 1967. J. Food Sci. *32*, 346.

Whitaker, J. R. 1959. Food Res. *24*, 37.

Whitaker, J. R. 1976. *Advances in Food Research* 22, 73.

Wilson, C. R., and W. H. Andrews. 1976. J. Milk Food Technol. *39*, 464.

Yesair, J., and O. B. Williams. 1942. Food Res. *7*, 118.

12

Cereals, Flours, and Pastas

Cereals

Wild grains and starchy roots were among the earliest human foods. It is likely that by 3000 B.C. all of the major grains except oats had been brought under cultivation. The only other cereal introduced in the past five millennia is triticale, a hybrid of wheat and rye recognized a century ago but only now coming into wide use.

Wheat is the most abundantly produced and used of all grains. However, many of the world's people survive almost entirely on a diet that is mostly rice. About 90% of all rice is grown in Asia. In the Western Hemisphere in pre-Columbian times corn (maize) was the staple grain. Even today it is by far the most important U.S. crop and second only to wheat in world production and consumption.

Cereals receive contamination from many sources—water, air, dust, soil, insects, rodents, birds, and humans. The flowering or fruiting grain tassle may be washed by rains or storms that carry various airborne microorganisms or their spores to the individual kernels deep within the grain head. Once in the developing seed, parasitic fungi may spread to other seeds or seed heads on the same plant or to other plants nearby. The crease in the kernel may fill with foreign matter or, if the grain falls to the ground, may be contaminated by manure. Such occurrences are most likely during wet or stormy weather. Weather conditions just prior to and during harvest determine to a large extent the moisture content of the stored grain.

Conditions Favoring Contamination

Since many of the agencies just named, including humans, may carry enteric organisms, it is easy for grain to become contaminated in the field or during handling or storage. Also, insects, rodents, and birds may foul the grain before or after it is threshed and while stored in bins inadequately protected from such vermin. Rusts and smuts attack the plant, the seed head, and the grain itself. Ordinarily, grain has sufficiently low moisture content to protect it from microbial attack. Bacterial spores, molds, rusts, and smuts may be present nonetheless. To control mold growth, grains having a moisture content exceeding 14% must be dried in the field or artificially after harvesting. However, the minimum moisture content that sustains microbial growth in the seed also depends on atmospheric temperature and humidity and the availability of nutrients.

Contaminating Microflora

When there is too much moisture *Helminthosporium* develops at the germ end of the kernel, causing "black point," or "smudge," a black or brownish discoloration of the seed coat. Since flour produced from such grain has black specks, wheat having the defect is downgraded (Harris and Sibbett, 1942). Rusts and smuts are well-known parasites of grains and are usually detected by the presence of undeveloped and shriveled grain, by odor, and by the presence of balls or masses of fungal mycelium.

Rusts. There are more than 300 different kinds of leaf and stem rusts. Probably the most destructive is black stem rust. This fungus, *Puccinia graminis,* requires two hosts for its complete cycle: the common barberry and the cereal grasses (Figure 12-1); the spores are spread by the wind. When rust sporidia, or

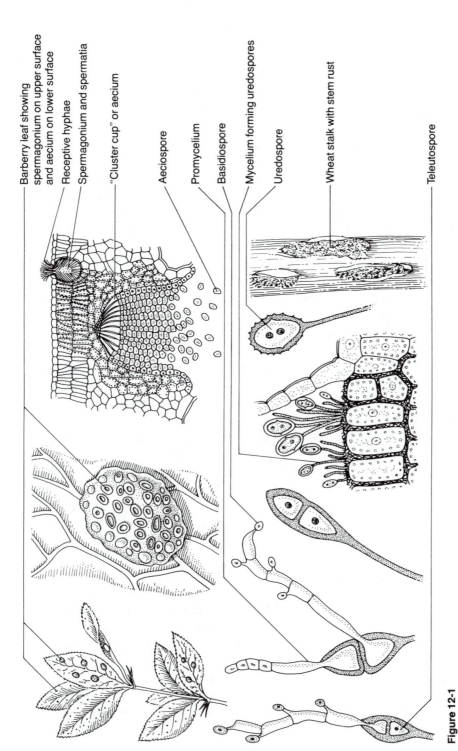

Barberry leaf showing spermagonium on upper surface and aecium on lower surface

Receptive hyphae

Spermagonium and spermatia

"Cluster cup" or aecium

Aeciospore

Promycelium

Basidiospore

Mycelium forming uredospores

Uredospore

Wheat stalk with stem rust

Teleutospore

Figure 12-1

Macroscopic and microscopic details in the life of the wheat rust *Puccinia graminis*. (Adapted from "Molds and Men," by Ralph Emerson. Copyright © 1951 by Scientific American, Inc. All rights reserved)

basidiospores, lodge on a barberry leaf they germinate and develop into + and − mating types of **spermagonia.** These develop on the upper epidermis of the leaf, where they produce receptive hyphae or + and − **spermatia.** The spermatia are exuded in nectar that attracts insects, which transfer the hyphae or spermatia to opposing mating types. The two types then fuse and form **aecia,** or "cluster cups," on the underside of the barberry leaf. The **aeciospores** produced within the cluster cups can infect wheat, barley, oats, rye and other grasses. The germinating aeciospores give rise to a mycelium that eventually forms a rust-red binucleate spore, the **uredospore,** which produces red pustules or the red stage on the leaves and stems of wheat and other grasses. After a large number of uredospores have developed within the pustule, they cause it to break and, once released, the spores spread to other plants—and frequently cause infections of epidemic proportions. Although these spores mature and germinate quickly on cereal crops, they are unable to infect the barberry.

Toward the end of the growing season, when the grain ripens, a much different two-celled spore, or **teleutospore,** arises from the mycelium produced by the mature uredospores. The teleutospore is surrounded by a thick cell wall, and when the pustules, or **telia,** that produce this type of spore dominate, a characteristic black pustule or black stage becomes evident. Initially each cell is binucleate, but later the two nuclei in the teleutospore fuse, and the spore overwinters as a dormant diploid uninucleate or karyogamous spore. Early in the spring the teleutospore germinates and produces a promycelium into which the diploid nucleus undergoes *meiosis* (reduction division), forming four nuclei separated by septa. From each of these four cells a haploid nucleated basidiospore is formed that, unlike the uredospore, can infect barberry but not cereal.

In the United States and Canada **black stem rust** has in the past destroyed as much as 300 million bushels of grain annually. In 1935 rust laid waste the farms of Minnesota and of North and South Dakota. Elimination of the barberry by eradication has greatly reduced the incidence of stem rust. However, even if the alternate host were completely destroyed, stem rust would not totally disappear, because in warmer regions or protected areas the fungus can survive in the uredospore stage and infect the grain directly. Also, when these spores are numerous, they may be blown to distant grain fields.

Black stem rust causes severe shriveling of the grain—in severe cases the crop is not even harvested. Grain that has been less drastically attacked is generally detected by its small size and low test weight. Rust infections are especially harmful to wheat, and the disease is most acute in low-lying damp areas. In recent years black stem rust has been held in check by development of wheat varieties resistant to the fungus.

Smuts. Smuts cause serious damage to cereal crops throughout the world. In the United States alone, these pathogens cost millions of dollars annually in losses. In fact, the huge economic losses resulting from infections by stinking smut of wheat, or **bunt** *(Tilletia caries* and *T. foetida),* equal those attributed to black stem rust of wheat. When bunt is present kernels are malformed, are filled with black spores, and have a fishy odor. In advanced stages, the spore masses become powdery and adhere to any grain that is nearby, causing it to become infected.

Loose smut of oats *(U stilago avenae),* loose smut of wheat *(U. tritici),* and **corn smut** *(U. maydis)* also have worldwide distribution and economic impact. In the smuts, when there is fusion of cells of compatible mating types, a heavy, extensive mycelium develops. Then masses of hyphae form and the protoplasts of each cell secrete a thick encompassing wall, converting the protoplast into closely packed globose teleutospores, or smut balls (sori). In some species of smuts teleutospores are cemented together, forming specialized spore balls in which some of the cells are fertile and others are sterile. The fungi that produce loose smut invade the flowering grain and develop in the ovary while the seed is maturing. When the seed germinates, the mycelium grows in the tissues of the seedling, eventually sporulating in the flower head, destroying the flower and replacing the grain with masses of smut spores.

Ergot. The fungus *Claviceps purpurea* occasionally attacks wheat and barley but is more commonly a parasite of rye (Figure 12-2), where it forms large purple sclerotia at the expense of

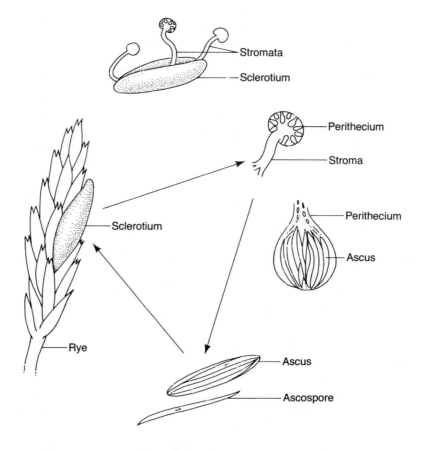

Figure 12-2
Life cycle of *Claviceps purpurea.*

the ripening grain seed, producing a disease known as *ergot.*
Adequate commercial cleaning operations remove these fungal
structures during milling. Both humans and domestic animals
have been poisoned by ergoty grain. Sometimes grazing animals are afflicted with gangrenous lesions of the legs when they
browse in grasses parasitized by species of *Claviceps.* Ergotism in
humans was common during the Middle Ages, often because
bread was made from poor quality rye flour containing ergot.
In A.D. 994 more than 40,000 people died from an ergot
epidemic (Kavaler, 1965). Improved methods of cleaning grain
before milling have resulted in the virtual elimination of er-

gotism from commercially milled flours. However, even today during famine, physicians have noted the gangrenous lesions of fingers and toes, arms and legs, ear lobes and nose, as well as vomiting, muscular pains, convulsions, and even the loss of these extremities and other symptoms typical of chronic ergot poisoning. As recently as 1951 a rash of St. Anthony's Fire occurred in Pont Ste. Espirit, France, when ergoty flour was used in locally produced bakery products. Five deaths were attributed to the mycotoxicosis.

During its growth cycle on grain, aerial conidia secrete a sweet, sticky "honey dew" that attracts insects, which transport the spores to the flowers of other plants. As these grains ripen, additional sclerotia form. When parasitized grain that is not harvested falls to the ground, the ergot remains dormant over winter. In the spring the sclerotia revive, producing a number of fleshy ascocarps or stalks with rounded masses of tissue (*stromata*) at their apex. Within each stroma, pear-shaped perithecia are embedded. Shortly thereafter, a number of asci, each containing eight needle-shaped ascospores, develop within the perithecia and, when mature, are forcibly discharged into the air and carried by the wind to other plants. Those lodging on developing rye ovules germinate, forming a mycelium and the cycle begins once more with the production of aerial spores. Such fungal infection can be controlled by planting fungicide-treated seed, cutting wild grasses and "volunteer" grain before they produce seed, rotating crops, and properly cleaning and milling grains. (See Chapter 23 for other mycotoxins in cereals.)

Handling of Grains

Grading. Much of the grain produced in the United States is officially graded. The United States Handbook of Official Grain Standards (1957) defines wheat as being "smutty" if it "has an unmistakable odor of smut, or . . . contains balls, portions of balls, or spores, of smut, in excess of a quantity equal to 14 balls of average size in 250 grams of wheat." If present, smut must be removed by scouring. The resultant weight loss of the wheat must be calculated as a percentage of the total grain after dockage.

Storage. Ripe grain of low moisture content (below 13%) can be kept for long periods without danger from too much respiration. However, when moisture content exceeds 16% molds and yeasts begin to grow, respiration becomes excessive, and the temperature in the storage facility increases greatly. Adequate ventilation of granaries, silos, elevators, and the holds of ships helps to dissipate the heat and to reduce anaerobic spoilage and foul odors. Applying fungistatic agents to grain effectively halts growth of the external microflora but has little influence on fungi in the subepidermal level (Milner and Geddes, 1955).

If sterile wheat has a moisture content of 19 or 22.5% and receives a dosage of .05 or 0.1% sorbic acid, respectively, mold grows sparsely if at all as Pomeranz (1957) demonstrated by inoculating such wheat with seven different mold species. With this treatment *Aspergillus ochraceus* and *Alternaria tenuis* will not grow. In wheat having .10% sorbic acid and 22.5% moisture *A. niger* grows sparingly, but fails to grow at all with .05% sorbic acid and 19% moisture. In wheat having 0.05% sorbic acid and 19% moisture *A. candidus* and *A. amstelodami* grow sparingly, but not at all with 0.1% sorbic acid and 22.5% moisture. The opposite is true for *Penicillium chrysogenum*. At either combination of fungistatic concentration and moisture *A flavus* grows moderately but not as well as the control. If 0.2% sorbic acid is added none of these fungi grows, even when the wheat has been dampened to 29% moisture. However, sorbic acid reduces seed viability, and when added to flour at levels of > 0.05% depresses loaf volume.

Flour

Steps in Making Flour

Milling wheat kernels into flour involves a series of treatments: The grain is cleaned, tempered, cracked, sized, purified, reduced, and milled; and the flour is "improved" and bleached.

Cleaning. Wheat should be dry and free of soil, filth, dust, debris, chaff, hair, weed, and other seeds, stones, cinders, insects, pieces of metal, and other materials. Generally grain is freed of foreign materials mechanically by magnetic separators and by passing through screens, seed separators, and scourers while aspiration collects the dust. Sometimes the wheat must be washed to remove mud or other materials clinging to it.

Tempering. Wheat is dampened or tempered so that the kernel breaks apart properly and the bran adheres in large flakes during cracking and milling. Grain that is not thoroughly clean may spoil during tempering. Water is added to bring the moisture content of the grain to approximately 20%. The dampened wheat is held in conditioning, or "tempering," bins for one to two days. Cleaning does not remove all bacteria and molds residing in creases of the grain, and tempering of the kernels may provide excellent conditions for growth (Ashby, 1939). Occasionally the moisture tempering bin is not sufficiently clean and as a result anaerobes, in particular the groups producing butyric acid, grow profusely in the bottom of the filled bins. They may reproduce sufficiently to produce a butyrous odor and spoilage of the wheat.

Breaking, Sizing, Reducing, and Purifying. Wheat milling operations are dry processes that separate the outer portions and germ from the starchy endosperm. The first "break" is made by feeding the grain between sets of grooved steel cylinders revolving rapidly but at different speeds. These shear or chop the kernel in such manner that the "first break flour," or endosperm, still adheres to the seed coat, or bran. These unsieved particles are fed into a second breaking roll and the flour is again sieved and separated from **semolina** and bran. Eventually, after passing through four or five break rolls and simultaneously being subjected to air aspirators or purifiers, the semolinas can be cleaned or separated from bran impurities. The cleaned semolinas are then gradually ground to fine size by the next series of reduction rolls and air aspirators. By suitably altering these operations, a wide variety or grades of flour may be made. For example, **whole-wheat** flour (also called graham or entire-wheat flour) contains the same proportions of unaltered natural constituents (with the exception of moisture) as the wheat from which it is made. **Straight** flour (also called 100% flour) is a milling term indicating all the flour from the wheat kernel after removal of the bran coats and germ but before its division into grades. Straight flour will make acceptable white bread and biscuits but is not generally marketed. **Patent** flour may be made from any class of wheat and comes from the more refined portion of the endosperm. How "long" or "short" a patent is depends on how much of the straight flour is included. Most patent flours include about 85% of the straight flour, but a very "short" patent used in fancy cake flour may include only 25% of the endosperm. The low amylase activity of patent flour can be overcome by adding malted wheat or barley flour. **Clear** flour is the less refined, bolted portion of the wheat meal recovered in the manufacture of patent flour. According to trade practice or demand, clear flour may be divided into first and/or second clears. The better clear flours are used in rye bread, the lower grades are used in dog biscuits.

Milling. Virtually all cereal crops are dry milled, except corn, which may be either wet or dry milled. In the United States the largest commercial activity using cereal grains is milling wheat into bread and pastry flours. From 25 to 30 billion pounds of wheat flour are produced annually in this country. Five classes of wheat are recognized: **hard red spring, hard red winter, soft red winter, white,** and **durum.** Flours from hard red spring and hard red winter wheats make the best "light breads." Flours from soft red winter and soft white wheats are better for making pastries, cookies, and crackers. Durum and red durum wheats are used to make flours for macaroni and noodles.

Dry milled corn is toughened by tempering with water to a 20–23% moisture content. After cleaning and scouring, the corn kernel is cracked and the bran peeled from the germ and endosperm. Usually, the endosperm is broken into two or more pieces during grinding and freed from the bran by scouring, screening, aspirating, and grinding. Or the kernel may be tempered to 17% moisture and placed in break rolls, where strong air currents remove the loose bran. Products of dry milling include corn germ, corn oil, hominy grits, brewer's grits, corn meals of varying screen size, corn flour, "dusting flour," a number of hot roll or flaked products such as corn flakes, brewer's flakes, paper-hangers' pastes, foundry flour, and stock feeds.

Wet Milling. Wet milled corn usually undergoes four principal steps: steeping, grinding, separating, and product recovery. After cleaning, the corn is transferred to wooden or stainless steel tanks. A countercurrent flow of water is sent through the tanks until the kernels of corn swell to maximum size. Steepwater, an aqueous solution containing SO_2 to control microbial growth, is used to loosen the hulls, soften gluten, help separate the starch from its protein matrix, and leach soluble carbohydrates, albumins, and globulins from the grain. Since the steepwater accumulates valuable nutrients, it is saved, concentrated, and used to make animal feeds or as a growth medium for fungi in the fermentation industries. Corn steep liquor is an excellent substrate for stimulating the production of penicillin and other antibiotics.

After steeping, the corn kernels are broken or shredded in degerminating mills. The starch suspension and extractives are recirculated in such manner that the germ is floated free and skimmed off in the overflow tank. The germ fraction is further processed to produce corn oil. Grinding in Buhr mills further frees the starch from the coarse bran. Rotating reels, screens, shakers, and washing devices free the starch and gluten from all remaining fine fiber. Tabling or centrifuging then separates the lighter gluten from the starch. The gluten and other protein by-products may be combined with the hulls and steepwater for use in animal feeds. Thorough washing, filtering, and drying yield a corn starch that can be used as such or can be hydrolytically converted to dextrins, corn syrup, and dextrose by cooking in the presence of acid. For syrup and dextrose production, the liquor is neutralized, filtered, and evaporated. Crude dextrose can be obtained by crystallizing the concentrate, but if the refined sugar is needed, impurities must be removed by centrifugation.

Cornmeal

Cornmeal often contains large numbers of bacteria (5,000 to 70,000 organisms/gram) and molds (1000 to 400,000) (Frazier, 1967). Their growth and survival depend on the moisture content of the stored meal and the storage temperature. Normally, in meals containing less than 13% moisture, bacterial and mold populations decrease rapidly. At 13% or more, xerophytic molds such as *Aspergillus repens* slowly produce growth balls. At 16% *Aspergillus flavus* becomes active (Thom, 1921) and is capable of producing mycotoxins. Streptococci have been isolated from samplings of water-ground corn meal. The strains isolated are plant-resident rather than human-resident types and, in all probability, derive from the corn used in making the meal.

Flour Terminology

According to the committee on terminology of the Millers National Federation (1972), flour is graded according to the milling process used. Thus, the grades of flour will be patent, clear, etc. By definition,

Flour, white flour, wheat flour, plain flour is the food prepared by grinding and bolting cleaned wheat other than durum wheat and red durum wheat; to compensate for any natural deficiency of enzymes, malted wheat, malted wheat flour, malted barley flour, or any combination of two or more of these, may be used; but the quantity of malted barley flour so used is not more than 0.25 percent. The flour is freed from bran coat, or bran coat and germ. . . .

Enriched flour conforms to the definition and standard of identity and is subject to the requirements for label statement of optional ingredients, prescribed for flour . . . except that it contains for each pound of flour between 1.66 and 2.5 mg of thiamin (B_1); 1.2–1.8 mg of riboflavin; 6–24 mg of nicotinic acid or nicotinic acid amide; and 6–24 mg of iron. . . .

Self-rising flour is prepared by mixing flour with a leavening agent.

Microbial Contamination

During milling the microbial content decreases. Patent flours have the fewest bacteria, straight flours have more, and lower flour grades have the most (Kent-Jones, 1937). More thermophilic bacteria are present in corn flour than in wheat flour (Yesair and Cameron, 1942). However, many of the bacteria and molds that survive the cleaning, tempering, breaking, and early stages of the bolting processes are removed with the bran and most are aspirated from the flour before the refined product is derived.

Bacteria decrease in number in stored flours, the rate being determined by the moisture content and storage temperature. Conversely, fungal contamination increases as storage is prolonged. In flours having a moisture content of 16% mold proliferation is excessive (Barton-Wright, 1938). Members of the genera *Penicillium, Aspergillus, Cladosporium, Mucor,* and *Rhizopus* are predominant. Most marketed flour has total bacterial counts ranging from 2750 to 19,500 per gram, spores of rope-forming bacteria from 49 to 85 per gram. Thermophilic counts are less than a thousand per gram, and only one or two are anaerobic (James and Smith, 1948).

Flour may contain sizable quantities of varied microbial flora. If the organisms are not destroyed by baking, the bread may become moldy or ropy. Certain bacterial spores are common in flour, e.g., *Bacillus mesentericus, B. subtilis* (Kent-Jones and Amos, 1930) and *Clostridium perfringens* (Aubertin et al., 1938; Farmiloe et al., 1954). Outgrowth of these organisms and of their capsular materials is responsible for rope in baked products (Barton-Wright, 1943).

Ready-to-Eat Cereals

Ready-to-eat cereals are cooked and dried during production. Since they are packaged under aseptic conditions, such cereals, properly stored, support no bacterial, yeast, or mold growth.

Pastas

There are two basic types of alimentary pastes, or pastas: macaroni and egg noodles. Both are made from a mixture of semolina—a hard, or durum, wheat—water, and salt. The main difference between them is that egg noodles also contain 5.5% egg solids by dry weight. This is equivalent to two eggs per pound of flour.

Macaroni

Macaroni is a generic term for a wide range of alimentary pastas including not only tubular macaroni (4–5 mm in diameter), spaghetti (1.5–2.5 mm), vermicelli (0.5–0.8 mm), and macaroni elbows but also many other shapes and sizes, such as alphabets, shells, ravioli, lasagna, and an assortment of unusual shaped pastas.

Pastas are produced in either of two ways: a dough of flour, salt, and water is passed through a series of rollers that knead it and roll it into a smooth, continuous sheet of required thickness before it is cut to desired size; or the unmixed ingredients are fed simultaneously through electronic metering devices where they are mixed at temperatures ranging from 40 to 60°C and extruded under pressure through forming dies of appropriate shape. The formed macaroni is cut by a rotating knife working against the face of the die and is collected for a preliminary drying, on a drying rack, receptacle, or a conveyor belt, to prevent sticking before it is transferred to the final dryer. During drying, large volumes of warm, dry air are circulated through the macaroni, reducing the moisture content from 25–33% to 13% or less. Since macaroni shatters if dried too rapidly, the usual drying time ranges from 4 to 24 hours, depending on the product. The dried macaroni is transferred to storage bins to await packaging.

Temperatures used to process commercial pastas are not high enough to completely destroy the microbial flora, but most surviving organisms are harmless saprophytes and are of little public health concern. Although extrusion, drying, and storage significantly reduce potentially pathogenic bacteria, several pasta products have been recalled because they were contaminated by salmonellae (Walsh, 1972; Walsh et al., 1974). Over 6% of 909 samples of noodles and macaroni examined by the U.S. Food and Drug Administration during selected periods from 1969 to 1971 were contaminated by salmonellae (Schwab, in private communication to Walsh et al, 1974).

Egg Noodles

Over 150,000 tons of egg noodles of many shapes and sizes are produced in the United States annually. About 70 percent of the product is made by a continuous system: durum flour, egg yolk or whole egg, and water are fed into a continuous sheet forming machine where they are mixed, kneaded, and rolled into a sheet of the required thickness. After the sheet has been cut or stamped to the desired size and shape, the noodles are automatically conveyed into a continuous drier. This product is unlikely to spoil and presents no microbial hazard if the ingredients are kept dry before use and are periodically monitored to show that the total microbial counts of staphylococci and salmonellae are maintained within satisfactory limits.

During storage at room temperature aerobic plate counts, staphylococci, and enteric streptococci (KF counts) on pastas and flat noodles decrease significantly. In the freshly prepared product, coagulase-negative staphylococci range from $10^{3.5}$ to $10^{5.5}$, but decrease to 10^3 in three to four months. In flat noodles and boccone pasta, viable aerobic counts (SPC) show similar reductions but persist at 10^4 for at least four months. Similarly, enteric streptococci initially 10^1 to $10^{1.5}$ in number, take less than two months to become too few to count in flat noodles and boccone pasta, but take four months to do so in eggshell pasta (Yeteran et al., 1974).

References

Ashby, G. K. 1939. J. Bacteriol *38*, 598.

Aubertin, E., A. Dangoremeau, E. Leuret, and F. Prechard. 1938. C. R. Soc. Biol. (Paris) *127*, 64.

Barton-Wright, E. 1938. Cereal Chem. *15*, 521.

Barton-Wright, E. 1943. J. Soc. Chem. Ind. (London) *62*, 33.

Farmiloe, F. J., S. J. Cornfold, J. B. M. Coppock, and M. Ingram. 1954. J. Sci. Food Agric. *5*, 292.

Frazier, W. C. 1967. *Food Microbiology.* McGraw-Hill Book Co., New York, N.Y.

Harris, R. H., and L. O. Sibbett. 1942. Cereal Chem. *19*, 403.

James, N., and K. N. Smith. 1948. Can. J. Res. C. *26*, 479.

Kavaler, L. 1965. *Mushrooms, Molds and Miracles.* The John Day Co., New York.

Kent-Jones, D. W. 1937. Analyst *62*, 649.

Kent-Jones, D. W., and A. J. Amos. 1930. Analyst *55*, 248.

Millers National Federation. 1972. Wheat foods in the marketplace. Flour and Feed Trade. 68.

Milner, M., and W. F. Geddes. 1955. In: *Storage of Cereal Grains and Their Products.* Amer. Assoc. Cereal Chemists, St. Paul, Minn.

Pomeranz, Y. 1958. Cereal Chem. *34*, 553.

Thom, C. J. 1921. Flora of corn meal. J. Agric. Res. *22*, 179.

U.S. Handbook of Official Grain Standards. 1957. Department of Agriculture. Washington, D.C.

Walsh, D. E. 1972. Macaroni J. *54* (6), 16.

Walsh, D. E., B. R. Funke, and K. R. Graalum. 1974. J. Food Sci. *39*, 1105.

Yesair, J. and E. J. Cameron. 1942. The Canner *94* (13), pt. 2, 92.

Yeteran, M., L. Chugg, W. Smith, and C. Coles. 1974. Food Technol. *74* (10), 23.

13

Sugars, Starch, and Sweetening Agents

Sugars

The word "sugar" comes from the Sanskrit *Sarkarah* (pebble). The sugar cane, *Saccharum officinarum,* was introduced to the Western world in the fourth century B.C. as a result of the campaigns of Alexander the Great in Eastern Asia. Cane is grown for its sugar throughout the tropics (see Figure 13-1). White, refined sugar is common to the European and American markets; crude, crystallized sugars are produced throughout the Far East.

The sugar grown and used in greatest quantity throughout the world, the disaccharide sucrose, $C_{12}H_{22}O_{11}$, is obtained from the sugar cane, which thrives only in the tropical and subtropical areas, and from the sugar beet, which thrives in the colder Temperate Zones of the Northern Hemisphere. The only other sugar refined extensively for commercial use as a sweetening agent is glucose, or dextrose, $C_6H_{12}O_6$, produced by the hot acid or enzymatic hydrolysis of starch. Other sources of sweetening agents are honey, maple sugar, which is obtained from the sap of the sugar maple *Acer saccharum;* palm sugar, in

Figure 13-1
Harvesting a field of sugar cane on a Caribbean island.

the East Indies; kaffir corn, a source of sorghum in Central Africa; skirret, a sweet root of the caraway family; licorice; and St. John's bread, which is the sweet pod of the carob tree. Sorghum is grown on a small scale in the Eastern United States.

Lactose, with 39% the sweetness of sucrose, is refined from whey, but it is not used extensively as a sweetening agent. A number of noncarbohydrate substances impart sweetness, or mask bitterness. Among them are the naturally occurring serendipity berry and katemfe, and the extract of the miracle fruit, which causes sour foods to taste sweet; others are the synthetic sweeteners such as saccharin or orthobenzosulfamimide, cyclamate, and aspartame, and the glycosidic substances such as glycorrhizin.

Cane Sugar

At the time of harvest the leaves of the cane are removed from the stalks, either manually by stripping, before the cane is cut, or, if the cane is cut by burning in the field after the leaves have wilted but while the stalks are still too moist to ignite. At the raw sugar mills the cane is washed and pressed through several—usually four—successive rollers to express the juice. Because the raw cane still contains valuable juice after several pressings, fresh water is added at the next-to-last crushing. The resulting mixture of juice and water is then passed over successive batches of cane emerging from the rollers, each time at an earlier phase. The juice is filtered to remove the fine **bagasse,** or particulate cane residue, is heated, and is clarified with slaked lime to neutralize the acid in the juice and prevent inversion of the sucrose. This process, known as defecation, yields a flocculent precipitate termed mud. The clarified juice, now containing 15 to 20% solids, is evaporated in several stages to a concentration of 70° Brix. The concentrated solution is seeded with finely pulverized crystals of sugar. The crystals that form are separated from the molasses in basket centrifuges, dried with a countercurrent flow of hot air (43.3°C), and bagged or stored in bulk.

Microbiology of Cane Juice. Bacilli now identified as strains of *Bacillus subtilis* and *B. cereus* (often termed *B. mesentericus* in the older literature) synthesize high molecular weight levan from the fructose portion of the sucrose molecule. In the fresh juice *Leuconostoc mesenteroides* produces copious quantities of transparent, slimy dextran, known as "frog spawn," and mannitol.

Strains of *Leuconostoc* spp. isolated from cane juice (McCleskey, et al., 1954) may be divided into four types: strains of type A occur typically in pairs and clusters and produce a semitransparent growth on an agar surface; strains of type C occur in pairs and short chains and produce a transparent slime; strains of types B and D produce an opaque, cartilaginous skin beneath which fluid forms. Many strains produce gas, which can be detected visually only in tubes sealed with a paraffin or agar overlay. All strains grow at 10°C, and many of the B and D strains initiate growth at pH 3.75, below the pH of the expressed cane juice.

Filamentous molds, including *Aspergillus repens, A. niger, Citromyces,* and *Monilia;* the yeasts *Saccharomyces cerevisiae, S. carlsbergensis, Pichia membranefaciens, P. fermentans, Candida krusei, C. guillermondi,* and *Torulopsis;* and the bacteria *Bacillus subtilis, B. subtilis* var. *aterrimus, B. cereus, Flavobacterium, Achromobacter, Escherichia,* and micrococci are associated with the deterioration of fresh juice. A caulobacterialike bacterium is reported to grow symbiotically with *Zygosaccharomyces* (now incorporated in the genus *Saccharomyces*), producing mucoid deposits on filters and in pipelines (Scarr, 1951). The initial filtration removes 99% of all microorganisms and the remainder are killed by the heat of evaporation.

Raw Sugars. The molds *Penicillium glaucum, Monilia* spp., *Aspergillus sydowi, A. niger, A. glaucum* (used in a general sense and not as a well-defined species), *Sterigmatocystis, Hormodendron, Stemphyllium* and a variety of yeasts are associated with the deterioration and fermentation of raw sugars. As early as 1613 water was recognized as the chief reason why raw sugar spoils (Browne, 1918). *Zygosaccharomyces nussbaumii, Z. major,* and *Clostridium saccharolyticum* are identified as vigorous agents of fermentation of Barbados sugar and high-grade Barbados cane syrup in barrels, producing gas and a rumlike flavor (Hall et al., 1937). Ethanol, furfural, acetone, and butanol have been identified in the distillates. Prior to 1900, when precautions began to be taken, flammable solvents would accumulate in confined spaces, and often caused serious explosions at raw sugar mills and on shipboard. Spores of filamentous molds secrete invertase, apparently even without growing (Kopeloff and Kopeloff, 1919). Yeasts may be present in the molasses films on raw sugar crystals (Scarr, 1951). Although the yeasts do not produce invertase, they acidify the traces of invert sugar in the adherent molasses films, and the acid catalyzes the hydrolysis of sucrose.

The Factor of Safety. The factor of safety is an expression of the water content in relation to the nonsucrose solids at and below which fermentation will not occur. Permissible water content (*X*) is calculated according to the formula

$$\frac{W}{100-S} = X$$

in which *W* equals the water content of the sugar and *S* the nonsucrose solids. Several versions of the formula are found in the literature. All arrive at the same value. Thus, if raw sugar contains 10% nonsucrose solids, the permissible water content is 3.3%; in highly refined sugar, with 1% nonsucrose solids, the water content should not exceed 0.333%.

Refined Sugar. The crystals of sucrose produced at the raw sugar mills are surrounded with adherent films of molasses which contain invert sugars, brown-pigmented material, and other impurities. In the refineries the crystals are suspended in a heavy sugar syrup, centrifuged, and affined, or washed. The crystals are then dissolved and the solution is treated with lime and carbon dioxide, filtered through diatomaceous earth and bone char to remove the remaining impurities and coloring matter, recrystallized, and dried.

By-products of the raw sugar mills are the bagasse, the sediment, or mud, and the molasses. The bagasse is used as a fuel and as material for a type of hardboard. The sediment is used as cattlefeed or is discarded. The molasses is used in yeast culture and alcoholic fermentation and is added to ensiled agricultural materials.

Beet Sugar

The cell juices of sugar beets contain about 19% sugar. At the refinery the beets are washed to remove the field soil and are sliced. The juices are extracted from the slices in batteries of counterflowing hot water, which kills and ruptures the cells to ensure complete extraction. Thereafter, the processes of clarification, concentration, and crystallization are similar to those used in the cane sugar industry. Modern practices enable beet sugar refineries to produce sucrose that can be distinguished from cane sugar only through exacting quantitative analysis of the ultraminute mineral impurities characteristic of each sugar.

Liquid Sugar

Liquid sugar is clarified cane juice concentrated to approximately 66.5 to 67.5% sucrose, or degrees Brix. The high sugar content and the exceedingly low content of minerals and nitrogenous substances restrict the growth of yeasts, the major microorganisms of spoilage. Contamination occurs most often in the intake hoses, valves, and couplings used to transfer the liquid from tank cars or trucks to the storage tanks, but also occurs through the gaskets and air vents. The surface of the liquid sugar can become diluted with atmospheric moisture. The absorbed moisture does not diffuse in the liquid sugar, but rather remains at the surface, there reducing the sugar concentration.

Liquid sugar storage tanks are equipped with filters at the air vents and with hot air blowers and banks of ultraviolet lights. Drained tanks are washed with hot water and chlorine solution. To prevent contamination, fill lines, pumps, outlets, and attachments should always be cleansed and sanitized with 200 ppm chlorine solution before use.

Molasses

Two types of molasses are produced commercially: *blackstrap* molasses, a by-product of raw sugar production, contains about 50% sugars, chiefly sucrose and invert sugars; *high test* molasses, evaporated cane juice with much of the sugar deliberately inverted through acid hydrolysis, contains between 72 and 86% sugars. Both types of molasses are major sugar sources for growing yeasts and may be used to produce ethanol.

Sorghum

Sorghum is made from the expressed juice of the grass plant *Sorghum.* Tropical in origin, the plant has been adapted to growth throughout the Temperate Zones. It has numerous small, glossy seeds. Varieties of the plant are used as pasture plants of fodder grain, or are made into syrup. Once the major sweetening agent in the home, sorghum is now produced in large volume commercially—consumers often misterm this product "molasses."

Sorghum is made by either the single-stage or the two-stage process. In the single-stage process the juice expressed from the plant between small rollers flows by gravity into a shallow cooking pan (Figure 13-2). The pan is baffled so that the juice is

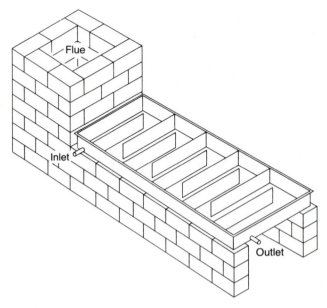

Figure 13-2
Diagram of a typical sorghum pan used in rural America.
By keeping the cane juice a longer time, the baffles favor
its concentration and also make it easier to remove the
coagulated impurities with skimmers.

kept longer, allowing it to reach proper concentration, and allowing coagulated solids, which rise as a scum, to be removed.
The flow rate is adjusted to yield a syrup with the desired concentration of sugar, which will vary between 60 and 75% according to the operator. In the modern, two-stage process the
partially concentrated juice is cooled and treated with malt diastase to digest the starches, and the cooking is resumed. The
starch content of the juice varies with the seasonal climatic conditions. The starch may result in a scorched flavor, coagulation
or jelling, and variation in color from amber to black.

Bacteria grow rapidly in frost-bitten and split, frozen cane,
leading to fermentations and foul odors. Because of the high
sugar content of the sorghum, yeasts are the only spoilage microflora. Sorghum produced in the home is often stored in

containers ranging in size from 1 to 10 gallons. During warm spring weather the yeasts form a dense cream-to-tan scum. Sorghum prepared for retail is bottled or canned while hot and is sealed.

Thermophilic Bacteria in Sugars

Three species of thermophilic bacteria found in sucrose are of major concern to the canning industries, because they may be incorporated in canned vegetables if sugar is added to the brines. They are *Bacillus stearothermophilus, Clostridium thermosaccharolyticum,* and *Desulfatomaculum (Clostridium) nigrificans.* Like many bacteria, they may occur in very low numbers in soils and will reproduce in exceedingly localized areas in canning plants when temperatures are high enough. In the sugar refineries they grow in the hot pressed juices, in hot wells, in pipes, and in sediment and storage tanks. More than 90% is removed from the juices during filtration, and many of the remainder stay in the molasses during the centrifugal separation of the sugar crystals. They grow only between 37 and 70°C, and grow best between 55 and 60°C.

Bacillus stearothermophilus

Despite its generic name, *Bacillus stearothermophilius (stearo-,* fatty; *thermo-,* heat; *philus-,* loving) is strongly fermentative under anaerobic conditions. The end products of fermentation are 75% lactic acid, the remainder being formic and acetic acids and ethanol in the rato 2:1:1. Because the bacterium lacks the enzyme formic acid hydrogen lyase, it does not decompose the formic acid to hydrogen and carbon dioxide. The spoilage is thus termed flat sour and cannot be detected in canned foods until the container has been opened.

The bacterium is nonmotile and spore-forming and requires glucose. It initiates growth at pH 5.0 and is strongly proteolytic in media containing casein. At pH 6.1 spores die only after heating to 100°C for 17 hours, but at 120°C they die in 11 minutes. The standard medium for its enumeration is dextrose-tryptone-BCP (brom cresol purple) agar incubated at 55°C. Upon incubation, two types of colonies may be observed. Those surrounded by a yellow zone, denoting acid formation, are enumerated as *Bacillus stearothermophilus*. The non-acid-producing colonies may arise from cells with weak fermentative ability. Reduction in the amount of tryptone, and a consequent reduction in the buffering capacity of the medium, increases the numbers of acid-producing colonies (Corpe and Frazier, 1950).

Cells of *Bacillus stearothermophilus* produce colonies 0.1 to 0.2 mm in diameter. Occasionally tiny, "pin point" colonies appear on plates incubated at 55°C. These are mesophilic, spore-forming bacteria, usually of the subgroup *Bacillus subtilis*, for which the temperature of 55°C is borderline for growth. In questionable circumstances, such colonies should be transferred to fresh media for incubation at both mesophilic and thermophilic temperatures.

Clostridium thermosaccharolyticum

The bacterium is commonly known as the thermophilic anaerobe, or "TA." Its cells produce lactic and butyric acids and the gases hydrogen and carbon dioxide. Fermentation produces "hard swells," cans with both ends strongly bulged, and if the side seams open, liquid is spewed as far as 40 feet. The bacterium is a very slender rod with swollen spores, with a tendency to become gram-negative with age. It is detected in sugars by inoculation of liver broth.

Clostridium (Desulfotomaculum) nigrificans

This bacterium, facetiously known as the "sulfur stinker," is a gram-negative rod with slightly swollen spores. It grows between 37 and 70°C. It is sensitive to acid and grows only at pH 6.0 and above. On rare occasions it is a spoilage agent of canned peas and corn, the only vegetables commonly packed in the United States that have this high a pH. It uses glucose, pyruvic and lactic acids, and ethanol as hydrogen donors to reduce sulfate to sulfide. Kernels of canned corn spoiled by this organism develop a bluish tip.

Sugar Standards for Thermophilic Bacteria

During the canning seasons of 1925 and 1926 American canners suffered extensive losses by spoilage of canned foods. Investigators traced the cause to thermophilic bacteria carried into the canning plant with the sugars. At that time microbiological standards for the numbers of thermophilic bacteria in sugars were established that remained virtually unchanged to the present and are applicable to both cane and beet sugars:

Total thermophilic spore count. Of 5 samples, no one sample may have more than 150 spores, and the average of 5 sequential samples must not exceed 125 spores per 10 grams of sugar.

Flat sour spores. Of 5 sequential samples, no 1 sample may have more than 75 spores, and the spores must average no more than 50 spores per 10 grams of sugar in 5 sequential samples.

Thermophilic anaerobes. Thermophilic anaerobes shall be present in not more than three (60%) of five sequential samples, and there shall be no more than four positive tubes of six inoculated from any one sample.

Sulfide-spoilage spores. Sulfide spoilage spores shall be present in no more than two (40%) of five sequential samples, and in any one sample they shall be present in no more than two of the six tubes inoculated.

The standard analytical practice is to dissolve 20 grams of sugar in 95 ml water, heat the solution rapidly and boil for 5 minutes to destroy the thermophilic vegetative cells. The solution is rapidly cooled and is returned to its initial volume by restoring water evaporated during heating. Ten ml of it are distributed among 5 plates that are poured with dextrose-tryptone-BCP medium. To detect the thermophilic anaerobes 40 ml of the solution are distributed among 12 tubes, 6 of chopped liver medium, and 6 of sulfite agar containing a small, acid-cleaned nail or iron strip to detect the sulfide spoilage bacteria. Both the liver and sulfite media must have been heated and be hot at the time of inoculation to maintain deeply reducing conditions. All media are incubated at 55°C in a moist incubator.

Experience has shown that the few spores permitted in sugars are not potential sources of spoilage bacteria in commercial canning processes. Also during storage, the viable spores appear to become gradually reduced in number, a process that has been termed "autosterilization" (Hall, 1939).

Bottlers' Standards

U.S. bottlers of carbonated beverages make up the only other industry that establishes microbiological standards for sugars. The bottlers are concerned with mesophilic bacteria, yeasts, and molds; all atmospheric contaminants of crystalline sugar during drying and subsequent handling. These contaminants used to enter the sugars through the mesh of the cloth shipping bags, but modern paper packaging and the use of liquid sugars have eliminated this source of contamination.

Industry standards stipulate that *crystalline* sugars shall bear no more than 200 mesophilic bacteria, 10 yeasts, or 10 molds per 10 grams of sugar. The last 20 samples of *liquid* sugars must average 10 yeasts or less per 10 grams dry sugar equivalent (DSE); 95% of the last 20 samples must have 18 yeasts or less; and only one of the last 20 samples may have more than 18 yeasts per 10 grams DSE if the other counts are within the tolerance. The standards for molds are identical, except that one of the last 20 sequential samples with more than 18 molds may be excluded. Mesophilic bacteria must average no more than 100 bacteria per 10 grams DSE, with 5% (one sample) of the last 20 permitted to have no more than 200, and of the last 20 samples, only one sample (5%) with more than 200 bacteria per 10 grams DSE may be excluded in the calculations if the counts of the other samples are within the tolerance.

Maple Syrup, Candy, and Honey

Maple Syrup and Sugar

Maple syrup and sugar are produced from the sap of the maple tree, *Acer saccharum,* across the northern tier of the United States, from the headwaters of the Mississippi River eastward into New England. Sap is generally harvested by inserting spouts into holes drilled just deep enough into the phloem and wood to support them and the pails hung from them. The trees are tapped in early spring when the sap begins to flow. The sap is gathered into barrels or tanks daily and concentrated at the cookers or pans to approximately 60% sugar. Modern harvesting methods include the use of more readily sanitizable metal spouts, pails, and tanks, and plastic hose to conduct the flowing sap directly to the storage tanks at the concentrators. As in sorghum making, the impurities are coagulated by heat and are removed by skimming.

The sap contains 2 to 3% sucrose with traces of invert sugars, proteins, minerals, and malic acid. Although sterile in the tree, sap becomes contaminated at the taphole, where microorganisms grow, as they do on the tapping and gathering equipment. The extent of contamination increases as the season advances. Bacteria growing in the tapholes seal the phloem vessels and reduce sap flow.

Amino acids appear in the sap as the season becomes warmer. As cells of *Leuconostoc* subsequently grow, sugar is reduced and flavor changes. *Enterobacter* produces a slimy or **ropy sap.** Other types of spoilage are **green sap,** caused by growth of fluorescent *Pseudomonas fluorescens;* **milky sap,** or a turbidity caused by excessive growth of bacteria; and **red sap,** attributed to both yeasts (Edson, 1910) and bacteria (Tanner, 1944).

"Buddy" sap, a flavor defect reminiscent of twig bark, appears late in the harvest season when the leaf buds enlarge. The buddy flavor, attributed to free amino acids (Wasserman and Willets, 1964), can be eliminated by diluting partially concentrated sap with water, inoculating it with *Pseudomonas geniculata* to a population of 5×10^6 cells per ml of sap, incubating for one to two days, and then reconcentrating the syrup. The bacteria are removed with the skimmings.

When bacteria grow in tapholes, sap flow often stops prematurely (Naghski and Willets, 1955). Transparent plastic containers in lieu of pails allow the sterilizing effect of sunlight to reduce microbial growth in the sap. Distribution of microbial growth on plastic containers has been tested with respect to the cardinal compass points: microbial counts ranged from a low of 11×10^4 per ml (western exposure) to a high of 6.8×10^7 per ml (eastern exposure), as compared with a count of 7.3×10^7 per ml of sap gathered in pails (Naghski and Willets, 1953).

The sugar content of maple syrups is borderline for the growth of osmophilic yeasts. Syrup in sealed containers has been pasteurized, so growth occurs only in opened containers. The yeast spores are substantially more heat resistant when heated in honey than when heated in broth (Fabian and Hall, 1933). *Aspergillus* spp. growing on maple candy bring about sucrose conversion, and the product becomes semiliquid (Fellers, 1933). The sap of the oil palm is used to make palm wine and crystalline sugar. The taphole in the tree must be cleaned regularly if sap flow is not to cease within a few days (Faparusi, 1974). In the warm climates, where oil palms abound, microbial counts reach 1×10^8 or more per gram of tissue within 36 hours. The microbial flora include several species of yeasts and a wide variety of both catalase-positive and -negative bacteria.

Candy

According to Cummins (1915), Cicero suggested that candy was a danger, in being a disseminator of disease. Such dangers have been moot, for Tanner (1944), in an excellent compendium of the literature to that time, includes not a single instance of illness associated with candies. In fact, salmonellosis has only recently (Morbidity and Mortality Weekly Report, 1974) been associated with a chocolate candy product. The candy, distributed throughout the United States and Canada, was the source of many cases of infections with *Salmonella eastbourne*.

Cummins found that both handmade and machine made dipped chocolates had low bacterial counts in both the centers and coatings. The bacterial death rates were directly proportional to water content. An inoculum of 4×10^5 cells of *Salmonella typhi* survived for 140 days in chocolate with 5% water content, and for 29 days in chocolate with 7% water content. *Bacterium (Bordetella) pertussis*, the causative agent of whooping cough (a rampant disease at that time), persisted between 10 and 36 hours, while *Mycobacterium tuberculosis* persisted for 10 days.

Fudge, fruit, coconut, and soft-centered candies usually contain more bacteria than do hard candies (Tanner and Davis,

1922). If taffies and fudge have high bacterial contents they probably were contaminated by handling and by organisms present in the air. Volatile oils such as menthol, clove, and cinnamon depress bacterial counts. Coliform bacteria may be found occasionally, but *Escherichia coli* has rarely been isolated from candies.

The malleability of the fondant of an enrobed candy limits the amount of sucrose that can be used in it. Not enough can be used to prevent growth of microorganisms in the finished product. Gaseous fermentations cause the chocolate robe to burst; the syrup exudes and then hardens. Among the suggested causes of blown chocolates were *Clostridium sporogenes* in egg albumen, *C. sporogenes*, *C. putrificum*, *C. aerofoetidum* and unidentified anaerobes in sugars, and clostridia in the starch used as molds (Weinzirl, 1927). Invertase prevents chocolate explosions, raising the cane and invert sugars to 79.1% (Paine et al., 1927).

Honey

Honey is the nectar of flowers concentrated to about 83% sugar content by the rapid wing movement of bees in the hive. The nectars initially contain from slightly less than one to slightly more than 2% sugars, principally glucose and fructose in fairly constant and equal ratio (Wykes, 1953). About 336,000 "bee trips" are needed to produce one pound of honey.

Although considered a pure, wholesome, and safe food, honey can be poisonous. According to Xenophon, a Greek historian of the fourth century B.C., soldiers of the Greek Army of the Ten Thousand, after consuming honey from the hills around the Black Sea, developed delirium, vomiting, and diarrhea. Poisonous honeys are made from the nectars of flowers such as the *Rhododendron ponticum*, *Datura stramomium*, and *Gelsemum*. Since 1976 honey has been incriminated in outbreaks of infant botulism. Spores of *Clostridium botulinum* that are present in honey fed to infants germinate in the body, producing botulinal toxin. It is now recommended that infants not be given honey (Morbidity and Mortality Report, 1978).

Nearly all honeys contain dormant bacteria, mold spores, and yeasts, but only the yeasts cause spoilage. In the hive the high temperature normally keeps ripening honey from fermenting, but high relative humidity can prevent proper ripening and fermentation can result. Lochhead and Heron (1929) recount a description of the antics of intoxicated bees from an article in the Russian literature. In this instance, the honey had an unusually high content of nitrogenous substances, apparently from the sap of the *Tillia* or lime tree.

Fresh honey rarely ferments if crystallization does not occur. If sugars in the honey have crystallized over winter, creating a disparity of as much as 9.1% water content between the crystals and the syrup, fermentation will occur as spring temperatures rise (Marvin et al., 1931). Although the high sugar content keeps ethanolic fermentation to a minimum, affecting only a small quantity of sugar, the yeasty flavor and the ethanol nevertheless render the honey unpalatable. It has been observed that among four strains of *Saccharomyces cerevisiae* cultured in 40% glucose broth, the quantity of gas produced was 4%, 16%, 26%, and 56% of the theoretical quantity (White and Munns, 1955). Honey with 17.1% water content does not spoil, despite a high population of viable yeasts. Honey with 18% moisture content does not support fermentation if the viable yeast population does not exceed 1000 cells per gram of honey, but moisture content of 20% or more presents the danger of spoilage.

Pasteurization is a practical means of preserving honey (Stephen, 1942). Strains of yeast are not equally resistant to heat, and a few highly resistant cells will remain that require severe heat to kill. Stephen's curves of survival show a point of inflection on the killing curve, suggesting that the relatively heat-sensitive strains die rapidly, the heat-resistant population more gradually. Stephen advocated that honey be pasteurized at 60°C for 30 minutes. Modern honey is flash pasteurized at 93°C.

When honey and other heavy sugar syrups are introduced into tap or distilled water blanks, yeasts plasmoptize, or burst. Therefore, dilution blanks and culture media are made with 50% honey or glucose to maintain the high osmotic pressure. Osmophilic yeasts may grow well in media containing 65% honey and poorly or not at all in media with less than 32% honey (Lochhead and Heron, 1929). The yeasts probably become adapted to high sugar concentrations and are not obligately osmophilic. They can be adapted to growth on customary culture media through serial, step-down (i.e., an incremental) reduction of the sugar content of the culture media.

According to Lodder (1970), only three species of yeast, all involved in spoiling honey, syrups, jams, and jellies, are capable of growing in 60% glucose-yeast extract medium: *Saccharomyces bailii* var. *osmophilus; S. bisporus* var. *mellis,* into which the yeast formerly known as *Zygosaccharomyces richteri* is merged; and *S. rouxii,* also known in earlier literature as *S. nussbaumeri, S. barkeri,* and *Z. mellis.* Of these species only some strains of *S. rouxii* grow at 37°C.

Starch

Starch is a complex polysaccharide of linear and branched chains that hydrolyze to maltose. It occurs very commonly in plants in nature, and large quantities of it are present in cereals and in the tuberous roots and underground stems of some plants. According to Brautlecht (1953), the Egyptians were using refined wheat starch as a food and in papyrus making before 3000 B.C. In the United States most starch comes from corn, rice, wheat, and the white potato. Major foreign sources include the root of the cassava plant *(Manihot utilissima),* sago starch from several species of palm, and arrowroot starch from the rhizome of maranta *(Maranta arundinacea).*

Starch is obtained from cereals by the following steps: the grains are first steeped in successive changes of mildly acidified water and are then coarsely ground, to facilitate separation of the bran, the endosperm, and the cotyledons that contain the starch. The starch is washed one or more times to free it of the residual nonstarch material and is deposited as a wet cake by centrifugation, sedimentation, or filtration. It is dried to about 16% moisture content. Nearly all starch is ground to a powder; however, pearl starch is unground, or lump, starch.

Potatoes, roots, and stems are rasped with fine teeth to rupture the plant structure and the cells. Thereafter, the successive washings, concentration into a cake, and drying, parallel the treatments of the corn starch industry.

All starch-making processes take several days' time. The initial slurries, rich in soluble carbohydrates, proteins, amino acids, minerals, and growth factors, are fertile media for the growth of the many kinds of bacteria, yeasts, and molds present on surfaces of the raw product. These microorganisms are suppressed by dry cleaning cereals and washing roots and stems, steeping in mildly acidified water, frequently changed, and using sulfurous acid and some form of chlorine. Results of plating common starches in routine laboratory exercises over the last 20 years suggest that the spore counts have been decreasing, probably reflecting improved starch refining practices. Some samples of starches have been free of bacteria. Unlike cane sugar, starch often contains sulfide-producing, sporulating anaerobes that produce large colonies in the sulfite agar used to detect the thermophilic species.

Most domestic and imported starch is used in the foods industries. Approximately 80% of the corn starch produced in the United States is converted into the dry sugar dextrose and into liquid sweetening agents such as the dextrins used by foods processors and corn syrups for retail. Selective use of mild acids and heat converts native or raw starch into soluble starch, the various dextrins, and dextrose.

The starch molecule is composed of amylose, or linear chains of *alpha*-D (1:4) bonds of several hundreds of glucose units; and amylopectin, which has branch points at the *alpha*-D (1:6) bonds. Cycloamylose, or the Schardinger dextrin, is a cyclic dextrin of 5 to 12 glucose units that is digested by the enzyme cycloamylase produced by *Bacillus macerans*.

Three enzymes are used to reduce starch to simpler carbohydrates:

Glucoamylase, secreted by *Aspergillus* spp., *Rhizopus*, *Saccharomyces diastaticus*, and *Clostridium acetobutylicum*, converts amylose, amylopectin, and maltose oligosaccharides directly into glucose.

Alpha-amylase, or alpha-D-1:4-glucan-4-glucanohydrolase, liberates *alpha*-maltose. The plant *alpha*-amylases are useful in malting and baking; the microbial *alpha*-amylases are used for desizing in the textile industry, for preparing starch slurries for coating paper, and for preparing liquid sweetening agents.

The *beta*-amylase, or *alpha*-D-1:4-glucan-maltohydrolase, releases maltose in the *beta* configuration. It is found in plants such as barley and wheat, the sweet potato, and in molds.

Neither *alpha*-amylase nor *beta*-amylase can hydrolyze starch molecules beyond the points of the branches. The remaining, undigested portion of the molecule is termed a "limit dextrin."

References

Brautlecht, C. H. 1953. Starch. Reinhold Publishing Co., New York, N.Y.

Browne, C. A. 1918. Indus. Engr. Chem. *10*, 178.

Corpe, W., and W. C. Frazier. 1950. Food Technol. *4*, 8.

Cummins, E. H. 1915. Amer. J. Public Health *5*, 1148.

Edson, H. A. 1910. Vt. Agr. Exp. Sta. Bull. 151.

Fabian, F. W., and H. H. Hall. 1933. Zentralbl. f Bakteriol. u. Parasitenk, II, Abt. *89*, 31.

Faparusi, S. I. 1974. J. Food Sci. *39*, 755.

Fellers, C. R. 1933. J. Bacteriol. *25*, 67.

Hall, H. H. 1939. Food Res. *4*, 259.

Hall, H. H., L. H. James, and E. K. Nelson. 1937. J. Bacteriol. *33*, 577.

Kopeloff, N., and L. Kopeloff. 1919. Indus. Engr. Chem. *11*, 845.

Lochhead, A. G., and D. A. Heron. 1929. Canad. Dept. Agr. Bull. 116.

Lodder, J., ed. 1970. *The Yeasts. A Taxonomic Study.* 2nd ed. North-Holland Publishing Co., London.

Marvin, G. E., W. H. Peterson, E. B. Fred, and H. F. Wilson. 1931. J. Agr. Res. *43*, 131.

McCleskey, C. S., L. W. Faville, and R. O. Barnett. 1954. J. Bacteriol. *54*, 697.

Morbidity and Mortality Weekly Report. 1974. *23*, 35, 37, 85, 89.

Morbidity and Mortality Weekly Report. 1978. *27*, 29.

Naghski, J., and C. O. Willets. 1953. Food Technol. *7*, 81.

Naghski, J., and C. O. Willets. 1955. Appl. Microbiol. *3*, 149.

Paine, H. S., V. Birkner, and J. Hamilton. 1927. Indus. Engr. Chem. *19*, 358.

Scarr, M. P. 1951. J. Gen. Microbiol. *5*, 704.

Stephen, W. A. 1942. J. Sci. Agr. *22*, 705.

Tanner, F. W. 1944. *The Microbiology of Foods.* Garrard Press, Champaign, Ill., 2nd ed.

Tanner, F. W., and E. Davis. 1922. Amer. J. Public Health *12*, 605.

Wasserman, A. E., and C. O. Willets. 1964. Food Technol. *18*, 1966.

Weinzirl, J. 1927. Amer. J. Public Health *17*, 708.

White, J., and D. J. Munns. 1955. J. Inst. Brew. *61*, 217.

Wykes, G. R. 1953. Biochem. J. *53*, 294.

14

Beverages

Fresh Fruit Juices

Fruit juices are merchandised in several different forms: fresh, frozen concentrate, canned, and dried. Fresh juices are of two general types: cloudy or pulpy, and clear or sparkling. Pulpy juices are prepared by squeezing the fruit in a manner that captures much of the tissue as well as the cell saps. Many citrus juices are so processed. Clear juices are prepared with enzymatic clarification and filtration to remove suspended matter and microorganisms. Apple juice, grape juice and berry juices are thus processed.

As their name implies, frozen concentrates are prepared by concentrating the juice to smaller volume. Large amounts of citrus and grape and tomato juice are prepared in this manner. Only a small proportion of fruit juices or combinations of juice and syrup are drum-, spray-, or freeze-dried and converted into powders that can be reconstituted (Strashun and Tolburt, 1954; Mylne and Seamans, 1954).

A large proportion of fruit juices such as grape juice, citrus juices, tomato juice, apple juice, pineapple juice, cranberry

Box 14-1

Several kinds of fruit are consumed as juices. The most-used are named below, in descending order of volume used per year.

Grapes. The grape is by far the most harvested of fruits, annual world production being over 60 million metric tons (63.28 in 1973). Most of the crop is used for wine production, only a very little for fresh juice. In the Mediterranean basin and in Southwestern Europe grapes have been cultivated for thousands of years. France, Italy, and Spain produce the most. They are followed by Hungary, Turkey, the Soviet Union, Algeria, Greece, Portugal, Romania, Yugoslavia, and the United States.

Citrus Fruits. Today the Mediterranean countries, the United States, Mexico, Brazil, Australia, and South America grow enormous quantities of citrus fruits—which are not native to any of these regions. In the twelfth century traders from India and Southeast Asia introduced them to the eastern Mediterranean, and European settlers transplanted them to the Western Hemisphere, Africa, and Australia.

The world's citrus crop is approximately 44 million metric tons (43.91 in 1974). The United States is the leading producer of oranges, lemons, and grapefruit, Mexico of limes. Ever-increasing amounts of oranges and grapefruits are being processed into juice, either fresh or canned, or frozen concentrate or dried concentrate.

Tomato Juice. Botanically the tomato is a fruit, although it is often considered a vegetable. Although most tomatoes are used whole, either fresh or canned, or are used as catsup, sauce, or puree, fresh or canned tomato juice is a major fruit beverage. The microbiology of tomatoes is considered in Chapters 4, on canning and 15, on fruits and vegetables.

Apple (Sweet Cider), Pineapple, and Other Juices. Fresh apple juice is a popular fruit drink but, like grapes, uses only a small portion of the total tonnage of apples harvested. When the pineapple fruit is processed into slices, chunks, or sections, the core and peripheral material would be wasted if they were not pressed to remove the entrapped juice, which is used either fresh, canned, or as frozen concentrate. Fresh fruit juices are also made from cranberries, pears, cherries, loganberries, blackberries, raspberries, strawberries, and guava or fruit nectars from peaches, pears, apricots, plums, and passion fruit.

Fruit Drinks. Various fruit drinks such as orangeade, lemonade, fruit punch, etc., exhibit wide variations in their formulas. Ordinarily these concoctions contain 15% or more of full strength fresh fruit, plus variable amounts of sugar to give a drink with 9° to 14° final Brix. Citric, malic, tartaric, or ascorbic acid may substitute in part or whole for the natural fruit acid.

juice, and blends or combinations of juices are stabilized by processing in localities near those where the fruit is grown.

Grape Juice

After Concord grapes are stemmed and pulped the total number of yeasts varies from 1200 to 60,000 per ml of pulp (Pearson, 1936). To prepare for storage the juice is flash heated at 80 to 85°C and flash cooled to about 0°C, then pumped into storage tanks. During storage, usually at –5.5 to –2.2°C, excess tartrates precipitate. After they and other argols precipitate, the juice is repasteurized to kill microorganisms and then bottled for sale. Such juice is essentially sterile. At times organisms—introduced principally from contaminated equipment, pipelines, container surfaces or from crevices, foam, or air—cause serious yeast or mold contaminations. Some yeasts of the genera *Saccharomyces, Hanseniaspora, Torulopsis,* and *Candida* have become adapted to grow more luxuriantly at 1°C than at 21°C (Lawrence et al., 1959). Cells of species of the first three genera produce alcohol, whereas those of *Candida* do not. Early in storage the fermentative types of yeast are the most common, but later in storage the nonfermenters often multiply the most readily (Pederson et al., 1959).

Citrus Juices

Citrus juices undergo quick microbial and chemical deterioration. Microorganisms in fresh juices are not highly heat resistant and are readily destroyed or inactivated by pasteurization. Although frozen concentrates are not usually pasteurized, their high acid and sugar content means that yeasts, molds, *Leuconostoc,* and lactobacilli are the organisms most likely to grow (Murdock et al., 1950).

Species of *Candida, Zygosaccharomyces, Hanseniaspora, Saccharomyces,* and *Pichia* are commonly isolated; *Trichosporon cutaneum* and *Torulopsis glabrata* also occur (Recca and Mrak, 1952). Few of these organisms are able to grow in 65°Brix or at 2.5°C; although *S. elongasporus,* first isolated from orange concentrate, can tolerate this temperature.

Apple Juice (Sweet Cider)

Sound, unwashed apples carry a very small load of micro-organisms (Forgacs, 1942), which typically consists almost entirely of weakly fermentative, nonsporing yeasts. At the processing plant apples are carefully sorted, but it is impossible to eliminate all rot. In apples that look sound rot may be localized around the core or may extend into the flesh. *Penicillium expansum,* cells of which form the mycotoxin patulin (see Chapter 23), is implicated in three-fourths of such moldy fruit (Figure 14-1), while *Aspergillus, Monilia* and *Mucor* are also implicated occasionally.

At the time of pressing the population varies widely according to the degree of internal or visible fungal damage (Marshall and Walkley, 1951). On sound fruits total yeast counts vary from 10^2 to 10^6 per apple, molds from about 10^3 to 10^5. "Apples with yeast or mold counts in excess of 2,000,000 and 200,000 respectively may be assumed to be damaged and with counts in excess of 5,000,000 the fruit is unsuitable for apple juice manufacture." In apple juice manufacture, the resident microbial population increases in direct proportion to the numbers of slightly damaged and bruised fruit used. When

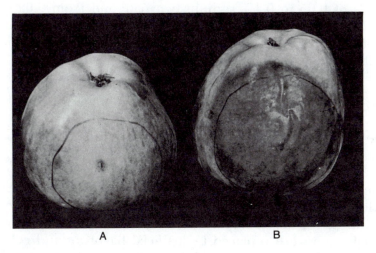

A   B

Figure 14-1
A. An apple nicked with an uninoculated scalpel; B. An apple nicked with a scalpel contaminated with *Penicillium expansum.*

apple juice is stored in bulk under carbon dioxide pressure for several months prior to bottling, microbiological control is essential to suppress growth of spoilage microorganisms.

Yeast species most often encountered are *Candida malicola, C. pulcherrima, C. krusei, Torulopsis* spp., *Rhodotorula* spp., *Cryptococcus* spp., *Sporobolomyces roseus, Debaryomyces kloeckera, Kloeckera apiculata, Saccharomyces bisporus, S. uvarum,* and *S. delbrueckii* (Clark and Wallace, 1954; Clark et al., 1954; Bowen and Beech, 1964). Distribution of *C. pulcherrima, C. krusei,* and *Kloeckera apiculata* is widespread on all varieties of apple. Conversely, *C. catenulata, Sporobolomyces roseus* and *Saccharomyces delbrueckii* are found only in isolated instances, suggesting that certain yeasts are peculiar to particular areas (Bowen and Beech, 1964).

Other Juices

Pineapple juice is easily pasteurized or, if canned, readily sterilized. It is far more acid (about pH 3.7) than its flavor suggests, because the high sugar content acts as a masking agent. Juice squeezed from ripe cranberries, which have a high benzoic acid content, does not spoil quickly.

Benzoates and sorbates are used in fruit juices, fruit drinks, and cider to control yeast, mold, and some bacterial growth. Levels of 0.05 to 0.1% sodium benzoate and from 0.025% to 0.075% sorbate ordinarily are used. Sodium benzoate is restricted to 0.1%, which is also a practical limit, because higher levels adversely affect flavor.

Carbonated Soft Drinks

The idea of carbonating water is quite old: in 1772 Joseph Priestley published *Directions for Impregnating Water with Fixed Air.* This practice was first tried as a way of imitating mineral waters and later as a way of making water more hospitable to added sweetening and flavoring agents such as sugars, syrups, lemon, ginger, and sarsaparilla (Riley, 1958). Eventually natural fruit flavors were replaced by artificial flavoring and coloring agents. Today, the carbonated beverage industry is a multimillion dollar business, manufacturing in this country alone more than 20 billion quarts of soft drinks per year. This is approximately the equivalent of one eight ounce bottle of "pop" per person every day.

Before 1945 many bottling plants served only their own localities and were of limited capacity. The buildings were of standard design and construction, which made environmental control of contamination difficult if not impossible. About 85 to 90% of the spoilage was caused by the presence of yeast in the syrup (Levine, 1939). The waters used in washing and bottling were variable in quality, and enough enteric pathogens survived to be a matter of concern. It was shown that against *Salmonella typhi* CO_2 had a lethal but relatively slow action; thus, before all viable cells could be killed the beverage could have been consumed.

Carbonation

If soft drink beverages have been properly bottled, the inhibitory action of acid and carbonation assures potability for a reasonable period. Decades ago it was found that microorganisms die more quickly in carbonated than in uncarbonated water or beverage (Gershenfeld, 1920; Koser and Skinner, 1922; Donald et al., 1924). It was concluded that bacteria could not survive more than 4.0 volumes of carbonation. Early workers (Donald et al., 1924) assumed that a beverage carbonated with sufficient pressure would not allow contaminating bacteria to multiply, and (Insalata, 1952) that increasing carbonation in turn increased its sterilizing effect on bacteria and yeasts. It is now known, however, that microbial spoilage in bottled or canned beverages is in fact prevented by the combined effects of high sugar concentration, acidity, degree of carbonation, and good plant sanitation. Each of these factors has received considerable individual study. Shillinglaw and Levine (1943) demonstrated that such edible acids as phosphoric, citric, acetic, lactic, glycollic, and tartaric, in a .02 N concentration—approximately that employed in many beverages—were effective germicides for *Escherichia coli* and *S. typhi,* and that adding CO_2 and sucrose slightly increased their germicidal efficiency. Although the low pH provides most of the killing effect, and is enhanced only slightly by carbonation (Eagan and Green, 1957), neither low pH nor carbonation alone is a sufficiently effective sterilizer to

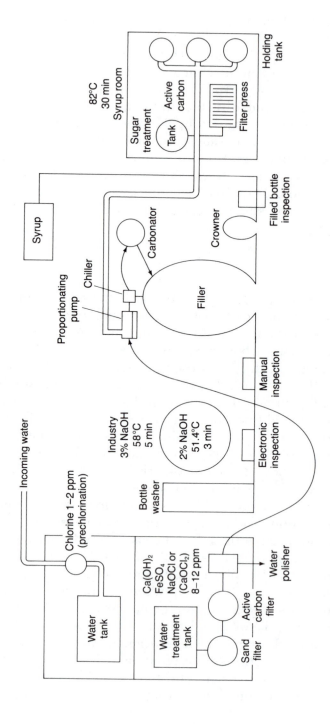

Figure 14-2

Flow diagram of processing stages in a beverage bottling plant. (Courtesy of B. Swaminathan.)

produce the observed powers of preservation. For example, to be germicidal to microorganisms, concentrations of sucrose must be 60 to 80% (Bell and Etchells, 1952; Erickson and Fabian, 1942; Tarkow et al., 1942), well above the range found in soft drinks. Without carbonation there is no lethality regardless of °Brix (Witter et al., 1958).

Modern bottling plants present good examples of aseptic packaging (Figure 14-2). The several ingredients enter the bottling machines through series of chemically sterilized pipes and meters. If sanitization is neglected, yeasts and molds may become established in the equipment; Sand et al. (1976) have isolated seven genera of yeasts from proportioning pumps in soft drink plants. Returnable bottles are washed, sterilized in hot (49.1°C) caustic solution, and rinsed; single-use containers are rinsed. The bottles and containers can become contaminated by exposure to the atmosphere before being filled. Water is chlorinated or, if it contains traces of organic compounds that react with chlorine to introduce objectionable flavors, is passed in shallow films under a battery of ultraviolet lamps to receive a dosage of 35,000 microwatts/cm (Woodroof and Phillips, 1974). Syrups, once a prominent vehicle by which yeasts got into bottled beverages, are now brought to a boil during manufacture (Levine, 1939). Syrups and concentrates must either be sterile or have total yeast counts low enough to result in autosterilization. Addition of 0.1% benzoate or sorbate effectively prevents growth of yeasts and molds in syrups and concentrates.

Box 14-2

Sodium benzoate is ordinarily used at a level of 0.03 to 0.05% as a preservative for carbonated beverages. Very few yeast strains can tolerate more than 0.03% benzoic acid. This chemical is not permitted to exceed 0.10% in the finished carbonated beverage, according to the Standard of Identity for Non-Alcoholic Beverages, Part 31, Title 21, Section 121.101 of the Code of Federal Regulations. As a food preservative, sodium benzoate at a level of 0.10% is considered GRAS (Generally Recommended as Safe) under the 1958 Food Additives Amendment to the Food, Drugs, and Cosmetic Act.

Ordinarily, sorbate is introduced in the syrup rather than in the final beverage. Occasionally an organism resistant to the usual concentration of benzoate or sorbate is found.

Woodroof and Phillips (1974) classify the microorganisms found in carbonated beverages into five groups:

1. the lactic acid bacteria;
2. the strongly and weakly fermentative yeasts;
3. the film yeasts;
4. "other microorganisms," including several molds, inert yeasts, and *Bacillus* spp.; and
5. the acetic acid bacteria.

Yeasts

Yeasts cause nearly all microbial spoilage of soft drinks (Mossel and Scholtz, 1964; Scharf, 1960). The indicators of yeast spoilage are visible sediment, turbidity, gas formation, and changes in odor, taste, and color. Yeasts can produce enough CO_2 to cause the bottle to explode during opening, with sometimes disastrous consequences. *Saccharomyces* and *Torulopsis* spp. are isolated and identified most frequently. Other yeasts that have been isolated from equipment and beverages are species of *Brettanomyces, Candida, Kloeckera, Hansenula, Lodderomyces, Pichia,* and *Saccharomycopsis.*

Bacteria and Molds

Lactobacilli and *Leuconostoc mesenteroides* are major spoilage bacteria. Lactobacilli produce lactic acid with unpalatable souring. *L. mesenteroides* develops dextran, the slime balls ranging in size from barely detectable to that of a large pea, although growth has been known to cover the bottom of the bottle.

Molds are occasional nuisances in bottled beverages, chiefly because they appear as foreign matter. Species of *Aspergillus* and *Penicillium* are isolated more frequently than are members of the family Dematiaceae and other members of the family Moniliaceae. They grow as delicate, fluffy, cottony white masses suspended in the liquid. From lack of oxygen, fruiting bodies cannot form. The molds grow slowly in bottled beverages and often are detected only after the beverages are in the mar-

Table 14-1
Sugar content, carbonation, % acid, and pH of various carbonated beverages.

Flavor	Sucrose Brix	Carbonation (gas volume)	Acid (%)	pH
Colas	10.5	3.4	0.09	2.6
Root beer	9.9	3.3	0.04	4.0
Ginger ale	9.5	3.8	0.10	—
Cream (vanilla)	11.2	2.6	0.02	—
Lemon and lime	12.6	2.4	0.10	3.0
Orange	13.4	2.3	0.19	3.4
Cherry	12.0	2.4	0.09	3.7
Raspberry	12.3	3.0	0.13	3.0
Grape	13.2	2.2	0.10	3.0

Source: Lachmann, 1975.

ketplace. Mold spoilage, frequently traced to improperly sanitized equipment, often is restricted to the products of a single bottling plant and can result in the recall of hundreds of cases of bottled beverages.

Pressure.

Gaseous carbon dioxide is introduced during bottling, the exact amount varying with the temperature at the time of bottling (Table 14-1) and the type of beverage. Most carbonated soft drinks receive about 3.2 to 3.8 atmospheres of CO_2, but soft drink bottles and cans must be able to withstand much higher internal pressures, since ambient temperatures in cars, heated rooms, and other holding areas where pop is stored often exceed 40°C.

Coffee, Cocoa, and Tea

Coffee

At least one-third of the world's peoples consumes beverages prepared from roasted coffee beans. More coffee bean beverage is consumed—either hot or cold—than is any other drink.

Since the sixteenth and seventeenth centuries, when coffee was introduced into Europe and America, world imports from the major producing countries—Brazil, Colombia, and various African nations, the Middle East, India and Indonesia—have expanded to almost 5 million tons (4.91 million metric tons in 1974). The United States consumes over 35% of world imports.

The coffee berry consists of an outer covering, or pericarp, enclosing two beans embedded in a mucilaginous parenchyma. The interior of the sound berry is sterile. To prepare, the beans are freed of the pericarp and mucilaginous substances by either the dry or the wet method. Berries for coffees of the best quality are hand-harvested at the proper stage of maturity; berries for coffees of lesser quality are stripped from the clusters and include mature, overripe, and immature berries.

In the *dry* method the harvested berries are dried in shallow layers for several weeks, during which they are occasionally turned over. The pericarp and mucilage are then removed mechanically. In the *wet* method, the berries are washed and the fresh fruit is broken in a pulping machine to remove the pericarp. The mucilaginous parenchyma is digested by the pectinolytic bacterium, *Erwinia dissolvens* in a period of about 24 to 36 hours; use of a commercial pectinolytic enzyme concentration reduces the time. Since the digestion is exocellular, the sugars liberated by pectinolysis become available to a wide variety of bacteria that play no part in coffee fermentation (Frank et al., 1965).

Next, the freed beans are dried to a moisture content of about 12%. With the classic method—drying them in the sun, over a period of several weeks—relative humidity and the frequency and duration of cloud covers affect how fast and how well they dry. The modern procedure is to dry the berries in counterflow drying chambers within four to six hours.

Because raw coffee beans are a dry organic material that comes to equilibrium with the moisture in the atmosphere, they must be shipped promptly. In warm, humid regions, such as the coastal regions of the coffee-growing countries, mold and insect damage result from prolonged storage. Cargoes of raw beans have been rejected upon arrival in American ports for such damage. Seizure and condemnation can be expensive: early in 1975, the Food and Drug Administration impounded 665,000 pounds of moldy coffee beans, worth an estimated $330,000.

For two centuries Sri Lanka and Indonesia (formerly Ceylon and Java, respectively) were major coffee producing nations. However, in 1869 the coffee rust *Hemileia vastatrix* attacked Ceylon plantations with devastating effect, and attacked Java coffee plantations a few years later. Long known in other coffee growing areas of the world, not until 1970 did coffee rust become widespread in Brazil—the first in the Western Hemisphere known to be infected.

Cocoa and Chocolate

Cocoa and chocolate are manufactured from the fruit of *Theobroma cacao,* a shrubby tree that is Amazonian in origin, and can grow no further than 20 degrees latitude from the equator. About 1.5 million metric tons (1.47 in 1974) of the fruit are produced annually. Each tree produces 20 to 30 woody pods that contain five tiers of eight cacao beans attached to a central placenta and embedded in a mucilaginous parenchyma. The beans are about the size of lima beans. If the pods have not been stung by insects or been knocked about by the wind, they are internally sterile when harvested. While men of Christopher Columbus' party were the first Europeans to see the cacao tree, probably Hernando Cortez brought the first beans to Europe. At one time an attempt was made, unsuccessfully, to restrict the term cacao to the raw bean and to designate the roasted bean as chocolate and its defatted pulverized product as cocoa.

Cacao pods are opened with a heavy knife and the beans are scooped out by hand (Figure 14-3). The beans are fermented in large heaps on the ground or in sweat boxes holding several hundred pounds of beans (Rohan, 1958; Martelli and Dittman, 1961). To ensure a uniform fermentation cure the heaped beans are turned over occasionally, and those in the boxes are transferred several times to other boxes.

Like coffee fermentation, the cacao bean is fermented to free it of the parenchyma through digestion by pectinolytic bacteria. The heat of fermentation kills the embryo and the bean becomes brown to purple because anthocyanin pigments are liberated and polyphenolic substances undergo enzymatic alteration. The wet beans become contaminated by contact with the hands, knives, and the sweat boxes and the ground and are further contaminated by Drosophila.

Figure 14-3
A heap of cacao pods fermenting on the ground.

The fermentation occurs in three stages over a period of five to seven days. During the first, or ethanolic, stage a variety of yeasts ferment the sugars, chiefly glucose ranging between 8 and 13% in the parenchyma. In the acetic acid stage the vinegar bacteria oxidize the ethanol to acetic acid. Some investigators attribute the death of the embryo to the acetic acid. During the last stage oxidative yeasts destroy the naturally present citric acid and raise the pH. During fermentation a great variety of gram-negative and positive bacteria and bacilli grow, but none is believed to play a consequential role in the cacao bean fermentations. The beans are then dried to a 6 to 8% moisture content, either by sun drying 6 to 21 days, depending upon the climatic conditions, or, more rapidly, by hot air in counterflow rotating drums. The beans are finally polished to remove residual dried parenchymal material, and are ready for market. If relative humidity is high the cacao bean, like the coffee bean, absorbs moisture from the atmosphere. Large shipments of moldy beans have been condemned and destroyed upon arrival in American ports. Some of the molds isolated from the beans are lipolytically active (Eyre, 1933). *Aspergillus fumigatus* attacks the cellulose of the seed coat and opens the way for other microorganisms to attack the cotyledons.

At the factory the cocoa beans are cleaned, broken, and shelled (Figure 14-4), and then roasted at between 95 and 134°C for 15 to 120 minutes, according to the desires of the roasters.

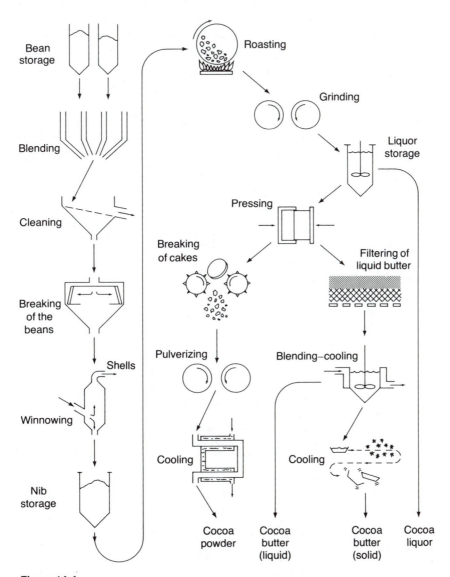

Figure 14-4
Flow diagram of steps in making cocoa products. (Courtesy Cacaofabriek De Zaan B. V., The Netherlands.)

Teas

The teas of commerce are prepared from the leaves of a shrubby plant, *Thea sinensis*, which grows in India, China, Taiwan, Japan, Korea, and similar geographic areas. There are two types of tea product: black tea, and green tea. During black tea manufacture the leaf undergoes a short (one to four hours) period of so-called fermentation, although no microbiological activity in fact takes place to prepare the dried leaves. Green teas are of leaves that have been withered, rolled, and dried rapidly to prevent changes in the catechins. Black teas are withered, rolled to break the cellular structure, and exposed to the air for several hours. During this time, catechins are converted to o-quinones, which are condensed to compounds of high molecular weight. Black teas are heated to about 82.2°C to inactivate the enzymes. The major enzyme in the leaf is polyphenol oxidase (Bokuchava and Skobelova, 1969). A blend of many compounds produces the aroma of tea. Several species of storage fungi, particularly *Aspergillus* spp. and *Penicillium* spp., grow on packaged teas kept insufficiently dry. Such growth is not only unsightly but also changes the flavor and makes the product unusable.

References

Bell, T. A., and J. L. Etchells. 1952. Food Technol. *6,* 468.
Bokuchava, M. A., and N. I. Skobelova. 1969. In: *Advances in Food Research.* 17, 215.
Bowen, J. F., and F. W. Beech. 1964. J. Appl. Bacteriol. *27,* 333.
Chatt, E. M. *Cocoa.* Interscience, 1965.
Clark, D. S., and R. H. Wallace. 1954. Canad. J. Microbiol. *1,* 145.
Clark, D. S., R. H. Wallace, and J. J. David. 1954. Canad. J. Microbiol. *1,* 145.
Donald, J. R., C. L. Jones, and A. R. M. MacLean. 1924. American Journal of Public Health *14,* 122.
Eagan, R. G., and C. R. Green. 1957. Food Res. *22,* 687.
Erickson, F. J., and F. W. Fabian. 1942. Food Res. *7,* 68.
Eyre, J. C. 1933. Annals Applied Biology *19,* 351.
Forgacs, J. A. 1942. Food Res. *7,* 462.
Frank, H. A., N. A. Lum, and A. S. Dela Cruz. 1965. Appl. Microbiol. *13,* 201.
Gershenfeld, L. 1920. American Food Journal *15,* 16.
Insalata, N. F. 1952. Food Engin. *24*(7), 84.
Lachmann, A. 1975. The Role of Sucrose in Foods. Intl. Sugar Res. Bethesda, Md.
Koser, S. A., and W. W. Skinner. 1922. Journal of Bacteriol. *7,* 111.

Lawrence, N. L., D. C. Wilson, and C. S. Pederson. 1959. Appl. Microbiol. *7*, 7.

Levine, M. 1939. Refrig. Eng. *37*, 157.

Marshall, C. R., and V. T. Walkley. 1951. Food Res. *16*, 448.

Martelli, H. L., and H. F. Dittman. 1961. Appl. Microbiol. *9*, 370.

Mossel, D. A. A., and H. H. Scholtz. 1964. *Annales de l'Institut Pasteur*, Lille, *15*, 11.

Murdock, D. I., J. F. Folinazzo, and U. S. Troy. 1950. Food Technol. *6*, 181.

Mylne, A. M., and U. S. Seamans. 1954. Food Technol. *8*, 45.

Pearson, C. S. 1936. Food Res. *1*, 9.

Pederson, C. S., M. N. Albury, D. C. Wilson, and N. L. Lawrence. 1959. Appl. Microbiol. *7*, 1.

Priestly, J. 1772. Philosophical Transactions, Royal Society, London.

Recca, J., and E. M. Mrak. 1952. Food Technol. *6*, 450.

Riley, J. J. 1958. *A History of the American Soft Drink Industry. Bottled Carbonated Beverages,* 1807–1957. ABCB, Washington, D.C.

Rohan, T. A. 1958. J. of Sci. Food and Agric. *9*, 542.

Sand, F. E. M. J., and A. M. van Grinsnen. 1976. Antonie von Leeuwenhoeck *42*, 523.

Sand, F. E. M. J., G. A. Kolfschoten, and A. M. van Grinsven. Brauwissensch. *29*, 294.

Scharf, J. J. 1960. Annales de l'Institute Pasteur, Lille *11*, 117.

Shillinglaw, C. B., and M. Levine. 1943. Food Res. *8*, 464.

Strashun, S. I., and W. F. Tolburt. 1954. Food Technol. *8*, 40.

Tarkow, L., C. R. Fellers, and A. S. Levine. 1942. J. Bacteriol. *44*, 367.

Witter, L. D., J. M. Berry, and J. F. Folinazzo. 1958. Food Res. *23*, 133.

Woodroof, J. G., and G. F. Phillips. 1974. *Beverages: Carbonated and Noncarbonated.* Avi Publishing Co., Westport, Conn.

15

Fruits and Vegetables

It may be recalled that all green plants possess an epiphytic microflora which normally subsists on the slight traces of carbohydrates, protein and inorganic salts which dissolve in the water exuding from or condensing on, the epidermis of the host.

THAYSEN AND GALLOWAY (1930)

Resident Microflora

In the quote above Thaysen and Galloway were referring to two classic studies: that of Burri (1903), in which leaves of apple, plum, bean, cabbage, cucumber, and onion plants were reported to have resident populations ranging from 300,000 to an excess of 14.4 million/gram; and that of Dueggeli (1904), in which the bacteria on plant surfaces were reported to be peculiar to the seeds and not commonly found in the soil. The major species cited were *Bacterium herbicola aurem* and *Pseudomonas fluorescens*. (Leben [1961] advanced the term "resident" to replace the less expressive terms epiphytic and commensal.)

310

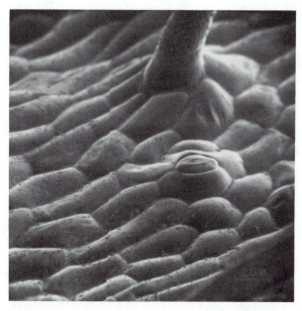

Figure 15-1
The inner surface near the tip of a soybean stipule.
Most of the small projections on the epidermal cells
are aggregates of bacteria. (Courtesy of Dr. Leben,
1969. Reproduced by permission of the National
Research Council of Canada from the Canadian
Journal of Microbiology, vol. 15, pp. 319–320,
1969.)

Most microorganisms on plants are bacteria, among which a
few genera and species of gram-negative bacteria are dominant
(Kroulik et al., 1955). Bacteria are transmitted primarily by the
seed (Stolp, 1952; Leben, 1961), but accompany the plant's
growing tip as its cells divide (Leben and Daft, 1966). The
bacterial cells that do not advance with the growing tip remain
isolated in place and form microcolonies as moisture permits
(see Figure 15-1). They do not exist as a continuous flow of
growth (Leben, 1965; Leben and Daft, 1967). Because plant
surfaces are adhesive and bacterial cell walls are mucilaginous,
much rainwater stays on the plant between rains. Table 15-1
gives numbers of bacteria on several vegetables. Cabbage and
lettuce grow by producing new leaves at the inner, growing tip.
Bacteria reach the newest leaves through the surface moisture
films on the surrounding leaves.

Table 15-1
Representative plate count populations (X 10^6) of several vegetables.

Vegetable	Minimum	Average	Maximum
Green beans	0.5	1–2	60
Cabbage, outer leaves	0.01		2
Cabbage, inner leaves	0.0		0.5
Lettuce, leaf	0.01	1–2	
Lettuce, head, outer leaf[a]	0.001	1–2	5
Lettuce, head, inner leaf[a]	0.0		
Lima bean pods	1.3		4.4
Southern peas, pods		5.0	

[a]Data from Nelson, 1974.

Different locations on the plant and different climatic conditions favor different resident microflora. On wheat, rye, and oats, for example, *Pseudomonas* spp., coliforms, and spore-forming bacteria are dominant in dry weather, the spherical and bacillary lactic acid bacteria during wet weather (Woeller, 1929). On the phyllospheres of plants *Flavobacterium* spp., *Chromobacter, Rhodotorula,* and *Sporobolomyces* spp. share dominance with *Pseudomonas* spp. (Last, 1961). Pathogens such as *Xanthomonas vesicatoria* (Leben, 1963) and *Erwinia* spp. are frequent saprophytic plant residents. In floral structures and on the succulent surfaces of vegetables are found *Enterobacter* spp.; typical and atypical nonfecal enterococci and lactic acid streptococci; *Aerococcus* spp.; pediococci; all species of the subgroups of *Streptobacterium* and *Betabacterium,* and of the genus *Lactobacillus;* yeasts; and molds. Members of the subgroup *Thermobacterium* and of *S. thermophilus* are seldom isolated. *S. bovis, S. equinus, S. salivarium* and the hemolytic, pyogenic streptococci appear not to exist on plant surfaces, although *S. bovis* and *S. equinus* have occasionally been isolated from harvested vegetables. Streptomycetes have not been reported as resident microflora. The colonies of *Streptomyces* spp. that occasionally appear on plates prepared from plant material probably arise from conidiospores transferred mechanically to the plant from the soil. Tables 15-2 and 15-3 are flow sheets for the rapid, tentative identification of gram-negative and gram-positive bacteria isolated from plants.

Table 15-2

Key to gram-negative rods found on plants.

OXIDASE POSITIVE

GLUCOSE

- Fermented (−): *Aeromonas*
- Not fermented
  - Yellow pigment
    - POLARLY FLAGELLATE (+): *Xanthomonas*[a]
    - (−): *Flavobacterium*
  - POLARLY FLAGELLATE (+): *Pseudomonas*[b]
    - (−): *Moraxella osloensis*

OXIDASE NEGATIVE

GLUCOSE

- Strongly fermented
  - PERITRICHATELY MOTILE (+): *Erwinia*, *Citrobacter*, *Enterobacter*, *Vibrio*
  - (−): *Klebsiella*
- Negative or weakly fermented
  - MOTILE (+): *Agrobacter*, *Achromobacter*
  - (−): *Acinetobacter*

[a] Some *Pseudomonas* spp. produce intracellular yellow pigments.

[b] Several species produce water-soluble fluorescein or pyocyanin. *Pseudomonas aureofaciens* produces soluble orange pigment; *P. lemonieri* and *P. indigofera* produce insoluble blue pigment; *P. stutzeri* produces light brown pigment on isolation; remaining species are colorless.

Table 15-3
Key to gram-positive bacteria found on plants.

SPHERICAL BACTERIA

CATALASE

+ FERMENTATIVE
  + Staphylococcus
  – Micrococcus

– CHAINS
    Streptococcus
  TETRADS
    ACIDITY
      BELOW pH 4.5 Pediococcus
      ABOVE Aerococcus

BACILLARY BACTERIA

CATALASE

– Lactobacillus

+ SPORES
  + Bacillus
  – DEFORMED CELLS OR PALISADES

PALISADES AND V FORMATION
  + FERMENTATIVE
    – Kurthia
    + Corynebacterium
  – Arthrobacter

COCCOBACILLARY
  + Microbacterium
  – Brevibacterium

Secondary Microflora

Secondary microflora on plants are microorganisms common to the soil and to human and animal bodies. They are transmitted to plant surfaces by contact with the soil or by wind, dust, rain, floods, irrigation waters, insects, birds, rodents, and humans. During harvest and after they are transmitted from machinery, trucks, reused packing crates, unsanitary conveyor belts, a dirty environment in the packing sheds, and directly or indirectly from human hands and clothing.

Ground-dwelling animals may contaminate plants in the field with intestinal bacteria. Bacteria from warm-blooded creatures do not survive in the soil if introduced in manures during seedbed preparation, but do survive on plants fertilized with night soil. They contaminate plants when fields are flooded, or are irrigated with polluted river waters. Ultimately these bacteria die because they are not competitive with the indigenous microflora, but salmonellae may persist for weeks to months in soil and probably on plants (see Chapter 22).

Pathogens

In the United States the pathogens of major concern on fresh fruits and vegetables are the salmonellae, shigellae, gastrointestinal viruses, *Entamoeba histolytica,* and *Ascaris* spp. The detection of pathogens on vegetables and their potential for contamination and persistence when introduced during irrigation has been demonstrated (Geldreich and Bordner, 1971).

As populations in rural areas have increased, so has contamination of streams receiving raw or partially treated sewage. When irrigation water is taken from below sewage outfalls before the self-cleansing action of the stream has been completed, intestinal microflora are conveyed to plants. Vegetables to be consumed should not be given overhead irrigation if the water is of questionable quality. Geldreich and Bordner (1971), review test data on streams from which irrigation water is taken: salmonellae occur in 53.5% of them when the fecal coliform *(Escherichia coli)* bacteria density averages 1000 viable cells/100 ml water or less a month, but they occur in 96.4% of those with higher monthly fecal coliform averages. For stream or river water to be used for irrigation, the National Advisory Committee on Water Quality established in 1968, the guideline figure of no more than 1000 fecal coliform cells/100 ml water.

Inner Plant Tissue

Unable to recover yeasts from the fluid of unbroken grapes, Louis Pasteur concluded that the inner flesh of grapes was sterile (Fernbach, 1888). After a study of 98 samples of 5 vegetables, Fernbach also concluded that healthy inner plant tissue was sterile. Although Fernbach recovered bacteria from tissue in 35% of the samples, he attributed their presence to contamination. It now seems that Fernbach may have been wrong— that the bacteria indeed are present within plant tissues. Immediately prior to World War II German investigators were engaged in controversy because extremely slow-growing bacteria were reported *within* the sprouts of germinated seeds, particularly in the sprouts of "Puff Bohne," a type of bean not grown in the United States. Meneley and Stanhellini (1972) report the recovery of cells of *Pseudomonas, Erwinia, Bacillus,* and *Xanthomonas* spp. from the inner flesh of potato, cucumber, carrot, celery, green bean, cauliflower, pepper, tomato, broccoli, and cabbage. Aerobic actinomycetes have been reported to be abundant in the phloem and to a lesser extent in the embryonic regions and floral structures of potato, Jerusalem artichoke, beet, carrot, parsnip, and turnips. Growths of *Bacillus megaterium* form threads of infection in the potato. Staphylococci have been reported in tomatoes, and nonfecal enterococci have been isolated from the ovules of corn and green beans. Fruits of the quince, apple, and cranberry frequently contain mold in the endocarp when its blossom end fails to close after fertilization.

Protective Plant Mechanisms

Although plants lack the reticulo-endothelial, antibody-producing mechanism that animals possess, they possess a variety of defensive mechanisms that keep microorganisms out of the susceptible subepithelial tissues, and that prevent invading microorganisms from growing. Subepithelial tissue can be invaded in several ways, among them insect stings, blown sand particles, wind damage such as knockage (violent contact of fruit with wood while on the tree), and abrasions. Dold and Witzenhausen (1953) have suggested that plant cells have a "vital substance" that inhibits microbial growth during life.

The *stratified epithelium* provides a continuous, relatively tough and impervious barrier that functions like animal skin. In lettuce the epithelium may be quite thin and fragile while in the potato it may be *cornified,* or *suberized.* In many fruits waxes that are soft and sticky or are hard complement the mechanical barrier by entrapping and immobilizing microorganisms.

Uninjured cells adjacent to damaged areas become *embryonic* and bring about rapid healing. If injured before the seed has matured, apples become cornified, sometimes observed as unpalatable scar tissue. The parenchymal cells of the damaged white potato become suberized if the temperature is below 20°C. Above this temperature soft rots caused by *Erwinia carotovora* and *Pseudomonas syringae* occur as well as ropy or gassy rots caused by *Clostridium* spp. These rots are frequently pink in color (Jones and Dowson, 1950). If relative humidity is low, the *concentration of soluble solids* of many fruits and vegetables raises osmotic pressure to the point where microorganisms of decay fail to grow. During dry weather, tomatoes on the vine and "ornamental" lemons that crack or split develop across the break a film consisting of sugars, amino acids, and salts. But at high relative humidity infections with *Oospora, Fusarium,* and *Rhizopus* occur.

The **internal** pH of vegetables ranges between 5.5 and 6.4, while that of most fruits ranges from pH 4.0 downward. The internal pH of fruits prevents the growth of microorganisms sensitive to acidic conditions. Most plants contain **antibiotic substances** that are demonstrable in the test tube in dilutions of expressed juices ranging from 1:10 to 1:150 of juice to water. A thermostable, crystallizable substance has been isolated from rye seeds and germfree bran that inhibits many microorganisms (El Shammaa, 1962), and a thermolabile substance has been found in cabbage juices that is inhibitory to several species of gram-negative bacteria (Pederson and Fisher, 1944).

Premature Maturation and Death

As soon as harvested, all succulent and actively respiring fruits and vegetables begin progressive changes termed *accelerated maturation*. Amylolytic and pectinolytic enzymes become active, browning reactions occur, and, in fruits, the titratable acidity may be reduced. Photosynthetic pigments may decrease, disappear, or be masked by anthocyanins. Vegetables and fruits with high respiratory rates rapidly oxidize soluble carbohydrates. Tissues can no longer regenerate the embryonic cells necessary to repair damage. However, harvested commodities do not die the instant they are excised from the plant or the root; proper storage may sustain the functions of living tissues for many weeks.

Storage

The objective of storage is to maintain the vitality of the stored commodities, keeping their appearance and texture attractive, and minimizing physiological and microbiological deterioration. Unlike other types of foods, vegetables and fruits are a very complex group, the members of which cannot be treated alike. Maximum storage life ranges from days, such as for mushrooms, to many months, such as for apples and potatoes. Lutz and Hardenburg (1968) comprehensively reviewed the necessary conditions for maximum storage life.

If they are to be properly stored, commodities harvested in warm weather must be rapidly cooled to remove their field heat. In general, respiratory rates are 6 to 10 times greater at 20°C than at 0°C. Most lettuce and some other vegetables are precooled by *vacuum cooling:* the vegetables are placed in chambers that can be evacuated, and cooling is achieved by evaporation of water at the surfaces. Rapid air currents are used to cool cane fruits and some vegetables that either lose attractiveness or decay rapidly if they remain wet. Leafy vegetables are packed with crushed ice. Large root vegetables and melons are top-iced during shipment; cold air from the melting ice flows through the stacks. Vegetables that are not damaged by contact with water, such as carrots, celery, and radishes, are cooled in washing waters or are *hydrocooled* in iced or refrigerated waters.

Duration of storage is influenced by *relative humidity,* air movement, the respiratory rates of the commodities, sanitation in storage rooms and packing sheds, controlled atmospheres, and the use of antioxidants. If too low, relative humidity causes wilting; if too high, it encourages molds to grow. The optimum relative humidity range is between 85% and 95%. Air circulation is essential to remove carbon dioxide, respiratory moisture, and gases such as ethylene oxide. Respiratory rates are expressed in terms of British thermal units (Btu)/ton/day. At 0°C the average rate of heat production, expressed as Btu/ton/day, is 750 for apples, 650 for cranberries, 1600 for celery; but it is about 10,000 for asparagus, 4400 for broccoli, 9000 for corn, and 14,000 for shelled green peas.

Use of **controlled atmospheres**—that is, deliberately decreasing oxygen and increasing carbon dioxide—increases the storage life of commodities such as apples by as much as one to two months. **Antioxidants,** such as sulfur dioxide for grapes and biphenyl for citrus fruits, decrease decay. Lutz and Hardenburg (1968) consider chlorine in wash waters of little use and ozone in a concentration as low as 0.5 ppm to be injurious to sensitive commodities such as lettuce and strawberries. Table 15-4 gives storage temperatures and lengths of storage life for a few representative commodities.

Many vegetables and fruits store well at temperatures just above freezing or between −1°C and to 0°C, but some suffer chilling injury. Lutz and Hardenburg (1968) list 26 commodities for which the lowest safe temperature is 2.2°C or above, and the lowest safe temperature of 18 of these is 7.2°C and upward. Below the recommended safe temperatures commodities undergo objectionable browning discolorations or pitting, which creates centers of decay.

Vegetables and fruits are divided into three groups according to susceptibility to freeze damage: beets, old cabbage, and rutabagas may undergo several light freezes during storage without harm; apples, new cabbage, cauliflower, and peas may undergo one or two light freezes; but most soft fruits, cucumbers, eggplant, and potatoes are damaged by a single light freeze.

Table 15-4
Recommended storage temperature and normal length of storage of selected fresh fruits and vegetables.

Commodity	Recommended storage temperature, °C	Length of storage
Apples	−1.11–4.44	3–8 months
Currants	−0.56–0	1–2 weeks
Raspberries	−0.56–0	2–3 days
Grapes	−1.11–0.56	3–6 months
Nectarines	−0.56–0	2–4 weeks
Tangerines	−0.56–3.33	2–4 weeks
Green beans	0	7–10 days
Cabbage, late	0	3–4 weeks
Sweet corn	0	4–8 days
Egg plant	7.22	7 days
Lettuce	0	2–3 weeks
Mushrooms	0	3–4 days
Potatoes, late northern	3.33–4.44	5–8 months
Rutabagas	0	2–4 months
Tomatoes, ripe	7.22–10.0	4–7 days

Source: Lutz and Hardenburg, 1968.

Storage Diseases

Storage diseases are microbially induced deteriorations that shorten the expected period of storage or destroy commodities in storage. The development of storage disease is hastened by mechanical injury to epithelial tissue; by bruising, asphyxiation, and sunscald, which damage underlying tissues; and by excess moisture. The agents of deterioration are divisible into three groups: those that are aggressive plant pathogens, infecting the commodity before harvest; those that are opportunistic pathogens, invading through insect stings or mechanical damage, and doing so either in the field or during storage or the steps in marketing; and those that attack weakening and overripe tissues. Cells of *Fusarium* spp. may invade before or after harvest, whereas cells of *Rhizopus* and *Oospora* spp. generally attack late in storage. Most field-infecting agents are totally or nearly inhibited by storage at or below 10°C. Meneley and Stanhellini (1972) suggest that many storage rots previously attributed to external contamination are in fact due to the soft rot bacteria present in healthy tissues that become activated at high temperatures.

Agents of Spoilage

Viruses. The viruses responsible for spoilage are primarily field pathogens that have infected the commodity, usually inconspicuously, and remain undetected during storage or marketing procedures. Leafy vegetables develop mosaics with unattractive blemishes; sweet potatoes develop internal cork; and pears develop stony pits. Unlike bacteria, yeasts, and molds, the viruses do not thrive during low-temperature storage.

Bacteria. Much vegetable spoilage is from bacterial decomposition. Growth occurs at sites of excision and at breaks in the epithelium. Pectinolysis leads to bacterial soft rot by members of the *Pseudomonas* and *Erwinia* spp. Cells of *Bacillus* spp. sometimes develop on decomposing surfaces. Streptococci and lactobacilli produce souring in masses of vegetables rich in sugars.

When vegetables are shredded for salad greens, surfaces and cell juices available for microbial action greatly increase, and rapid wilting and alteration of flavor are evidence of bacterial activity. Greens can be kept fresh longer if, before shredding, they are washed in water containing 50 to 200 ppm available chlorine. The chlortetracycline antibiotics, while effective in extending freshness (Shapiro and Holder, 1960), are both expensive and illegal. Ascorbic and citric acids at 100 ppm and above in the cooling waters prevent brown spots on celery (Johnson, 1974), but they are ineffective as bactericides.

Yeasts. Freshly harvested fruits and vegetables have a mixed yeast flora of oxidative, or nonfermentative, and fermentative types. Yeast populations are carried from the soil, their natural habitat, to plant surfaces, leaf surfaces, the phyllospheres, and the nectaries of flowers by wind, rain, insects, and pollen grains. Commonly encountered on fruits and vegetables are the oxidative yeasts *Rhodotorula* and *Sporobolomyces* spp., both of which are red-pigmented; species of *Cryptococcus* other than the highly virulent pathogen *C. neoformans;* and fermentative yeasts of *Torulopsis, Pichia, Candida,* and *Hanseniospora* spp. It is interesting that although very few sporulating yeasts are isolated from flowers (Phaff et al., 1966), they become dominant yeast flora in fermenting fruit juices.

Yeasts rarely cause spoilage of vegetables—and indeed rarely attack even sweet potatoes and yams, despite their relatively high free sugar contents. The more rapidly growing bacteria and molds are fierce competitors. Although a few yeasts are amylolytic, none of these appear to be spoilage agents of fruits. Most yeasts require simple sugars, usually present in fruits as glucose and disaccharides, but cannot use the more complex carbohydrates. They seldom proliferate on the surfaces of green fruits, and cannot penetrate them. They proliferate in the nectaries of flowers; in tissues exposed through mechanical damage, such as by windfall or knockage; and in juices exposed by split skin or liberated through crushing or the pectinolytic collapse of overripe tissues.

Bell and Etchells (1956) detected polygalacturonase in 20 of 61 species of yeasts among the genera *Debaromyces, Candida, Endomycopsis, Hansenula, Rhodotorula,* and *Zygopichia,* all of which were isolated from brines of cucumber salt stock, and from *Saccharomyces fragilis* and a variant of *S. cerevisiae.* Polygalacturonase produced by the yeasts accounts for the softening of finished cucumber pickles. In rare instances, yeasts are found among the spoilage agents where cellobiose and glucose are liberated by the enzymes of cellulolytic fungi.

Molds. Both fungal field pathogens and normally nonpathogenic molds cause spoilage. The distinction between field pathogens and harvest spoilage fungi is sometimes more arbitrary than real. Members of *Diplodia* spp. are typical plant pathogens that invade the stem of harvested fruits and vegetables. Members of *Fusarium* and *Oospora* spp., generally considered storage molds, grow on exposed and succulent tissue such as that in the cracks of tomato. Although members of the *Mucorales* spp. are not considered field pathogens, they will grow on sheltered fruit surfaces in the field.

Most molds produce very light spores easily wafted into the atmosphere. Although all vegetables bear fungal spores, root vegetables may be heavily contaminated with the spores of the soil saprophytes.

Various species of fungi may be cellulolytic, amylolytic, proteolytic, or pectinolytic; some molds are more than one of these. Very low storage temperatures materially retard or completely inhibit all molds; but unfortunately, some fruits and vegetables stored at these temperatures suffer cold damage not associated with microbial deterioration. Mold growth is directly related to the relative humidity, which affects the available moisture (a_w) at the surface. Thus, commodities stored in the lower range of humidity for the fungus may exhibit no signs of fungal growth or deterioration during short storage, but may be heavily molded after protracted storage. Some molds attack a relatively narrow variety of commodities, while others, such as *Botrytis*, *Colletotrichum, Rhizoctonia,* and *Sclerotinia* spp., attack a wide variety of fruits and vegetables. Although saprophytic molds are said to primarily degrade dead tissue, opportunistic fungi often establish growth on such tissue. Table 15-5 lists molds that commonly cause spoilage of vegetables and fruits.

Table 15-5
Common spoilage molds of harvested fruits and vegetables.

Alternaria	Firm, black rots of citrus, drupe and pome fruits; nail head spot of vegetables and tomato; calyx rot of bell peppers; stem end rot of grapes
Aspergillus	Dry, "charcoal" rots of vegetables; storage mold of improperly dried or stored grains and cereals; black rot of onion
Botrytis	Gray mold rots; infections occur after harvest at stem ends and injured areas; wet weather rot; wide host range
Ceratostomella	Black rot of pineapple, sweet potato

continued next page

Table 15-5 *continued*

Cladosporium	Olivaceous to black rot of cucurbits, drupes, figs; storage fungus attacking aging and damaged produce
Colletotrichum	Anthracnose or bitter rot; brown surface with rotting of internal tissues; wide host range
Diplodia	Field pathogen following citrus, cucurbits, pome fruits and sweet potatoes into storage; browning and shrivelled end of watermelon
Endoconidiophorum	Black rot of sweet potatoes
Fusarium	Conspicuously tufted white to pink stem end and damage rots of all vegetables; may follow bacterial soft rot
Glomerella	Bitter rot of apple
Mucor	Wound infections; non-souring rot; grows below 10°C
Oospora	Storage rots; sour and acrid rots of citrus, tomato, and potato
Penicillium	Blue mold and green mold rots of citrus; *P. expansum* of apple; adventitious on aging and wet commodities
Peronospora	Downy mildew of grapes and currants
Phoma	Dry brown rots of beets and tomatoes
Phornopsis	Rot of lima beans, soybeans and eggplant
Phytophthora	Field infection continuing into storage; water-soaked, rusty-tan blemishes; leathery and brown spots
Pythium	Damping-off of many young seed plants
Rhizoctonia	Widespread field pathogen accompanying fruits into storage; crown and dark "charcoal" rots
Rhizopus	Late storage rot; "leak" in cane fruits; waterbag of tomato; often distinguished by sour odor; little growth below 10°C
Sclerotinia	Also Monilinia; probably most damaging fungus of stone fruits such as peaches and cherries; deeply penetrating brown rots with little surface mycelium; field infecting agent
Septoria	Leaf spot of pear, raspberry, tomato, late blight of celery
Trichoderma	Green mold and black rot of *Brassicae*
Trichothecium	Pink mold of fruits, cucurbits

Source: Duckworth, 1966.

Detection of Mold in Finished Products

Because molds seldom produce chemical or organoleptic changes detectable in the finished product, small fruits and tomatoes that are heavily molded can nevertheless be made into finished products, such as jams, preserves, catsup, and chili sauce, acceptable to the consumer. However, the practice is esthetically and toxicologically undesirable because the raw products are decomposed and the molds may produce mycotoxins.

Since the mold filaments remain intact during cooking, they can be observed in suitable preparations under the microscope (Howard, 1911; Howard and Stephenson, 1917). The product to be observed is mounted on a special slide, the Howard mold slide (Figure 15-2), which consists of a central circular or rectangular well surrounded by a moat, two bars to support the cover slip at an elevation of 0.1 mm above the surface of the well, and the cover slip itself. The material is examined through a microscope with a standardized field diameter of 1.382 mm. The volume of each field examined is 0.0015 mm^3 of material.

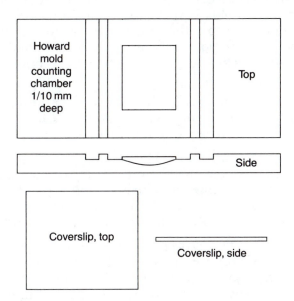

Figure 15-2
Howard mold counting chamber and coverslip.

An ocular disc fitted to the microscope divides the field into 1/6 diameters. A field is positive for mold if one filament or the combined lengths of two filaments or of three portions of filament are equal to or exceed 1/6 the field diameter. Usually, 25 fields are examined, the number of fields containing mold is recorded; and the result is expressed in percent positive fields. Molds are known to establish themselves on fruits in the field, but their presence is not taken per se as a sign of contamination. However, government guidelines establish the permissible use of the raw product and the graded quality of the finished product that reflect different levels of tolerance among different fruits.

Molds are distinguished from plant tissues and cells by one or more of the following observable properties:

1. *Flow of protoplasm.* In genera of Mucorales spp., and especially in *Rhizopus* spp., granules flow in the filaments of living cells toward the growing tip in preparations of fresh fruits.

2. *Segmentation.* Except for members of the genera of Mucorales, molds have septate filaments. Some molds have delicate septa that are not easily found, but the boundaries of most cells are readily observed in the filaments. Trichomes, or plant hairs, are not septate.

3. *Granulation.* Mold cells contain granules, although in some molds, such as of *Oospora* spp., the granules are so small as to be inconspicuous. Plant tissue cells and trichomes are never granulated internally.

4. *Parallelism of side walls.* The observable sides of mold cells and filaments are continuously parallel. Exceedingly long trichomes may seem to have parallel sides, but they taper to a needle point at one end and may flare at the site of attachment at the other end.

5. *Focal plane of the side walls.* Both sides of the mold filament are visible for the entire length. Epithelial cells frequently appear to be mold filaments, but one or another of the ends disappears, giving the impression of rolling under the cell.

6. *End of the filament.* The end of the mold filament is always blunt and roughly square, while trichomes are needle-pointed at one end.

Accurate mold counting requires that plant tissue cells be recognized and identified. The more prominent cells are the large, flat, irregular epithelia. Epithelial cells frequently exhibit false parallelism for as much as half of their circumference. Although similar in appearance to epithelial cells, parenchymal cells have thinner cell walls. Trichomes may be simple or branched; their major and the minor tips are fusiform, or needle-shaped; they are not granulated; and their sidewalls do not exhibit parallelism. Tracheid cells are spirillar in shape, resembling a coiled spring. The spirals may be very short or very long. Clusters of very narrow, tubular cells with comparatively thick sidewalls are the supporting cells of the vascular bundles. Ribbonlike, twisted cells are fibers of cellulose from the lens paper used to clean the glassware.

A single Howard mold count procedure does not give precise results; the standard deviation decreases as the total number of fields observed increases. Spot samples are less reliable than thoroughly mixed samples of volumes at or above 20 ml. Not all analysts obtain normal distribution curves (Vas et al., 1959), and the reliability of even trained analysts varies from day to day. Mold counts of a finished product not only correlate poorly with the mold count of the raw product but also vary greatly among themselves because of inherent variabilities in the raw product (Esselen, 1948). Washing the raw product with suitable detergents removes surface molds (Haynes et al., 1953) but not deeply penetrating molds.

The Howard Mold Count Method

The Howard mold count method has been adapted to determine the amount of mold in the finished products of many commodities such as fruits, juices, and vegetables. All plant material has a resident mold population, some of which accompanies the material through processing. Mold counts above normal or background level are interpreted to signify either unsound raw material or unsanitary processing.

All tomato processing plants use the method to determine the quality of tomatoes used. Tomatoes are easily bruised and cracked during harvesting and post-harvest handling, and become even more susceptible to cracking as they ripen and the firm tissue softens by action of self-contained pectinase.

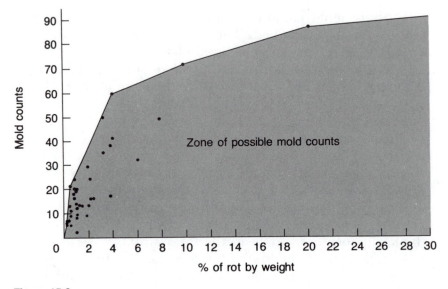

Figure 15-3
Relation, in factory samples, between percentage, by weight, of rot and mold
count. (Howard and Stephenson, 1917.)

Cracked and broken tomatoes are immediately contaminated
with molds of *Geotrichum candidum,* which thrive and bring
about rapid spoilage. Howard and Stephenson (1917) showed
that as little as 2% of decayed tomatoes by weight always yields a
mold count of at least 37%, and that 4% of decayed fruit by
weight yields a mold count of not less than 60% positive fields
examined. (See Figure 15-3.)

Fresh tomatoes to be examined are put in a Waring blender or passed through a cyclone, a conical perforated sleeve
equipped with a motor-driven inner rotor. The rotor forces the
tomato pulp through the performations and the skins and
seeds emerge at the end of the sleeve. Canned tomatoes are
drained, the flesh is pulped with a cyclone, and the pulp and
juice are examined separately. Tomato juice to be examined
needs no preparation. Tomato puree and catsup are adjusted
with water to a refractive index of 1.3448 to 1.3454 at 20°C.

The Compendium of Methods for the Microbiological
Examination of Foods (Speck, 1976) gives specific procedures
for examining other fruits, juices, and some canned foods.

Ades and citrus concentrates, pineapple juice concentrate, and fresh grape pulp are mounted directly on the Howard cell. Apple butter, cranberry sauce, jams, berry fruit preserves, heat-processed grape pulp, and catsup are diluted with a stabilizer made with sodium carboxymethylcellulose, pectin, or algin. Apple pomace and canned or frozen raspberries, blackberries, and strawberries are pulped in a Waring blender or by passage through a cyclone before dilution with stabilizer. Capsicums, garlic powder, infant foods, and tomato soup are treated with alkali to dissolve starches. Undiluted citrus juice to be examined is centrifuged and the precipitate is resuspended in water and stabilizer.

Machinery Mold

Geotrichum candidum is also called the machinery mold. Its cells grow on surfaces of harvesting machinery that are wetted by plant juices. In processing and bottling plants it grows on all surfaces that are wetted by the materials being handled, such as conveyers, metering and filling devices, flumes, and valves. It also thrives in refuse not removed immediately from the processing and sorting areas. It grows as a tenacious slime from which fragments break to enter the product. In heavily contaminated plants or in areas where processing refuse accumulates it becomes conspicuous by its acrid odor. To be detected, sufficient amounts of the mold must be accumulated. This is done by draining liquids or washings of solids on a fine sieve, staining the material remaining on the sieve with crystal violet, and examining it on rot fragment slides or hemocytometers.

Characteristics of Fruit and Vegetable Types

Vegetables

Ovular and Podded Vegetables. To consume English and Southern, or field, peas and lima, butter, and shellie, or October, beans, the pods are removed and the green ovules in the intact pod are prepared and eaten. The ovules are surface sterile. They may

Figure 15-4
A mobile viner harvesting Southern or black-eye peas. The
harvested peas are gathered in lug boxes beneath the cover
immediately above the wheels. This viner will harvest about three
acres and gather about 4500 pounds of peas per day. Note the
harvested field in the background.

contain a variety of bacteria, probably carried into the ovule
with the growing pollen tube. Usually only one species of bacteria is present. It is highly questionable whether any of these
bacteria play a role in the health or disease of plants, or in post-
harvest spoilage.

Stationary or mobile viners (Figure 15-4) separate the podded ovules from the pods. During use, the operating surfaces
of the viners become covered with thick films of cell juices
mingled with dirt in which bacteria grow freely. During the
relatively rapid passage over the operating surfaces, the vined
vegetables acquire bacterial loads ranging from the low
thousands to millions per gram. The bacteria continue the
logarithmic rate of growth on the vegetable surfaces. Moreover,
although fungi are a very minor problem at this stage, molds
develop on the films when the equipment is not operational and
has not been cleaned.

All modern harvesting operations are performed in the field,
and the peas or beans are transported to processing plants in
hoppers, crates, or in bulk on trucks. The rapidly respiring,

thin-coated, succulent English peas must be processed rapidly to prevent intrinsic and microbial deterioration (Talburt and Legault, 1950). Rapidly developing anaerobic conditions within the container encourage the growth of lactobacilli. Lima and butter beans and Southern peas, harvested at a later stage of maturity than are English peas, are less subject to rapid intrinsic and microbial deterioration. *Leuconostoc mesenteroides,* aerococci, and pediococci become dominant, especially if the peas and beans are moist, or become so. *Achromobacter* spp. cause shelled lima beans to become sticky. Various species of *Uromyces,* or rists, cause the reddish-brown discoloration, spotting, and streaking on legume pods, particularly on those of green beans; making them unmarketable. Some species of *Phomopsis* cause serious rot of lima beans and soy beans.

A significantly higher percent of machine-harvested beans undergoes breakage than do hand-harvested beans. The damaged surfaces develop water-soaked, soft rot (Figure 15-5), which spreads to sound beans during storage. Growth of mold frequently follows bacterial soft rot.

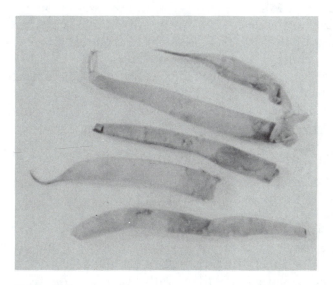

Figure 15-5
Soft rot of beans, evident 24 hours after harvest. These beans were crushed and broken during machine harvesting.

Corn. Corn remains in the husk until the beginning of processing. The husking, cutting, desilking and transporting equipment rapidly become coated with cell juices that support prolific growth of bacteria. Bacteria remain on corn a much shorter time than on podded vegetables because corn is rapidly moved into the blanchers (Wolford et al., 1965).

Leafy Vegetables. Spinach, chard, kale, mustard, dandelion, lettuce, and celery are high in water content and low in firm tissue, have high respiratory rates, and are easily bruised, broken, or crushed during harvesting and packing. Intact surfaces prevent rapid microbial decomposition after harvest, but bacteria proliferate where the leaf has suffered mechanical breakage. *Erwinia carotovora* is an important spoilage organism of leafy vegetables. The vegetables are spoiled almost exclusively by bacterial soft rot, except for that caused by *Sclerotinia sclerotiorum,* the cause of lettuce leaf drop and celery rot. These vegetables are well adapted to cooling with waters and storage when packed with crushed ice.

Cabbages, cauliflower, broccoli, and brussel sprouts store well in dry atmospheres. The storage life may be limited by the softening of the tissues, as seen in cauliflower, or yellowing of brussel sprouts. The vegetables are susceptible to field infection by *Alternaria* spp., which begins as an inconspicuous, tiny spot termed nail-head spot. The spots enlarge and coalesce into large unsightly black semisoft blemishes on the outer leaves (Figure 15-6). Occasionally *Phoma oleracea* is responsible for serious attacks.

Sound, mature cabbage may store well for several months, but as storage progresses *Erwinia carotovora* causes bacterial soft rot and *Rhizopus* spp. cause mushiness.

Root Vegetables. All common root and underground stem vegetables such as beets, turnips, salsify, parsnips, carrots, and white and sweet potatoes are firm, with an outer coating of dense, sometimes suberized, epithelium. At maturity all have low respiratory rates. Dehydration causes a shriveled appearance,

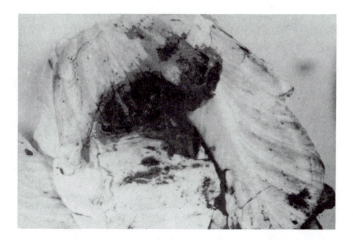

Figure 15-6
Nailhead or black spot of cabbage caused by *Alternaria* spp.
The tiny black spots, initial sites of infection, enlarge
rapidly and coalesce to form large, rotted black areas such
as that in the center of the leaf.

and considerable loss on the fresh market. Fungal deterioration
develops during prolonged storage, especially when there is
inadequate ventilation, high temperature, or both.

Irish or white potatoes are susceptible to bacterial soft rot
caused by *Erwinia, Pseudomonas,* and *Clostridium* spp. Bacterial
soft rot from *Clostridium* spp., is characterized by gassy, malodorous spoilage. At the relatively high temperatures during harvest and shipment the bacteria grow more rapidly on
damaged surfaces than cells grow to repair the damage. *Alternaria* spp. produce a purple-to-brown superficial cortical rot.
Alternaria solani causes early potato blight, *Phytophthora infestans*
late blight. In 1846 *P. infestans* brought about the great potato
famine in Ireland—a famine so severe that thousands starved
and thousands more emigrated to the United States. *P. infestans*
survives in the tubers, and the fungus grows into the new tissues sprouting from the potato and sporulates on portions of
the plant above ground. Production of sporangia is much influenced by temperature and humidity.

Several species of *Fusarium* produce a dark-brown-to-black, dry tuber rot that penetrates deeply during the low temperature of storage. Fusaria are believed to be field parasites and not wound-infecting agents. Western potatoes are susceptible to jelly end rot produced by *Rhizoctonia* spp., which may also produce hard, black masses of sclerotia. The commonly seen black heart of winter potatoes is darkening caused by freezing injury or by bruising, which causes production of melanin.

The sweet potato is one of the few commodities, if not the only commodity, in which a virus infection contracted in the field continues to develop after harvest and during storage. The virus causes a condition known as internal cork. Sweet potatoes are also susceptible to black rot caused by *Endoconidiophora fimbriata,* which may be stopped by briefly heating the potatoes to between 40.6 and 43.3°C. Fungi of *Rhizopus* spp. cause mushy, soft rot. Fungal filaments intertwining among many potatoes cause them to cling together, a condition known as *nesting.* Other fungi produce firm black dry rot, shriveling, and blackening through the formation of pycnidia or sclerotia. Harvested sweet potatoes are cured by storage for several weeks at or above 21.1°C, to harden and dry the surface tissue, after which they are stored at 12.8°C.

Cucurbits

Cucurbits—melons, pumpkin, squash, gourds, cucumbers—mature during warm to hot weather, deteriorate rapidly after harvest, and have a relatively short storage life. In general, the surface is relatively thin and is easily pitted and abraded or damaged during handling. In the pits, tiny recessed areas, grow the bacteria which cause soft rots and the rapidly growing *Fusarium* and *Rhizopus* spp. Cucurbits maturing in middle to late fall develop a hardened surface during the final stages of maturation that allows prolonged storage far into the winter months under cool, dry conditions.

Alternaria is a general agent of storage decay among cucurbits. *Cladosporium* and *Diplodia* spp. are reported to cause stem end rot of cantaloupe and watermelon, respectively. Blossom end rots of squash and watermelons are caused by *Choanephora* and *Macrophomia* phaseoli respectively. *Mycosphaerella* spp. infect melons and squash in the field and growth continues to produce black rots after harvest. *Penicillium* spp. and bacteria cause soft rots of cantaloupe.

Tomatoes

In every field of tomatoes some mold grows. It is minimal during dry, perfect harvest seasons. If rains follow prolonged dry periods, tomatoes enlarge rapidly and develop cracked skins. If weather after the rains is dry, with low relative humidity, the soluble solids form a glaze and no spoilage occurs. During more humid seasons, *Fusarium* spp. and *Geotrichum* sp. grow in the cracks, the latter microbe producing a peculiarly acrid odor. The fruit fly *Drosophila* deposits its eggs in the infected areas and the newly hatched flies distribute cells of *Geotrichum* to freshly cracked fruit. Tiny chewing beetles also distribute these molds to internal tissue via their mouthparts. Cells of *Rhizopus* spp., probably deposited during insect puncture, pectinolyze the contents of the ripened tomato on the vine, converting it to a "bag of water." Field infections by *Phoma* continue to develop in the harvested tomato, producing concave, watery areas; *Alternaria* spp. are responsible for the black "nail-head spot," while *Collectotrichum* spp. produce anthracnose and cause shallow, dark-brown-to-tan, concentrically ringed spots. *Rhizoctonia cinerea* commonly cause rot during wet weather, particularly of tomatoes touching the ground.

Garlic and Onion

During storage *Erwinia* and *Botrytis* spp. attack garlic and onion. *Fusarium* spp. cause bulb end rot. Members of *Pseudomonas cepacea* are frequent spoilage agents of onions. Cells of *Aspergillus alliaceus,* a specific agent of decay producing black mold rot, are seldom seen on other commodities.

Small Fruits

Cane Fruits and Strawberries. Cane fruits and strawberries have minimal, delicate epithelial tissue. In normal climatic conditions green and ripening berries have little mold, but mold grows rapidly on fully mature berries and spreads from box to box. The fruits have high respiratory rates, and wooden berry boxes easily accumulate moisture, raising the relative humidity within them. The boxes themselves may be the source of mold spores. The newer plastic containers with netted bottoms are more sanitary than wooden boxes and in addition facilitate air movement among the fruit within.

During wet weather sugars leach through the epithelium of the ripening berry and cane in direct proportion to amount of rainfall and relative humidity. The sugars support the growth of both harmful and harmless fungi. On the strawberry these can be seen beneath the sepal—but only after picking. On cane fruits they can be seen within the cup and sometimes on the surfaces themselves while still on the plant. Although nearly 20 species of molds have been isolated from cane fruits, the most frequently seen are the green-to-black *Cladosporium* spp., which are unsightly but do little damage, and *Botrytis,* which cause a soft, watery rot. *Rhizopus* spp. develop in overripe berries that fall to the ground spontaneously. These berries, however, are not harvested. The fruit fly deposits enormous numbers of eggs in overripe and leaking fruit, and the adult flies and later the larvae become a major means of spreading mold spores. During the warm weather usually prevailing during cane fruit harvesting, the life cycle of the fruit fly is as brief as eight days.

Members of *Phytophthora* and *Rhizoctonia* are wet weather fungi of strawberries that invade berries that touch the ground, by penetrating the unbroken skin. *Botrytis* is a gray mold that turns the invaded area brown. *Rhizopus* spp. invade ripe, overripe, or waterlogged berries; the berry structure collapses and a condition known as leaking occurs. *Dendrophora* spp. cause stem end rot and *Colletotrichum* spp. cause bitter rot in strawberries.

Cane fruits and strawberries are among the few commodities in which the quality of the raw or processed product is determined visually, by the Howard mold count. The count of freshly harvested berries is directly related to the climatic conditions during ripening. The mold count ranges between 4 and 8% for dry-weather berries and 48% or more for wet-weather berries, with a harvest season of frequent rains and high relative humidity. Thorough washing removes roughly half the mold, which consists of surface-resident fungi that have not penetrated the epithelium.

Grapes. Various fungi of the genera *Alternaria, Aspergillus, Botrytis, Dematium, Guignardia, Mucor, Penicillium, Phomopsis, Plasmapara,* and *Sclerotinia* are often observed on grapes. Some fungi, notably *Penicillium* and *Mucor* spp., grow primarily on berries with split skins. The native American downy mildew *Plasmapara viticola* is of historical interest: accidentally introduced into France on American grape stock imported to resist the root aphid *Phylloxera,* the fungus attacked the stem, leaves, and fruits of the *vinifera* grape plants. During wet years it was very destructive until, in 1882, Millardet developed what is called Bordeaux mixture. Even today vineyard attacks by *P. viticola* are disastrous unless *vinifera* grapes are sprayed.

To prevent the spread of *Botrytis* it is standard commercial practice to fumigate stored grapes at 10 day intervals with 0.25% sulfur dioxide (Harvey, 1956; Harvey and Pentzer, 1960), and to add sodium bisulfate to packaged grapes during storage and long-distance shipment (Pentzer, 1939). *Botrytis cinerea* is the most common contaminant of the mature fruit. Often when grapes are left on the vine as long as possible to ripen adequately *B. cinerea* penetrates the healthy skin, to obtain nourishment from the juice. When very humid weather is followed by less humid, grapes thus attacked become partially dehydrated. In the past, such sugar concentration was mistakenly considered desirable, since grapes too sour for wine were thereby made sweet enough for the purpose. The "noble mold" was encouraged to grow on white wine grapes to be used for French Sauternes and other concentrated wines. The oxidative enzymes of the mold cause red wines made from infected grapes to be weak and off-colored. As it grows *B. cinerea* reduces the tannin content, destroys anthocyanins, and lowers the total phenol content. In its perfect stage the fungus is identified as *Sclerotinia sclerotiorum,* which is associated with the production of psoralen compounds.

Black rot, caused by *Alternaria* spp., generally first attacks a single grape, which becomes shriveled and mummified. Mucors, cottony and hairy, grow in profusion over the entire cluster; such attacks appear to be extensive but do only superficial damage.

A serious problem of grapes during storage, transportation, and at the market is stem end rot, often caused by *Phomopsis* spp., but also caused by *Aspergillus, Aureobasidium, Botrytis,* and

Penicillium spp. It is of interest that in 2 of 33 wines tested aflatoxin was detected (Schuller et al., 1967). Presumably, contamination arose from infection of white wine grapes that were harvested late.

Cranberries. Molds invading the fruit cavity of the ripening cranberry may grow to produce spoilage after harvest. Wet storage, high temperatures, long holding periods, and sweating encourage molds to grow. The common molds and the spoilage they produce are *Godronia cassandrae,* stem end rot; *Guignardia vaccinii,* early rot; *Glomerella cintulata,* bitter rot; *Sporonema oxycocci,* ripe rot; *Diaporthe vaccinii,* fruit storage rot; and *Centhospora lunata,* black rot. Often the mold grows only internally and the fruit may remain firm in texture but develop a brownish discoloration; or the mold may digest the contents of the skin.

Large Fruits

Most large fruits are harvested while quite green to enable them to withstand the rigors of transportation and storage. As such, they are not readily bruised. Enzymatic changes leading to browning, softening, and changes in flavor occur in bruised areas, and fungal rots, initiated at the bruised sites, penetrate and spread the spoilage. Often rots are initially firm and marked by conspicuous brown discolorations, but soften during later stages, at which time fruiting structures characteristic of the mold develop over the affected surfaces.

Stone Fruits. Monilinia (Sclerotinia) fructicola causes brown rot of peaches, plums, and other stone fruits. Conidia or ascospores invade blossoms, twigs, or leaves, where they germinate and produce mycelia. From these growths come long, branched conidiophores bearing chains of conidia, which wind and insects often carry to other trees. During wet years or humid weather *M. fructicola* causes enormous losses of marketable fruit. During storage of ripe peaches, brown rot often reduces the number of sound fruits by 5 to 25% overnight. As the green fruit ripens, the fungus enters via insect punctures, hair sockets, or wounds. The fungal mycelium softens and invades the surrounding tissue, penetrating the entire fruit, which eventually shrivels and mummifies and, if not removed, may be a source of brown rot in the next season. Other fungal agents of spoilage during storage are *Botrytis, Rhizopus, Alternaria* and *Penicillium* spp.

Pomes. Most spoilage of apples in storage is caused by *Penicillium expansum.* This soft brown rot spreads rapidly from apple to apple. *P. expansum* produces the mycotoxin *patulin,* which is found in apple products such as ciders if pressed from molded fruit. Apples going to the presses are now given rigorous inspection to eliminate rotten fruit.

Pink rot of apples, generally following scab infection, gets its name from the masses of pink conidia of *Trichothecium roseum.* *Sphaeropsis malorum,* the imperfect stage of *Physalospora cydomiae,* causes black rot; *Volutella fructi,* dry rot; and *Glomerella cingulata,* bitter rot and anthracnose. *Rhizopus* spp. develop on weakened, overmature, or damaged fruits, not on sound apples. *Botrytis* and *Penicillium* spp. (pinhole rot) destroy pears. Apples and pears are protected by being wrapped individually in wrappers containing mineral oil and copper sulfate.

Citrus. Penicillia are the major spoilage agents of citrus fruits. *P. digitatum,* often called the "green mold" or "green rot," causes serious losses of lemons and oranges, while *P. italicum,* the "blue mold" or "blue rot," attacks grapefruit and oranges, but both species attack all citrus fruits. From a small, localized infection *P. digitatum* grows rapidly, taking several days to cover the entire surface of the fruit with a dense, olive-colored layer of mycelium and powdery conidia. Fruit tissues become soft and the sections lose their integrity. Eventually the fruit becomes watery and rotten. *P. italicum* is blue-green and velvety, with tufted conidial masses covering the fruit rind.

Other organisms beside penicillia attack citrus. *Diplodia natalensis* causes destructive stem end rot of citrus; *Phytophthora citrophthora* produces brown rot of lemon. *Botrytis* and *Oospora* may develop on fruit after lengthy storage. *Sclerotinia* is a cottony contact decay. *Alternaria,* entering at ripening or ripened buttons of lemons, sometimes causes serious losses. To retard fungal deterioration citrus fruits may be treated with sodium orthophenylphenate or with diphenyl.

References

Alexopoulos, C. J. 1962. *Introductory Mycology.* 2nd ed. John Wiley & Sons, New York.

Bell, T. A., and J. L. Etchells. 1956. Appl. Microbiol. *4*, 196.

Burri, R. 1903. Zentralbl. f. Bakteriol. u. Parasitenk. II. Abt. *10*, 756.

Dold, H., and R. Witzenhausen. 1953. Zentralbl. f. Bakteriol. u. Parasitenk. I. Orig. *160*, 217.

Duckworth, R. B. 1966. *Fruits and Vegetables.* Pergamon Press, New York.

Dueggeli, M. 1904. Zentralbl. f. Bakteriol. u. Parasitenk. II. Abt. *13*, 56.

El Shammaa, Z. A. 1962. Zentralbl. f. Bakteriol. u. Parasitenk. I. Abt. Orig. *183*.

Esselen, W. 1948. Fruit Prod. J. *28*(1), 6.

Fernbach, M. A. 1888. Ann. Inst. Pasteur *2*, 567.

Geldreich, E. E., and R. H. Bordner. 1971. J. Milk Food Technol. *34*, 184.

Harvey, J. M. 1956. Phytopathol. *46*, 690.

Harvey, J. M., and W. T. Pentzer. 1960. U.S. Dept. Agric. Handbook *189*, 1.

Haynes, R. D., H. Harlin, J. O. Mundt, and R. Stokes. 1953. Food Engr. *27*(3), 50.

Howard, B. J. 1911. U.S. Dept. Agric. Bureau Chem. Circ. 68.

Howard, B. J., and C. H. Stephenson. 1917. U.S. Dept. Chem., Bull. 581.

Johnson, C. E. 1974. J. Food Sci. *39*, 678.

Jones, D. R., and W. J. Dowson. 1950. Ann. Appl. Biol. *37*, 563.

Kroulik, J. T., L. A. Burkey, and H. G. Wiseman. 1955. J. Dairy Sci. *38*, 256.

Last, F. T. 1961. Mycol. Trans. *44*, 300.

Leben, C. 1961. Phytopathol. *51*, 553.

Leben, C. 1963. Phytopathol. *53*, 778.

Leben, C. 1965. Canad. J. Microbiol. *11*, 677.

Leben, C., and G. C. Daft. 1966. Canad. J. Microbiol. *12*, 1119.

Leben, C., and G. C. Daft. 1967. Canad. J. Microbiol. *13*, 1151.

Lutz, J. M., and R. E. Hardenburg. 1968. *The Commercial Storage of Fruits, Vegetables, and Florist and Nursery Stock.* U.S. Dept. Agric. Agric. Hdbk. No. 66.

Meneley, J. C., and M. E. Stanhellini. 1972. Phytopathol. *62*, 778.

Nelson, F. A. 1974. Amer. Soc. Microbiol. Ann. Meeting. Abstr.

Pederson, C. S., and P. Fisher. 1944. N.Y.S. Agric. Exp. Sta. Tech. Bull. 273.

Pentzer, W. T. 1939. Blue Anchor *16*(7), 2.

Phaff, H. J., M. W. Miller, and E. M. Mrak. 1966. *The Life of Yeasts.* Harvard Univ. Press., Cambridge, Mass., p. 108.

Schuller, P. L., T. Ockhuizen, T. Werringoer, and P. Marquardt. 1967. Arzneimittel-Forsch. *17*(7), 888.

Shapiro, J. E., and I. A. Holder. 1960. Appl. Microbiol. *8*, 341.

Speck, M. L. 1976. *Compendium of Methods for the Microbiological Examination of Foods.* Amer. Public Health Association. Washington, D.C.

Splittstoesser, D. F., M. Wilkinson, and W. J. Harrison. 1972. J. Milk Food Technol. *35*, 399.

Stolp, H. 1952. Arch. f. Microbiol. *17*, 1.

Talburt, A. A., and R. R. Legault. 1950. Food Inds. *22*, 1021.

Thaysen, A. C., and L. D. Galloway. 1930. *The Microbiol. of Starch and Sugars,* Oxford Univ. Press, London.

Vas, K., I. Fabri, N. Kutz, A. Lang, T. Orbangi, and G. Szabo. 1959. Food Technol. *13*, 318.

Woeller, H., 1929. Zentralbl. f. Bakteriol. u. Parasitenk, II. Abt. *79*, 173.

Wolford, E. R., A. D. King, Jr., and H. D. Michener. 1965. J. Milk Food Technol. *28*, 183.

16

Dairy Products

Fluid Milk

Milk is the natural secretion of the mammary glands of animals. The milk begins to flow at the birth of the young and normally continues throughout nursing of the newborn. Not only is milk the nursling's principal food, but in its natural, unaltered state it protects the health of the young, if nursling and nursing mother are of the same species (Jeliffe and Jeliffe, 1971). The newborn of domesticated dairy cattle are weaned shortly after birth and the milk is used for human consumption. Milk is produced in the lacteal glands (LG in Figure 16-1), from which it flows through tubules into storage areas termed cisterns, CG. During milking the orifice normally closed by the sphincter muscles of the teat releases the milk.

Milk may be clarified, homogenized, and enriched with vitamin D. Clarification is a mild centrifugation of the milk to remove some of its mucins and leucocytes and some of the bacteria that adhere to them. High-speed centrifugation,

Box 16-1

In the United States most milk comes from cows, and a small amount comes from goats. In other parts of the world milk is taken from the domesticated animal prevalent in the area, such as the sheep, goat, or water buffalo. The average American dairy cow produces about 45 pounds of milk per day and about 15,000 pounds of milk during 9 months of lactation. By contrast, the white mouse, milked for studies of immunological factors, produces about 0.7 ml milk.

The United States Public Health Service defines milk as "the lacteal secretion practically free of colostrum, obtained by the complete milking of one or more healthy cows, which contains not less than 8.25 percent milk solids-not-fat and not less than 3.25 percent milkfat" (Schalm et al., 1971). Colostrum, the milk produced during the first few days of lactation, differs from ordinary milk in being richer in leucocytes and components from the blood that furnish protective factors from the mother, and having lower milkfat and casein contents. The content of milk fat (lipid) produced by various mammals differs considerably (see Table 16-1).

A working average for some breeds of cattle is 3.25% milk fat, but some animals produce less. Dairies adjust the fat content of milk for the purpose of uniformity. For dietary purposes, some consumers prefer milk with 2% milk fat, or skim milk with the fat removed. Recombined milk is skim milk to which fats of vegetable origin are added.

Table 16-1
General Composition of milk from different mammals.

Mammal	Composition, %					
	Water	Protein	Lipid	Lactose	Ash	Solids
Goat	86.9	3.7	4.1	4.5	0.8	13.1
Sheep	83.5	5.2	6.2	4.2	0.9	16.5
Water buffalo	76.9	6.0	12.5	3.7	0.9	23.1
Horse	89.1	2.7	1.6	6.1	0.5	10.9
Pig	83.9	7.2	4.6	3.2	1.1	16.1
Reindeer	66.3	11.1	18.7	2.7	1.2	33.7
Whale	62.3	12.0	22.2	1.8	1.7	37.7
Rat	68.5	11.3	14.8	3.9	1.5	31.5
Elephant	81.3	2.5	9.1	8.6	0.5	18.7
Human	87.5	1.1	4.5	7.0	0.2	12.8
Cow	87.0	3.1	4.1	5.0	0.8	13.0

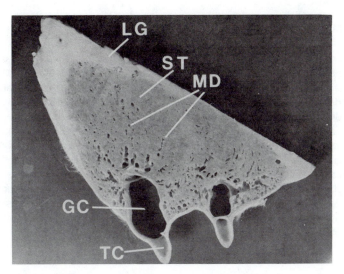

Figure 16-1
Cross section of the bovine udder. LG: lymph gland; ST:
secretory tissue; MD: milk ducts; GC: gland cistern;
TC: teat cistern. (Courtesy of E. W. Swanson, University
of Tennessee.)

suggested as a means of removing pathogens and saprophytic
bacteria from milk, has not gained favor by either the regulatory agencies or the dairy industry. Homogenization is a process
that shears the fat droplets—which range from as large as 20
microns in diameter to 5 microns or less—to stabilize the suspension of fat in the milk and to prevent the cream layer from
forming. By shearing microcolonies, homogenization increases
the plate count of milk but does not increase the bacterial numbers. Vitamin D is added to milk as a sterile concentrate.

Casein, the protein predominant (2.5% by weight) in cow's
milk, is a colloidal mixture of phosphoproteins that precipitate
from milk at room temperature at the isoelectric point of pH
4.6. When acted on by rennin the molecule splits into two
molecules of paracasein, which immediately react with calcium
to form a soft coagulum. Lactalbumin and serum globulin are
coagulated by boiling and are precipitated from whey, the fluid

---

Box 16-2

Composition of milk.   Milk, a fluid of heterogeneous composition, contains over 100 identified constituents (Kon and Cowie, 1961). Its average moisture content is 87% water. All constituents are surprisingly constant in quantity except milk fat and water; the percent of water changes inversely with the amount of triglycerides, which range between 3.2 and nearly 5%. The solid components consist of colloidally dispersed and water-soluble proteins; nonprotein nitrogenous compounds; lactose; traces of other sugars, triglycerides, sterols, and carotinoids; Vitamins A, D and E; the water-soluble vitamins; minerals; and somatic cells. The lacteal glands synthesize casein, lactalbumin, lactose, lactoglobulin, and the triglycerides; the blood secretes the other components through the membranous lining of the lacteal glands.

---

residue of milk after the casein has been removed in cheesemaking. The albumins are used in making such whey cheeses as ricotta, primost, gjetost, mascarone, and Schottenziger. Lactoglobulin and serum globulin are stable to heat and to acid treatment. The nonprotein, nitrogenous constituents include between 0.1 and 0.01% ammonia, amino acids, urea, purines, pyrimidines, and hippuric acid. Enough nonprotein constituents are present to enable the indefinite growth of species of bacteria such as the lactic acid bacteria, which do not elaborate proteolytic enzymes.

Nearly all of the lipid fraction of milk consists of a mixture of triglycerides. The sterols, carotinoids, and Vitamins A, D, and E account for less than 0.05% of the weight of milk. The fatty acids of the triglycerides range in size from 4 to more than 20 carbon atoms. Less than 2.5% of the acids have branched chains. The fatty acids are synthesized from acetic acid, produced in the rumen by bacterial fermentation.

Palmitic and stearic acids, the major fatty acids, are nearly 40% of the total fatty acid content. Oleic and vaccenic acids, each with one unsaturation group, account for nearly 30%. The others are, in descending order of quantity, caproic, capric, and caprylic acids. A small quantity of fatty acids has from two to five double bonds, which are easily oxidized, producing peroxides that contribute to the flavor of oxidative rancidity in milk and its products. The bond sites are a ready point of attack by bacteria, which split fatty acids into small molecules that in turn contribute to the flavor of some types of cheese.

Lactose is a disaccharide, beta-D-galactopyranosyl-1:4-beta-D-glucopyranose, fermented by streptococci and lactobacilli to lactic acid; the coliform bacteria and some *Erwinia* spp. also ferment this disaccharide to lactic acid and a mixture of fermentative end products, including gas. *Pseudomonas* spp. oxidize lactose to the slightly bitter lactobionic acid.

During cheesemaking relatively little of the lactose is fermented and most of the sugar is found in the whey. The lactose is reclaimed for pharmaceutical preparations and foods, or bacteria such as *Lactobacillus bulgaricus* may ferment it to lactic acid to be reclaimed for industrial use. Lactose brings about intestinal disturbances in people not accustomed to milk. Such lactose intolerance is widespread in many of the developing nations. Milk sent in food aid programs to these countries is treated with the yeast enzyme lactase, to convert lactose into the constituent monosaccharides. The traces of other sugars probably account for the transient reduction of litmus produced in litmus milk by non-lactose-fermenting, plant-resident streptococci.

Filled and Imitation Milks

In filled milk the natural lipid fraction has been replaced by another fat product—usually coconut oil, although safflower, corn, or cottonseed oils have also been used. Imitation milk is a beverage made to resemble milk but containing no dairy products. Sodium caseinate or soybean protein may be used as protein sources—although the former is of dairy origin, it is recognized by the Food and Drug Administration as a chemical rather than a dairy product. Imitation milk is lower in fat, protein, and solids than is cow's milk (Table 16-2); it also has no lactose and contains far less calcium and magnesium than does cow's milk. Since milk meets most of a growing child's calcium and protein needs, exclusive use of an imitation milk such as the one in Table 16-2 could have significant adverse nutritional

Table 16-2
Composition of imitation and cow's milk compared.

Measurement	Imitation[a]	Cow
Fat, %	2.85	3.70
Protein, %	0.75	3.50
Total Solids, %	10.83	12.6
Calcium, mg/1	21.0	1230
Magnesium, mg/1	6.3	120
pH	7.4	6.6
Acidity, %	0.03	0.17

Source: Kosikowski, 1968.

[a] Product also appeared to have no lactose as evidenced by absence of a typical lactic acid fermentation when inoculated with 1% active starter culture and incubated at 21°C for 20 hours.

consequences. Because lactose stimulates calcium absorption in the young, the absence of this milk carbohydrate from the imitation product is considered undesirable. Conversely, milk substitutes are useful for persons who are lactose intolerant.

Other products such as imitation ice cream and imitation cheese are also on the market. These, like natural dairy products, are perishable foods and should be manufactured under the same standard of sanitation required for the production of all high-quality foods. In refrigerated imitation milks that have no lactose and also contain excessive numbers of bacteria, proteolytic and lipolytic degradation are likely to occur and produce undesirable bitter, fruity, and rancid flavors. The bacteriological condition of the raw ingredients affects the product's quality, and care should be taken to insure that the pasteurized product comes in contact with only clean, sanitary equipment surfaces.

Microbiology of Raw Milk

Bacteria inevitably enter the milk during milking and later during handling. The dairy industries have made remarkable progress in herd health, in sanitation, in handling and storing milk, and in mechanization. Much of the improvement in milk quality became possible after 1945, mechanization and mechanical refrigeration were developed and soaps were replaced by effective detergents. Undoubtedly, the closed milking system, in which the milk is not exposed to the atmosphere with its bacteria and debris, has contributed more to the quality of modern milk than has any other development. As a result of these changes, the "natural" flora of raw milk has changed from what it was when different production methods were used. Prior to 1945 most of the bacteria present in raw milk were gram-positive acid-producing types. Today, because mechanical refrigeration cools milk to 4.5°C in ½ to 2 hours, gram-negative, psychrotrophic organisms (chiefly pseudomonads) are dominant.

Because, in nature, the young take milk, their source of nourishment, directly from the nursing mother, it is probable that milk in the udders of animals in the wild state is sterile. However, once milk leaves the udder it is totally unprotected from microbial attack. Because modern dairy animals have been selected and bred to produce milk, their udders are hypertrophied. Bacteria from the exterior pass through the sphincter muscle to reside chiefly in the storage areas. Figures commonly stated are 15,000 bacteria per ml of foremilk, or the first milk to be drawn, and none in the final milk taken from the cow. Relatively few of the bacteria passing into the udder can grow, because anaerobiosis, high body temperature, and the presence of inhibitory substances are inimical to their development. Most bacteria residing in the saprophytic state are members of *Micrococcus* spp. and a peculiar, tetrad form of *Staphylococcus albus,* known in the past as *Gaffkya* and *Tetracoccus.*

Improved conditions of milk production make possible very high quality raw milk with less than 5000 bacteria per ml. Milk taken aseptically from normal udders may run from 300 to 1000 bacteria (mostly micrococci) per ml. When such milk comes in contact only with clean, properly sanitized equipment and is maintained below 10°C, it will be delivered to the dairy processing plant with a very low bacterial count. A low bacterial count, however, is no guarantee that the milk has been produced under sanitary conditions, since refrigeration can prevent growth of many types of bacteria that may have entered milk from an unclean environment.

Cattle on Grade A dairy farms are housed in clean stables and yards. They are bathed as needed and the flanks are clipped of hair. They are milked in a milking parlor, a facility separate from the stable. Prior to each milking the udders and teats are washed and sanitized. The milk is drawn by the teat cups of the milking machine, which operates on the principle of vacuum and release, and the milk of each animal is drawn immediately from the pail of the machine through glass pipes to the bulk storage tanks. Here it is cooled to 4.5°C or less within two hours. Between milkings all equipment is cleansed, without disassembly, with modern detergents and sanitizers. At the time of marketing the milk is pumped into stainless steel or glass-lined tanks for transport.

Formerly cows were milked by hand; body and udder surfaces were cleansed infrequently or dry cleaned (that is, brushing with the dry hand); the milk drawn into open-top pails received contamination from the flanks, teat and udder surfaces, strainers, cans, and the atmosphere. It is believed that much, if not all, contamination with *Mycobacterium bovis*, *Escherichia coli*, and the enterococci came from the atmosphere. The cleansing agents then in use did not provide the state of cleanliness achieved with modern detergents and milk was cooled in 10-gallon cans immersed in water. In consequence, milk was constantly contaminated with *Streptococcus lactis*, which grew during the prolonged cooling period. Consumers accepted that milk would sour, and objected only if it soured too quickly. *E. coli* produced indole and skatole, and so common was its presence in milk that the fecal flavor was believed to be the normal flavor of milk. Occasionally, during the summer months, *Alcaligenes viscolactis* produced a viscous slime. The adhesiveness of this slime is remarkable: often it could be drawn into a thread several feet long.

Milk stone deposits of calcium salts and precipitated proteins, forms on surfaces of utensils. The deposits are not removed by soaps and alkaline cleansing agents and can be removed only by periodic scrubbing with scouring agents. The milk stone gradually thickens to form a porous layer that is a haven for bacteria, which can be a major source of reseeding milk with *Streptococcus lactis* and, during the summer months, with thermoduric bacteria, chiefly *Bacillus* spp. Modern dairy detergents prevent the formation of milk stone through a combination of chemical and physical reactions that emulsify fats, penetrate and deflocculate protein films, sequester calcium ions, and dissolve the salt residues.

The bacterial count of much raw milk today ranges from the low hundreds to low thousands per ml. Contaminants include a number of airborne microorganisms from the soil. Species of *Pseudomonas, Achromobacter, Flavobacterium, Erwinia, Micrococcus, Rhodotorula,* and fungi are most common, but not all contaminants grow in refrigerated milk. Those that grow are *psychrotrophs* although commonly called *psychrophiles* by the dairy industry. (Eddy [1960] proposed the term *psychrotroph,* suggesting that microbiologists restrict use of the term psychrophile to bacteria with an optimum temperature below 20°C and use the term psychrotroph for those able to grow at 5°C and below, whatever their optimum temperature; in this sense, the bacteria with high counts in refrigerated raw or pasteurized milk are psychrotrophs.) Psychrotrophic bacteria are natural soil inhabitants and are widespread in the farm milk-producing environment, especially in well and municipal water supplies. Because these gram-negative types are not pathogenic, municipalities do not add enough chlorine to water to eliminate them. Therefore, equipment washed with usual water supplies must be sanitized prior to use in the milking process to eliminate these organisms. Species of *Pseudomonas, Achromobacter, Alcaligenes,* and *Flavobacterium* are most often encountered, but psychrotrophic *Bacillus* and *Clostridium* species also have been isolated.

After a prolonged lag period of 47 to 72 hours pseudomonads grow rapidly to produce several characteristic changes in flavor, described as stale, oxidized, bitter, or rancid. The stale flavor is difficult to describe more precisely. The oxidized flavor may be approximated by immersing a bit of corrugated cardboard in milk. Bitterness is associated with proteolytic activity. When amino acids oxidize, the released ammonia reacts with free fatty acids to form a soap and soapy flavor. Rancidity is the result of lipase acting on milk fat.

The American Public Health Association suggests that milk leaving the farm should have no more than 100,000 bacteria per ml of raw milk, as determined by the standard plate count—ordinances in some communities stipulate much lower permissible counts. The limitations are far from stringent: contamination must be excessive to reach 50,000 bacteria per ml. Sanitary practices and temperature control make satisfactorily low bacterial counts readily attainable, but it must be remembered that milk with an initial count of 5000 per ml will have 160,000 per ml in only five bacterial generations.

Determining Microbiological Quality of Milk

Procedures for determining the microbiological quality of milk are found in *Standard Methods for the Examination of Dairy Products* (American Public Health Association, 1978). The publication specifies in detail the formulation and preparation of culture media, the standardization of supplies, and equipment and procedural manipulation.

Standard Methods Agar—composed of pancreatic digest of casein, yeast extract, glucose, and agar—is used to enumerate bacteria in raw and in pasteurized milk and to detect thermoduric bacteria in laboratory-pasteurized milk incubated at 32°C for 48 ± 3 hours. Plates are incubated at 55°C to enumerate thermophilic bacteria, and at 10°C for seven days to enumerate psychrotrophic bacteria. Anaerobes, very slow-growing bacteria, and nutritionally fastidious bacteria are not detected.

Coliform bacteria are enumerated with any of several selective media. If the numbers of coliform bacteria are very low, as they may be in pasteurized milk, the multiple tube procedure is used and tables for Most Probable Numbers are consulted. Most of these bacteria appear to be airborne contaminants originating with soil, plants, or water and many fail to grow on selective media at 37°C. Standard Methods specify that coliform media be incubated at 30°C.

Dye-reduction tests, which measure the rate of bacterial metabolism, are most useful for grading milk to be used in manufacturing dairy products. The numbers of bacteria in such milks tend to be high and the bacteria are rapidly metabolizing. Dye reduction tests are not reliable if bacterial numbers are under 100,000 per ml or if the bacteria are gram-negative contaminants.

Both methylene blue and resazurin, the dyes commonly used, measure the oxidation-reduction potential and are reduced, i.e., change color, when the dissolved oxygen is exhausted. Methylene blue changes from blue to white. Tubes with 10 ml milk and 1 ml dye solution at a final concentration of 1:250,000 are incubated in water at 36°C and examined at 0.5 hour, i.e., at the first half-hour, and hourly thereafter for 5.5 hours or longer. Resazurin passes through two color stages: from purple to the vivid pink resorufin and then to the white or colorless dihydroresorufin. Resazurin is commonly used as a triple reading test with the tubes of milk-dye mixture examined at 0.5 hour and at 1.5 and 2.5 hours.

Direct microscopic examination (DME) yields information regarding the types of bacteria and the apparent reasons for high bacterial content. It is also used to monitor the health of the udder. In the DME procedure 0.01 ml milk is spread over a delineated area of one cm² on a slide, fixed with a mild heat treatment, and stained with the Levowitz-Weber stain, which contains a fat solvent and methylene blue. The slide is examined through a microscope in which the field is calibrated to one of several specified diameters. A constant is obtained from the field diameter, volume of milk, and area on the slide, and, by multiplication with the average number of cells per field, yields a direct conversion.

Pairs and short chains of spheres usually signify *Streptococcus lactis* and indicate slow cooling rates. Large rods, occurring singly or in chains, and yeasts indicate atmospheric and utensil contamination. Long chains of streptococci are *S. agalactiae,* a causative agent of mastitis. Clusters of small rods suggest fecal material. Brown-to-black, irregular-to-amorphous, particles in milk are dirt.

Normal milk contains between 500,000 and 800,000 leucocytes or somatic cells per ml. Numbers above this range, coupled with the presence of *Streptococcus agalactiae* or staphylococci, suggest mastitis. High counts of somatic cells also occur if colostrum is added to the milk, if the lactation period is prolonged beyond the normal span, or if trauma has occurred.

Mastitis

Prior to 1945 *Streptococcus agalactiae* was believed responsible for 70% or more of the incidence of mastitis. Being quite drug sensitive, especially to penicillin, this bacterium has assumed less significance as a causative agent since antibiotics were introduced for mastitis treatment. With the advent of penicillin, *Staphylococcus aureus* has replaced *S. agalactiae* as the major causative agent, and *Escherichia coli, Corynebacterium* spp., *Pseudomonas aeruginosa,* and *Klebsiella* spp. account for many cases. Bacteria that ferment lactose to acid and gas can cause acute illness. During the summer months, *P. aeruginosa* is most often encountered, its presence being indicated by the blue color of the milk that results from elaboration of the organism's pigments.

Solutions of penicillin or aureomycin, the antibiotics most commonly used in treating mastitis, are injected into the udder through the opening in the teat canal and distributed throughout the lacteal system by vigorous manipulation of the udder. Although the antibiotic remains in the system for several days, the milk soon looks normal again, misleading many dairy farmers, who then market milk containing antibiotics. Hence, if an animal is treated with penicillin, current regulations prohibit the marketing of milk from six consecutive milkings or for three days, or, when other antibiotics are administered, for as many as eight consecutive milkings or four days. The regulations were based on several considerations, among them:

1. the esthetic value, or the marketing of milk of a diseased animal;

2. the possible sensitization of a consumer to that antibiotic—although this is a nebulous consideration, of which there are no known confirming occurrences;

3. reaction by a sensitized consumer, a very real consideration, because many consumers have been sensitized as the result of medication; and

4. the lethal effect of the antibiotic on the bacteria used in making cultured milks and cheeses. (As little as 0.5 μg penicillin per ml milk destroys *Streptococcus lactis*. Milk thus affected is termed "dead milk" and is discarded, with economic loss.)

Milk is examined routinely for the presence of penicillin. Discs of 0.5 mm diameter are touched to the surface of the milk and wetted by capillary action, drained, and placed on seeded agar plates. The agar is an enriched medium receiving approximately 2.5×10^7 spores of strain ATCC 6633 *Bacillus subtilis* prior to preparation of the plate. After an incubation period of 14 to 25 hours zones of incubation surrounding the discs indicate the presence of an antibiotic substance. In raw milk, antibiotic substances may be naturally occurring or may have been produced by bacteria such as *Streptococcus lactis,* which produces the antibiotic nisin (Hurst, 1972; Duthie et al., 1976) or acetic acid (Sorrells and Speck, 1970).

Heating the milk to 82°C for two to five minutes destroys the heat-labile antibiotic substances. This heat treatment is more severe than pasteurization, which may not completely destroy the labile substances. If further confirmation of the presence of penicillin is desired, duplicate sets of discs may be treated with the enzyme penicillinase.

During infection the permeability of the alveoli of the udder is altered, and components of the bloodstream infiltrate the milk in abnormal amounts. If blood vessels are ruptured erythrocytes enter the milk. The synthesizing activities of the gland are suppressed, reducing casein and milk fat content. Fermentative bacteria produce acid that precipitates casein as clots in the teat canal. If the first few streams of milk are squirted onto a black cloth it retains any clots present, thus supplying one means of detecting mastitis in the barn.

Laboratory tests to detect mastitis involve the direct microscopic examination of milk and chemical tests that measure increases in mucin and leucocytes. Of the latter type of test, the modified Whiteside and California mastitis tests are based on the formation of a viscous mass when a sample of milk is mixed with normal NaOH. The Wisconsin mastitis test also employs NaOH and measures the increase in viscosity or the time of flow through a standard orifice. A membrane filter DNA method (Bremel et al., 1977) allows colorimetric measurement of somatic cells based on indole staining of the cell DNA. The increase in catalase, as measured by the amount of oxygen released from hydrogen peroxide, is an indirect determinant of the numbers of leucocytes that possess the enzyme. The procedural details of these tests are found in *Standard Methods for the Examination of Dairy Products* (American Public Health Association, 1967 and 1972). The term "somatic cells" is considered more appropriate than "leucocytes" to describe those body cells present in milk obtained from an inflamed udder. Whereas the Whiteside, California, and Wisconsin screening tests only determine the presence of somatic cells, the direct microscopic somatic cell count gives an actual count of cells present.

A completed positive diagnosis of mastitis requires isolation and identification of the infectious agent. Isolation is also often necessary for antibiotic sensitivity testing, so that the proper treatment can be prescribed.

Pathogens in Milk

Pathogens that can occur in milk are divided into three groups, according to source—the animal, humans, or the environment. Pathogens transmitted directly or indirectly from the milking animal are *Mycobacterium bovis, Brucella* spp., *Coxiella burnetti,* salmonellae other than *S. typhi, Staphylococcus aureus,* and *Leptospira pomona.* Humans are the most probable source of *Streptococcus pyogenes, Salmonella typhi, Shigella* spp., *Corynebacterium diphtheriae,* and possibly viruses. Contaminants from the environment are *Bacillus anthracis, B. cereus, Clostridium perfringens,* and *C. botulinum.* The environmental contaminants are not known to cause milk-borne diseases. Indirectly, they may cause illnesses in improperly handled foods in which milk is an ingredient. The bovine type *S. agalactiae* is not a known human pathogen; the *S. agalactiae* responsible for infections of the vagina and of newborn infants is of a different serotype that has not been implicated in bovine disease.

Bovine milk contains a number of antibacterial substances such as lysozyme, lactoferrin, agglutinins, complement-mediated specific antibodies, polymorphonuclear leucocytes, and lactenin. Lactenin is composed of three substances found in milk that, acting together, are bactericidal: lactoperoxidase, thiocyanate, and hydrogen peroxide (Reiter et al., 1976). These substances are generally distributed in human milk and other body secretions where they operate as defense mechanisms (Reiter and Oram, 1967). The lactenin system has been shown to kill cells of *Escherichia coli, Salmonella typhimurim,* and *Pseudomonas aeruginosa;* it may help prevent enteric infections in neonates. Collectively these antibacterial substances are minor defense mechanisms in humans and animals. They probably do little to inhibit the spread of disease through market milk and other dairy products.

Fermented milk products are well known to be less susceptible than unfermented products to spoilage and hence inhibit or destroy any pathogens present. Starter bacteria—especially *Streptococcus diacetylactis* and *Leuconostoc cremoris*—produce substances that kill pathogens (Daly et al., 1972; Sorrells and Speck, 1970). To work effectively as inhibitors or destroyers of pathogens, these substances may require not only that pH be lowered but also that hydrogen peroxide, lactic and acetic acids, nisin, D-leucine, and other, unidentified substances be produced (Lawrence et al., 1972). In a fermented product the time required by the starter bacteria to kill pathogens varies according to the numbers and types of pathogens present.

Mycobacterium bovis became a pathogen of major importance as dairy herds increased in size, the dairy industry became concentrated, a free exchange of cattle occurred among herds through purchase and sale, and confinement of dairy cattle in stables brought them close together and facilitated the spread of the bacterium from one animal to another. Testing of dairy cattle was advocated as early as 1912 (Rosenau) and subsequently, systematic control programs were initiated in the United States around 1918. The control program consists of injecting tuberculin into the animal. Development of a red wheal in 48 hours indicates an infected animal, which then is removed from the herd.

Dairy animals in the United States and in many other countries must be tested for tuberculosis at least once every three years—the minimum time that the disease takes to pass from the closed, or noninfective, to the open, or infective, state. In the open state the tubercular pustules in the lungs rupture to discharge the bacilli into the alveoli. The bacilli either are coughed into the mouth and leave the animal in the saliva, or are swallowed to leave with the feces. Bacilli in the intestinal tract also pass through the intestinal wall and may be transmitted to the milk through the udder.

Bovine tuberculosis occurs principally in children, and most of its victims die before the age of 15 years. Many suffer tuberculosis that begins in the intestinal tract and disseminates by the bloodstream. Occasionally, it localizes in the maxillary lymph glands to form an impacted enlargement. During World War II, refugees from Eastern Europe expressed amazement at not seeing children with impacted glands in the United States.

Q fever, a disease of unknown etiology when it first appeared in 1935, was designated "?" on medical reports, which speedily gave it its present name. The causative agent is the rickettsia *Coxiella burnetti,* an obligate parasite of cattle and humans. It resides in the placental tissues of cattle in an asymptomatic condition and is discharged into milk through the lacteal membranes. Common human symptoms are pneumonia, evidence of hepatitis without overt symptoms of clinical jaundice, pericarditis, and proteinurea. *C. burnetti* enjoys a dubious distinction: it is the most heat-resistant nonsporulating pathogen known.

Brucella abortus of cattle, *B. suis* of swine, and *B. melitensis* of goats are agents of infectious abortion, of brucellosis, or Bang's disease, in animals, and of brucellosis, or undulant fever, in humans. The brucellae reside in the udders of animals and pass through the lacteal membranes directly into the milk. Swine brucellae are transmitted to animal handlers. Cattle acquire the porcine species when cattle and swine share common feedlots and yards. In man the more common symptoms are daily fluctuation in temperature and malaise and exhaustion during the latter part of the day. The bacteria are carried throughout the body by the blood and may settle in any organ to simulate an infection of that organ. Most incidents in the United States now occur among veterinarians and among travelers returning from countries where milk is not commonly pasteurized.

Incidents of brucellosis have been reduced among dairy and beef cattle because infected animals are now detected and removed from the herds. To detect the presence of brucellosis in a herd the milk ring test is applied to mixed herd milk. Antibodies from the bloodstream of infected animals adsorb on the surfaces of the fat droplets. When milk and stained, killed cells of *Brucella abortus* are mixed the antibodies react with the stained cells, which are carried upward into the cream layer. If the herd is free of brucellosis the fat-free layer retains the stained cells.

Streptococcus pyogenes is the pathogen that humans most often transmit directly to milk. It is responsible for throat infections commonly known as strep throat, septic sore throat, and, simply, sore throat. It leaves the body with nasal and buccal discharges and, in the barn, enters milk exposed to coughing, sneezing, poorly aimed expectoration, and the use of the hand in lieu of the handkerchief. Some strains are highly virulent to humans. When raw milk was a common market commodity, outbreaks traced to a single dairy farm worker occasionally involved several hundred individuals with a mortality rate as high as 5%. Unlike typhoid fever, milk-borne streptococcosis often occurred during the winter months when the incidence of throat infections is high. During the summer months, outbreaks have been associated with the sale of pastries containing whipped cream. In many communities ordinances now prohibit the sale of such foods unless they are maintained under continuous refrigeration. The streptococci do not grow at 10°C.

Staphylococcus aureus, an organism of ever-increasing incidence as a pathogen of milk, is a species for which more than 80 strain-specific bacteriophages are known. Some phage types cause illness specifically in humans and rarely in cattle; other types characteristically attack cattle but not humans. Animal strains invade the udder through the sphincter muscle and produce numerous pimples or pustules on the surfaces of the lacteal system. And like pimples, the pustules mature and break during the agitation of milking, releasing staphylococci into the milk. If the milk is promptly cooled and maintained at 10°C or below there is no human consequence. If, however, the milk reaches that temperature too slowly, the staphylococci grow and produce enterotoxin with subsequent human intoxication. During cheesemaking the toxin is adsorbed on the casein, where it remains stable throughout storage, causing intoxication on consumption.

Typhoid fever and shigellosis occur exclusively in humans, but the causative bacteria grow freely in warm milk, to which they are transmitted either directly by humans with undesirable hygienic practices, or indirectly by polluted well waters. Milk has not been incriminated in the spread of viral infections or infections caused by yeasts and molds.

Pasteurization

Louis Pasteur popularized the process of selective heating to destroy food-spoilage microorganisms. During his studies on wine in 1864 Pasteur found that acetic acid bacteria could sour wine when excessive oxygen was present, and that holding the bottled wine for a short time at 50°C destroyed these organisms. First used to stabilize the composition and flavor of beer and wine, pasteurization was applied to milk in Denmark in about 1870 by Fjord. The principal objective of pasteurizing milk was to destroy the nonsporulating pathogens and to render the milk safe for consumption; however, as many as 99% of the saprophytes are also destroyed.

At one time all milk was sold directly by the producer to the consumer, who either was a neighbor without a cow or lived on a milk route and for whom milk was ladled from a dip tank. The first glass bottles were made in about 1885, but the dip tank was in use in some areas as late as 1912. Milk-borne diseases then attacked only the families of producer and consumer; typhoid fever—often traced to bacteria in contaminated well water—was a common milk-borne disease.

As cities grew in size, so did the demand for milk among people distant from producers. The first rail shipment of milk into New York City was in 1835. This practice introduced, in the larger cities, the middleman, who commingled the milk of many producers into a common tank, thereby distributing pathogens in the milk of one producer into the milk of many. Some of the pathogens grow readily in milk, and in large cities of the United States and Europe as many as 28% of the children died at less than a year in age from gastroenteritis, 85% of them dying in July and August. Records of the day associated the mortalities with the consumption of raw milk contaminated by *Shigella dysenteriae* (Rosenau, 1912).

Public health concerns resulted in sporadic attempts to heat milk as early as 1878, but Soxhlet, in 1886, was the first to systematize and popularize the heating and special care of milk for infant use. In the United States Jacobi introduced heating of milk in the tenements of New York, and impetus to commercial pasteurization was given when philanthropic associations established pasteurized milk depots in New York. In 1898 the exclusive use of pasteurized milk at Randall's Island Hospital for Children in New York City reduced infant mortality from 42 to 20% in the first year. Soon after, the process became widely used in most cities. In 1899, Theobald Smith determined that the tubercle bacillus was destroyed in milk held at 60°C for 15 minutes; this became the basis for the holding method of pasteurization. Equivalent exposures for bacterial destruction were subsequently established as 60°C for 60 minutes, 62°C for 25 minutes, 66°C for 4 minutes, 72°C for 15 seconds, 76°C for 2.4 seconds, 80°C for 0.4 second, and 85°C for 0.04 second. These treatments were determined to kill *Mycobacterium tuberculosis* and from 90 to 99% of all other microorganisms present in raw milk. By 1924 a number of cities, including New York, Chicago, and Philadelphia, had compulsory pasteurization laws.

Meanwhile, also to assure the availability of milk of good quality, the Certified Raw Milk Program came into being, and persisted through several decades of this century. Such milk was produced and distributed under the scrutiny of veterinarians and sanitarians. Also during this time, dairy distributors resorted to illegal practices to maintain an acceptable product. They commonly used such chemicals as borax and boric acid, sold under the name of Aseptine; formaldehyde, sold under the name of Freezene; salicylic and benzoic acids; fluorides; common salt; sodium bicarbonate; and hydrogen peroxide. Some chemicals harmed the infant; sodium bicarbonate merely neutralized lactic acid; and common salt had no merit. Some cheesemakers have resisted pasteurization of milk to be used for cheese manufacture because the required heating and holding destroy milk enzymes important in full flavor development during cheese ripening. Pasteurization is also reported to delay draining and help weaken the cheese curd. In addition, if the milk used contains many anaerobic spore formers (e.g., *Clostridium sporogenes*), heat may activate the spores, resulting in bloated or gassy cheese with poor flavor. To avoid these problems, cheesemakers may "pasteurize" milk by treating with hydrogen peroxide to kill coliform bacteria and other gram-negative rods. Food grade hydrogen peroxide (35%) is added to milk at 0.2% and held at about 49°C for 20 to 40 minutes. After cooling to 38°C, excess catalase is added to decompose the residual peroxide and the cheese is then made in the usual manner. To make plant operation speedier and less costly, a flash method also has been used to treat cheese milk. In this process, clarified milk is treated in the pasteurizer surge tank by adding 0.02% hydrogen peroxide (100%) by means of a flowmeter. The milk passes through the short-time pasteurizer, where it is heated to about 52°C for 25 seconds and is then cooled to cheese-ripening temperature. While the cheese vats are being filled excess catalase is added to hasten decomposition of the peroxide. When the milk is free of peroxide, as determined by the p-phenylenediamine or the potassium-iodide–starch tests, starter is added to start cheese manufacture.

The Standard Milk Ordinance and Code defines pasteurization as

the process of heating every particle of milk or milk product to at least 62.6°C and holding it continuously at or above this temperature for at least 30 minutes, or to at least 71.5°C and holding it continuously at or above this temperature for at least 15 seconds, in equipment which is properly operated and approved by the health authority: Provided, *that milk products which have a higher milk fat content than milk and/or contain added sweeteners shall be heated to at least 67.7°C, and held continuously at or above this temperature for at least 30 minutes, or to at least 74.3°C, and held continuously at or above this temperature for at least 15 seconds:* Provided further, *that nothing in this definition shall be construed as barring any other pasteurization process which has been recognized by the United States Public Health Service to be equally efficient and which is approved by the State Health Authority.*

The present ordinance went into effect in 1956. The temperatures specified are slightly higher than temperatures specified in 1924, because the extreme heat resistance of *Coxiella burnetti,* the Q fever rickettsiae, has been recognized. Temperatures of 61.7°C for the long-time low-temperature, or kettle, process, and 71.3°C for the short-time high-temperature, or continuous flow, process, had been based on the thermal destruction time of *Mycobacterium bovis.* The bacterium is actually destroyed at

61.7°C in 15 minutes, but the time was doubled because the relatively crude pasteurizing equipment then in use made it possible for mycobacteria to survive in foam and in dead end valve connections.

In addition to the low-temperature holding and short-time high-temperature methods of pasteurization, also in use are ultrahigh-temperature processes by plate heat exchange or steam injection. These are thermal processes that have holding times of two seconds or less with holding temperatures of 88 to 132°C—specific recommendations being 88°C for 1.0 second, 90°C for 0.5 second, 94°C for 0.1 second, 96°C for 0.05 second, and 100°C for 0.01 second.

Milk pasteurized at ultrahigh temperatures has better keeping quality because more bacteria are destroyed—although shelf life is affected more by types of organism present than by their numbers. For example, milk treated at 97°C for 3 seconds has very few organisms, even after holding for 2½ weeks at 7°C, but the same milk pasteurized at 80°C for 16 seconds has a SPC of thousands of organisms per ml after holding for 16 days at the same temperature (Speck, 1961). Occasionally milk supplies may contain spores of *Bacillus* spp., which are activated at ultrahigh temperatures and grow out rapidly; they can spoil the pasteurized product more quickly than when the short-time high-temperature method is used. Table 16-3 shows keeping quality of milk pasteurized at ultrahigh temperatures. The data also illustrate how low temperatures extend the milk's shelf life. In retail outlets milk is usually kept at about 7°C. At that temperature high-quality raw milk pasteurized by the short-time high-temperature method and given minimal post-pasteurization contamination by psychrotrophs can be expected to have a shelf life from 7 to 10 days.

Pasteurized Milk Quality Tests

The adequacy of pasteurization is determined by tests to detect the enzyme phosphatase. Details for conducting the test are described in *Standard Methods for the Examination of Dairy Products.* All raw milk contains the enzyme (Burgwald, 1939). Because this raw milk enzyme has a greater thermal resistance

Table 16-3
Relationships of initial total and psychrotrophic counts, pasteurization conditions, and storage temperatures to shelf life of ultrahigh-temperature pasteurized milk.

Milk lot	Raw			Bacteria	Storage life, weeks[b]			
	SPC/ml.	PPC/ml.	Condition	SPC/ml.[a]	0°C	2°C	4°C	7°C
A	150,000	3200	220 F, 16 sec	<30	13+	13+	9+	3
			200 F, 16 sec	120	7	4	3	1
			220 F, 0.5 sec	110	10	6	4	1
			210 F, 0.5 sec	130	9	4	3	1
E	87,000	4700	220 F, 16 sec	80	13+	13+	11+	4+
			200 F, 16 sec	100	8	5	5	1
			220 F, 0.5 sec	130	12+	4	4	1
			210 F, 0.5 sec	120	6	6	4	1
I	14,000	7600	220 F, 16 sec	120	15	14	7	3
			200 F, 16 sec	130	10	7	4	2
			220 F, 0.5 sec	120	13	7	4	2
			210 F, 0.5 sec	140	11	3	3	2
M	35,000	1600	220 F, 16 sec	<30	20	20	20	18
			200 F, 16 sec	32	12	7	6	4
			220 F, 0.5 sec	39	12	10	6	5
			210 F, 0.5 sec	31	10	7	5	4

Source: Finley et al., 1968.
[a] In all samples <30 psychrotrophic organisms/ml were detected.
[b] Acceptability determined by flavor score of 35.0 or higher and SPC and PPC less than 1,000,000/ml.

than do pathogens, the phosphatase test is used extensively to determine causes for underpasteurization and to detect contamination of pasteurized milk with raw product. The combination of time and temperature required for complete inactivation are very nearly those established for the pasteurizing process. In the tests commonly used, disodium phenyl phosphate is added to a milk sample. After a brief incubation, a dye, 2,6-dichloroquinonechloroimide (CQC; indophenol blue), is added. The CQC reacts with any enzymatically freed phosphate to form a compound with a blue color. This compound is extracted with *n*-butanol, which is then compared visually with a color chart or measured with a spectrophotometer. If a decision about a large total production volume is based on only a small sample, the phosphatase test can lead to false negative reactions. Also, because of the sensitivity of the test, readings from blends of overheated and underheated milks can be falsely negative. Similarly, if microbial phosphatase is present, especially in cream, the phosphatase test will be falsely positive, because the phosphatase is more heat resistant than the milk enzyme. If the positive milk sample is reheated at 62.8°C for 30 minutes and no decrease occurs, the test should be regarded as false positive.

Bacterial quality tests on pasteurized milk are designed to detect numbers and/or types of bacteria that might spoil the product before it reaches the consumer. In the past, coliform counts on pasteurized dairy products were used extensively as an indication of the sanitary state of post-pasteurization processing conditions. Because these bacteria do not survive pasteurization, their presence in a pasteurized product indicates contamination from unclean equipment or some other source. They should be absent from retail dairy foods. Pasteurized milk maintained continuously at or below 4.4°C has a normal storage life of 15 to 30 days. Spoilage of milk is associated with the growth of psychrotrophic, gram-negative bacteria of *Pseudomonas, Alcaligenes, Achromobacter,* and *Flavobacterium* spp. Usually the same species and strains are found that cause spoilage of raw milk. They bring about the changes in flavor described earlier. Rancidity may occur if bacterial lipase has been produced in the raw milk supply. Much of the enzyme remains stable during pasteurizing and is slowly active during the storage of pasteurized milk.

The Standard Plate Count. The Standard Plate Count (SPC) with incubation at 7°C for 10 days is the generally accepted reference method for determining psychrotrophic colony counts in refrigerated milk (Thomas, 1969). But because it is time-consuming other methods have been used for routine analyses in commercial quality control programs. These include selective or differential media such as crystal violet agar, penicillin agar, medium for counting oxidase-containing colonies, determination of SPC after incubation at 7°C for five days (the

Moseley test), and quick-count methods. Plating these types of organisms with any medium requires at least two precautions: care should be taken to cool the agar media to 45°C or lower, to prevent destruction of some psychrotrophs; the choice of diluent deserves thought, and 0.1% (w/v) peptone-saline is widely used.

For many years agar media containing 0.5 ppm of crystal violet were used to enumerate gram-negative bacteria. However, some gram-positive types such as streptococci grow under these conditions, while many gram-negative organisms are inhibited by the dye. Another crystal violet based medium (Olson, 1963) frequently used to enumerate psychrotrophs in pasteurized milk is prepared by adding 1.0 ppm of crystal violet to SPC agar before sterilizing and adding 50 ppm of 2,3,5-triphenyltetrazolium chloride just before pouring the plates. Each plate receives 0.5 ml of milk sample before the agar is added; once solidified, the plates are incubated at 32°C for 48 hours. A piece of onionskin paper is placed under the plates on a colony counter and the red colonies counted and expressed as the CVT count per ml. Table 16-4 gives data showing the relationship of CVT and other counts to shelf life. There is no correlation between any of the counts and the keeping quality of the product; thus, CVT counts are useful only in detecting contamination. Numbers in pasteurized milk exceeding 100 per ml indicate excessively contaminated equipment.

Table 16-4
Relationship between various bacterial counts and shelf life of commercially pasteurized milk.

Number of samples	Number positive for coliforms	Bacterial counts per ml[a]			Average shelf life, days
		SPC	CVT	PPC[b]	
5	1	1800	83	13	7.2
6	3	4800	86	28	8.3
8	3	1300	30	12	12.0
5	1	2100	6	3	13.2
2	0	1100	11	14	19.0
5	3	3700	22	2	21.2

Source: Olson, 1963.

[a] Logarithmic averages.
[b] Psychrotrophic plate count: SPC with incubation at 7°C for 10 days.

Table 16-5
Minimum time required for one bacterium to reach indicated population at 7°C.

Generation time, hours	Approximate number of days required to reach SPC of at least			
	100	15,000	1,000,000	100,000,000
4	1	3	4	5
8	3	5	7	9
12	4	7	10	14
18	6	11	15	21
24	7	14	20	27

Source: Olson, 1969.

Psychrotrophic counts of various types fail to correlate with shelf life because sample size is small and because the temperature at which the product is held is not known by the tester. Depending upon the particular organism and the temperature, psychrotrophs may have generation times ranging from 4 to 16 hours. If generation times of a single cell are 38, 12, and 4 hours at 1, 4, and 10°C respectively, the SPC after 7 days would theoretically be 24, 16,000 and 4.4×10^{12} per ml. Table 16-5 indicates how the influence of generation time on theoretically predicted shelf life can be determined. From the data it can be seen that samples of milk with undetectable contamination by SPC could be spoiled in 10 to 12 days when counts of a million or so per ml are reached.

The Mosely Test. A practical shelf life test of the quality of pasteurized dairy products is the Mosely test. In this test, fresh product is held at 7°C for five days and the SPC determined before and after storage. Any increase in count during the five days is of psychrotrophs that have entered the product from unclean equipment; they grow at 7°C but grow better at 32°C, at which temperature they can be enumerated in two days. The product should be tasted on the fifth day to determine if undesirable flavors and odors are developing. The Mosely test is useful in dairy processing plants to measure the effectiveness of cleaning and sanitizing procedures; trouble spots within plants may be located by analyzing samples taken at different steps in processing. The Mosely test can be made more sensitive by

Table 16-6

Oxidase reaction of *Pseudomonas* and other known bacteria.

Organism	Oxidase reaction	Number of colonies by	
		SPC	Oxidase count
Pseudomonas fragi	+	404	419
P. fluorescens	+	186	186
P. mucidolens	+	119	133
P. putrefaiens	+	215	221
P. viscosa	+	178	178
Streptococcus lactis	−	55	0
Escherichia coli	−	120	0
E. aerogenes	−	102	0

Source: Hankin and Dillman, 1968.

extending the time of incubation to 7 or even 10 days. A highly significant correlation exists between the Mosely count and keeping quality (Randolf et al., 1964).

Psychrotrophic organisms in pasteurized dairy products may also be detected by the oxidase test (Hankin and Dillman, 1968). Being highly aerobic, *Pseudomonas* spp. and related bacteria have an active cytochrome oxidase. This is taken advantage of in a plating technique that incorporates α-naphthol and p-aminodimethylaniline oxalate into a medium. These reagents are coupled in the presence of cytochrome oxidase to form indophenol blue. After incubating at 32°C for 48 hours the plates are flooded with oxidase reagent and the blue colonies are counted after 10 to 15 minutes. Since psychrotrophs of other genera—such as *Alcaligenes* spp.—may give a negative oxidase test, oxidase counts are likely to be lower than total number of psychrotrophs present. However, the test is as rapid as the SPC. Table 16-6 shows the relationship between oxidase and SPC results in known organisms.

Dried Milks

Nonfat dry milk and dry whole milk are used as reconstituted fluid milks in the home and for cheesemaking and as dry ingredients in baked foods, confections, infant formulas, cake mixes, ice cream, and sausage. They are made from freshly skimmed or whole milk. Dry buttermilk, the fluid residue of sweet cream butter production, is used in some pancake mixes and also as animal feed. Dry whey, prepared from the fluid obtained during cheesemaking, is used primarily as an animal feed. Malted milk is a dry mixture of whole milk, ground barley malt, and enzymatically degraded wheat flours, to which salts or dried yeast may be added for stability, flavor, and nutritional value.

All products are dried to a moisture content of 5% or less, expressed or determined on the fat-free basis. They are hygroscopic and to be preserved must be packaged in plastic or metal, which renders them impervious to the absorption of moisture from the atmosphere. If the absorbed moisture reaches 10 to 11% at the surface, molds grow.

Milks are dried by spray drying (described below) or by roller or drum drying, either in normal atmosphere or in a vacuum system. In roller drying processes the milk is sprayed to form a film on the surfaces of heated, rotating drums. Nonfat and dry whole milk powders are now generally prepared by the spray process, which uses milder heat treatments (Figure 16-2) than the roller dryer process. If dry whole milk is desired, the milk is clarified to remove mucins and body cells, and homogenized to shear the fat droplets; if dry skim milk is desired, the milk is passed through a centrifuge (termed a separator in the dairy industry). After centrifugation about 60% of all microorganisms in the whole milk are found in the cream effluent and the remaining 40% in the skim milk layer. Such concentrations of bacteria are observed also in the production of coffee and whipping creams and in butter fat concentrates to be used subsequently in the manufacture of ice cream and butter.

Milk to be spray dried is preheated to inactivate milk enzymes and to reduce the numbers of bacteria. The milk then passes to a vacuum pan, a chamber under high vacuum, where it is concentrated to half its original volume or less, and passes to the spray drier. The droplets, entering as a fine mist, lose 90% of the water during the first 10% (or approximately one foot) of vertical drop. Thereafter, the concentration of milk solids helps protect surviving bacteria. Air enters the bottom of the drier at a temperature of 121°C or higher and leaves at approximately

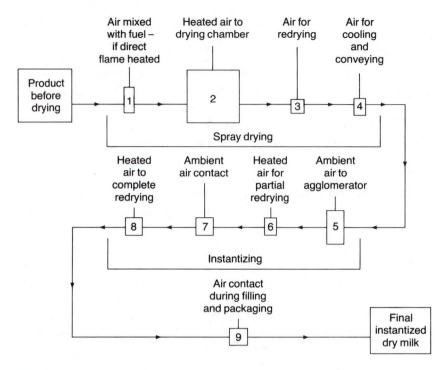

Figure 16-2

Diagram illustrating points of air contact with product during manufacturing. Sizes of square indicate relative magnitude of contact. (Courtesy of J. Dairy Sci.)

71°C. The evaporational effect keeps the temperature of the droplets materially below that of the heated air. The spray dried milk is globular and thus when observed microscopically is readily distinguished from the less easily soluble, flaky particles of drum dried milk.

Not enough heat is applied to spray dried milks to totally destroy any pathogens present. In fact, there may be stages, termed "critical points," in the process at which pathogens may be introduced and actually grow. During outbreaks of salmonellosis involving dry milk, 21 different serotypes of *Salmonella* spp. were isolated during a 6 month period (National Communicable Disease Center, 1967; Blackburn and Ellis, 1973). Miller et al. (1972) point out that each drying process and plant, regardless of commodity being dried, has unique points at which salmonellae may be introduced into the product. Such points, once recognized, receive constant in-plant monitoring to prevent future instances of food-borne illnesses.

Milks entering drying processes are now pasteurized at the outset. Microbiological quality of dried milks is achieved by the standard plate counts for mesophilic, thermophilic, thermoduric, and psychrotrophic bacteria, hemolytic streptococci, staphylococci, yeasts, and molds. Since bacteria killed during pasteurization may be observed in the stained preparation, direct microscopic observations must be interpreted with care.

Butter

Butter was probably used first only for cosmetic and medicinal purposes (Guthrie, 1923), and came to be used extensively as a food less than 2000 years ago. Butter is microbiologically more stable than most other dairy foods, except some hard cheeses. Three types of butter are generally produced: sweet cream butter, sweet cream starter butter, and cultured or ripened cream butter. Table 16-7 lists the traditional butter-making steps for sweet cream butter, indicating how each step influences microbial stability. Ripened cream butter is made from pasteurized cream inoculated with starter bacteria (*Streptococcus lactis, S. cremoris, Leuconostoc cremoris, S. lactis* subsp. *diacetylactis*) and incubated overnight at 21–22°C until a pH as low as 4.7 is reached. The cream is then churned as usual but is not salted because salt and acid—especially if more than 0.075 μg of copper per gram of butter is present (Pont and Rogers, 1962)—favor deterioration of the fat during storage, which results in oily and fishy flavors. Ripened cream butter is made extensively in Europe, especially in Scandinavian countries. Ripening the cream increases the keeping quality of the unsalted butter by suppressing growth of undesirable bacteria. Sweet cream starter butter is made from cream inoculated with starter but held at low temperatures before churning to keep acidity from increasing during the holding period. The butter is thus given desirable culture flavor with no sacrifice of keeping quality due to elevated acidity levels. As much as 10% starter inoculum may be used depending on the holding

Table 16-7
Influence of butter manufacturing steps on the microbial stability of sweet cream butter.

Step	Function	Effect on microbial stability
Cream pasteurization	Destroys pathogens; reduces microbial load by 99.9%	Increases because of reduced numbers of microorganisms
Cool to 2 to 5 C and hold overnight	Allows formation of fat globule clusters, minimizing fat losses during churning	Minimal
Warm to churning temperature (8 to 12 C)	Favors cohesion of fat globules	None
Churn	Promotes separation of buttermilk from the continuous fat phase as butter granules are formed	Increases because organisms are concentrated in the buttermilk
Drain buttermilk	Removes much of the buttermilk	Increases because organisms are removed in the buttermilk
Wash	Removes most of remaining buttermilk	Increases as above
Salt and work	Incorporates water and salt in small droplets and breaks up fat globules, beginning the continuous fat phase	Increases because availability of moisture to microorganisms is limited and the salt content is elevated

temperature and species of organisms used (McDowall et al., 1960).

Butter, especially the sweet cream variety, may be held for as long as 12 months at the usual storage temperatures of −23 to −29°C without bacteriological or chemical deterioration. Under these conditions cultured cream butters may deteriorate chemically, and culture flavor compounds, especially diacetyl, are lost during storage. Therefore, most cream made into butter to be stored for several months before consumption is not cultured, and in the United States cultured cream butter, even of the sweet variety, is virtually unavailable.

Pasteurization destroys all microorganisms important in butter spoilage, but a variety of undesirable types may enter the product at some later stage if the buttermaking operation has any unsanitary features. Hence, conditions during removal of the buttermilk and working the butter properly are especially important. If neither underworked or overworked, butter contains about 10^{10} moisture droplets per gram, most less than 5 microns in diameter. Thus, for sweet cream butter of good microbiological quality, the majority of the moisture droplets are sterile and bacteria trapped in the fat are inhibited by lack of moisture.

The replacement of wooden by aluminum churns (Figure 16-3A) has greatly improved the microbiological quality of butter, since metal churns are much easier to clean, sanitize, and maintain free of cracks and crevices that could harbor organisms. Around 1900, soon after the continuous centrifugal separation of cream was developed, allowing improved butter quality and uniformity, continuous churning equipment was perfected (Figure 16-3B). Today much of the butter made in the United States is produced by several different continuous methods that generally involve oiling off the cream and separating a butterfat-serum mixture containing 87 to 99% fat. The mixture then passes through a vacreator, chiller, texturator, and printer. Figure 16-3 B shows one type of continuous churn, with the separator on the far left and the texturator, inclined upward, on the right.

Psychrotrophic bacteria pose a keeping quality problem for butter stored under refrigeration (5 to 10°C). These gram-negative types of *Pseudomonas*, *Alcaligenes*, *Achromobacter*, and *Flavobacterium* spp. enter the product after pasteurization from improperly cleaned and/or sanitized equipment and from butter wash water. Much drinking water, whether from wells, streams, or reservoirs, contains these types of organisms; delivered at the tap they are not a public health concern, but in buttermaking such water must receive an additional chlorination treatment (5 to 10 ppm for 15 minutes) to insure that these bacteria do not contaminate the product. Because of their proteolytic and/or lipolytic abilities, certain microorganisms can bring about characteristic flavor defects in refrigerated butter. *Pseudomonas fragi, P. fluorescens,* and possibly other gram-negative bacteria are able to hydrolyze the milk fat, liberating short-

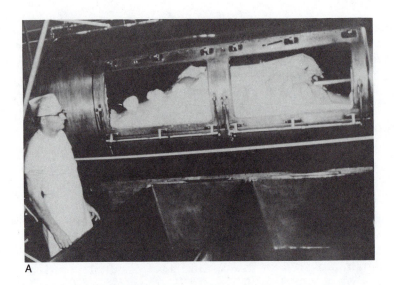

A

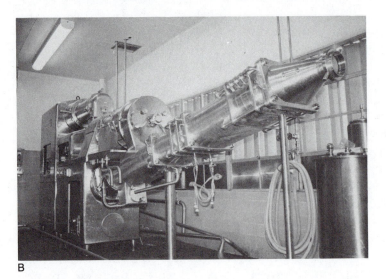

B

Figure 16-3
A. Aluminum butter churn; B. Continuous butter churn. (A.
courtesy of Tillamook County Creamery Association, Tillamook,
Oregon; B. courtesy of Dairyman's Cooperative Creamery
Association, Tulare, California.)

chain fatty acids that lead to a rancid flavor. Putrid flavors may result from proteolysis carried out by *P. putrefaciens*. *P. nigrifaciens* may cause black discoloration on the surface of butter, and *P. mephitica* may cause skunklike odors. A cheesey flavor defect, sometimes called surface taint, also is caused by *P. putrefaciens*.

Numbers of psychrotrophs in butter may be determined as indicated for raw or pasteurized milk. The proteolytic members of this population may be determined (American Public Health Association, 1978) as well as those that are lipolytic (Mourey and Kilburtus, 1976).

If coliform bacteria are present in butter there is cause for concern about its microbiological quality. Although their absence is in itself no assurance of high quality, their presence suggests poor production conditions. Since coliforms are relatively sensitive to salt and freezing and may die off in stored butter, enumeration of enterococci is a better test for butter quality (Saraswat et al., 1965). More than 10 enterococci per gram of butter is regarded as excessive.

The yeast and mold count, especially with butter made in wooden churns, has been used extensively to determine microbiological quality. For this purpose media containing antibiotics (Overcast and Weakley, 1969) are superior to acidified media (Jarvis, 1973), and a count of less than 10 per gram is desirable.

Cheese

The Old Testament contains references to the use of cheese as a food. Sometime between the domestication of the cow, around 9000 B.C., and early Biblical times (5000 to 3000 B.C.), the art of making cheese curd from fresh milk was developed. It was the practice in early civilizations to store water in animal skins, especially that of the stomach, and no doubt milk was similarly stored. When stomach lining enzymes, especially rennin, or chymosin, and pepsin, act in combination with the milk's natural bacterial flora curd forms (i.e., casein precipitates). Rennin, still used extensively in cheesemaking, is prepared as a crude extract, called rennet, from the fourth stomach (abomasum) of unweaned calves. Rennin attacks κ casein, which functions in milk as a protective colloid to keep the other

Box 16-3

From its beginnings cheese making has been an important part in
each of the world's major civilizations. By 2000 B.C. writings from the
ancient East reveal that cheese was a cherished food and often used
to trade for other items. The Greeks believed cheese to be of divine
origin and offered it in sacrifice to their gods. The Romans developed
cheesemaking skills and especially favored one type, made on the
island of Cyprus, as a delicacy. Soldiers in the early Christian centuries, through the time of Genghis Khan, carried cheese with them.

At the time of the Crusades, in the eleventh through the thirteenth
centuries, cheese was a well-established important food throughout
the East, and crusaders returning to Europe brought cheese of various
types with them as well as information about their manufacture. By the
Middle Ages a variety of cheeses, such as Roquefort, Swiss, Gouda
and Sapsago, were being made in Europe, where cheese making was
then under protection of the church. From this heritage come such
cheeses as Port du Salut, Trappist, and Oka.

Cheese was first introduced into North America with the early
settlers from England, who brought both cheese and cows on shipboard. Both cottage and cheddar cheeses were made in the home,
and a great variety of other types were also produced, especially in
Wisconsin, by immigrants who longed for their familiar European varieties. In 1851 Jesse Williams, a dairy farmer at Rome, New York,
opened the first cheese-making factory in the United States, beginning
a great industry. With an annual cheese production of almost 1.5 million tons, today the United States leads the world, followed by France,
about 0.75 million tons; and Russia, 0.5 million tons. World cheese
consumption is increasing by approximately 250,000 tons annually,
and, according to the National Cheese Institute, the worldwide total of
cheese produced in 1974 was almost 6 million tons.

casein fractions (α, β, σ) in suspension. During cheesemaking
rennet converts κ casein molecules from a molecular weight of
30,000 into two molecules with molecular weights of 22,000
and 8000. Lactic acid in the milk helps casein to precipitate by
reducing the pH to below the isoelectric point (pH 4.7) and
eliminating the stabilizing effect of repelling negative charges
on the casein molecules:

$$\left(R \begin{smallmatrix} \diagup COO \\ \diagdown NH_2 \end{smallmatrix} \right)^{-}_n + n\ H^+ \longrightarrow \left(R \begin{smallmatrix} \diagup COOH \\ \diagdown NH_2 \end{smallmatrix} \right)_n$$

pH 6.8   pH 4.7

Greater cheese production and consumption throughout the world has caused rennet shortages and increased prices. Mixtures of rennet and pepsin (1:1) have alleviated this situation, but searches for rennet substitutes from plants and microorganisms continue (Sardinas, 1972). In cheese manufacture, rennet is added (about 3 ounces per 1000 pounds of milk) either at the same time as the starter bacteria or after an initial ripening period of 30 to 60 minutes. Soft cheeses such as cottage and bakers' cheese may be made without rennet. After the addition of rennet and/or starter, most cheeses are made by the same basic process, which includes: incubation to allow curd formation, cutting the curd, stirring and heating (cooking), draining the whey, salting and pressing the curd, and ripening. During the last step, variations of time, temperature, and humidity are major factors in growth of the microorganisms that contribute to the unique flavor and texture of each cheese. These variations in ripening environments are the source of the several hundred types of cheese now produced throughout the world. Similarly, the final texture of the cheese will have been partly determined by its cooking temperature; hard cheeses are generally cooked at 35°C or higher, soft cheeses at 32–33°C.

Cheeses are usually classified by texture or degree of hardness achieved after ripening:

1. Very hard grating cheeses:
    Romano, Parmesan, Sapsago

2. Hard cheeses:
    a. Ripened by bacteria, without eye formation:
        Cheddar, Colby, Jack, Caciocavallo, Edam, Gouda, Kasseri, Provolone
    b. Ripened by bacteria, with eye formation:
        Gruyere, Swiss, Emmentaler
    c. Unripened whey cheeses:
        primost, mysost, gjetost

3. Semisoft cheeses:
    a. Ripened principally by bacteria:
        brick, Münster
    b. Ripened by bacteria and surface microorganisms:
        Limburger, Port du Salut, Schloss
    c. Ripening contributed by blue mold growing inside the cheese:
        blue, Roquefort, Stilton, Gorgonzola

4. Soft cheeses
   a. Ripened from the surface inward:
       Camembert, Brie, Bel Paese
   b. Unripened:
       cottage, cream, bakers', ricotta

Cottage Cheese

Cottage cheese is an unripened skim milk curd to which salt and sweet or cultured cream is added prior to marketing. A perishable food, it should be kept under refrigeration (5°C) until consumed. In the United States it must contain at least 4.0% butterfat, but even at that level it has only about 210 to 220 calories per cup (200 grams).

The bacteria usually present in mixed-strain cottage cheese starters are the acid-producers *Streptococcus lactis* and *S. cremoris* and the flavor-producer *Leuconostoc cremoris*. *S. diacetylactis* is not used in cottage cheese cultures, because it may produce gas (carbon dioxide), causing the curd to float and mat during cooking.

To prepare the bulk starter most cottage cheese manufacturers inoculate the starter milk with a commercially prepared frozen, concentrated culture. Usually one of three combinations of time, temperature, and percent culture inoculum is used: 4 to 6 hours at 32°C with a 5 to 8% inoculum (short set); 8 to 9 hours at 28°C with a 2 to 2.5% inoculum (intermediate set); or 12 to 16 hours at 22°C with 1 to 2% starter (long set). When the curd has been cut, cooking is essentially the same for all methods, taking a little longer to reach the final temperature with the intermediate and long set methods.

Next, the desired amount of culture and rennet are thoroughly mixed into the milk. During setting, the vat temperature is carefully controlled to remain uniform throughout the mass of milk. Cutting the curd accelerates separation of whey from curd; the acidity at cutting is a determining factor in final cheese quality. Cutting acidities may range from 0.44 to 0.60% and are influenced by solids-not-fat. However, the best criterion of proper cutting acidity is pH of the curd, i.e., pH 4.6–4.7.

Box 16-4

Common defects of cottage cheese and some of their causes:

1. Mealy curd: cooling too fast; particles touching hot surfaces.
2. Matted curd: pH too high or not enough acid at cutting.
3. Shattered curd: overpasteurization; too much acid at cutting; rough handling of curd; too low in solids.
4. Rubbery curd: cooking temperature too high.
5. Weak pasty curd: too much heat treatment of skim milk; cooking temperature too low; pH too low at cutting; or too much acidity before and during cooking.
6. Acid or unclean flavor: acid too high at cutting; cooked too fast; contamination during and after manufacture; poor quality skim milk and/or starter.
7. Gelatinous curd: spoilage bacteria; alkaline wash water.
8. Poor shelf life: unclean equipment; failure to keep cream or curd cold (5°C).
9. Bitterness: contamination by psychrotrophic bacteria; cooked too fast.
10. Gassiness: contaminated milk; use of gassy starters containing *Streptococcus lactis* subsp. *diacetylactis.*
11. Sediment in cheese vat at cutting: clumping of starter bacteria by agglutinins in milk (most common with mastitic and early lactation milk)
12. Medicinal flavor: use of chlorinated wash water high in organic matter.

The cut curd is cooked to heat the curd-whey mixture to a temperature that will expel enough whey to achieve the proper texture and moisture content in the final cheese. So that each curd particle can begin to become firm ("repair") and take shape, cooking should not begin until 15 or 20 minutes after the curd has been cut. The final cooking temperature is usually from 49 to 55°C. If the curd is heated too fast a film forms on the surface of individual particles and prevents proper whey expulsion; a sour and/or bitter cheese results. When a temperature of 39 to 41°C is reached starter bacteria cease producing acid. A cooking temperature of at least 52°C is required to

ensure the death of any contaminating salmonellae. Psychrotrophic spoilage bacteria such as *Pseudomonas fluorescens, P. fragi,
P. viscosa* and *P. putrefaciens* do not survive cooking at 49°C for
30 minutes, and a temperature of 55°C for at least 18 minutes is
required for destruction of *Escherichia coli.* The cooked cottage
cheese curd is free of psychrotrophic spoilage organisms, and
additional curd-handling steps must be carefully controlled to
prevent contamination from these and other undesirable microorganisms.

After cooking, the whey is drained and washed in water that
may or may not be chlorinated (5 to 10 ppm) and acidified.
Acidification (pH 5.0) of the water potentiates the action of the
chlorine in killing microorganisms and a retention time of at
least 1 to 2 minutes is needed after chlorination before adding
to the cheese; it also helps maintain the low pH curd. The
volume of water added should be about the same as the volume
of whey removed; it should be added slowly and the curd allowed to equilibrate to the water temperature during a 10-
minute stirring period. The final curd temperature should be 6
to 7°C after draining. If the cheese plant wash water is high in
organic matter from algae, decaying leaves, etc., chlorophenol
compounds will be produced and will cause a medicinal flavor.
Charcoal filtration is an effective means of solving this problem.

Flavor, body, and texture defects are common in cottage
cheese. A product that maintains fresh flavor for 12 days at 7 to
10°C will keep for at least 3 weeks at 4°C. The dominant spoilage organisms of cottage cheese are *Pseudomonas fragi,* which
produces an obnoxious flavor and an odor of overripe strawberries, and molds of *Geotrichum* spp., which digest the casein
and grow as a velvety film on the surface.

Cheddar Cheese

This cheese originated in the village of Cheddar, Somerset-
shire, England. Most Cheddar cheese is made from pasteurized
milk; if made from raw milk, the cheese must be aged at least 60
days at temperatures not less than 1.7°C before consumption,
to bring about the death of any pathogens which may have been
introduced into the cheese.

Vats containing 10,000 to 30,000 pounds of milk are commonly used, yielding 1000 to 3000 pounds of cheese. Curd

formation is begun by adding starter bacteria and rennet to the milk. Sufficient starter bacteria are added to bring the numbers to about one million per ml of milk. The starter produces an acidic reaction which favors the action of the pepsin in the rennet, favors the curdling of the casein at pH 5, helps expel the whey, protects against the putrefactive bacteria, and helps the curd particles fuse.

Starters may be mixed strains of *Streptococcus lactis* or a mixture of *S. lactis* and *S. cremoris*. The streptococci remain active in the curd after pressing, and reach their maximum population, about one billion per gram of cheese, in 48 hours. Thereafter, their numbers slowly decline. *Lactobacillus casei* and *L. plantarum* appear adventitously and, after several months of curing, may constitute 99% of the bacterial population. *Streptococcus faecalis* and *S. durans* have been used experimentally as starters.

Beginning in the 1890s cheddar cheeses were wrapped in cheesecloth and coated with paraffin wax to reduce dehydration and to limit the surface growth of molds. Today cheddar cheese is wrapped in plastic film, which prevents dehydration and growth of mold. The cheese may be cold-cured at 0°C for 3 to 12 months, or warm-cured at 10 to 15°C for 13 to 18 weeks. During this period the sharpness and intensity of flavor increase.

Numerous investigations (Bills and Day, 1966; Bills et al., 1965; Day, 1967; Lawrence, 1966; Lawrence et al., 1972; Reiter et al., 1966; Sandine et al., 1972) have been conducted to identify the bacteria that determine flavor and aroma. Much research has been done on *Streptococcus lactis*. Because this bacterium is the host of strain-specific bacteriophages, starter cultures are made with mixed strains. To eliminate fruity and musty flavors, strains with a pronounced hexose monophosphate pathway of carbohydrate metabolism are avoided. *S. lactis* subsp. *diacetilactis* also contributes these flavors and, by producing carbon dioxide from the citrate in the milk, contributes to openness, or gas pockets in the cheese. *Leuconostoc* spp. produce carbon dioxide and also ethanol, which becomes esterified with hexanoic and decanoic acids, producing a fruity flavor. *S. lactis* var. *maltigenes* forms methyl isobutanal and methyl butanal from leucine and isoleucine, imparting a malty flavor. Adventitious microorganisms such as *Pediococcus*, *Micrococcus*, and enterococci have relatively little influence on the course of the fermentation or curing, although the enterococci may be responsible for the high content of tyramine which is found in some lots of cheese.

Endoenzymes liberated by bacterial autolysis produce proteolytic and lipolytic activity; the balance of that activity is the source of the typical flavor and aroma of a cheese. The primary role of the lactobacilli appears to be to convert casein to soluble nitrogenous components. As much as 35% of the casein may be solubilized in curing.

Swiss Cheese

Swiss cheese and cheeses of its type, are distinctive by having holes, or eyes. The eyes are produced by species of *Propionibacterium*, which ferment lactic acid to CO_2 and to propionic and acetic acids during ripening. Produced anaerobically within the cheese and trapped there, the gas forms the eyes. Figure 16-4 shows the series of reactions carried out by the propionibacteria (Allen et al., 1964) during the fermentation of lactate.

Most of the Swiss cheese made in Switzerland (its country of origin) is manufactured in the Emmen Valley and is therefore called Emmentaler cheese. There many family-owned cheesemaking enterprises produce it in 200-pound wheels by an age-old traditional process. Figure 16-5 shows the steps in Emmentaler cheesemaking as practiced in Switzerland.

Starters for Emmentaler cheese are rod- and coccus-shaped bacteria as well as the eye-forming organisms. The rod cultures may be *Lactobacillus helveticus, L. bulgaricus,* and *L. lactis;* the coccus is *Streptococcus thermophilis,* which, together with the rod, is selected for growth and its ability to produce acid during cooking (at 50 to 55°C). The starters are added as milk or whey cultures, usually in a 1:1 ratio at about 0.1% by volume of the vat milk. Eye formers may or may not be added, because, although made from clarified raw milk with a low bacterial count, the cheese will contain some propionibacteria. When added, *Propionibacterium freudenreichii* var. *shermanii* is used, in amounts ranging from one to 25 ml of broth culture in 1000 pounds of milk. Frozen concentrates of starter and eye-forming bacteria are commercially available for use in inoculating bulk starters or direct addition to the vat milk, as the propionibacteria are.

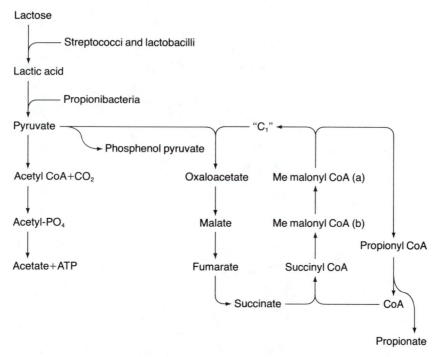

Figure 16-4

Reactions by which propionibacteria in Swiss cheese produce CO_2, proprionate, and acetate from lactic acid. Me-malonyl CoA is methyl-malonyl-Coenzyme A and (a) and (b) represent two isomers produced during the reaction.

Emmentaler produced by the traditional process has a tough, dry surface called a rind, but most of the Swiss cheese produced in the United States is made by a process that prevents rind formation. Starters for rindless Swiss cheese include rod and coccus types as well as propionibacteria. *Streptococcus lactis* or *S. cremoris* is frequently used to ensure rapid early acid production during milk coagulation and early in the cooking phase, up to about 40°C. The *S. thermophilus* then produce acid up to about 50°C, after which the lactobacilli grow, especially during pressing and thereafter. The eye formers grow rapidly in the warm room during the two to seven weeks the cheeses are held there. Thus, there is a succession of four different bacteria—each essential to producing the final cheese.

A

B

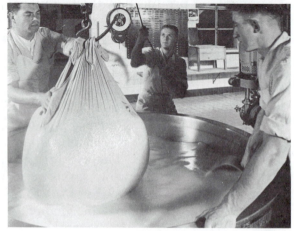

C

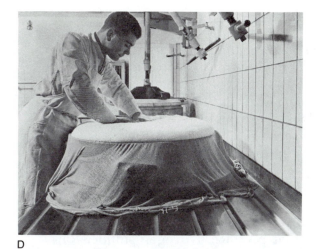

D

E

F

Figure 16-5

Steps in Swiss Cheese manufacture as practiced in Switzerland: A. After renneting, the curd is cut (by "harping") before being cooked in the copper kettle; B. Inserting the cheesecloth to remove the curds; C. Removing the curd; D. Placing the curd in the wheel-shaped hoops; E. Pressing the curd overnight prior to dry and brine salting; F. Turning the cheese in the warm (20 to 25°C) fermentator room; Here the wheels also are washed and lightly salted three times a week. (Courtesy of Switzerland Cheese Association, Inc.)

Gruyere cheese (also named for the area of its origin in Switzerland) is similar to Emmentaler but usually has less moisture, fewer and smaller eyes, and a somewhat stronger flavor because some surface ripening occurs during curing.

Emmentaler cheese does not acquire its typical flavor until after eye formation is extensive. The 6 to 10 months in the curing cellar allows sufficient lipolysis to liberate small amounts of butyric, caproic, caprylic, and capric acids. Free amino acids are believed responsible for the sweetness typical of well-aged Swiss cheese.

Cheese becomes defective through growth of butyric acid and species of clostridia that produce hydrogen sulfide. Raw milk contaminated with dust or manure will contain these organisms—even if the milk is pasteurized, lactate fermentation by the clostridia creates gas, and off-flavors may occur in the ripening cheese. Milk hygiene is the best means of control, although in Europe the antibiotic nisin is also used effectively as a control (Hirsch et al., 1971).

Blue Cheese

Several blue-veined cheeses are made throughout the world, such as Roquefort in France, Gorgonzola in Italy, Stilton in England, Danablue in Denmark, and blue in the United States. Blue cheese may be made from raw, heated, or pasteurized milk. A common practice is to separate cream from raw milk, bleach the cream with benzoyl peroxide, then homogenize it and recombine it with the skim milk to provide a fat content of about 3.8%. The bleaching provides a white cheese against which the blue mold *(Penicillium roqueforti)* will appear distinct and attractive. Homogenization is done to rupture the fat globule membrane so that natural milk lipase can attack the milk fat more extensively and liberate free fatty acids, which are important in Blue cheese flavor.

The treated milk is pumped at 30°C into a vat of appropriate size and inoculated (2 to 5%) with a lactic streptococcal culture composed of several strains of *Streptococcus lactis* or *S. cremoris*. Also added (at 2 to 5 grams/1000 pounds of milk) are *Penicillium roqueforti* spores that have been previously prepared by growing the mold on bread cubes that are then dried and

Box 16-5

Roquefort cheese deserves special mention because of its long and unique history, having been produced in the first century A.D., if not earlier. All Roquefort cheese is produced from ewe's milk and is manufactured in the village of Roquefort in Southwest France. An underground stream keeps natural limestone caves in the area at a constant temperature and humidity. Moist air rises from the caves through horizontal crevices, issuing through the floor of a high plateau. Over 30 million pounds of this cheese are produced each year by the villagers, working cooperatively raising the sheep, milking the ewes, making the cheese, and curing it in the caves. On August 31, 1966, the parliament of Toulouse passed a law stipulating that the name Roquefort could be applied only to cheese cured in the natural caves at Roquefort.

ground into a fine powder. Dried mold spores also may be added to the cheese curd as the hoops are being filled when salt (1.0 to 1.5%) is added. After inoculation and ripening to allow titratable acidity to increase by about 0.05%, rennet is added. When the milk is firmly coagulated (1 to 1½ hours), the curd is cut and allowed to "repair" for about 15 to 20 minutes. When the whey is partially removed the curds are dipped with cheesecloth into a concave draining form. The curds are then placed in perforated cheese hoops about seven inches in diameter on a draining table and turned frequently during the next one to two hours, to encourage draining and matting. The next day the cheese hoops are removed and, for the next 1 to 1½ weeks, the cheeses are individually dry salted each day in a cold room (5°C)—or they may be held in refrigerated brine for two to three days. The cheese may be dipped in paraffin wax, punched with needles to admit air, and held in the curing room at 10 to 12°C for three months. If not waxed, it is removed at one month intervals and brushed lightly to remove excessive mold growth.

During Blue cheese manufacture the lactic starter bacteria function primarily to produce cheese curd and a clean acid flavor; the typical flavor and aroma of the mature cheese are produced mainly by *Penicillium roqueforti*. The low pH (4.5 to 4.7), high salt content (5%), and reduced oxygen conditions of the cheese favor growth of the mold, which reaches its maximum within two weeks after the aeration holes are punched in the cheese. Within three months the pH increases to about 6.2, and active proteolytic enzymes produced by the mold make the cheese become progressively softer. Free fatty acids continue to be liberated, especially caproic, caprylic, and capric acids; a number of ketones, especially methyl-n-amyl ketone, also are produced and contribute to the flavor. Within 14 to 20 days unwaxed cheese will develop a reddish-brown surface flora consisting of micrococci and *Brevibacterium linens,* which also contribute to the desirable flavor of the cheese.

The most common defects in blue cheese are insufficient or excessive blue mold and excessive surface growth of the *Brevibacterium linens.* Waxing is practiced to control the latter, but it is generally agreed that ripening conditions that promote optimum growth of surface flora yield better-flavored cheeses. *Penicillium roqueforti* requires iron for good growth and pigment development, and a food grade iron salt may have to be added to the milk. A leakage or water exuding defect is also sometimes encountered and is attributable to any of a number of manufacturing variables (Pedersen et al., 1971).

Camembert Cheese

Camembert cheese is a relatively soft, perishable cheese with a short shelf life. It has been manufactured in France since 1791 (Moquot, 1955). Legend has it that Napoleon Bonaparte named the cheese after a small hamlet in the district of Orne, France, where he first tasted it.

The cheese is made from whole raw or pasteurized cow's milk (Lombard and Bester, 1975). The milk is warmed to 32°C and lactic acid culture *(Streptococcus lactis)* and rennet are added. When the curd has formed it is cut into small cubes, which undergo a short holding for shrinkage and are then dipped into hoops four to five inches in diameter. The hoops are placed on mats for draining, and metal disks are placed on the cheese to exert a mild pressure. After the cheese has drained and formed it is immersed for 30 minutes in saturated salt brine to develop the firm rind. On the second day the cheese is sprayed with a suspension of spores of *Penicillium camemberti.* The cheese is cured at 12°C in an atmosphere of 90% relative humidity to enable the mold to grow. After 8 to 10 days it is wrapped in foil and cured in a cold room for an additional 20 days. During this time proteolytic enzymes diffuse from the surface into the interior, softening the cheese. The cheese is considered marketable when a small residual core remains in its center. The proteolytic enzymes remain active during the marketing period, and if not consumed the cheese is overcured, or unpleasantly soft and ammoniacal. *Brevibacterium linens* and *Candida, Geotrichum,* and *Mucor* spp. may appear adventitiously, or they may be added to the surface of the cheese. Products of their metabolism help the flavor develop.

Parmesan Cheese

One of several very hard cheeses made from partially skimmed milk and grated for use in cooking, Parmesan cheese has been made for centuries in Italy and is now also manufactured extensively in the United States. Cultures of *Lactobacillus bulgaricus* or similar organisms, and *Streptococcus thermophilus* are used to inoculate the milk at 37°C and rennet is added at the same time. When a firm curd has formed, it is cut and cooked out to firmness at 46 to 54°C. Curds are then dipped, drained in cheese cloth, hooped, and pressed for about 20 hours at 15 to 18°C. Then the cheese is held for several days at 15°C and, after removal from the hoops, is brine salted for about two weeks. It then is dried on shelves or in the sun for 7 to 10 days and cured for about a year at 12°C in a room with 80 to 85% relative humidity. The cured cheese is very hard and keeps indefinitely; it has a moisture content of about 30% and a fat content of 28%.

Sapsago Cheese

Sapsago is another hard grating cheese that has been made in Switzerland for hundreds of years. Its greenish color and pungent odor and flavor are due to a powdered alpine clover (*Melitolus coerulea*) that is in the cheese. Partially soured skim milk is put into a round kettle and stirred while heated to boiling. Cold buttermilk is slowly added with continuous stirring and heating. Sour whey is added until the casein precipitates and settles. Then the curd is collected by dipping with cheese-cloth, spread on a draining table, salted, placed in wooden hoops, and pressed under heavy pressure at 15 to 16°C for several hours. The curd is under lighter pressure during a five week ripening period, after which it is mixed with salt and dried clover at the rate of 5 and 2.5 pounds respectively per 100 pounds of curd. A homogeneous paste of the mixture, produced by stirring, is added to small cylindrical forms and cured for about five months at 15 to 20°C. Sapsago cheese has about 38% moisture, 5 to 9% fat, and 4 to 5% salt.

Mozzarella Cheese

Mozzarella is a mild, soft cheese that may be eaten fresh, but a variety having less moisture (Pizza cheese) is used in cooking. It originated in Italy and was made from buffalo milk but is now made from cow's milk. Whole milk is pasteurized and cooled to 35 to 37°C and inoculated with starter consisting of *Streptococcus thermophilus* and *Lactobacillus bulgaricus*. Rennet is added and the milk is allowed to coagulate to a firm set in 30 to 35 minutes. The curd is cut but not cooked and the whey drained. The curd is hooped into block-shaped hoops and allowed to mat and drain in a warm room until a pH of 5.2 to 5.3 is reached. Then the blocks of cheese are heated in hot water, pulled and stretched to develop desirable body characteristics, and molded into the familiar ball shapes. Curd-stretching machines are also now widely used to work the heated cheese. It then is lightly brine-salted and is ready for consumption. Whole-milk Mozzarella cheese contains 18% fat and 54% moisture, whereas Pizza cheese contains 24% fat and 47% moisture.

References

Allen, S. H. G., R. W. Kellermeyer, R. L. Stjernholm, and H. G. Wood. 1964. J. Bacteriol. *87*, 171.

American Public Health Association. 1967. *Standard Methods for the Examination of Dairy Products*. 12th ed. New York.

American Public Health Association. 1978. *Standard Methods for the Examination of Dairy Products*. 13th ed. Washington, D.C.

Bills, D. D., and E. A. Day. 1966. J. Dairy Sci. *49*, 1473.

Bills, D. D., M. E. Morgan, L. M. Libby, and E. A. Day. 1965. J. Dairy Sci. *48*, 1168.

Blackburn, B. O., and E. M. Ellis. 1973. Appl. Microbiol. *26*, 672.

Bremel, R. D., L. H. Shultz, F. R. Gabler, and J. E. Peters. 1977. J. Food Protection *40*, 32.

Burgwald, L. H. 1939. J. Dairy Sci. *22*, 853.

Daly, C., W. E. Sandine, and P. R. Elliker. 1972. J. Milk Food Technol. *35*, 349.

Day, E. A. 1967. The chemistry and physiology of flavors. In: *Symposium on Foods*. Schultz, H. W. *et al.*, ed. Avi, Westport, Conn.

Duthie, A. H., C. G. Woelfel, K. M. Nilson, and H. V. Atherton. 1976. J. Milk Food Technol. *39*, 774.

Eddy, B. P. 1960. J. Appl. Bacteriol. *23*, 189.

Finley, R. D., H. B. Warren, and R. E. Hargrove. 1968. J. Milk Food Technol. *31*, 382.

Guthrie, E. S. 1923. *The Book of Butter*. Macmillan Co., New York, N.Y.

Hankin, L., and W. F. Dillman. 1968. J. Milk Food Technol. *31*, 141.

Hirsch, A., E. Grinsted, H. R. Chapman, and A. T. R. Mattick. 1951. J. Dairy Res. *18*, 205.

Hurst, A. 1972. J. Milk Food Technol. *35*, 418.

Jarvis, B. 1973. J. Appl. Bacteriol. *36*, 723.

Jelliffe, V. B. and E. F. P. Jelliffe. 1971. The uniqueness of human milk. A symposium. Amer. J. Clin. Nutr. *24*, 968.

Kon, S. K., and A. T. Cowie. 1961. *Milk: The Mammary Gland and Its Secretion*. Academic Press, New York.

Kosikowski, F. W. 1968. J. Milk Food Technol. *31*, 174.

Lawrence, R. C. 1966. N.Z. J. Dairy Technol. *1*, 122.

Lawrence, R. C., L. K. Creamer, J. Gilles, and F. G. Mortley. 1972. N.Z. J. Dairy Sci. and Technol. *7*, 32.

Lombard, S. H., and B. H. Bester. 1975. So. African J. Dairy Technol. *7*, 141.

McDowall, F. H., J. A. Singleton, and B. S. Le. Heron. 1960. J. Dairy Res. *27*, 91.

Moquot, G. 1955. J. Soc. Dairy Technol. *8*, 17.

Mourey, A., and G. Kilburtus. 1976. J. Appl. Bacteriol. *40*, 47.

National Communicable Disease Center. 1967. U.S. Department of Health, Education, and Welfare. *Salmonella Surveillance Report No. 61* (June).

Olson, J. C. 1963. J. Dairy Sci. *46*, 362.

Olson, J. C. 1969. J. Milk Food Technol. *32*, 323.

Overcast, W. W., and D. J. Weakley. 1969. J. Milk Food Technol. *32*, 442.

Pedersen, D. H., E. R. Vedamuthu, G. W. Reinbold, and C. J. Washam. 1971. J. Dairy Sci. *54*, 1615.

Pont, E. G., and W. P. Rogers. 1962. Austr. J. Dairy Technol. *17*, 173.

Randolph, H. E., T. R. Freeman, and R. D. Peterson. 1964. J. Milk Food Technol. *28*, 92.

Reiter, B., T. F. Fryer, M. E. Sharpe, and R. C. Lawrence. 1966. J. Appl. Bacteriol. *29*, 231.

Reiter, B., V. M. Marshall, L. Bjorck, and C. G. Rosen. 1976. Infection and Immunity *13*, 800.

Reiter, B., and J. D. Oram. 1967. Nature *216*, 328.

Rosenau, M. J. 1912. *The Milk Question*. Houghton Mifflin.

Sandine, W. E., C. Daly, P. R. Elliker, and E. R. Vedamuthu. 1972. J. Dairy Sci. *55*, 1031.

Saraswat, D. S., G. W. Reinbold, and W. S. Clark, Jr. 1965. J. Milk Food Technol. *28*, 245.

Sardinas, J. L. 1972. In: *Advances in Applied Microbiology, 15*. D. Perlman ed., Academic Press, New York, N.Y.

Schalm, O. W., E. J. Carroll, and N. C. Jain. 1971. *Bovine Mastitis*. Lea and Febiger.

Sorrells, K. M., and M. L. Speck. 1970. J. Dairy Sci. *53*, 239.

Speck, M. L. 1961. J. Milk Food Technol. *24*, 358.

U.S. Public Health Service. 1967. *U.S. Public Health Service Bull. 229*, Washington, D.C.

17

Meat and Meat Products

The earliest human meat eaters were hunters, but gradually, over centuries or millennia, they took to keeping wild animals for fattening and slaughter. From this practice came animal husbandry, some 10,000 years ago, when nomadic tribesmen of central and southern Asia domesticated cattle, using them for milk and meat. By the rise of the great civilizations in the Middle and Near East, beginning some 5000 years ago with the Egyptians, agriculture was flourishing and horses, cattle, sheep, and goats were grazed. There was already significant trade not only in their flesh but in their by-products as well—hides, wool, milk, and cheese. Rabbits and swine were domesticated and were important sources of meat. Large supplies of fish harvested from the Mediterranean and Black seas supplemented local catches.

Box 17-1

The United States meat industry is the world's largest. Some 36 million cattle, 85 million swine, 9 million sheep, and 2.5 million calves, as well as small numbers of goats and horses, are slaughtered each year, providing more than 40 billion pounds of meat and meat products. Americans annually spend two-fifths of their food dollars for meat and poultry, and consume about 178 pounds of red meat per capita per year. Meat from cattle, hogs, poultry, and sheep supplies at least one-third of the human need for protein and fat, about 0.3 of thiamin and iron requirements, as well as 0.4 of the niacin and 0.2 of the caloric and vitamin A needs. Meats are nutritious not only for humans, but for a wide variety of microorganisms as well. Hence, it is most difficult to prevent or retard spoilage of meats and meat products.

Meat Inspection

History

As long ago as 2000 B.C. the Egyptians introduced supervisory laws regulating inspection and sanitation during slaughter. Biblical laws governing "kosher" (Hebrew for fit or proper) slaughter date from about 1300 B.C., the time of Moses: only that animal flesh that is defined as clean can be used according to the ancient laws set forth in the Old Testament. People of the Jewish religion continue to abide by these Biblical tenets; cattle, calves, sheep, and poultry intended for Jewish consumption are killed and bled by men ordained under the Mosaic law to perform such duties.

Today nearly every country in the world regulates the preparation and handling of meat; and animals are inspected for wholesomeness and graded for quality. In the United States federal inspection was initiated in 1891. However, Upton Sinclair's description of meat handling in a Chicago packing house, in his novel *The Jungle,* and the rejection of American meats on the international market stimulated sufficient public concern so that in 1906 the Wholesome Meat Act of 1906 was passed in Congress. This Act, virtually unchanged for the next 60 years, except for broadening to include inspection of horse meat by passage of the Horse Meat Act of 1919, was rigidly enforced by the federal government. With enactment of the Imported Meat Act of 1930 wholesomeness, cleanliness, and truthful labeling of imported meats were subjected to equal scrutiny and to an inspection program equivalent to that in the United States.

In 1962 the Talmadge-Aikin Provision allowed federal inspection to be conducted by state meat inspectors. Throughout this period laws regulating meat products moving in intrastate channels varied greatly in accordance with the quality of differing state meat inspection programs. In 1963 the United States Congress asked the United States Department of Agriculture (USDA) to assess the quality of the various programs, but the report was kept confidential. When an exposé published by a reporter for the Des Moines Register revealed that "sanitation and inspection quality were often horrendous and sometimes worse," further regulatory action was taken and the Wholesome Meat Act of 1967 was passed. This law specifies that all states must have inspection systems equal to federal inspection. Today, except for the custom slaughtering of game animals or for animals slaughtered for family use on the farm, almost all carcasses are inspected by a veterinarian or lay inspector in accordance with the requirements of the Food Safety and Quality Service (FSQS) of the USDA. Except for three states that elected to allow federal inspectors to operate intrastate, inspections of slaughter carcasses may be made by state meat inspectors in accordance with FSQS requirements.

Zoonoses

There are some 70 zoonoses, or diseases that can be transmitted from livestock to humans. The most important are tuberculosis, brucellosis, salmonellosis, anthrax, rabies, leptospirosis, listeriosis, erysipeloid, staphylococcal disease, pasteurellosis, streptococcosis, tetanus, tularemia, actinomycosis, moniliasis, histoplasmosis, coccidioidomycosis, cryptococcosis, geotrichosis, and foot-and-mouth disease (rarely transmitted to humans), Q fever, cow pox, balantidiasis, toxoplasmosis, taeniasis (including cystocercosis), hydatidosis, and trichinosis. Veterinary inspection is designed to prevent or interrupt this mode of transmission (see Table 17-1).

Table 17-1

Carcasses found by FSQS to demonstrate infectious disease or parasitic conditions upon postmortem inspection, fiscal year 1976 (abridged).

| | Number of carcasses condemned | | | |
Cause of condemnation	Cattle	Calves	Sheep & lambs	Swine
Infectious diseases:				
Actinomycosis, Actino-bacillosis	833	2		10
Anaplasmosis	308			
Caseous lymphadenitis			11,637	
Coccidioidal granuloma	23		1	1
Swine erysipelas				3,165
Tuberculosis, nonreactor	86	1		4,996
Tuberculosis Reactor with lesions	60			
Miscellaneous	184	79	8	54
Inflamatory diseases:				
Enteritis, gastritis, peritonitis	5,087	1,617	289	9,129
Eosinophilic myositis	3,878	9	28	28
Mastitis, mammitis	1,012	2	4	43
Metritis	1,932	5	161	904
Nephritis, pyelitis	3,825	257	639	2,661
Pericarditis	4,585	76	223	1,350
Pneumonia	13,398	4,368	6,912	12,553
Miscellaneous	462	249	404	424
Parasitic conditions:				
Cysticercosis	135		189	4
Sarcosporidiosis			3,695	
Stephanuriasis				2,322
Miscellaneous	88	3	133	74
Septic conditions:				
Abscess, pyemia	11,582	534	2,116	28,136
Septicemia	7,749	2,318	828	3,878
Toxemia	3,535	384	627	2,738
Miscellaneous	169	154	32	53
Other:				
Arthritis, polyarthritis	1,468	2,786	2,102	19,332
Asphyxia	56	272	41	2,727
Bone conditions	21	1	2	38
Contamination	1,135	36	316	5,111
Icterus	477	2,133	1,688	9,884
Uremia	1,364	45	1,265	874
Miscellaneous general	4,445	1,564	374	5,308

Source: Food Safety and Quality Service, U.S. Dept. of Agriculture, 1977.

Inspection, Antemortem and Postmortem

All inspections are made by a veterinarian or are conducted under his supervision. Two types of animal inspection are made: *antemortem* and *postmortem*. The latter is much more exhaustive and detailed than the former.

Regulations require that the live animal be observed while in motion and at rest. A dead or dying animal, one that is comatose, or evidences parturition, displays symptoms of disease, has an abnormal temperature ($> 41.5°C$), or exhibits other abnormal symptoms that would render its meat unfit for food is tagged "U.S. Condemned" and is eliminated from the food chain. In addition, seriously crippled or immature animals, "downers" (prostrate animals unable to rise to their feet), positive tuberculin test reactors, those with eye cancer, and brisket edema are tagged as "U.S. Suspect," as are swine with vesicular exanthema or minor erysipelas. All animals so tagged are subjected to thorough postmortem inspection.

So that postmortem inspection can proceed as the animal progresses along the dressing line, the movement of carcass and viscera is synchronized, and duplicate numbered tags assure positive identity of carcass, head, and viscera. When any abnormal glandular condition or other unusual condition in the meat is noted, a "U.S. Retained" tag is affixed and that carcass or part is removed from the normal flow of product until a complete inspection is possible. Upon final inspection, all carcasses or parts found to be unfit for food are tagged "U.S. Inspected and Condemned" and thereafter handled in such manner that they cannot be introduced into edible products. Generally they are put in sealed rendering tanks, chemically denatured, incinerated, or are segregated from the normal product flow, cooked (tanked), and otherwise handled, in quarantine rooms, for inedible use.

Fresh Meats

Huge numbers of organisms are associated with the animal's hide, hoofs, and hair, with the gut, and with excrement. Because dirt is an important source of contaminants that cause meat spoilage, handling methods before, during, and after slaughter determine to a large extent the ultimate storage life of meats.

Resident Flora and Contaminants

Resident Flora on Hides. Microorganisms abound on the hides, skins, and wool of cattle, swine, and sheep, and are an important source of contamination. Counts of 10^5 to 10^7 aerobic bacteria per cm^2 are commonplace on cattle hides (Empey and Scott, 1939a), and numbers ranging from 10^5 to 10^{12} have been observed on hogs (Jensen and Hess, 1941); similarly heavy microbial loads are present on unwashed wool. It may seem paradoxical that beef prepared in tropical climates has a potential storage life some two to three weeks longer than has similar meat prepared in temperate zones, or that in subtropical regions, beef prepared in summer has a storage life several days longer than beef prepared in winter. The answer is that the meats in the tropics and the summer-prepared meats are less contaminated by low-temperature mesophiles. Because of this difference in bacterial temperature response, smaller numbers of bacteria grow in tropical soils than in temperate soils, and thus fewer are found on cattle, and, ultimately, on meat. Empey and Scott (1939a) considered meat from animals living in hot climates to be better suited for cold storage than that from beasts living in colder regions.

Various procedures have been advocated for reducing microbial loads on animal hides, hoofs, and hair. A forced water spray of 20–25 gallons over the whole hide area of cattle removes about half the initial microbial load. Using 0.025, 0.05, 0.10 and 0.15% chlorine solutions 15 minutes before processing animals has produced microbial kills of 82, 90, 94.3, and 95.3% respectively. (Empey and Scott, 1939a). A hot water wash (74°C) of the outside of hog carcasses before and after evisceration produces inconclusive results (Shotts et all., 1969). However, still hotter (75 to 90°C) water rinses of carcasses reduce microbial numbers by one log unit (Patterson, 1971).

Two novel techniques for sanitizing carcasses were recently developed.

1. Hog carcasses are treated after slaughter with a mixture of 1.5 M acetic and propionic acids in a 60:40 w/w ratio (pH 2.3); total microbial numbers are reduced by two log units without detriment to the carcass (Reynolds and Carpenter, 1974).

2. U.S. Patent No. 3,114,450, issued to Swift and Co., the meat processors, covers an intermittent chlorination (Clor-Chil) process for beef and pork: the hot carcass is sprayed with no more than 200 ppm chlorinated water and while in the cooler receives an intermittent spray of chlorinated water for a period of from four to eight hours. The spraying is programmed to maintain a relative humidity of 97 to 99% for pork and 94 to 95% for beef. The process both significantly reduces microbial numbers and prevents customary shrinkage. Interim guides issued by FSQS stipulate that the chlorine must not exceed 200 ppm, that the temperature of application be limited to $< 27°C$, and that the spray time be limited to 2 minutes per 30 minute period.

Flora in the Gut. Huge numbers of myriad varieties of microorganisms are swallowed with food and water and other materials introduced into the gastrointestinal tract, but few escape the tract and find their way into the rest of the body. Except for the enteric pathogens, organisms present in water, soil, or food that are swallowed rarely cause difficulty; the vast majority are killed or inhibited within minutes after ingestion. Inimical to the survival of most organisms are enzymes in the saliva and in the gut, gastric acidity, bile salts, the viscosity and alkalinity of the bile, anaerobic conditions, and high mammalian body temperatures. Even among salmonellae, hundreds of thousands and even millions of organisms are needed to cause illness (McCullough and Eisele, 1951a, b, c, d). To produce infection, from 10^3 to 10^5 more cells are required when introduced by way of the mouth than when injected into the peritoneal cavity (Sarles et al., 1951).

The lower small intestine and the colon contain astronomic numbers of bacteria. While the thick mucous membranes that line the gut wall hold this microflora in check, living and dead bacteria are often from 15 to 19% of the total weight of fresh fecal material (Lissauer, 1906).

Tissue Contamination. The intact skin and the animal's defense mechanisms effectively prevent bacterial penetration, but infection takes place where the skin is broken or when the local inflammatory response is insufficient or too slow. There is much controversy about the source of tissue contaminants (Ayres, 1955); there is adequate support for the hypothesis that a few microorganisms may be present in certain of the living tissues (Burn, 1934; Haines, 1937; Adamson, 1949). Bacteria in the deep tissues reach the lymph nodes via the lymphatic capillaries (Lepevetsky et al., 1953) by "fixed" macrophages. Two hours after slaughter, 40% of the lymph nodes contain organisms (Reuter and Trukenbrod, 1964). Almost a fourth of these are cocci (24.2%); about one of 20 (4.8%) is a heat-resistant spore. Ordinarily the lymph nodes serve as filters where the bacteria are engulfed and digested. However, when microorganisms are too numerous or are invasive they continue to multiply, causing swelling and soreness of the nodes. Bacteria belonging to the genera *Enterobacter, Alcaligenes, Bacteroides, Clostridium, Corynebacterium, Escherichia, Flavobacterium, Micrococcus, Proteus, Pseudomonas, Serratia,* and *Streptococcus* were recovered from 15 of 23 samplings of lymph nodes from beef chuck and rounds—but, by contrast, were isolated from only 3 of 23 marrow samples and 2 of 23 muscle samplings (Lepevetsky et al., 1953).

Glycogen Reserve

When animals are slaughtered and circulation stops, the pH of muscle falls steadily as the glycogen reserve is converted to lactic acid. The ultimate pH is proportionate to the amount of glycogen present at the time of death (Bate-Smith, 1948). If the animal is exercised (fatigued) shortly before slaughter, much of its muscle glycogen has already been converted to lactic acid and other glycolytic intermediates and is lost during exsanguination. Among hogs that were rested and fed, Callow (1949) found that the ultimate pH of the psoas muscle was pH 5.58, as compared to 5.87 for a control group that was not rested; and Ingram (1949) demonstrated that a slight acidification of the medium markedly reduced the growth rate of a wide variety of bacteria. From these observations Bate-Smith (1948) concluded that when meat reached

its ultimate pH with an excess of glycogen still remaining, the growth of microorganisms may not cause an alkaline shift . . . until the glycogen is completely exhausted. . . . In other words, from the point of view of resistance to bacterial growth there cannot be too much glycogen in the muscles.

Steps in Processing

Skinning. During dressing and processing the lean and cut surfaces of the meat often become heavily contaminated. Cattle and sheep are usually skinned before they are eviscerated, while swine receive a preliminary "scald" to help loosen the hair, making it easier to singe and rub free of the hide. Microorganisms are first transferred from the hide to the underlying tissues in the first stage of skinning, being most numerous just below the incision of the hide and least numerous in the areas furthest from the incision (Empey and Scott, 1939a). Average populations at this juncture are 10^4 to 10^5 per cm^2. The blade introduces some organisms directly when it incises the exposed tissues, while other organisms may be transferred during separation of the hide from the carcass. The hands, arms, and clothing of workers transfer additional contaminants to the carcass, in the presence of moisture, blood, and tissues. By the end of a day's operation as many as 3 billion per gram or 30 million per cm^2 of these organisms may be in material scraped from a skinner's clothing.

Scald Tank. Scald tank water is usually maintained at a temperature of 63 to 65°C—adequate for reducing the number of psychrophilic and pathogenic bacteria but not for many of the mesophilic organisms or thermophilic bacteria and spore-forming organisms. Because bits of soil, fecal material, blood, and other foreign matter on the feet, hair, and hides of animals slough into the water, millions of microorganisms are present in every milliliter of the water after the first few carcasses have traversed the tank (Ayres, 1955). As a more sanitary procedure than the scald tank, Danish meat packing plants are arranged to have hog carcasses pass through steam condensate from a boiling water bath. The microbiological advantages of this procedure, however, remain to be determined.

Cutting. Material adhering to saw blades, knives, cleavers, pans, and other utensils, swab cloths, tables, and conveyor belts is heavily populated by microorganisms and mechanically inoculates newly exposed cut flesh.

After the animal has been skinned and eviscerated the carcass is chilled to free it of body heat, firm the flesh, and retard microbial and chemical changes. Hog carcasses are chilled for a much shorter time than beef. In some instances pork is cut before it has lost animal heat; such dismembering is called "hot cutting."

The temperature in the cooler is kept as low as possible to quickly and adequately cool the thickest parts of the carcass. Also, a relative humidity of about 90% is maintained to prevent excessive shrinkage. If beef carcasses are aged for a week or more, humidity must be carefully monitored to prevent excessive microbial proliferation. Air circulating in chill rooms contains many bacteria, molds, and yeasts. An electrostatic sampler placed in a beef cooler to monitor the air for contamination by mold spores indicated a range of from 882 to 4500 spores per plate using an exposure time of 2 minutes (Richardson et al., 1954).

Of the organisms isolated by several investigators from fresh meat held at chill room temperatures (Table 17-2), many are casual or adventitious residents and only a few find the substrate and temperature to their liking. Since the principal sources of these organisms are soil, water, air, and humans, transplanted cells must be able to adapt quickly to an entirely different environment. It has been surmised that growth on the surface of the whole or quartered animal is so limited because the extensive covering layer of fat and connective tissue offers a poor supply of nutrients (Haines, 1933). A further restriction is that many of the transient forms, being mesophiles, grow scantily at low temperature and are poor competitors with psychroduric spoilage bacteria (Ayres, 1951).

Stored, unfrozen beef suffers primarily only surface spoilage (Moran, 1935; Moran and Smith, 1929), and the deep flesh of beef stored for two weeks at 5°C suffers only negligible increases in the number of organisms.

Microbial Spoilage

If cutter and canner grade beef are maintained, uncut, at a temperature of approximately 4 to 5°C during overnight shipment, aerobic microbial loads on the lean (uncut) surface of knuckles, inside rounds and outside rounds generally range from 10^5 to 10^7/cm^2 or from 10^6 to 10^7/gram. Total numbers on the cut surfaces of similar pieces sliced at the packing house (cut surfaces) and packaged just before shipment usually range from 10^4 to 10^6/cm^2 or 10^5–10^6/gram (Ayres and Adams, 1953) (see Table 17-3). Organisms present on sliced areas were introduced there at the time of slicing, whereas surface flora on the intact carcass tissue develops when the animal is skinned.

Much of the surface of the uncut beef carcass is covered with a layer of fat and connective tissue upon which little or no bacterial growth occurs during normal hanging (Haines, 1933). High counts of flora on pieces of knuckle range from 10^6 to 10^7 per g on cut surfaces and from 10^5 to 10^6 for intact surfaces. Although it is possible to select areas for swabbing or sectioning inside and outside rounds having little or no fat or connective tissue, this is not the case for knuckle cuts. Also, the white fibrous tissue that covers much of the knuckle musculature provides poorer nutrient qualities than do lean cut surfaces. Consequently, microorganisms transferred to these new surfaces during cutting outgrow those surviving on the native tissue during time of shipment.

Microorganisms introduced from the air or water or by handlers and miscellaneous sources in the packing plant, together with those entrenched in the lymph nodes or transferred from the hide or viscera, are redistributed by comminution. The more finely the meat is cut, the more widely the microflora are

Table 17-2
Generic distribution of microorganisms isolated from surface of meats.

Genus represented	Fresh beef				Fresh pork sausage
	(Brooks & Hansford, 1923)	(Haines, 1933)	(†Empey & Scott, 1939a)	(Mallman et al., 1940)	(Sulzbacher & McLean, 1951)
Bacteria					
Pseudomonas		+	+		+
Micrococcus		+	+	+	+
Gaffkya				+	
Sarcina				+	+
Neisseria					+
Lactobacillus					
Microbacterium					+
Alcaligenes					+
Achromobacter[e]		+	+	+	+
Flavobacterium			+	+	+
Escherichia					+
Aerobacter[e]					+
Paracolobactrum					+
Proteus		+			+
Salmonella					
Bacterium				+	+
Bacillus				+	+
Xanthomonas					+
Diplococcus				+	
Streptococcus		+			
Corynebacterium					
Diphtheroids					
Hemophilus					
Serratia					
Azotobacter type		+			
Streptomyces					
Actinomyces		+			
Molds					
Zygorrhynchus					
Mucor	+				
Thamnidium	+				
Rhizopus	+				
Penicillium	+		+		
Sporotrichum	+		+		
Aspergillus					
Geotrichum[a]				+	
Cladosporium			+		
Alternaria			+		
Monascus					
Monilia[b]			+		
Yeasts					
Torulopsis[c]	+				
Rhodotorula			+		
Wardomyces[d]	+				

[a] Odium, Oospora, and Geotrichoides included.
[b] Candida, Mycotorula, and Blastodendrion included.
[c] Also Torula and Cryptococcus.
[d] Debaryomyces isolated from sausage by Mrak and Bonar, 1938.
[e] Moraxella-Acinetobacter and Enterobacter respectively.

Cured meats (Yesair, 1936)	Sausage (Jensen & Hess, 1941)	Frankfurters (Ogilvy & Ayres, 1953)	Chicken (Gunderson et al., 1947)	Ayres et al., (1950)	Cod (Dyer, 1947)	Haddock (Stewart, 1932)
	+	+	+	+	+	+
	+		+	+	+	+
			+			
	+		+	+		
			+	+		
	+	+				
		+	+		+	
			+		+	
	+		+	+	+	+
	+		+	+	+	+
			+	+		
			+	+		+
			+	+		
	+		+	+	+	
			+	+		
		+		+		
			+			
			+	+		
			+			
			+			
			+			
	+					+
			+	+		
			+			
		+				
+		+				
+						
+		+		+		
		+				
+		+				
			+	+		
+		+		+		
+						
+		+	+			
		+	+	+		+

Table 17-3

Incidence of total viable organisms and of spores growing aerobically or anaerobically in packaged raw beef.

Kind of growth	Unit measured	Surface examined	Number of organisms		
			Smallest	Usual range	Largest
Total aerobic	No./cm²	Uncut	100	100,000–10,000,000	20,000,000
		Cut	<100	10,000–1,000,000	3,300,000
	No./gram	Uncut	43,000	1,000,000–10,000,000	400,000,000
		Cut	17,000	10,000–1,000,000	68,000,000
Total anaerobic	No./cm²	Uncut	100	1000–100,000	300,000
		Cut	<100	1000–100,000	100,000
	No./gram	Uncut	3200	100,000–1,000,000	160,000,000
		Cut	3200	100,000–1,000,000	70,000,000
Spores growing aerobically (facultative)	No./gram	Uncut	200	1000–100,000	14,000,000
		Cut	900	1000–100,000	1,900,000
Spores growing anaerobically (putrefactive)	No./100 grams		<0.06	0.7–6.0	140

dispersed; comminuted meat often has counts of 10^6 to 10^7. Weinzirl and Newton (1914) long ago forcefully illustrated the dilemma of establishing meaningful standards for ground meats: they proposed a bacteriological standard of 10^7 organisms/gram for ground meats after observing that if an earlier standard, suggested by Marxer (1903), of not over 10^6/gram were in effect, most ground meats would be condemned.

Spore-Forming Organisms. Cells that grow anaerobically are consistently fewer in number than those that grow in an atmospheric environment—cells that reproduce without free air making up about 20% of the total flora. A surprisingly large number of bacteria survive 20 minutes of heating at 80°C and remain capable of reproducing in atmospheric conditions. This group, the aerobic (or facultative) spores, may play a significant role in the spoilage of canned meats (Niven, 1961).

Because viable putrefactive anaerobic spores are enumerated by use of a Most Probable Number (MPN), results are not directly comparable with plate counts. In no case did the putrefactive anaerobic spore count from 99 separate samplings of packaged raw beef exceed 1.4 spores per gram, and in only 11 samples were there more than 0.06 spores per gram (Ayres and Adams, 1953).

Fresh beef trimmings have higher counts of putrefactive anaerobic (PA) spores—the mean being 1.8 to 11 per gram—than primal cuts have—the mean of which is 0.7 to 6.0 per 100 grams of meat (see Table 17-4).

An extensive study was made of the incidence of mesophilic clostridial spores in 2358 samples of raw beef, pork, and chicken in processing plants in the United States and Canada (Greenberg et al., 1966); mean putrefactive anaerobe levels in beef and pork were 3.0 per gram and in chicken were 2.5 per gram. Of the samples tested, 77% contained three or fewer PA

Table 17-4

Comparison of spore counts obtained on two lots of fresh beef trimmings.

Sample number	MPN spores/gram of meat	
	Lot 1	Lot 2
1	0.80	0.70
2	5.5	46
3	0.23	5.5
4	0.39	1.1
5	2.3	2.5
Total	9.22	55.80
Mean	1.8	11

Source: Steinkraus and Ayres, 1964. Reprinted from Journal of Food Science, Vol. 29, p. 87, 1964. Copyright © by Institute of Food Technologists.

Table 17-5

Incidence of mesophilic putrefactive anaerobic (PA) spores, including those of *Clostridium botulinum,* in raw beef, pork, and chicken.

Type of sample	No. of samples	Total no. of PA spores isolated	Mean PA spores/gram	No. of Botulinal isolates
Beef				
Bloody neck area	298	2929	3.277	0
Trimmings	326	2742	2.803	0
Total	624	5671	3.029	0
Pork				
Bloody neck area	319	3655	3.820	0
Trimmings	337	2308	2.317	0
Total	656	5963	3.030	0
Chicken				
Anterior	373	2673	2.390	0
Posterior	379	3071	2.700	1
Giblets	326	2349	2.403	0
Total	1078	8093	2.500	1
Total, all types	2358	19,727	2.789	1

Source: Greenberg et al., 1966.

spores per gram. PA counts were significantly higher in the neck area than in the trimmings (see Table 17-5). No spores of *Clostridium botulinum* were recovered from beef and pork, and only one PA isolated from chicken (a posterior sample) was a *C. botulinum* (type C). Statistical analyses of the data revealed that the counts of PA spores varied with season—being highest in autumn and lowest in spring.

Off-Odor and Slime. Upon prolonged storage, chilled meats develop off-odor and become slimy. Glage (1901), the first to recognize this defect, characterized the organisms as "aromabakterien," and described the odor that developed during their early growth as rather pleasant. The typically rancid, sweetly aromatic esterlike odor has also been described as a "dirty dishrag" odor. Since this aroma occurs only a few hours before large numbers of minute, translucent, moist colonies appear, it must be considered a sign of incipient spoilage. In comminuted meats off-odor is always evident when microbial populations exceed 10^7 cells/cm^2. The colonies first resemble droplets of moisture, but as they become larger they develop a creamy white color and coalesce to form a uniform sticky or slimy layer. If the atmosphere is sufficiently saturated with moisture, sliminess occurs when the microbial load exceeds 3 to 6×10^7/cm^2 (Haines and Smith, 1933; Ayres, 1960a and b). According to Scott (1936), bacterial growth on beef muscle "only became manifest as slime at relative humidities of 99% and above." When atmospheric moisture is lower, colonies remain small and discrete and look almost as clear as water.

When the initial microbial load exceeds 10^5 organisms/cm^2 on meat stored at 5°C sliminess can be detected within six days (Figure 17-1), but when the initial load is less than 10^2 organisms/cm^2 it does not occur until the twelfth day. The time that elapses before slime occurs on sliced beef varies inversely with the storage temperature: at 0°C, sliced beef with a microbial population of 5×10^2 organisms became slimy after 20 days

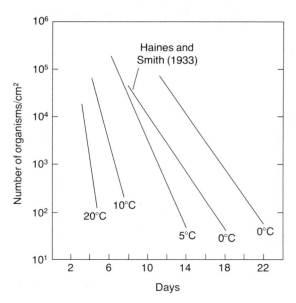

Figure 17-1
Relation of initial numbers of organisms to the time elapsed before slime appears on slices of beef held at 0, 5, 10, or 20°C. (Ayres, 1960. Reprinted from Food Technology, Vol. 25, pp. 1–18, 1960. Copyright © by Institute of Food Technologists.)

of storage; at 5°C, with an initial population of 6×10^2, within 12 days; at 10°C, with an initial population of 3×10^2 cells, in 7 days; and at 20°C, with an initial load of 150 bacteria (1.5×10^2), within 3 days.

On meats stored at 0 or 5°C the numbers of bacteria decline during the first few hours (see Figure 17-2). At these temperatures, it appears, only the spoilage organisms survive or grow—and do so only during the early part of storage—and only psychroduric organisms persist.

Microbial behavior during storage is affected not only by the size of the initial contamination but also by the nature of the

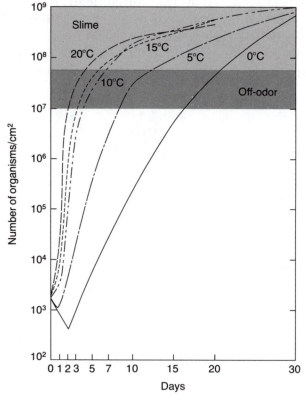

Figure 17-2

Growth rates of organisms on beef stored at 0, 5, 10, 15, and 20°C. (Ayres, 1960. Reprinted from Food Technology, Vol. 25, pp. 1–18, 1960. Copyright © by Institute of Food Technologists.)

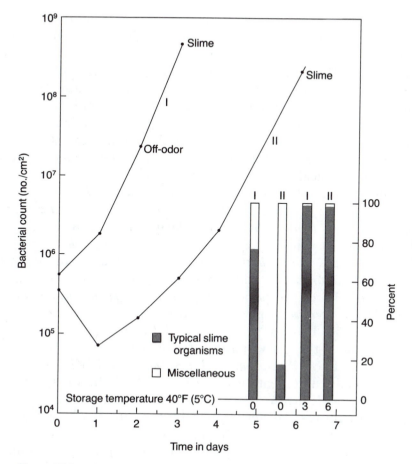

Figure 17-3
How type of initial flora of (I) frozen ground beef trimmings and (II)
experimental fresh ground beef affects their storage lives. The vertical
bars in the right-hand corner give the percentage distribution of
organisms by type at different storage times. (Ayres, 1951.)

organisms involved. As Figure 17-3 shows, two samples of
ground beef having almost identical initial loads developed
off-odor and slime at different rates. One sample (I) was prepared from frozen trimmings; the other sample (II) was made
to order from fresh chunks of chuck and plate. In the 50 representative colonies examined from subcultures, organisms of the

spoilage type (i.e., pseudomonads) were 78% of the isolates from the frozen ground trimmings but only 18% of the isolates from the freshly ground meat. On samples of the latter type (II), over 80% of the flora comprised many gram-positive micrococci, sarcinae, bacilli, and short gram-negative rods. Both lots of meat remained packaged in cellophane until they developed off-odor and/or slime; the trimmings (I) spoiled in three days, the fresh ground (II) took six days. Once the slime was evident, more than 98% of the flora from both lots of meat comprised typical spoilage microorganisms.

The genera *Pseudomonas, Micrococcus, Microbacterium, Corynebacterium, Moraxella-Acinetobacter, Flavobacterium, Enterobacter,* and *Penicillium* are regularly found in meats at the beginning of the storage period. Initially, *Achromobacter, Flavobacterium, Enterobacter* spp., spore-forming organisms, chromogenic bacteria, and molds and yeasts collectively make up 20 to 50% of the population, but *Pseudomonas, Micrococcus,* and *Microbacterium* spp. are the predominant genera (Ayres, 1951, 1955). During storage the flora changes. Except for *Acinetobacter*—the growth rate of which is reduced at $\leq$ pH 5.7—growth of *Pseudomonas, Enterobacter* and *Microbacterium* spp. is maximal in meat juice medium with pH between 5.5 and 7.0 (Gill and Newton, 1977). Maximal cell densities of aerobic spoilage cultures are determined by the rate at which oxygen becomes available to the cells. Flora on meat kept at 0, 5, and 10°C consists primarily of pseudomonads. Having generation times about 30% shorter than those of other species on meat surfaces, the pseudomonads have a growth advantage at temperatures between 2 and 15°C. In addition, they are relatively insensitive to the presence of other species and therefore dominate the aerobic spoilage flora of meat. On meat stored at 15°C or higher, the pseudomonads and micrococci are approximately equal in number. By the time meat develops off-odor or becomes slimy, the flora consists of motile gram-negative rods, usually species of the genus *Pseudomonas* (Ayres et al., 1950). Simultaneously, the protein becomes highly hydrated and tacky to the touch, the free amino acids and nucleotides are much reduced, and the pH rises to 8 through ammonification (Jay, 1972). Strains of *Achromobacter, Alcaligenes,* and *Aeromonas* (see Brown and Wiedemann, 1958; Ayres, 1960a) have occasionally been recovered from off-odor meats. *Flavobacterium, Bacillus, Clostridium,* and *Streptococcus* spp. sometimes occur at 10°C or lower, but enterobacteria and staphylococci are seldom recovered from raw meats stored under such conditions.

Lactics and Gram-Positive Rods. Small nonmotile gram-variable (usually gram-positive) short rods (0.5–0.7 × 1.0–2.0 μm) have been variously identified as belonging to the genus *Lactobacillus* (Kirsch et al., 1952), *Microbacterium* (Ayres, 1951; McLean and Sulzbacher, 1953; Ogilvy and Ayres, 1953; Weidemann, 1965), *Corynebacterium* (Brown and Weidemann, 1958; Kraft et al., 1966), *Butyribacterium,* and *Propionibacterium* (Sutherland et al., 1976). Their ability to hydrolyze starch and produce catalase suggests that they belong to the genus *Microbacterium* rather than *Lactobacillus* or *Arthrobacter.* However, it is likely that some of the isolates belong to the genus *Corynebacterium,* because they use glucose oxidatively and produce L-lactic acid rather than by metabolizing glucose through the Embden-Meyerhof pathway.

The water-holding capacity (WHC) of pre-rigor meat is very high immediately after slaughter, but decreases rapidly, reaching minimal value in a day or two. After another 24–48 hours have passed, the WHC begins to increase and continues to increase slowly with age (Hamm and Deatherage, 1960). Hence, Jay (1964a and b), using aging comminuted beef (hamburger), plotted decreases in extract release volume (ERV) by the filter-paper press method. Jay (1966) also suggested that measurement of free water is a rapid technique for determining microbial spoilage of meats.

Vacuum-Packaged Boxed Meats. Meat formerly handled and transported as hanging eviscerated carcass halves in refrigerated railroad cars and trucks is increasingly being vacuum packaged and boxed as primal cuts. Today about half of all choice beef is vacuum packaged at the processing plant. This not only reduces freight costs—90 carcasses can be boxed and loaded in railroad cars that could hold only 55 carcasses of swinging beef—but also markedly lowers labor costs, improves inventory and product control, reduces trim loss, and prolongs the sanitary life of the product.

Microbial changes of vacuum-packed beef and lamb differ from those of meats wrapped with films permeable to oxygen (Sutherland et al., 1975a; Carpenter et al., 1976). The myoglobin of the surface flesh fails to take on the desirable scarlet color that it acquires when oxygenated in the presence of air. More important, however, microflora growth is retarded in meats packaged anaerobically or vacuum packaged. Since pseudomonads have a higher affinity than other bacteria for O_2, their growth is restricted by O_2 limitation and presence of CO_2 (Kraft and Ayres, 1952), and the characteristic off-odor and slime associated with these organisms does not develop. The dominant microbial floras in stored meats usually are gram-positive, fermentative rods (Ayres, 1951; Kraft et al., 1966; Patterson and Gibbs, 1974). Although some workers characterize the dominant lactic acid organisms as belonging to the genus *Lactobacillus* (Kirsch et al., 1952; Seideman et al., 1976a and b), others consider that while their biochemical reactions may be similar to those of *Lactobacillus* spp., except for catalase activity, they may be members of genera such as *Microbacterium* (Ayres, 1951; Sulzbacher and McLean, 1957; Ogilvy and Ayres, 1953), *Corynebacterium* and *Arthrobacter* (Kraft et al., 1966). Some of the gram-positive, catalase-negative rods usually termed lactobacilli may be strains of *Butyribacterium* or *Propionibacterium* (Sutherland et al., 1976), which can metabolize lactic acid to volatile acids.

When vacuum-packaged meats have been tested for the prevalence of *Microbacterium thermosphactum* (McLean and Sulzbacher, 1957), findings have been somewhat at variance—possibly due in part to differences in initial contamination, type of meat tested (for example, lamb vs. beef), and competitive growth rates. If *M. thermosphactum* is initially present in large numbers on beef, it is a major contaminant, making up 25% of the aerobic flora of meat spoilage (Barlow and Kitchell, 1966; Roth and Clark, 1975). On the other hand, only a low proportion (about 5%) of these organisms occurred in the spoilage flora of steaks cut from vacuum-packaged beef and stored under atmospheric conditions (Sutherland et al., 1975a). *Microbacterium thermosphactum* may be prevalent on mutton (Barlow and Kitchell, 1966; Newton et al., 1977), whereas *Lactobacillus* spp. may prevail on beef (Seideman et al., 1976a and b).

Under anaerobic conditions *Lactobacillus* spp. grow faster than *M. thermosphactum* and *Enterobacter* spp. do; and in lightly contaminated meats lactobacilli proliferate rapidly and markedly inhibit growth of *M. thermosphactum* (Roth and Clark, 1975). Conversely, *Lactobacillus* spp. are detected only in atmospheres where competing organisms grow more slowly (Newton et al., 1977). In vacuum-packaged meats, whereas *Pseudomonas* spp. are prevalent between the seventh and fourteenth days, after 21 days lactobacilli predominate (Carpenter et al., 1976). Because lactobacilli reproduce in the presence of small concentrations of oxygen, and because they lower the pH and apparently have antagonistic effects on other bacteria, they may significantly prolong the storage life of vacuum-packaged meats (Beebe et al., 1976; Seideman et al., 1976).

Green Discoloration. Meat stored at 1 to 2°C under low oxygen tensions occasionally becomes discolored, and a bright green exudate may be formed. The pigment results when myoglobin is converted to sulfmyoglobin in the presence of H_2S, which has been formed by *Pseudomonas mephitica* (Nicol et al., 1970). *P. mephitica* grows and produces H_2S only when the oxygen tension is low (about 1%) and the pH of the meat is 6.0 and above.

Salmonellae. Reports vary concerning the incidence of salmonellae on meats (Table 17-6). Although contamination varies with sanitation of the meat, the presence of viable salmonellae on carcass meats is of a low order, i.e., < 3 to 18/cm² of meat. As could be expected, the larger the portion of carcass analyzed, the greater is the probability that salmonellae will be recovered. Samplings representing quadrants of entire sides of beef carcasses (i.e., swabs from complete surface areas of inside and outside forequarters and hindquarters) provide a higher proportion of salmonella-positive results (Weissman and Carpenter, 1969) than when smaller areas are tested. These organisms may survive for at least six weeks on meats held at 0 to 2°C (Patterson and Gibbs, 1974). Diebel and Goepfert (1977) consider that these organisms present no public health hazard if the holding temperature does not exceed 10°C (Table 17-7).

Table 17-6
Recent reports of samplings of beef for contamination by salmonellae.

Investigators	Date reported	Total no. of samples	No. having salmonellae	% contamination
Weissman and Carpenter	1969	50	37	74
Fomin and Simmons	1972	92	16	18
Childers et al.	1973	93	0	0
Duitschaever et al.	1973	213	0	0
General Accounting Office	1974	71	0	0
Ladiges and Foster	1974	100	0	0
Manickam and Victor	1975	67	4	5.9
Pivnick et al.	1976	218	8	3.7
Smith et al.	1976	112	3	3
Field et al.	1977	112	3	3
Foster et al.	1977	150	0	0
Swaminathan et al.	1978	36	4	11.1

Bone Stink. The packing industry identifies bone stink, bone taint, and ham souring as a putrefactive type of spoilage in the deep tissue of large pieces of meat, such as the hindquarters of pork and beef. Organisms implicated are: *Clostridium putrefaciens, C. histolyticum, C. sporogenes, C. tertium, C. novyi,* and *C. putrificum (C. lentoputrescens)* (Tucker, 1929; Moran and Smith, 1929; Haines and Scott, 1940; Mundt and Kitchen, 1951).

Discoloration. On meats *Penicillium* and *Cladosporium* spp. attack and penetrate the superficial layers of connective tissues or of fat covering the musculature and produce discoloring spots. *Penicillium* spp. attack stored meats more often than *Cladosporium* spp. do. On meats stored for several weeks members of both genera may produce discoloring areas ranging from yellow to black. Black spots on chilled beef and mutton shipped to

Table 17-7
Recent reports of critical temperatures at which salmonellae grow or die gradually in refrigerated, packaged ground beef.

Investigator(s)	Date	Temperature °C	Temperature °F	Growth	Gradual destruction
Patterson and Gibbs	1977	0–2	32–35.6	—	+
Goepfert and Kim	1975	0.6	33	—	+
Brown	1977	1.1	34	—	+
Kennedy and Oblinger	Personal communication	1.7	35	—	+
Goodfellow et al.	1977	1.7	35	—	+
Deibel and Goepfert	1977	4.4	40	—	+
Goepfert and Kim	1975	4.5	40.1	—	—[a]
Brown	1977	5.6	42	—	+
Goepfert and Kim	1975	8	44.6	—	—[a]
Deibel and Goepfert	1977	10	50	—	+
Goodfellow et al.	1977	10	50	+	+
Davidson and Witty	1977	12	53.6	+	
Goepfert and Kim	1975	12.5	54.5	+	

[a] Neither growth nor destruction noted.

England from Argentina were identified as being due to *Cladosporium herbarum* (Massee, 1912). This mold can grow and produce black spots on meat kept at temperatures several degrees below freezing. The organism affects not only beef and mutton but also pork, veal, lamb, and rabbit meat. Such discolorations are much less common than formerly because animal carcasses are no longer aged in the packing plant and spend shorter time in transit. Aspergilli are recovered from meats stored at 10°C or higher.

Whiskers. On rare occasions in meat storage rooms, mycelia of various members of the *Mucorales* are observed on the surface of meats; isolates of *Thamnidium, Mucor,* and *Rhizopus* spp. produce an extensive whiskery, airy, or cottony gray-to-black growth.

Yeast Spoilage. Yeasts seldom cause spoilage of fresh red meats, being only a small part of the initial microbial population and growing more slowly than most bacteria. Yeasts cause spoilage of refrigerated meats only when bacterial numbers have been restricted. Yeasts may reach large populations when meat has been pasteurized by ionizing radiation, or when antibiotics such as the chlortetracyclines have been used as bacteriostatic agents. Three asporogenous genera of yeasts—*Torulopsis, Candida,* and *Rhodotorula*—occasionally are recovered from refrigerated meat (Ayres, 1960b). They are not known to have much role in the spoilage of fresh meats (Eklund et al., 1942; Tarr et al., 1952). Spoilage of chilled beef (at −1.1°C) is caused by growth of a mixture of bacteria, yeasts, and molds on the beef surface (Empey and Scott, 1939b). In general, yeasts can grow on much drier surfaces than can bacteria (Scott, 1936 and 1957). If, during meat storage, atmospheric relative humidity decreases from 99.3 to 98.0%, the growth characteristics of spoilage yeasts change little or not at all. Slime is present at 99% relative humidity, and the yeast population ranges from 2 million to 10 million cells per cm^2, depending on the size of the individual cells. The small discrete nodules that are transparent at 97 and 98% relative humidity become an opaque white at 96%. The characteristic yeasty odor is most pronounced on the moist muscle. At 97% relative humidity growth is only slightly retarded but at 96% both the lag period and generation time for *Candida* spp. increase. *Candida* spp. fail to grow at 91% relative humidity. Growth of all yeasts becomes irregular at vapor pressures below 94% saturation.

Asporogenous yeasts grow at −1.6 to −1°C on beef (Lea, 1931; Vickery, 1936a, b). Some of these yeasts are lipolytic and produce appreciable lipolysis of beef fat. Many yeasts readily attack fats—possibly greatly increasing spoilage of food (Ingram, 1962).

Molds and yeasts can grow on lamb and mutton carcasses stored at −5°C (Haines, 1931; Lea, 1938). Colonies of these organisms are observed after about 7 weeks of storage, becoming fairly numerous on the inside of the flanks after 12 weeks. A satisfactory storage temperature for controlling microbial growth on lamb and mutton is −10°C. The fat of lamb and mutton does not appear to be attacked when the carcasses are held for three days at a mean temperature of 12°C after storage.

Processed Meats

Processed and fresh meats often harbor different groups of microorganisms. The predominant flora of raw meats—including pseudomonads, *Aeromonas, Flavobacterium, (Achromobacter), Bacterium* spp. gram-positive spore formers, *Corynebacterium, Arthrobacter, Microbacterium* spp., enterobacteria, and chromogenic micrococci—are not the most common on processed meats. Table 17-8 gives the principal microbial groups commonly found; those in capitals are the primary offenders. Yeasts may form a slimy layer on improperly refrigerated packaged frankfurters. When newly processed, meats such as weiners and other sausages, bacon, and country cured hams show little hospitality to molds, but after these meats have aged, molds frequently gain ascendancy and may cause spoilage.

Although *Bacillus* spp. are common spoilage microorganisms in canned hams (Niven, 1955), most canned meat spoilage is

Table 17-8
Principal microbial groups isolated from cured meats.

Bacteria	Fungi
MICROCOCCUS	DEBARYOMYCES
STAPHYLOCOCCUS	Torulopsis
Sarcina	Rhodotorula
Neisseria	Candida
STREPTOCOCCUS	Trichosporon
LEUCONOSTOC	PENICILLIUM
Pediococcus	Mucor
Bacillus	Rhizopus
LACTOBACILLUS	ASPERGILLUS
MICROBACTERIUM	Monilia
Lactobacillus (catalase +)	Cladosporium
Clostridium	Alternaria

attributed to the group of bacteria known as putrefactive anaerobes. These organisms produce heat-resistant spores that can germinate and decompose meat under the anaerobic conditions that prevail within the sealed container. Generally, nonacid food, such as meat, is given a heat process calculated to destroy spores of *Clostridium botulinum,* an extremely toxigenic putrefactive anaerobe.

Sterilization periods advocated by canning-industry laboratories are based on the assumption that heavy loads of putrefactive anaerobic spores of high heat resistance are present before heat processing—an assumption not supported by observations in the field. Because canned meats subjected to lower processing values are known to have more acceptable flavor and texture, and possibly also better nutritive values, many shelf-stable canned cured meats receive a heat process of $F_0 = 0.1{-}0.6$ (Chang et al., 1974). Packers use only cooking processes that sterilize the product without making it less appetizing. Most nonacid canned vegetables are processed to sterilizing values in excess of $F_0 = 2.78$ (Ecklund et al., 1942); packers of luncheon meat using processing values lower than this figure rely on the bacteriostatic effect of curing agents and a rigid program of plant sanitation to prevent losses through spoilage. The possibility that the spores associated with the product can germinate must be fully considered before a lower processing value is adopted.

The various manufacturers of canned meat products apply widely differing processing schedules; yet there is no concrete evidence from the field that pork luncheon meat products in 12 ounce or 40 ounce tins are underprocessed (Subsistence Research and Development Laboratory, 1948). However, a similar item merchandized in 3 pound and 6 pound rectangular cans is generally considered to be understerilized, because were it given the heat believed necessary to destroy spores, the meat would have suffered excessive rendering and purging, becoming unpalatable and having low drained weight. Although luncheon meat packed in large containers does not have the pronounced canned or "overcooked" flavor associated with the sterilized item, its usefulness is restricted by the necessity of refrigeration.

Table 17-9

Comparison of frequencies of various spore counts in fresh pork trimmings from different plants.

Range of MPN spore count per gram of meat	Number of samples containing spore count				
	Plant A	Plant B	Plant C	Plant D	Total for all plants
Below 0.18	2	1	2	18	23
0.18–1	2	2	4	12	20
1–2	4		3	1	8
2–3	3	1	1		5
3–4			2		2
7–8		1			1
18–23			1	1	2
51				1	1
Total samples	11	5	13	33	62

Source: Steinkraus and Ayres, 1964. Copyright © 1964 by the Institute of Food Technologists.

Usually, fewer than 0.18 PA per gram are recovered from fresh pork trimmings, although the maximum number in any one sample may be much higher. When 62 samples from 4 different packing plants were analyzed (Steinkraus and Ayres, 1964), the MPN spore counts per gram of meat ranged from ≤ 1.0 spore per gram for almost 70% of the samplings (43) to over > 4 spores per gram for only 6%. As Table 17-9 shows, the spore count is usually quite low and differs little among the packing plants. PA spore counts in cured pork trimmings are similar to those in fresh pork trimmings. Of samples from 2 packing plants, 16 (70% of 23 samples) had fewer than 1.0 PA spores per gram and only 2 samples had more than 3 spores per gram. The heat process given 6 pound cans of pork luncheon meat is considered insufficient to destroy spores. As may be seen in Table 17-9, the spore count in this product approaches that of cured pork trimmings. As with cured trimmings, most of these samples contain 0.18 to 1.0 PA spores per gram.

Only very small populations of putrefactive anaerobic spores are present in commercial meat trimmings (Burke et al., 1950; Ayres and Adams, 1953; Silliker et al., 1958; and Greenberg et al., 1966). Hence, when the stability of heavily inoculated canned cured meats is studied, the results contradict the actual situation—a long history of satisfactory commercial results using heating schedules far less severe than such results imply are necessary. Probably much more significant are the curing agents sodium chloride, sodium nitrate, and sodium nitrite. Commercial mixed cures contain much smaller amounts of sodium nitrate and sodium nitrite than of sodium chloride. NaCl concentrations are limited by flavor considerations—most consumers judge levels above 3.5% in cured meats to be too salty. Nitrate and nitrite are used to improve color and flavor and for the bacteriostatic effect. Investigators agree on the value of sodium chloride in checking PA spoilage of canned meat, and regulatory agencies consider sodium nitrite and sodium nitrate necessary to prevent the production of botulinal toxin in canned ham, bacon, and in some processed meat, poultry, and fish products.

The preservative value of curing salts has been recognized for some time (Tanner and Evans, 1933a, b, c). Early work by Tanner and Evans revealed that species and strains of microorganisms differ in susceptibility to the same salt, and that differences in substrates affect how similar organisms tolerate various levels of the same salt. The hydrogen ion concentration of the medium used to test the salt has an important influence upon the results (Tarr, 1941). Sodium nitrite appears to be an effective preservative in acid media. As more spores are added the curing agents become correspondingly less effective in preventing spoilage (Stumbo et al., 1948).

Curing Agents

Sodium Chloride. If sodium chloride alone is used to prevent rapid spoilage of pork, the critical concentration needed is 4%. When an inoculum of a few hundred spores of *Clostridium* spp. (PA 3679) per gram is added to pork, 4% NaCl extends storage life for a short time, 5% NaCl extends it at least 6 months, and 6%

Table 17-10
Spoilage of inoculated[a] pork containing different levels of NaCl.

Sodium chloride		Percent spoilage with incubation times at			
— Added	Found by analysis	1 week	4 weeks	6 months	10 months
%	%				
0	0.1	100	—	—	—
2.0	2.1	100	—	—	—
4.0	4.1	0	100	—	—
5.0	—	0	0	0	100
6.0	6.1	0	0	0	0
8.0	8.2	0	0	0	0

[a] Analyzed to contain 175 PA 3679 spores per gram.

Source: Bulman and Ayres, 1952.

or 8% NaCl extend it at least 10 months (Table 17-10). Lower levels of NaCl are sufficient to preserve meat in which 50 PA 3679 spores per gram are used (Table 17-11). Length of the incubation period also must be considered in evaluating the preservative properties of sodium chloride. No samples containing 3.7% NaCl spoil during the first two months of incubation, but eventually all of them spoil.

Nitrate and Nitrite in Meats. As early as 1200 B.C. sodium nitrate was in use in ancient China and in Mideastern civilizations to develop the red color and the flavor in cured meats. To do either effectively, however, the nitrate must be reduced to nitrite by microbial action. And it delays spoilage only in concentrations of several percent (Table 17-12). In canned hams bacilli have reduced nitrate to acquire oxygen for growth, and the hams have then developed putrefactive spoilage (Greenwood et al., 1940).

To achieve rapid cure, sodium nitrite is added to hams, shoulders, moist canned comminuted meats, bologna, wieners, Thuringer and summer sausages, and other products at the legal limit of 156 μg per gram of meat. It, too, develops color

Table 17-11

Spoilage of inoculated[a] pork containing different levels of NaCl.

	Sodium chloride			Percent spoilage with incubation times at 37°C (98.6°F) of				
Trial	Added	Found by analysis	No. of tubes	1 week	2 weeks	4 weeks	9 weeks	16 weeks
	%	%						
1	0	0.1	6	100				—
	3.5	3.6	6	0	0	0	100	—
	4.0	4.0	6	0	0	0	0	—
	4.5	4.6	6	0	0	0	0	—
Uninoculated pork	0	0.1	6	33	67	67[b]	100	—
2	0	0.1	5	100				
	3.5	3.7	5	0	0	0	0	100
	4.0	4.4	5	0	0	20	20	40
	4.5	4.6	5	0	0	0	0	0
Uninoculated pork	0	0.1	5	100	—	—	—	—

[a] Approximately 50 PA 3679 spores per gram.
[b] Four samples spoiled; other two doubtful.

Source: Bulman and Ayres, 1952.

Table 17-12

Spoilage of inoculated[a] cooked pork containing different levels of NaNO$_3$.

Sodium nitrate		Percent spoilage with incubation times at 37°C (98.6°F) of			
Added	Found by analysis	1 week	2 weeks	4 weeks	10 weeks
%	%				
0	0	100	—	—	—
2.9	3.1	100	—	—	—
3.4	3.8	0	100	—	—
3.9	4.1	0	0	0	0

[a] Approximately 50 PA 3679 spores per gram.

Source: Bulman and Ayres, 1952.

and flavor; it also inhibits anaerobic bacteria—including the putrefactive anaerobes as well as *Clostridium botulinum*—and prevents warmed-over flavor of pork (Gray, 1976). Inhibition of anaerobes is extremely essential: none of the meats is heated to sterility; in the many meats that are moist the anaerobes could otherwise grow, causing putrefactive spoilage; and the meats are commonly subjected to temperature abuse, despite prominent labels stipulating that they be kept under refrigeration at all times. The level of nitrite added to meat drops rapidly during the first few days of cure, and thereafter it decreases more slowly to persist to about eight weeks. By this time the meats presumably have been consumed.

The major role of nitrite in curing meats is to prevent growth of germinating spores. The prevention may be through the action of an inhibitory substance (Pivnick et al., 1970; Chang et al., 1974), or by rendering unavailable a substance required in metabolism. The inhibiting action of nitrite has been related to its chelating action with iron (Tompkin et al., 1978a and b). The chelating action may explain the Perigo effect (Perigo et al., 1967): nitrite added to a medium before it is heated becomes much more toxic to spores than when added to the medium after heating. If, however, meat is added to the heated medium, the inhibitory effect is lost (Johnston et al., 1969). Chelation may also explain the relationship between the numbers of spores and the level of nitrite required to prevent formation of toxin. Meats inoculated with a small number of spores, 210 per gram, become toxic if the curing pickle is formulated with 120 μg (120 ppm) nitrite, but not if the level of nitrite exceeds this amount. If the inoculum consists of 19,000 spores, more than 340 μg (ppm) must be added to the curing pickle to prevent formation of toxin (Christiansen et al., 1974).

Because nitrosamines form when meats containing nitrite are heated, a search is underway for inhibitors to replace nitrite in the cure. No such factor has been found, and there is at present no assurance that *Clostridium botulinum* will not grow in the present array of cured meats if nitrite is omitted. The nitrosamines are formed when nitrite reacts with diamines and tertiary amines in heated meats (they have not been detected in raw meats). They are seldom present at levels greater than 25 μg per gram of meat (Sebranek and Cassens, 1973), and in many trials with cured meats they are either not detected or are detected at the lower end of the scale, in the region of 10 ng per gram of meat. Nitrosodimethylamine and nitrosopyrrolidine appear regularly in cooked bacon (Wassermann et al., 1978). The nitrosamines are carcinogenic, and commonly cause carcinomas of the liver in rats and of the liver and other organs in other animals. The threshold level for the induction of carcinomas in rats lies between 0.3 mg (300 μg) and 1.0 mg (1,000 μg) per kilogram of body weight (Preussmann et al., 1976).

The quantity of nitrosamines in cooked cured meats is directly related to the amount of nitrite used in the cure. Much current research is devoted to determining the absolute lowest level of nitrite that will ensure a safe product, because meats cured without nitrite and stored at 27°C invariably spoil. When minimal levels of nitrite are used, sodium ascorbate and sodium erythorbate are added to obtain the cured meat color (Christiansen, 1979). Also, at 0.2%, sorbic acid either inhibits or contributes to the action of nitrite (Busta and Sofos, 1979; Sofos et al., 1979).

Cured Hams

More pork is sold cured than fresh. The principal cuts that are cured are hams, shoulders, briskets, sides (called gammon in the United Kingdom), bellies, jowls, and feet. Although the proportion is much smaller, a considerable quantity of corned and chipped beef and lesser amounts of salted mutton are also merchandized.

Curing Methods

Four general procedures are used to distribute the curing ingredients throughout the meat: dry curing, brine immersion (pickling), artery pumping, and stitch pumping. In dry curing, the oldest and simplest method, the mixture of curing agents is rubbed into the surface of the meat and the meat is held at low temperature (2.2 to 4.4°C) until the curing agents have penetrated to the center. With large or fat hams penetration is often slow and undependable, and is much slower from the skin side than from the lean side. In brine immersion (or cover picking curing), which has also been used for hundreds of years, the trimmed cuts are placed in barrels or in a large wooden tank and sufficient cold (2.2 to 4.4°C) curing solution is added to submerge the meat. Cuts near the surface are weighted to keep them submerged. Because the brine penetrates slowly in this method, large pieces may spoil—particularly if the brine is too weak (low salt concentration), excessively diluted by meat juices, contaminated by putrefactive or other bacteria, or maintained at insufficiently low temperature. Obviously, the higher the storage temperature, the more rapidly the brine penetrates but, unfortunately, the more quickly the microorganisms proliferate.

Except for limited use in the processing of country-cured hams, dry curing was abandoned commercially because of its high incidence of internal spoilage—5 to 7% (Jensen and Hess, 1941; Mundt and Kitchen, 1951). The slow loss of heat and the uneven penetration of salts permitted putrefactive anaerobic growth to take place and cause "ham souring" or "bone stink" before the curing agents could inhibit growth of the anaerobes. Among several species of *Clostridium* isolated from tainted country-cured hams—including *C. bifermentans, C. mucosum, C. parabifermentans, C. septicum, C. paraputrificum,* and *C. putrefaciens*—only *C. putrefaciens* is able to grow within 8 weeks at 7.2°C (Mundt and Kitchen, 1951). Because the meat does not absorb sufficient concentrations of curing agents to limit anaerobic growth (Rust and Olson, 1973), it is essential that hams be properly cooled and maintained at low temperature during curing and ripening.

Because of the high percentage of failures of the dry curing and brine immersion methods, artery pumping was introduced to cure meats more rapidly. In this technique the femoral artery near the surface and slightly to the side of the ileum bone is located. A hollow needle is inserted into this artery and—at a pressure of about 45 psi—curing solution is pumped into the ham until it has gained approximately 10 to 15% of its original weight. The cure is forced into various parts of the ham through the vascular system. However, the distribution of cure is by no means uniform. In recent years, stitch pumping has come into use. In this procedure a long needle is inserted into the meat; from the several holes along the length of the needle, curing solution is forced into the interior of ham, shoulder, brisket, or belly. A variant technique, called continuous brine injection, has also come into use. It employs gangs of injection needles. With today's instrumentation, multiple punctures are closely spaced and each needle penetrates to proper depth or until it makes contact with bone. The pressure applied to each needle is automatically adjusted to prevent damage to the needle or the meat. Care must be employed to keep needles and conveyor belts sanitized, because insertion of contaminated needles into the meat interior potentially introduces microorganisms along the entire pathway. Also, the excess brine, which often is recirculated, may become contaminated and diluted by meat juices and serve as a source of spoilage organisms.

In 1960 a patent was issued in Britain (Holmes, 1960) wherein individual bacon slices or ham slices are exposed to curing ingredients and the curing process is completed in evacuated hermetically sealed plastic retail packages. This process has not received wide acceptance in the United States to date.

Smoking Meats. Smoke deposited in and on meats alters their flavor and appearance. Many cured meats are also smoked, but country-cured hams and salt-cured bacons often are aged without smoking. Some consumers prize the changes brought about by smoking, others do not.

Smoke may penetrate the meat (1) while it is hanging in a smoke-filled chamber and exposed to the vapors arising from the combustion of slowly burning sawdust (usually from a hardwood) or (2) in the form of a liquid percolate that perfuses into the meat. The first method has the advantage of causing considerable drying from the heat and of causing a glazed or "case-hardened" surface crust to form that acts as a barrier to microbial penetration. Moreover, the formaldehyde and phenolic and cresylic resins that accumulate on the surface have antibacterial properties, deterring rapid spoilage, as does the increased salt level in the interior of the ham caused by surface dehydration.

Contaminants of Hams

A motile, homofermentative species of *Lactobacillus* is a common contaminant in ham curing brines and on the surface of cured unprocessed ham (Deibel and Niven, 1958). However, meats cured by arterial or needle pumping and held for no more than a few days are seldom contaminated by microorganisms. On hams, shoulders, and bacons that receive long dry cures or immersion in cover-pickle, a varied and luxuriant microbial population develops. During brine curing the meat juices and the soluble proteins are drawn from the muscle and mingle with the brine, providing a rich source of nutrients for micrococci, lactobacilli, leuconostocs, pediococci, and aerococci. Souring of cured meats results from extensive growth of these organisms. While they are not "heat tolerant" in the sense that the spore-forming anaerobes are, they can survive mild heat processes, and can be destroyed only by an internal temperature of at least 67°C (Deibel et al., 1961).

Vacuum packaging and curing agents—particularly sodium chloride and nitrite—change the balance of flora from almost equal proportions of lactobacilli (30%), streptococci (30%), and micrococci (40%) to predominantly lactic acid bacteria (70%) at the expense of the streptococci and micrococci. Reduced oxygen availability, while limiting for pseudomonads, flavobacteria, and achromobacters, appears to favor growth of lactic acid bacteria—especially *Lactobacillus* and *Streptococcus* spp. On bacon held at 5°C for 6 days the flora is composed almost entirely of lactobacilli (75%) and streptococci (25%)—proportions that persist for at least 24 days (Hansen and Riemann, 1962).

Enterococci are found on canned hams (Ingram and Hobbs, 1954; Barnes, 1956; Barnes et al., 1956). *Streptococcus faecalis* was once thought to be restricted to the human gut, whereas *S. faecium* is widely distributed among animals. Inasmuch as fecal contamination of food from either source is highly undesirable, Mossel et al. (1957) state that no practical reason exists for differentiating among the two. Yet this view may be questioned in the light of results obtained by Mundt et al. (1958), who, on a modified MRS medium, readily isolated enterococci similar to, though not identical with, *S. faecium* from plant sources from geographical areas largely devoid of humans and large mammals.

Products with high salt content are often too dry for most microbial growth other than that of *Staphylococcus aureus* and aspergilli. The amount of moisture that is available for growth can be expressed in terms of water activity (refer back to Figure 3-1). When pure water is used as a reference (or an $a_w = 1.00$), the amount of salt present displaces part of the water and reduces the a_w proportionally. With cured meats such as ham, bacon, cured and/or smoked pork, sausage and frankfurters, a_w values of from 0.87 to 0.93 prevail. These levels are below those necessary for the growth of spore-forming aerobic bacteria, pseudomonads achromobacters, salmonellae, and *Clostridium botulinum*. Normal salting of cured meats eliminates these organisms. However, staphylococci are able to survive and grow well on cured meats; when these products are heated to an internal temperature of 57.2°C to ensure death of *Trichinella spiralis*, staphylococci persist as the primary contaminants. Improper handling or storage, contaminated equipment or surfaces, and inadequate heating provide the opportunity these organisms need for growth and toxin production. Generally this form of foodborne illness is associated only with ham that has remained in the temperature range 10°C to 46°C for at least four hours (Niven, 1961).

The microbiology of aged hams differs significantly from that of packing-house hams. By virtue of their high salt content, country-style hams may lose 10% to 25% of their weight during

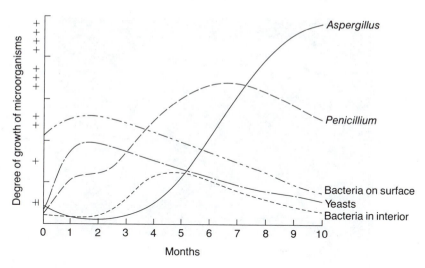

Figure 17-4

Changes in microbial flora during curing and ripening of "country-cured" hams. (Ayres et al., 1974.)

curing and aging. Ordinarily such hams remain raw throughout storage, and because their proteins are not heat denatured, the flesh is less susceptible to staphylococcal attack. While micrococci and lactobacilli are proliferating, so are the yeast populations (i.e., *Debaryomyces, Pichia, Mycoderma, Torula,* and *Rhodotorula* spp.). With further a_w reduction—as with country-cured hams—molds supplant these organisms and become the dominant flora of aged cured meats (Figure 17-4). The amount of mold growth varies considerably. Hams stored in damp rooms, caves, or cellars or wrapped in paper during ripening generally exhibit abundant surface growth, whereas unwrapped hams or those held at a low relative humidity (about 65%) exhibit very little mold development (Leistner and Ayres, 1968).

Molds of the genera *Cladosporium, Penicillium, Scopulariopus,* and *Aspergillus* are the predominant fungi isolated from ripening hams. Other genera found occasionally are *Alternaria, Rhizopus, Paecilomyces, Oospora* and *Epicoccum. Cladosporium* spp., in addition to being present on a large number of aged hams investigated, cause undesirable, large deep-seated "black spots" that cannot be washed away but instead must be cut out or

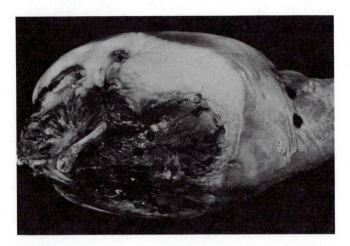

Figure 17-5
Growth of *Cladosporium* on ham. (Leistner and Ayres, 1968.)

trimmed from the meat (Figure 17-5), a treatment which results in significant economic loss. *Scopulariopsis* spp. appear as superficial dusty white, amorphous spots or blotches on the skin. These spots—unlike those produced by *Cladosporium*—cause little real damage except for the moldy appearance of the meat cut.

The dominant fungal genera on country-cured hams are *Penicillium* and *Aspergillus*. Penicillia predominate early in ripening, aspergilli predominate on the mature product. The opinion of some producers of country-cured ham that high-quality hams exhibit a dark green moldy surface (caused by xerotolerant aspergilli) is fallacious; more likely, such mold growth merely indicates that the product has been stored at a low a_w, permitting a long ripening period during which flavor develops (Leistner and Ayres, 1968).

Debaryomyces spp., having the greatest tolerance to salt and having high tolerance to organic acids, are the yeasts most often present in brines. These characteristics and their ability to grow well at low temperatures account for their prevalence in foods preserved by salting and brining (Mrak and Bonar, 1939). *Debaryomyces membranaefaciens* var. *hollandicus* causes film to form on brines used for hams, beef tongue, bacon sides, and Canadian bacon. *D. kloerkeri,* a non-film-forming yeast, also is found in subsurface brine samples (Costilow et al., 1954). Yeasts develop readily on vacuum-packed bacon slices (Ingram, 1962).

Bacon from sugar-fed pigs is usually more acid than other bacons (Ingram, 1962). When the pH of Wiltshire bacon is below 5.6, bacterial growth decreases but yeast growth does not (Gibbons and Rose, 1950). Because yeasts are not strongly proteolytic or odoriferous, they do not necessarily cause rapid spoilage. However, whereas small numbers of them in hams, bacons, or curing brines are of little consequence, internal growth of yeasts can cause spoilage.

Sausage

Sausages are made of ground or finely chopped, seasoned meats stuffed into casings. In the United States about 15% of all meat processed is used in sausages. Some sausages are skinless, made by stuffing the ingredients into an artificial (cellulose) casing that is stripped from the meat after a thin outer protein film has coagulated. For a few fresh, cooked, or dried sausage meats, such as fresh pork sausage and pemmican, no casing materials are used at all; instead, these meats are pressed into cakes or strips. With a few exceptions, fresh and uncooked smoked sausages are heated before serving, whereas cooked, dry, and semidry sausages are ready to eat without further preparation (see Table 17-13) (Smith et al., 1975).

Box 17-2

Meat was made into sausage (from the Latin *salsus,* "salted") as early as Babylonian times. In the thirteenth century B.C. the Chinese were reported to have used salt to preserve their meat. And the art of sausage making was later well known to the Greeks and Romans. During the Middle Ages local sausage preparations were identified with their place of origin—e.g., bologna with Bologna, Italy; Genoa salami with Genoa; romano with Rome; and frankfurters with Frankfurt, Germany. In the United States Lebanon sausage is named for Lebanon, Pennsylvania, and coneys for Coney Island, in Brooklyn, New York.

Table 17-13
United States consumption of various types of sausage.

Sausage classification	Percent of consumption

Cooked-smoked emulsion sausages
  Frankfurters
  Bologna
  Vienna sausage
	50

Loaves and luncheon meats
  Olive loaf
  Pickle and pimiento loaf
  Pressed ham
  Head cheese
	18

Fresh or fresh and smoked sausages
  Fresh pork sausage
  Bratwurst
  Brockwurst
  Mettwurst
	15

Cooked-gelled sausages
  Liver sausage
  Braunschwieger
	11

Fermented sausages
  Dry
    pepperoni
    salami
  Semidry
    cervelat
    Thuringer
	6

Source: Smith et al., 1975.

Fresh Sausage

Fresh pork sausage, the most widely manufactured fresh sausage, has very limited shelf life because it is made from pork trimmings and has a very high fat content (up to 50%). Although some spices are added, its salt content is insufficient to delay microbial spoilage—caused by many of the same organisms that cause spoilage of comminuted beef (i.e., hamburger). However, the most common spoilage in refrigerated pork sausage (kept at 2 to 10°C) results from the growth of *Microbacterium* spp. and lactic acid bacteria. The bacteria that cause souring are a heterogeneous group and not confined to the genera *Lactobacillus, Streptococcus,* and *Leuconostoc.* Some of the microorganisms growing on fresh pork sausage are catalase-positive and can be assigned to the genera *Arthrobacter, Corynebacterium,* and *Microbacterium.* Although these organisms can cause meat to spoil by souring (which does not always occur), their dominance serves the useful purpose of preventing survival of salmonellae and other enterobacteria. Nonfecal coliforms, *Escherichia coli,* and salmonellae, often recovered from fresh sausage, seldom cause incipient spoilage. However, because the presence of salmonellae constitutes a potential health hazard, state or federal regulatory agencies planning to impose microbial standards must assess not merely the total quantity of organisms but also the quality of contamination. In processed meats 50,000 *E. coli* or salmonellae are of much more serious consequence than 500 million lactic acid bacteria—even though such numbers of the latter could cause souring or produce slime.

Cooked Sausage

Bacterial discoloration, or greening, an all too common defect of cooked sausage (wieners and large sausages), is also observed on luncheon meats and, far less often, cooked hams. Niven (1951) categorizes three types of greening:

1. Surface discoloration—resulting from contamination of the sausage surface after heat processing; followed by holding conditions that allow extensive growth of the contaminating bacteria.

2. Green cores—appearing on the cut surface a few hours after the sausage is sliced; caused by faulty heat processing and poor refrigeration of the finished product.

3. Green rings—resulting from an unusually high bacterial count of the sausage mix before heat processing.

The microorganisms responsible for these defects belong to the genera *Lactobacillus* and *Leuconostoc.* These organisms have certain distinctive characteristics, including an ability to:

1. grow at low temperature (3.3°C);

2. tolerate high concentrations of salt (up to 10%);

3. produce H_2O_2 in the presence of air;

4. produce gas;

5. grow equally well with or without air;

6. grow throughout a sausage mix or sausage;

7. grow under relatively acid conditions;

8. ferment sugars, producing acid or acid and gas;

9. do not reduce nitrates to nitrites;

10. do not produce catalase, the enzyme that destroys hydrogen peroxide.

The summer months, when temperatures are high, are most conducive to outbreaks of greening, especially if the surfaces of sausages are not kept dry. Hence, moist surfaces, caused by "sweating," are undesirable. Contaminants are transferred from one product to another when spoiled products are returned to the sausage kitchen, when faulty back-slopping practices are permitted, or when plant equipment or workers' hands or clothing are contaminated. After an outbreak of greening a thorough and complete cleaning and disinfecting of the entire operation is essential.

Yeasts

Among the microflora on packaged frankfurters, yeasts (Ogilvy and Ayres, 1953), lactic acid bacteria, and micrococci (Drake et al., 1958) predominate. Most often found are members of the genera *Debaryomyces*, *Candida*, *Torulopsis*, and *Trichosporon*, but also occasionally isolated are *D. kloeckeri*, *D. hansenii*, *D. subglobosus*, *D. nicotianae*, *C. lipolytica*, *C. zeylanoides*, *C. calinulata*, *T. gropengiesseri*, and *Trichosporon pullulans* (Drake et al., 1958).

Debaryomyces that assimilate nitrite have been isolated from cold meat cuts, which appear to be a good source of them (Wickerham, 1957). Because these yeasts have limited ability to use protein, nitrite is the likely source of nitrogen for growth; and because they tolerate high osmotic pressure, they are able to

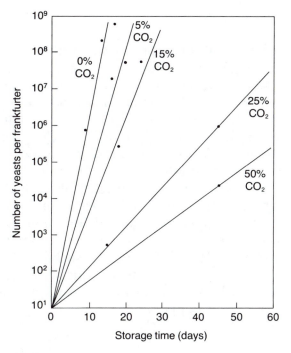

Figure 17-6
Effect of carbon dioxide on the growth of yeasts on frankfurters stored at 7.2°C. (Ogilvy and Ayres, 1953. Reprinted from Food Technology, Vol. 18, pp. 121–130, 1953. Copyright © by Institute of Food Technologists.)

grow in meat brines. When frankfurters are irradiated, the surviving flora consists principally of yeasts and of bacteria of *Bacillus* spp. *Debaryomyces subglobosus* and *Torulopsis candida* are commonly isolated (Drake et al., 1958).

As carbon dioxide becomes more concentrated in the atmosphere in which frankfurters are stored, yeast growth is increasingly retarded. Figure 17-6 shows that as little as 5% CO_2 reduces yeast proliferation; 15% CO_2, more than doubles the storage life before yeast populations exceed a million cells. Growth becomes progressively slower as the gas level is increased: during 45 days of storage with 50% CO_2, yeast numbers are still in the 10^4 range; at 75 or 96% CO_2 no yeasts are found (Ogilvy and Ayres, 1953).

Molds

The spores of mycelial fungi, or molds, are abundant in the air and during sausage manufacture readily contaminate pickling solution surfaces, curing room walls, and sausage casings. Favorable temperature, humidity, and food sources aid the growth of threadlike vegetative mycelial strands. Early in their development molds of the genera *Penicillium, Aspergillus, Fusarium, Oospora, Monilia,* and *Monascus* appear as white, feltlike or fuzzy white splotches but as they mature and form spores they often assume various bright colors such as yellow, green, blue-green, blue, pink, red, purple, or brown (Yesair, 1928). Fungi such as *Cladosporium* and *Alternaria* spp. become dark gray or even black when given sufficient time to develop. The vegetative strands of other molds, such as *Mucor, Thamnidium,* or *Rhizopus* spp., are initially almost transparent but soon take on a whiskery or cottony light gray appearance, later becoming dark grey or peppery black.

References

Adamson, C. A. 1949. Acta. Med. Scand. *227,* 1.

Ayres, J. C., W. S. Ogilvy, and G. F. Stewart. 1950. Food Technol. *4,* 199.

Ayres, J. C. 1951. Iowa State Coll. J. Sci. *26,* 31; (also) Proc. Third Conf. on Res., Amer. Meat Institute, Chicago.

Ayres, J. C. 1955. Microbiology of Meat Animals. E. M. Mrak and G. F. Stewart, eds. In: *Advances in Food Research. 6,* 109, Academic Press Inc., New York.

Ayres, J. C. 1960a. Food Res. *25,* 1.

Ayres, J. C. 1960b. J. Appl. Bacteriol. *23,* 471.

Ayres, J. C., and A. T. Adams. 1953. Food Technol. *7,* 318.

Ayres, J. C., L. Leistner, M. Sutec, P. E. Koehler, M. T. Wu, N. A. Halls, E. Strzelecki, and F. Escher. 1974. Proc. IV Int. Congress Food Sci. and Technol. Vol. III, 218.

Barlow, J., and A. G. Kitchell. 1966. J. Appl. Bacteriol. *29,* 185.

Barnes, E. M. 1956. J. Appl. Bacteriol. *19,* 193.

Barnes, E. M., M. C. Ingram, and G. C. Ingram. 1956. J. Appl. Bacteriol. *19,* 204.

Bate-Smith, E. C. 1948. Physiology and Chemistry of Rigor Mortis, with Special Reference to the Aging of Beef. In: *Advances in Food Research. 1,* 1, E. M. Mrak and G. F. Stewart, eds. Academic Press, New York.

Beebe, S. D., C. Vanderzant, M. O. Hanna, Z. L. Carpenter, and G. C. Smith. 1976. J. Milk Food Technol. *39,* 600.

Brooks, F. T., and B. A. Hansford. 1923. Mould Growths upon Cold-store Meat. Dept. Sci. Ind. Res. Food Invest. Bd (Brit.) Special Rept. No. 17.

Brown, A. D., and J. F. Weidemann. 1958. J. Appl. Bacteriol. *21,* 11.

Brown, W. L. 1977. National Provisioner, Nov. 12.

Bulman, C., and J. C. Ayres. 1952. Food Technol. 6, 255.

Burke, M. V., K. H. Steinkraus, and J. C. Ayres. 1950. Food Technol. 4, 21.

Burn, C. G. 1934. J. Infect. Dis. 54, 395.

Busta, F. F., and J. N. Sofos, 1979. Proceedings, 39th Inst. Food Technol. Mtg., St. Louis, Mo., June 10–13.

Callow, E. H. 1949. J. Roy. Sanit. Inst. 69, 35.

Carpenter, Z. L., S. D. Beebe, G. C. Smith, K. E. Hoke, C. Vanderzant. 1976. J. Milk Food Technol. 39, 592.

Chang, P. C., S. M. Akhtar, T. Burke, and H. Pivnick. 1974. Canad. Inst. Food Sci. and Technol. J. 7, 209.

Childers, A. B., E. E. Kealey, and P. G. Vincent. 1973. J. Milk Food Technol. 36, 635.

Christiansen, L. N., 1979. Proceedings, 39th Inst. Food Technol. Mtg., St. Louis, Mo., June 10–13.

Christiansen, L. N., R. B. Tompkin, A. B. Shaparis, T. V. Kueper, R. W. Johnston, P. A. Kautter, and O. J. Kolari. 1974. Appl. Microbiol. 27, 733.

Costilow, R. N., J. L. Etchells, and T. N. Blumes. 1954. Appl. Microbiol. 2, 300.

Davidson, C. M., and J. R. Witty. 1977. J. Inst. Canada Sci. Technol. Aliment. 10, 23.

Deibel, R. H., and J. M. Goepfert. 1977. Report to Amer. Can Co., Barrington, Il., 12/8.

Deibel, R. H., and C. F. Niven, Jr. 1958. Appl. Microbiol. 6, 323.

Deibel, R. H., C. F. Niven, Jr., and G. D. Wilson. 1961. Appl. Microbiol. 9, 156.

Drake, S. D., J. B. Evans, and C. F. Niven, Jr. 1958. Food Res. 23, 291.

Duitschaever, C. L., D. R. Arnott, and D. H. Bullock. 1973. J. Milk Food Technol. 36, 375.

Dyer, F. E. 1947. J. Fish. Res. Bd. Can. 7, 128.

Ecklund, O. F., H. L. Roberts, and H. A. Benjamin. 1942. Food Ind. 14, 62.

Empey, W. A., and W. J. Scott. 1939a. Australia Coun. Sci. Ind. Res. Bull. No. 126.

Empey, W. A., and W. J. Scott. 1939b. Australia Coun. Sci. Ind. Res. Bull. 129.

Field, R. A., F. C. Smith, O. O. Deane, G. M. Thomas, and A. W. Kotula. 1977. J. Food Protection 40, 385.

Fomin, L., and G. C. Simmons. 1972. Queensland J. Agric. and Animal Sci. 29, 79.

Foster, J. F., J. L. Fowler, and W. C. Ladiges. 1977. J. Food Protection 40, 790.

General Accounting Office. 1974. Report to Congress. Publication B-164031 (2). Comptroller General, U.S., Washington, D.C.

Gibbons, N. E., and D. Rose. 1950. Canad. J. Res. F. 28, 438.

Gill, C. O., and K. G. Newton. 1977. J. Appl. Bacteriol. 43, 189.

Glage, F. 1901. Z. Fleisch.-u. Milchyg. 11, 131.

Goepfert, J. M., and H. U. Kim. 1975. J. Milk Food Technol. 38, 449.

Goodfellow, S. J., and W. L. Brown. 1977. Report to Amer. Meat Institute, Natl. Independent Meat Packers Assoc., New England Meat Dealers Assoc. and Meat Trade Institute.

Gray, J. I. 1976. J. Milk Food Technol. 39, 686.

Greenberg, R. A., R. B. Tompkin, B. O. Bladel, R. S. Kittaka, and A. Annelis. 1966. Appl. Microbiol. 14, 789.

Greenwood, D. A., W. M. Urbain, L. B. Jones, and W. L. Lewis. 1940. Food Res. 5, 625.

Gunderson, M. F., K. D. Rose, and M. J. Henn. 1947. Food Ind. 19, 1516

Haines, R. B. 1931. J. Soc. Chem. Ind. 50, 223T.

Haines, R. B. 1933. J. Hyg. 33, 175.

Haines, R. B. 1937. Microbiology in the Preservation of Animal Tissues. Dept. Sci. Ind. Res. Food Invest. Bd. (Brit.) Special Rept. No. 45.

Haines, R. B., and W. J. Scott. 1940. J. Hyg. 40, 154.

Haines, R. B., and E. C. Smith. 1933. The Storage of Meat in Small Refrigerators. Dept. Sci. Ind. Res. Invest. Bd. (Brit.) Special Rept. No. 43.

Hamm, R., and F. E. Deatherage. 1960. Food Res. *25*, 533.

Hansen, N. H., and H. Riemann. 1962. Die Fleischwirtschaft *14*, 861.

Holmes, A. W. 1960. Unilever. U.K. Patent No. 848,014.

Ingram, M. 1949. J. Roy. Sanit. Inst. *69*, 39.

Ingram, M. 1962. In: *Recent Advances in Food Science*, Vol. 2. J. Hawthorne and J. M. Leitch, eds., Butterworths, London.

Ingram, M., and B. C. Hobbs. 1954. J. Roy. Sanit. Inst. *74*, 1151.

Jay, J. M. 1964a. Food Technol. *18*, 1633.

Jay, J. M. 1964b. Food Technol. *18*, 1637.

Jay, J. M. 1966. Appl. Microbiol. *14*, 492.

Jay, J. M. 1972. J. Milk Food Technol. *35*, 467.

Jensen, L. B., and W. R. Hess. 1941. Food Res. *6*, 273.

Johnston, M. A., H. Pivnick, and J. M. Samson. 1969. Canad. Inst. Food Technol. *2*, 52.

Kirsch, R. H., F. E. Berry, G. L. Baldwin, and E. M. Foster. 1952. Food Res. *17*, 495.

Kraft, A. A., and J. C. Ayres. 1952. Food Technol. *6*, 8.

Kraft, A. A., J. C. Ayres, G. S. Torrey, R. H. Salzer, and G. A. N. da Silva. 1966. J. Appl. Bacteriol. *29*, 161.

Ladiges, W. C., and J. F. Foster. 1974. J. Milk Food Technol. *37*, 213.

Lea, C. H. 1931. J. Soc. Chem. Ind. *50*, 215T.

Lea, C. H. 1938. Rancidity in Edible Fats. Dept. Sci. Ind. Res. Food Invest. Bd. (Brit.) Special Rept. No. 46.

Leistner, L., and J. C. Ayres. 1968. Die Fleischwirtschaft *48*, 62.

Lepevetsky, B. C., H. H. Weiser, and F. E. Deatherage. 1953. Appl. Microbiol. *1*, 57.

Lissauer, M. 1906. Arch. Hyg. *58*, 136.

McCullough, N. B., and C. W. Eisele. 1951a. J. Infect. Dis. *88*, 278.

McCullough, N. B., and C. W. Eisele. 1951b. J. Immunol. *66*, 595.

McCullough, N. B., and C. W. Eisele. 1951c. J. Infect. Dis. *89*, 209.

McCullough, N. B., and C. W. Eisele. 1951d. J. Infect. Dis. *89*, 259.

McLean, R. A., and W. L. Sulzbacher. 1953. J. Bacteriol. *65*, 428.

Mallmann, W. L., L. Zarkowski, and M. Ruster. 1940. The Effect of Carbon Dioxide on Bacteria with Particular Reference to Food Poisoning Organisms. Mich. Agr. Expt. Sta. Bull. No. 489.

Manickam, R., and D. A. Victor. 1975. Indian Vet. J. *52*, 44.

Marxer, A. 1903. Fortschr. Vet. Hyg. *1*, 328.

Massee, G. 1912. J. Hyg. *12*, 489.

Moran, T. 1935. J. Soc. Chem. Ind. *54*, 149T.

Moran, T., and E. C. Smith. 1929. Post-mortem Changes in Animal Tissues. The Conditioning or Ripening of Beef. Dept. Sci. Ind. Res. Food Invest. Bd. (Brit.) Special Rept. No. 36.

Mossel, D. A. A., H. M. J. Van Diepen, and A. S. De Bruin. 1957. J. Appl. Bacteriol. *20*, 265.

Mrak, E. M., and L. Bonar. 1939. Zentralbl. f. Bakteriol. u. Parasitenk Abt. II, *100*, 289.

Mundt, J. O., and H. M. Kitchen. 1951. Food Res. *16*, 233.

Mundt, J. O., A. H. Johnson, and R. Khatchikan. 1958. Food Res. *23*, 186.

Newton, K. G., J. C. L. Harrison, and K. M. Smith. 1977. J. Appl. Bacteriol. *43*, 53.

Nicol, D. J., M. K. Shaw, and D. A. Ledward. 1970. Appl. Microbiol. *19*, 937.

Niven, C. F. 1951. Sausage Discolorations of Bacterial Origin. Amer. Meat Institute Foundation Bull. 13.

Niven, C. F. 1955. La Symposium International de Bacteriologie Alimentaires. Annales de L'Institut Pasteur.

Niven, C. F. 1961. Microbiology of Meats. Amer. Meat Institute Foundation Circular, Chicago, Il.

Ogilvy, W. S. and J. C. Ayres. 1953. Food Res. *18*, 121.

Patterson, J. T. 1971. J. Food Technol. *6*, 63.

Patterson, J. T., and P. A. Gibbs. 1974. 20th European Meeting of Meat Res. Workers, Dublin, Sept. Abstracts and Communications.

Patterson, J. T., and P. A. Gibbs. 1977. J. Appl. Bacteriol. *43*, 25.

Perigo, J. A., E. Whiting, and T. E. Bashford. 1967. J. Food Technol. *2*, 377.

Pivnick, H., I. E. Erdman, D. Collins-Thompson, G. Roberts, M. A. Johnston, D. R. Conley, G. LaChapelle, U. T. Purvis, R. Foster, and M. Milling. 1976. J. Milk Food Technol. *39*, 408.

Pivnick, H., M. A. Johnston, C. Thacker, and R. Loynes. 1970. Canad. Institute Food Sci. and Technol. J. *3*, 103.

Preussmann, P. D., I. Schmaehl, G. Eisenbrand, and R. Port. 1976. Proc. 2nd Int. Sympos. Nitrite Meat Prod., Zeist, Pudoc, Wageningen.

Reuter, H., and I. Trukenbrod. 1964. Die Fleischwirtschaft. *44*, 310.

Reynolds, A. E., and J. A. Carpenter. 1974. J. Ani, Sci. *38*(3), 515.

Roth, L. A., and D. S. Clark. 1975. Canad. J. Microbiol. *21*, 629.

Richardson, J. H., V. J. Del Guidice, and C. K. Weisman. 1954. Appl. Microbiol. *2*, 177.

Rust, R. E., and D. C. Olson. 1973. Meat Curing Principles and Modern Practice, 32 pp. Koch Supplies Inc., Kansas City, Mo.

Sarles, W. B., W. C. Frazier, J. B. Wilson, and S. G. Knight. 1951. *Microbiology: General and Applied*. Harper, New York.

Scott, W. J. 1936. Australia J. Coun. Sci. Ind. Res. *9*, 177.

Scott, W. J. 1957. Water Relations of Food Spoilage Microorganisms. *Advances in Food Research*. 7. E. M. Mrak and G. F. Stewart, eds. Academic Press, New York.

Sebranek, J. G., and R. G. Cassens. 1973. J. Milk Food Technol. *36*, 76.

Seideman, S. C., Z. L. Carpenter, G. C. Smith, C. Vanderzant, and C. Hoke. 1976a. J. Milk Food Technol. *39*, 740.

Seideman, S. C., C. Vanderzant, M. O. Hanna, Z. L. Carpenter, and G. C. Smith. 1976b. J. Milk Food Technol. *39*, 745.

Shotts, E. B., W. T. Martin, and M. M. Galton. 1969. Proc. Annual Meeting, U.S. Livestock Sanitary Assoc., Minneapolis, Oct. p. 309.

Silliker, J. H., R. A. Greenberg, and W. R. Schack. 1958. Food Technol. *12*, 551.

Smith, F. C., R. A. Field, D. D. Deane, and G. M. Thomas. 1976. J. Ani. Sci. *43*, 247.

Smith, G. C., G. T. King, and Z. L. Carpenter. 1975. Laboratory Manual for Meat Science. 1st ed. Howard Kemp Printing, Inc., Houston, Tx.

Sofos, J. N., F. F. Busta, and C. E. Allen, 1979. Proceedings 39th Inst. Food Technol. Mtg., St. Louis, Mo., June 10–13.

Steinkraus, K. H., and J. C. Ayres. 1964. J. Food Sci. *29*, 87.

Steinkraus, K. H., and J. C. Ayres. 1964. J. Food Sci. *29*, 87.

Stewart, M. M. 1932. J. Marine Biol. Assoc., U.K.N.S. *18*, 35.

Stumbo, C. R., C. F. Gross, and C. Vinton. 1945. Food Res. *10*, 85.

Subsistence Research and Development Laboratory. 1948. Q.M. Food and Container Inst. for Armed Forces, Chicago, Il.

Sulzbacher, W. L., and R. A. McLean. 1957. Food Technol. *5*, 7.

Sutherland, J. P., P. A. Gibbs, J. T. Patterson, and J. G. Murray. 1976. J. Food Technol. *11*, 171.

Sutherland, J. P., J. T. Patterson, and J. G. Murray. 1975a. J. Appl. Bacteriol. *39*, 227.

Swaminathan, B., M.A. B. Link, and J. C. Ayres. 1978. J. Food Protection. *41*, 518.

Tanner, F. W., and F. L. Evans. 1933a. Zentrabl. f. Bakteriol. u. Parasitenk Abt. II, *89*, 44.

Tanner, F. W., and F. L. Evans. 1933b. Zentrabl. f. Bakteriol. u. Parasitenk Abt. II, *89*, 48.

Tanner, F. W., and F. L. Evans. 1933c. Zentrabl. f. Bakteriol. u. Parasitenk Abt. II, *91*, 1.

Tarr, H. L. A. 1941. Nature. *147*, 417.

Tarr, H. L. A., B. A. Southcott, and H. M. Bissett. 1952. Food Technol. *6*, 363.

Tompkin, R. B., L. N. Christiansen, and A. B. Shaparis. 1978a. Appl. and Environ. Microbiol. *35*, 59.

Tompkin, R. B., L. N. Christiansen, and A. B. Shaparis. 1978b. Appl. and Environ. Microbiol. *35*, 863.

Tucker, W. H. 1929. Studies in Clostridium putrificum and C. putrefaciens. Institute of American Meat Packers, Chicago.

Vickery, J. R. 1936a. J. Coun. Sci. Ind. Res. *9*, 107.

Vickery, J. R. 1936b. J. Coun. Sci. Ind. Res. *9*, 196.

Wassermann, A. E., J. W. Pensabene, and E. G. Piotrowski. 1978. J. Food Sci. *43*, 276.

Weidemann, J. F. 1965. J. Appl. Bacteriol. *28*, 365.

Weinzirl, J., and E. B. Newton. 1914. Amer. J. Public Health. *4*, 413.

Weissman, M. A., and J. A. Carpenter. 1969. Appl. Microbiol. *17*, 899.

Wickerham, L. J. 1957. J. Bacteriol. *74*, 832.

Yesair, J. 1928. The Cause and Prevention of Molds on Meat and Meat Products. Unnumbered Bulletin, Institute of American Meat Packers, Chicago.

Yesair, J. 1936. Color Control and Conservation of Sausage and Cured Meats. Unnumbered Bulletin, Institute of American Meat Packers, Chicago.

Fish and Shellfish

Fish and shellfish make up a significant portion of the world's food consumption (Box 18-1). Finfish, or vertebrate fish, reside in all waters except the most acidic and salinic, under a wide variety of temperatures except the hottest, and subsist on a wide variety of foods. The indigenous microorganisms of the waters are those that are metabolically most active at the prevailing temperature. The bacteria may be at the beginning of a food chain that ends variously with the large, carnivorous freshwater and marine life, fish-eating birds, or humans (Figure 18-2).

Box 18-1

Of the five finfish species that yield the largest catches, three are used for fish meal rather than for human food (Figure 18-1). Fish and shellfish were among the first animal foods to be eaten by humans, supplying protein when other animals were unavailable or in short supply. The chief methods of harvesting fish—the baited hook, the spear, the trap, the seine, and the purse set—are age-old, and have changed little throughout history. Primitive South Americans, for example, "poison" fish by asphyxiating them and causing them to float. Today, some cultivated fish are driven from the rearing ponds into concrete channels and stunned electrically before dressing. Shellfish are harvested by digging and dredging. Most fish and shellfish are still captured in the wild state, but increasing quantities are being cultivated in inland ponds and estuarine areas.

442

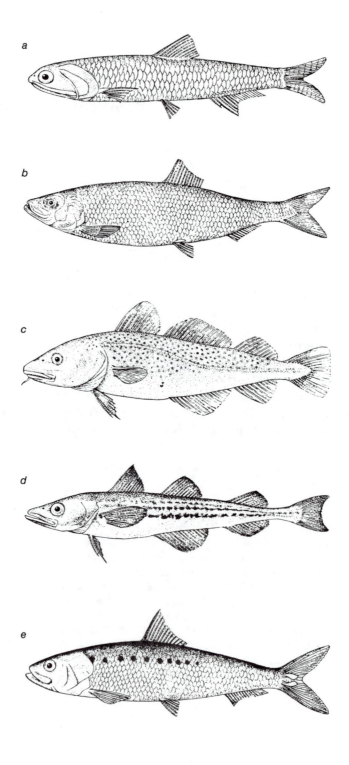

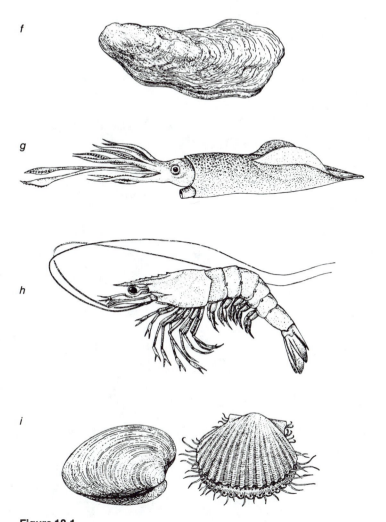

Figure 18-1
Largest catches of individual fish species according to the FAO
Fisheries Statistics Yearbook, 1975. The Peruvian anchoveta (a)
with a catch of 3.9 million metric tons; the Atlantic herring (b)
with a catch of 3.2 million tons; the Atlantic cod (c) with a catch of
3.0 million tons; the Alaska walleye pollack (d), and the South
African pilchard (e). No single invertebrate species (right) is
harvested in similar quantities. Taken as a group, oyster species (f)
totaled 0.8 million tons in 1975; squids (g), 0.7 million tons;
shrimps and prawn (h) 0.6 million tons; clams and cockles (i) 0.6
million tons. (From "The Food Resources of the Ocean" by S. J.
Holt. Copyright © 1969 by Scientific American, Inc. All rights
reserved.)

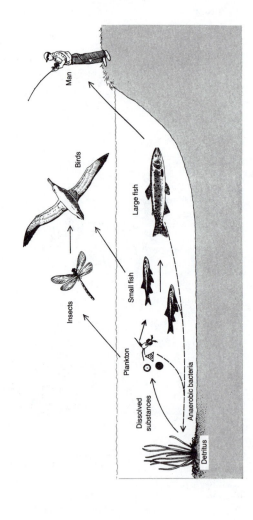

Figure 18-2

Aquatic food chain. (From "Mercury in the Environment" by Leonard J. Goldwater. Copyright © 1971 by Scientific American, Inc. All rights reserved.)

Microorganisms, adsorbed on the surfaces of the fish, do not affect the fish during life, but after death saprophytic and commensal residents invade the flesh and bring about its decomposition.

Body Structure

Fish organ systems are roughly comparable to those of the terrestrial vertebrates. The outer dermal layer of the skin, where the scales attach, is a protective barrier. The inner dermis, or corium, houses blood vessels, nerves, and tubular, flask-shaped glands that secrete a slippery mucus, a lubricating agent. During life the mucus, a natural residential site for bacteria, is constantly produced and constantly sloughed off as the fish moves through the water.

Few fish possess salivary glands. Because many species of fish have a vestigial, nonfunctional gallbladder and bile duct, the selective bile salts are not present in the fish gut. The gut is lined with glands supplying mucus and powerful proteolytic enzymes, which become autolytic after death and hasten decomposition. For this reason, most fish are gutted or dressed at sea; fish, such as herring, that are not gutted are quickly brought to shore after capture or packed in salt.

Ocean fish are divided into the **demersal,** or **flat** fish, such as the cod, skate, and dogfish, which are low in fats, and the **fatty** fish such as herring, pilchard, and mackerel. The water content of fish flesh ranges from 53 to 89%, and averages 80%. The flesh contains less than 1% carbohydrate, and, unlike red meats, there is therefore no protective reduction of pH of the flesh upon death. Lactic-acid-producing bacteria are seldom agents of spoilage—except in **kamaboko,** a fish product to which sugar has been added (Niscolo and Frank, 1966).

It has been the custom in the fishing industry to use wood for catch boxes, bins, holds, dressing surfaces, decks, scoops, and cutlery handles. With age the wood becomes pitted and the recesses become repositories for watery refuse and microorganisms. Washing with water alone does not satisfactorily clean fish containers (Spencer, 1959; 1960; Bhadra and Bose, 1975). As a result, all fish and flesh coming into contact with such wood are seeded with microorganisms. Seawater is used for washing gutted fish and also for cleansing equipment and decks at sea. Only in recent decades has stainless steel replaced wood and the use of detergents, microbicides and refrigeration become common.

Microbiology of Living Fish Surfaces. The microbiology of living fish surfaces has been much studied, but practically all studies have been made on ocean fish, few on freshwater fish. No definitive statements can be made as to numbers or identity of the microflora associated with the slime coat of fishes. Numbers reported range from near zero to many thousands per gram or cm^2 of skin surface (Avery, 1950; Zobell, 1961). Some genera of the bacteria are isolated from the surfaces frequently, others only occasionally. Principal genera are *Moraxella, Flavobacterium* (which may include *Cytophaga,* although this genus is not mentioned in the earlier literature), and *Pseudomonas.* In one study the surface slime of mackerel was reported to contain *Bacillus,* 24%; *Achromobacter,* 22%; *Pseudomonas,* 18%; *Proteus,* 18%; *Flavobacterium,* 4%; and *Micrococcus,* 4% (Reed and Spence, 1929). In another study of that fish species the bacteria isolated were, in descending order of frequency of occurrence, *Achromobacter, Pseudomonas, Flavobacterium, Sarcina, Kurthia, Lactobacillus,* and *Streptococcus* (Kiser and Beckwith, 1944). *Aeromonas, Vibrio,* and coryneform bacteria are also present. Numbers of these organisms may depend upon the amount of organic matter in the waters in which the fish live (Evelyn and McDermott, 1961; Horsley, 1973). *Vibrio* spp. occur only on fish caught in marine (salt water, ocean) environments. Fish caught far from any coast carry no bacteria signifying human pollution (Shewan and Hobbs, 1967); nor do fish caught in unpolluted waters (Spencer and Georgala, 1957). Most bacteria recovered from polar waters are those adapted to the lower temperatures, in contrast to the bacteria on fish taken from tropical waters.

A high percentage of all surface residents are proteolytic—a fact that tends to counter the common statement that fish surfaces are not selectively hospitable to bacteria (Spencer, 1961). An alternative explanation, however, is that the ocean environment itself selectively favors proteolytic types, because it abounds in proteinaceous degradable materials. Chitinoclastic

Table 18-1
The bacterial load on the gills of wild and hatchery salmonids.

	Estimated number of organisms $\times 10^2$/g wet weight of gills	
Fish	Mean value	Value range
Wildfish		
Salmo aquabonito	613	320–1200[a]
Salmo clarki lewisi	1900	1300–2400
Salmo gairdneri	6100	21–46,000
Salvelinis fontinalis	22,000	44–36,000
Oncorhynchus nerka[b]	16	11–22
Hatchery cultured fish[c]		
S. gairdneri	1473	17–9000
O. keta	3430	9–36,000
O. kisutch	2500	1300–4800

[a] Plated on trypticase-soy agar.
[b] These fish were spawners.
[c] No countable organisms were recovered on anaerobic plates of samples from hatchery cultured fish.

Source: Adapted from Trust, 1975.

bacteria are known to occur in fish slimes. A relatively high percentage of bacteria involved in the nitrogen and sulfur cycles is found in ocean waters but not on fish surfaces, possibly because these autotrophic bacteria will not grow on customary laboratory culture media. Similarly, luminescent bacteria occur in the ocean waters and have been observed growing on fish (Zobell, 1961). In earlier times luminescence was the cause of great dismay to the superstitious housewife, whose salted fish stored in her rootcellar seemed to fix her with a stare.

As indicated in Table 18-1, bacterial populations removed from the gills range from 1.6×10^3 to 2.2×10^6. Numbers on skin surfaces may be high, and include both gram-positive and gram-negative bacteria (Table 18-2) and may reflect the proportionate populations of the bacteria in the environment (Trust, 1975).

Intestinal Microflora. The intestinal microflora ranges from 0 to 64 million per gram of contents. The intestinal microflorae of herbivorous fish reflect their consumption of plant foods, whereas

Table 18-2
Frequency of isolation of bacterial species from the gills of salmonids.

Gram-positive species	Number of isolates		Gram-negative species	Number of isolates	
	Wildfish	Cultured		Wildfish	Cultured
Bacillus	49	27	Achromobacter	18	2
Brevibacterium	34		Acinetobacter	25	26
Coryneforms	56		Aeromonas	65	10
Erysipelothrix	6		Alcaligenes	5	2
Micrococcus	25	9	Arthrobacter	20	
Sarcina	5		Cytophaga	82	82
Staphylococcus	2	2	Enterobacter	23	
Streptococcus	10		Flavobacterium	13	43
Streptomyces	3		Pseudomonas	89	83
Yeast	8		Xanthomonas	4	
			Vibrio	27	3

Source: Trust, 1975.

proteolytic bacteria are found in the contents of carnivorous fish. The microflorae are responsible for the weedy odor of ungutted fish. Bacteria not indigenous to the waters of the fish occur only through pollution, and such organisms are transient.

Spoilage

Freshly caught fish have a shining, iridescent surface with bright, characteristic colorations and markings. The surfaces are covered with a thin, whitish but transparent, smooth, homogeneous layer of mucus or slime. The eyes are clear, bright and full, with a prominent jet black pupil and transparent cornea; the gills are bright pinkish red; the skin and flesh are firm, moist, and elastic. A delicate fresh-fish odor is typical, although sometimes it is described as a feedy odor, one associated with the feed of the fish. In the marketplace fresh fish odors are masked by the bacterial growth on decomposing refuse and slime. Although the fish begins to deteriorate as soon as it is removed from water, the spoiled condition develops gradually.

Fresh whole fish are decomposed by several types of enzymatic and microbial activities. While the fish is still in a state of rigor, autolytic enzymes begin breaking down nitrogenous compounds in areas adjacent to the visceral cavity. Meanwhile, bacterial decomposition of the slime layer, the gills, and the intestinal tract gets underway and, after rigor is complete, proceeds rapidly. In late stages of spoilage, the highly unsaturated fats—which are especially abundant in fatty fish—become oxidatively rancid. The exact point at which fish are deemed spoiled is difficult to determine. Among the more noticeable physical changes are the shrunken and sunken eye, cloudiness and milkiness of the pupil, and opacity of the cornea. These changes are the source of the adage that guides shoppers: "Look the fish in the eye." Fish that are promptly refrigerated to 0°C have an estimated average acceptable market life of 21 to 30 days. Fish refrigerated to 7°C remain edible for about seven days.

Little bacterial activity occurs until well after rigor mortis has reached maximum (Reay and Shewan, 1949), after one to two days in fish stored in ice. The rate of spoilage is related directly to time and temperature; each successive degree of reduction in temperature during storage lengthens the storage life more than the previous degree of reduction. Birdseye (1929) stated that haddock, mackerel, and sole lose fresh flavor when the plate count is between one and four million per gram of flesh, and soon thereafter the fish become sour and "fishy." When spoilage is incipient the numbers of bacteria are approximately 10^7 per cm² of surface (the same figure for red meats and poultry on the point of spoilage). Not all bacteria present bring about spoilage. *Pseudomonas, Flavobacterium,* and *Achromobacter* are the genera most commonly implicated.

The eventual spoilage of fresh fish packed in crushed ice is aerobic, but if air is excluded—if, say, the fish are packed close together or pressed against slime-soaked pen boards—conditions become anaerobic and a characteristic foul or "bilgy" odor becomes evident, although the fish look and feel fresh (MacCallum, 1955). The bacterial load on surfaces of wooden fish pens, before cleaning, uniformly exceeds 10^7/in² (Levin,

1971). Hand scrubbing with untreated harbor waters appears to clean the surfaces but leaves a heavy load of microorganisms. Fish holds and pen boards used for storing fish at sea must be efficiently sanitized to maintain high quality of product and to prevent the development of bilgy fish (Castell, 1954).

The duration of market quality in fish has been improved chiefly by use of sanitizable equipment and of detergents, sanitizers, and chlorinated waters in washing both on shipboard and in onshore plants, and by rapidly cooling dressed fish by immersion in water refrigerated to nearly freezing or packing in ice. The inclusion of sodium acid phosphate in the salt to reduce the surface pH, and of sodium nitrite, chlorine, calcium hypochlorite, and antibiotics in the ice has been proposed to prolong the edibility of fish (Boyd et al., 1956; Firman et al., 1956).

Yeasts. Yeasts make up less than 1% of all microorganisms on fresh salmon from the North Pacific (Snow and Beard, 1939). Among the identified strains are species of *Debaromyces, Torulopsis, Candida, Rhodotorula, Pichia,* and *Cryptococcus* (Ross and Morris, 1965). Yeasts commonly occur on the gills and slime layers of marine fish (Morris, 1973), including *Rhodotorula infirmo-minuta* occurring most frequently, and *Candida scottii, Cryptococcus diffluens, C. laurentii, Debaromyces hansenii, Rhodotorula glutinis, R. mucilaginosa, Torulopsis formata, T. sake,* and *Trichosporon pullulans.* Morris found that all cultures grew at 4 and 10°C but none at 30°C, suggesting that these strains of normally mesophilic yeasts are adapted to growth at low temperatures.

Seafoods and Foodborne Disease. Microbial pathogenic contaminants are possible threats to human health. In the United States most fish-related foodborne illnesses are traced to *Salmonella, Staphylococcus,* and *Escherichia* spp., *Clostridium perfringens, C. botulinum* E., *Vibrio parachemolyticus,* enteroviruses, fish and shellfish toxins, and parasitic worms (Center for Disease Control, 1974, 1975, 1976, 1977).

A variety of salmonellae have been isolated from 24 species of fish found in a single market, and from 5 of the 42 waters from which the fish were taken (Gulasekharam et al., 1957). Salmonellae on contaminated fish taken from a river in Canada survived processing of smoked whitefish, infecting over 300 persons with salmonellosis (Gangarosa et al., 1968). The same researchers state that 2% of fish caught in Central Africa and up to 19.6% of fish caught in rivers harbor salmonellae in a balanced host-parasite relationship. Of 597 fresh and frozen seafood products in San Francisco retail markets, analysis indicated 7.9% to be positive for *Staphylococcus aureus,* 4.7% for *Escherichia coli,* and 2% for *Clostridium perfringens;* in these trials, neither salmonellae nor *Vibrio parahemolyticus* were recovered (Foster et al., 1977).

Fish may be carriers of waterborne *Erysipelas, Leptospira, Pasteurella, Aeromonas, Pseudomonas, Vibrio,* and *Mycobacterium* spp. Only the first two genera have no members considered to be pathogens of fish (Brown and Dorn, 1977). Although *Vibrio parahemolyticus* has been recognized for at least a quarter century (Fujino, 1953), it was not linked with disease outbreaks in this country until 1971. Various species of *Vibrio* are associated with numerous disease outbreaks in marine fishes (Vanderzant and Nickelson, 1973).

If fish and shellfish are promptly chilled on the fishing vessel and properly processed, the resident bacterial flora presents little human health hazard. Sporadic incidents of paralytic shellfish poisoning (Ray, 1971), ciguatera, and scombroid fish intoxications occur in this country (Center for Disease Control, 1977) and puffer fish poisoning occurs frequently in Japan (see, in Chapter 21, the discussion of tetrodotoxin). The only fish parasite of public health significance in the United States is *Diphyllobothrium latum,* the broadfish tapeworm (see, also in Chapter 21, the discussion of Diphyllobothriosis). Elsewhere, any of a variety of flukes of fish or shellfish that are present may be transferred to humans during handling and processing.

Chemical Determinants of Spoilage. During postmortem autolytic and microbial activity, numerous chemical changes occur in fish flesh, resulting in liberation of end-products of digestion or of metabolism. A number of these end-products have been suggested as indicators of spoilage. Many compounds, however, cannot practically be well detected. Bacterial plate counts, for example, are time-consuming. Free fatty acids can be determined only in the fatty fish, and then only if *Achromobacter* is not the dominant agent of decomposition. Methyl mercaptan, dimethyl sulfide, dimethyl trisulfide, and 3-methyl-l-butanol have only recently been detected when fish flesh is decomposed by *Pseudomonas putrefaciens, P. fluorescens,* and *Achromobacter* (Miller et al., 1973a, b) and detection of their presence may become practical through the use of gas chromatograms.

Among the components used or suggested for use are mono-, di-, and trimethyl amines, indole, volatile acids, hydrogen sulfide, oil acidity, steam-volatile oxidizable substances, ammonia, free fatty acids, noncoagulable nitrogen, and tyrosine (Farber, 1952; Cooks and Ritchie, 1938; Bradley and Bailey, 1940). Hypoxanthine appears able to determine spoilage of albacore, increasing to as many as 1500 μ/g during deterioration (Crawford and Finch, 1968).

The nonodorous trimethylamine oxide (TMAO), a common component of marine fish but not of freshwater fish, is reduced to the odorous trimethylamine (TMA) characteristic of stale fish according to the equation,

$$TMAO + H^+ \text{-------} \blacktriangleright TMA + acid\ residue + CO_2 + H_2O$$

If the hydrogen donor is lactic acid, acetic acid is formed (Tarr, 1939). The alkali-sensitive enzyme ceases to function at pH 8.0, which is achieved when urea, another common component of fish flesh, is decomposed (Elliott, 1952). In fatty fish, fat rancidity masks the odor of TMA (Tarr and Ney, 1949).

The measurement of volatile reducing substances (VRS) was developed to determine spoilage in protein foodstuffs (Land et al., 1944). In tests for VRS, purified air is swept through a homogenate of food material into a volume of standardized potassium permanganate. Values are expressed as microequivalents (ME) reduction of the permanganate. The VRS test is applicable to fish that have been held at sea during periods of sluggish catch and in which the changes in decomposition have not proceeded to the point of organoleptic detection. Such fish do not yield a canned product of good quality. For example, tuna fish with ME values of 15 or less are acceptable for canning; those with ME values between 15 and 20 are borderline; and those with ME values above 20 are not acceptable for canning (Farber, 1952).

Most microbial and autolytic decomposition of fish has only economic and esthetic consequences, but sometimes public health is also involved. If scombroid fish like tuna decompose either in the marketplace or before canning, scombroid poisoning may result (further discussed in Chapter 21). Several bacteria capable of decarboxylating histidine form the intoxicating agent histamine. The normal histamine content of fresh tuna is no more than 6 mg/100 g of flesh; a count of more than 10 mg/100 g indicates decomposition (Geiger, 1944). Values ranging between 100 and 275 mg have been reported in instances of obvious intoxication. Miller et al. (1973a) identified a volatile compound produced by *Pseudomonas perolens* in otherwise sterile fish muscle; this compound is the source of the musty potatolike odor common in chilled fish muscle.

Fresh fish are sometimes rejected because of an environmental rather than a spoilage factor. Salmon residing in tidal muds (Thaysen, 1936) and catfish, carp, crappie, and other fish captured in inland waters of the United States during the summer months acquire an objectionable earthy or muddy taint. The taint, which would appear to be the oil geosmin, produced by streptomycetes growing in the muds, can be removed by placing live-caught fish in clean, flowing waters.

Preservation of Fish

Fermentation and Pickling. Fermentation is practiced widely in the Far East to preserve fish and fish protein, both major sources of protein for coastal and inland populations. In a typical acidic fermentation the cut, salted, and drained fish *(burong dalag)* or shrimp *(burong hyphon)* is mixed with cooked rice and water. The acid fermentation is completed in 2 to 10 days (Orillo and Pederson, 1968).

Throughout the Philippines and Indochina, the preservation of fish by salting and drying is limited by the high price of salt (Van Veen, 1953). Whole small fish, often below edible size, are preserved in quantity as a fish sauce—produced, however, by a process of digestion rather than fermentation. The microbiology of the process is unknown, but may be caused by autolytic enzymes rather than by bacteria. Whole fish are kneaded, salted, pressed, and sealed either in large vats in commercial processes or, if made at home, in vessels that are then buried in the ground. During the digesting process, several months long, all the nitrogen is liberated as amino acids, and methyl ketone is formed, imparting a cheesy flavor. The product is marketed as a clear, supernatant fluid.

Numerous species of fish are pickled directly in vinegar, with salt, spices, and herbs added as desired. Because during pickling some acetic acid is neutralized through reaction with the calcium salts in the bones, bacterial spoilage has been associated with the use of poor-quality vinegar. The pickled product is perishable, and if not consumed immediately it should be refrigerated, pasteurized, or kept under anaerobic conditions. Lang (1935) found *Clostridium botulinum* in the slime and washings from empty sardine cans.

Lightly Salted Smoked Fish. Lightly salted smoked fish are a perishable delicacy. As with many other food items, the history of smoked fish preparation has two phases. Before the early 1960s the amounts of salting and smoking varied among processors. Fish marketed in translucent, brown, moisture- and oxygen-permeable paper became dehydrated and developed a brown surface discoloration. Packing in Mylar film prevented both dehydration and discoloration.

Although *Clostridium botulinum* type E was identified by Gunnison et al. in 1936, occurrence of this organism in fish from the Great Lakes was unsuspected until 1963 (Bott et al., 1964), when a much-publicized outbreak of botulism in Huntsville, Alabama, and Nashville, Tennessee, was traced to a shipment of vacuum-packed smoked whitefish chubs (Communicable Disease Center, 1963). Through a tragic series of errors, high moisture content coupled with anaerobiosis and lack of refrigeration resulted in the growth of *Clostridium botulinum* type E in the fish. Allegedly the fish were held for one day at 1.7°C (35°F) before being smoked for 5 hours at 82.2°C (180°F) and then re-stored at 1.7°C for a day before vacuum packaging. There is no documentation of refrigeration during shipment. Ten days after being smoked the fish arrived at their destinations and the next day were displayed for retail sale. A day later two persons who had eaten fish from this shipment contracted botulism and died. From intensive studies after this outbreak it was learned that *C. botulinum* type E is a common resident of fresh and estuarine waters and bottom muds, and is on the surfaces of captured fish and can be transferred to their flesh during processing (Foster et al., 1977). Because of the incident described above, the Food and Drug Administration set regulations designed to prevent further outbreaks (Olsen, 1969). During brining the salt content must exceed 3% in the water phase of lean muscle; during smoking every portion of the fish must be heated to 82.2°C for 30 minutes to destroy the spores of *C. botulinum* type E; the fish must be cooled to refrigeration temperature within two hours of removal from the smokehouse. Refrigeration must not be interrupted throughout shipment and until sold. Packages must be dated, and are legally salable for only 14 days after production. Present GMPs for smoked fish are: hot smoking at 82°C for a minimum of 30 minutes if the water phase salt (WPS) of the finished product is 3.5% or above; or 66°C (150°F) treatment for 30 minutes if the WPS is at least 5.0%. A storage temperature of 3.3°C (38°F) or lower is also specified for smoked fish (Federal Register, 1970).

Heavily Smoked Fish. Applying smoke to preserve fish, geese, and other wildlife is an ancient art still practiced in some parts of the world, such as in the Pacific Northwest, Eastern Canada, and countries surrounding the North Sea. Dehydration of the fish during smoking helps control microbial growth. Smoking and drying may require as many as seven days. The smoldering wood forms and releases chemical constituents that react with the food and inhibit most microorganisms. Moreover, most consumers prize highly specific flavor characteristics formed in the product. Because smoking accelerates oxidative rancidity, it is used only on lean fish.

Aerobic sampling of smoked salmon from roadside retail counters along the Pacific Northwest coast yielded several varieties of microorganism at levels ranging from 0.13 to 2200 ×

$10^3/g$; most were staphylococci and micrococci, although in some samples *Bacillus* and *Pseudomonas* spp. and yeasts were predominant (Lee and Pfeifer, 1973). While the gram-positive cocci were able to grow in 10 to 25% NaCl, most did not multiply at 4°C and were readily inactivated by 66°C or 82°C. These organisms were considered to be post-processing contaminants.

Salt Preservation. Salt (NaCl) is used to preserve large quantities of fish by brining, wet salting, or salting and drying. Processers throughout the world make much use of solar or sea salt, which is rich in a variety of halophilic and haloduric bacteria. The numbers of bacteria in solar salts vary according to the sources of the seawater; counts of 200,000/g are reported in the salts used in Russia (Petrowa, 1933), less than 100,000/g in salts used in the Atlantic region (Shewan, 1938). The relatively few bacteria causing spoilage of salted fish have been identified as *Serratia* (see Box 18-2), *Pseudomonas salinaria* (Harrison and Kennedy, 1922), *Sarcina* spp. (Hess, 1942), and several species of *Micrococcus* (Hanzawa and Takeda, 1931). All of the above bacteria are obligately halophilic and grow best at warm temperatures. They produce proteolysis of the flesh and red pigments on their surfaces.

Herring, a fatty fish, is preserved in saturated salt brines. Within a few days it acquires a characteristic cured flavor, made possible by action of autolytic enzymes (Shewan, 1937). In cold climates properly preserved and submerged herring are kept through the winter months. Surfaces projecting above the

---

Box 18-2

The bacteria were termed *Serratia* because of the red pigment. Unlike the fermentative genus *Serratia* in the family Enterobacteriaceae, the red-pigmented bacteria found in solar salt and products preserved with solar salt do not ferment carbohydrates. They are probably members of the genus *Halobacterium,* but as with *Achromobacter,* their identity is in doubt. Halophiles occur rarely in human foods. When solar salt is used these organisms may cause spoilage.

liquid develop a rusty red growth of bacteria and of haloduric, filamentous molds, especially as temperatures rise in spring.

Wet salted fish and fish to be salted and dried must receive sufficient salt for a minimum of 8% salt to reach the deepest layers of tissue within five days. Much of the loss of salt-dried fish occurs during drying, if salt incrustation is too slow or case hardening is too fast (Finn, 1941). The halophilic mold *Sporendonema epizoum (Torula epizoa* in the older literature) causes most spoilage of salted fish destined for tropical and subtropical countries, where refrigeration is rare. To prolong quality maintenance, salted cod can be treated with salt containing sorbic acid or its potassium salt, or can be dipped into a sorbic acid solution (Boyd and Tarr, 1955).

In warm climates fish can be preserved by a quick salting process whereby a mixture of ground fish and salt receives extreme pressure and forms into cakes. The presscake of shark taken in the Gulf of Mexico has a salt content of 22.6% and a water content of 45% (by contrast, heavily salted commercial fish have about 20% salt and 49% water) (Del Valle and Nickerson, 1968; Del Valle and Gonzales-Inigo, 1968). The protein-coagulating action of the salt makes the cakes cohesive and causes them to retain their shape when boiled in water.

Sardines. Most sardines are canned. Although the bacterial counts of freshly captured sardines are low, the bacteria can grow rapidly if the fish are mishandled before processing. The surfaces of the packing tables present the greatest hazard (Highland and Williams, 1944).

Molluscs

Although markedly different in external appearance, snails, slugs, and whelks; clams, mussels, and oysters; and octopuses, cuttlefish, and squids are all soft-bodied (i.e., *Mollusca)* invertebrates and fundamentally resemble one another more than they resemble the crustaceans or the finfishes.

Shellfish make up a good proportion of seafoods. Squids and octopuses are popular foods in Oriental and Mediterranean

Box 18-3

Several bivalve shellfish are eaten by humans. Most important are several species of *Ostrea,* the oyster; *Mya arenaria,* the soft shell clam; *Venua mercenaria,* the quahaug or hard-shell clam; and *Rangia,* the Gulf Coast clam; the mussels, *Mytilus* and *Modiolaria;* and the scallop, *Pecten* sp. Oysters are sessile residents in shallow estuarine waters; clams move by means of a muscular foot; and the scallop swims by snapping its two deeply grooved earlike shells with its edible adductor muscle. Generally, mussels attach themselves to rocks and other surfaces by threadlike secretions. Shellfish thrive in brackish, estuarine waters where fresh and salt waters mix, and growth rates are proportionate to the amounts of organic matter suspended in the waters.

In the United States, to conserve natural oyster beds and ensure continuing supplies, the several coastal states have taken legal possession of all offshore beds. Many of the northern oyster beds suffer from periodic floods, which introduce excessive quantities of fresh water into the estuaries, contributing to failure of the oyster culture.

Under the supervision of a state commission, seed oysters can be obtained from natural beds. The bottoms of the beds are dredged and the small, immature oysters are transplanted to special beds for further growth. A mature oyster may produce thousands of eggs during her lifetime. The immature oysters, or "spats," are cultivated in beds located between high and low tidelines. The spats have a natural tendency to attach themselves to rocks, shells, or other objects. It is common practice to provide them with shucked oyster shells. The shell to which the spat attaches itself is called the "clutch." When oysters reach maturity, at approximately 18 months, they free themselves from the clutch, which may then be used by the next generation of spats.

Oysters feed by pumping water through the gills, which strain out the food. The food consists of algae, bacteria, and detritus. Oysters begin feeding at about 5°C and at the optimal temperature—around 30°C—a mature oyster will pump several liters of water through the gills each hour.

countries, although not very popular in the United States. Over a million metric tons of squids, cuttlefish, and octopuses are caught annually. Once caught they decompose quickly. Surprisingly few definitive studies have been made of postmortem microbiological changes in these animals.

Every year over a million metric tons of shrimps and prawns are harvested, as is a similar quantity of clams, mussels, and scallops. About 750,000 tons of oysters are harvested annually while the yearly catch of crustaceans other than shrimps and prawns (i.e. crabs, lobsters, etc.) is a little over a half million tons.

Oysters

The microflorae of the oyster are typical of the waters in which they grow. *Saprospira* and *Cristispira* spp. are the only known commensal residents. In 1973 aerobic plate counts of freshly harvested oysters from Galveston Bay, on the Gulf Coast of Texas, ranged from 2.3×10^4 to $3.0 \times 10^7/g$, whereas water samples are usually $10^2/ml$ of *Vibrio, Aeromonas,* and *Moraxella* spp. *V. parahemolyticus* was isolated from 39 of 66 water samples and from 9 of 30 sediment and water samples (Vanderzant et al., 1973c). Species of *Aeromonas* and *Moraxella* were predominant in oysters on the retail market; *V. parahemolyticus* was detected in only one of eight refrigerated retail oyster samplings.

Other commonly encountered seafood isolates include *Arthrobacter, Pseudomonas, Micrococcus, Flavobacterium-Cytophaga,* and *Acinetobacter* (Lee and Pfeifer, 1974). At 35°C, 68, 51, 42, 33 and 29% respectively of these organisms failed to grow, while at 5°C, 40% of *Flavobacterium-Cytophaga,* 26% of *Acinetobacter,* and 12% of *Arthrobacter* failed to grow. At 35°C, 46% of *Moraxella* failed to grow, compared to only 9% at 5°C. Between 25 and 37°C, from 1.6 to 3.7 logs difference in microbial counts may be observed (Vanderzant et al., 1973c). High percentages of *Vibrio* (22%) and *Aeromonas* (21%) occur in Gulf Coast oysters while North Sea fish contain less than 1% of each (Shewan, 1971) and neither genus is present on Dungeness crab, frozen shrimp, or iced fillet of sole (Lee and Pfeifer, 1974).

Large quantities of oysters are consumed on the half shell. Moreover, cooking them, as for oyster stew, impairs their delicate flavor and texture if done too long or at too high a temperature. However, because oysters thrive in waters that may be polluted with human wastes, they raise public health concerns. This was reflected as early as 1905 in an editorial appearing in the *Journal of the American Medical Association* (Clem, 1969). It was not until 1925, however, when a widespread outbreak of typhoid fever occurred with 1500 cases and 150 deaths, that controls of shellfish were initiated. Despite surveillance, outbreaks of typhoid occasionally occurred (Wise et al., 1948; Wilson and McCleskey, 1951). Now all coastal states where shellfish are harvested participate with the federal government and with industry in the National Shellfish Sanitation Program (NSSP), administered since 1968 by the Food and Drug Administration.

A 24 month survey, completed in 1974, in which 539 samplings of the Eastern oyster *Crassostrea virginica* and the overlying waters were collected, indicated that salmonellae were not detected in oysters from approved growing areas. However, the routine analysis of shellfish-growing waters for the presence of pathogens remains an impractical task for state and federal officials charged with guaranteeing the quality of shellfish (Andrews et al., 1975a). The NSSP has traditionally measured the total coliform group to assess the bacteriological quality and safety of shellfish-growing waters. At the 1974 NSSP national workshop, a proposal was made to substitute a fecal coliform standard for the total coliform standard. Andrews et al. (1975a) report that a higher percentage of salmonella-positive shellfish samples was harvested from waters conforming to the standard of <70 total coliforms/100 ml than from the same waters that met a proposed fecal coliform standard of <14 organisms per ml.

Enteroviruses, *Vibrio parahaemolyticus,* and *Pseudomonas aeruginosa* are real or potential pathogens associated with oysters and other shellfish. In recent years numerous outbreaks of enteroviral illness have followed consumption of raw oysters, and have stricken many persons. Often shellfish are implicated as the vehicle for food-associated hepatitis outbreaks (Cliver, 1971). Enteroviruses have been detected in oysters taken from a contaminated estuary (Metcalf and Stiles, 1968). Viruses are more stable in the oyster than are bacteria, but at present no rapid microbiological method is known to indicate their presence or to ensure the safety of the oyster consumer (Fugate et al., 1975). Oysters taken from approved Gulf Coast waters sometimes contain enteroviruses. Echovirus 4 and poliovirus 1 were recovered from 1 of 17 Texas samplings and poliovirus 3 from 1 of 24 samplings from Louisiana. Two of 24 oysters that contained *V. parahemolyticus* also yielded virus. Poliovirus 1 was also found in frozen shucked oysters imported from Japan.

Vibrio parahaemolyticus is most abundant in inshore and estuarine areas when the ambient temperatures rise seasonally to levels permitting growth, i.e., above 15°C. This halophilic organism is found in water, sediment, plankton, and fin fish, but in North America is most abundant in molluscan shellfish and in waters high in organic matter (Liston and Baross, 1973). Similar regular cycles coinciding with plankton blooms have been observed in the Chesapeake Bay (Colwell, 1973). It survives, unchanged in number, in oyster shellstock being stored for eventual purification to be served (Johnson et al., 1973).

Pseudomonas aeruginosa has been isolated from nearly half of all oysters and 74% of all mussels sampled in southern France. Eight bacteriophage serotypes were detected among the strains, with Type P_3 accounting for 26% of the isolates. At the time of the study, Type P_3 predominated among nosocomial infections in the hospitals in the district of Poitiers. The author suggested that shellfish are a fomite for transmitting *P. aeruginosa* into the hospitals and for colonizing intestinal tracts (Denis, 1975).

Shellfish-growing waters are classified as clean, moderately polluted, or heavily polluted. Harvesting of bivalves is prohibited in heavily polluted waters. In 1965 an estimated two million acres of shellfish waters off the Atlantic Coast were closed to harvest, and 8.2 million acres of waters were approved. The several areas prohibited to shellfish harvest receive vast quantities of sewage sludge and other types of human waste. Shellfish taken under the jurisdiction of the State of Massachusetts must be harvested by bonded diggers, and all clams must be taken to shellfish treatment plants for self-cleansing (Fox, 1953).

Oysters harvested in mildly polluted waters are purified by natural and by artificial means. When transplanted to clean waters, they will become relatively free of coliform bacteria in one or two days. The rapidity of cleansing depends largely on the conditions in which the oysters are placed. Under good feeding conditions, the rate at which water is passed through the gills determines minimal necessary cleansing time.

To be artificially purified, oysters are placed in water containing 6 ppm chlorine, at which concentration they hibernate until the chlorine has been dissipated. Adding calcium hypochlorite to the water reduces the irritation caused by free chlorine so that the mollusks carry on normal activities.

Japanese investigators have proposed that bacteria can be removed from oysters and clams by the process of electrophoresis (Tanikawa et al., 1967). The bivalves are placed on a plastic net between the cathode and the anode and 5 to 200 volts is circulated for 50 minutes, achieving a 2 to 4 log reduction in microbial content.

The numbers of bacteria in freshly shucked oysters ranges widely. The genera represented include *Achromobacter, Pseudomonas, Escherichia, Serratia, Flavobacterium, Micrococcus, Streptococcus, Lactobacillus, Enterobacter, Proteus, Bacillus* and *Clostridium*. Freshly shucked oysters usually remain sound for 10 to 20 days, depending on the initial numbers of bacteria and on degree and duration of refrigeration.

During storage *Achromobacter* and *Pseudomonas* are among the most metabolically active, and at the point of spoilage may number 10 million per gram of meat, and much of the flesh weight is lost as fluids (Lartique et al., 1960). Increases in volatile acids (Beacham, 1946), indole, and TMA are too inconsistent to be useful indices of spoilage. Freshly shucked oysters have a pH 6.2 to 5.9. At pH 5.8 oysters are considered "off," at pH 5.7 "musty," and at pH 5.2 or less sour or putrid. Because Southern oysters demonstrate seasonal variations in pH, sour odor and drop in pH do not always correlate (Gardner and Watts, 1957). Southern oysters produce more carbohydrate in May than in August, and acid formations may contribute to the differences (Lee and Pepper, 1956). Whatever the cause, a clear relationship has been demonstrated between the pH of oyster liquor and spoilage (Simidu and Hibiki, 1957). A pink yeast, of either *Rhodotorula* or *Sporobolomyces* spp. (Phaff et al., 1952), grows in refrigerated oyster liquor, imparting an unsightly pink tinge. The yeast grows in the shell piles at the shucking house when unsanitary conditions prevail, and when the shells are deposited on the oyster beds they inoculate the water (Hunter, 1920).

High coliform counts have been observed in shucked oysters, but *Escherichia coli* is seldom present in oysters harvested in clean waters (Perry, 1939). Prior to World War II, oysters were one of the few foods in which the nonfecal lactose-fermenting bacteria were often present without the concomitant presence of *E. coli*. When these bacteria are found in well waters, sanitarians rightfully include them along the potential indices of pollution. Subsequently, this concept was extended to other foods as well, although the association between the nonfecal coliform bacteria and pollution was obscure. Hunter (1939) stated:

In the sanitary control of food production and handling there is, of necessity, considerable groping for bacterial indices of pollution and potential danger to health. A knowledge of the food product which permits an explanation of the sources of contamination is fully as important as a knowledge of the identity of the coliform organisms detected in it, and it is not unusual to encounter misconceptions and differences of opinion related to the applicability of coliform bacteria as indices of pollution of food.

Later investigators (Raj and Liston, 1961) believe that the large heterogeneous flora in frozen seafoods tends to discredit the specificity of the fecal coliform test.

Clams

In a 24 month microbiological survey of 214 samples of the hardshell clam, or quahaug, *Mercenia mercenaria,* none of the samples meeting a total coliform of <70/100 ml of water or a fecal coliform value of <14/100 ml of water contained salmonellae (Andrews et al., 1975a). The range of total coliform MPN of the waters at which *Salmonella* spp. could be recovered was from 490 to 11,000, whereas for fecal coliforms it was from 33 to 2300. Salmonellae were not detected on any quahaug samples meeting the wholesale market quality standard of 230 fecal coliforms/100 g of shellfish and total plate count of 500,000 organisms/g of shellfish (Public Health Service, 1965).

The fecal coliform standard, determined by the Elevated Temperature Plate Count procedure of Hefferman and Cabelli (1967), requires 72 hours to complete (APHA, 1975). When used to determine the sanitary quality of clams, this procedure underestimates the densities of fecal coliforms by about 10% compared to the standard MPN procedure (Varga et al., 1977).

Scallops

The microbiology of scallops is similar to that of other molluscan shellfish. The microflora of scallops is of less public health significance than that of oysters, because the inedible siphons,

Table 18-3
Distribution of different bacterial groups during iced storage of whole queen scallops.

Days in ice	No. of isolates	Pseudomonas	Moraxella-Acinetobacter	Flavobacterium-Cytophaga	Coryneforms	Gram-positive cocci	Entero-bacteraceae	Unclassified	No growth[a]
0	101	16	22	12	17	22	2	3	7
2	106	16	52	10	22	3	0	2	1
3	100	20	54	9	10	3	0	3	1
5	100	22	76	0	0	0	0	0	2
6	101	3	91	2	1	3	0	0	1
8	101	8	91	0	2	0	0	0	0
12	102	6	87	1	1	1	0	0	6
14	102	8	80	0	0	0	0	0	14

[a] Refers to primary isolates not viable on subculture.

Source: Thomson et al., 1976.

digestive tract, and gills are the organs most susceptible to microbial contamination, but only the adductor muscle (the large muscle that closes the shell) is used for food.

Bacterial counts in harvested scallop muscle increase from 10^3 to 10^5 per gram in 8 to 9 days. *Moraxella-Acinetobacter* comprise 22% of the initial microflora, the remainder consisting of *Pseudomonas*, gram-positive cocci, coryneform bacteria, and *Flavobacterium-Cytophaga* spp. After six days the *Moraxella-Acinetobacter* group accounts for more than 90% of the total flora (Table 18-3).

Snails

Snails are gastropod mollusks with a spiral protective shell that live on land and in water. Some species are eaten as food. The Moroccan snail, an imported delicacy, harbors salmonellae (Andrews et al., 1975b). To prevent growth of competing bacteria and to ensure recovery of the salmonellae, snail homogenate is added directly to the selective media without preenrichment.

Enrichment in tetrathionate broth with brilliant green has proven to be superior to that using selenite-cysteine broth. Salmonellae have been isolated from the garden variety of snails, which are not consumed as food (Mueller, 1965).

Crustaceans

The edible crustaceans differ from finfish and other shellfish in having a number of specialized appendages including paired antennae and legs; some species also have large pincers. All have a calcified cuticle or carapace over their dorsal surface. Shrimps, lobsters, and crayfish have additional thin plates of hard cuticle covering a segmented abdomen and tail but with a softer and thinner shell covering the ventral surface.

Shrimps

Once caught, shrimp die quickly and almost immediately begin to deteriorate, remaining organoleptically pleasing for no more than four days (Campbell and Williams, 1952). Microbial counts on whole shrimp examined immediately after being emptied from trawler nets typically range from 10^3 to 10^5 per gram (Green, 1949). Partially digested, rapidly decomposing food forms a dark liquid that exudes from the stomach and intestines of whole iced shrimp (Fieger and Novak, 1961). On the ship deck, the shrimp often are smeared with slime and exuded intestinal contents from other marine life taken at the same time. The heads of freshly captured shrimp should be removed, the surface slime and digesta washed away, and the beheaded shrimp packed in ice until they reach market. Removal of the heads extends the storage life by two days (Lantz, 1951). The heads account for about three-fourths of the bacteria (Fieger, 1950). Often several hours elapse before the shrimp are beheaded and iced. During this period, the shrimp may be held at temperatures ranging from near freezing to about 40°C. Shrimp are extremely susceptible to bacterial and enzymatic degradation (Lapin and Koburger, 1974), and begin to develop off-color and off-odor if held for 3 hours at 30, 37, and 44°C, for 6 hours at 20°C, or for 24 hours at 10°C; after 6 to 24 hours at 30 to 44°C there is a large increase in bacterial count (Cobb et al., 1977).

The microbial flora of shrimp reflect the microbiological characteristics of the water from which they are harvested (Cobb and Vanderzant, 1971). Gram-negative bacteria, particularly *Flavobacterium-Cytophaga,* represent most organisms isolated from fresh rock shrimp taken off the East Coast of Florida (Koburger et al., 1975). In freshly harvested white shrimp with low counts—i.e., with bacterial loads of $<10^4/g$—spoilage odors appear after 25 to 37 days of storage on sterile ice; for brown sea shrimp, similar odors appear in 20 to 30 days, and in brown boat-shrimp after 0–15 days (Cobb et al., 1973). As spoilage occurs, value of amino nitrogen decreases and that of total volatile nitrogen increases. Some losses of both forms of nitrogen occur during iced storage as a result of the continual washing the shrimp receives from the melting ice (Box 18-4).

Almost all the bacteria initially present in freshly caught shrimp are of the genera *Micrococcus, Moraxella-Acinetobacter, Achromobacter, Pseudomonas, Flavobacterium,* and *Bacillus* (Campbell and Williams, 1952; Vanderzant et al., 1970). Dur-

Box 18-4

Shrimp trawls generally contain an assortment of shrimp, fish, and crabs, as well as mud and debris. In the United States shrimp landings greatly exceed those for any other shellfish in tonnage. Most of the shrimps—brown, white, pink, seabobs, royal red, and rock—are harvested in South Atlantic and Gulf waters, but northern shrimp landings—off Alaska, the West Coast, and New England—have increased greatly during recent years. Also, as many shrimp are imported to this country every year as are harvested annually.

ing storage at 8°C, an initial count of 25,000 may rise to as many as 12 million bacteria per gram in 16 days. Similar flora are found on prawns (Magar and Shikmahmud, 1956). As storage continues, the percentages among the genera change drastically (Table 18-4). The microbial flora on pond-reared brown shrimp *(Penaeus aztecus)* initially consists primarily of species of *Aeromonas, Pseudomonas,* and *Vibrio;* after three or four weeks of refrigerated storage coryneforms, acinetobacters, and pseudomonads become dominant (Vanderzant et al., 1973a). *Pseudomonas putrefaciens,* the predominant hydrogen-sulfide-producing organism always present in low numbers in fresh shrimp, increases rapidly until it represents a significant percentage of the total population (Lapin and Koburger, 1974). In nitrogen-packed shrimp, it reaches high numbers by the eighth day of storage. With Gulf shrimp, *Achromobacter* spp. grow markedly while pseudomonads and flavobacteria decrease as percentages of the total population. Micrococci, bacilli, and miscellaneous organisms disappear completely or are so far outnumbered as to be undetectable by normal plating procedures. Shrimp also become contaminated with bacteria not common to ocean waters, such as streptococci and coliform bacteria, because of the unsanitary conditions of the boats (Virgilio et al., 1970).

The increase in total volatile nitrogen is thought to be a better index of declining shrimp quality than are bacterial counts or changes occurring in pH, amino nitrogen, hydrolysis of water-

Table 18-4

Changes in bacterial flora of Gulf shrimp during storage in crushed ice.[a]

Days held in storage	Achromobacter	Bacillus	Flavobacterium	Micrococcus	Pseudomonas	Miscellaneous
0	27.2	2.0	17.8	33.6	19.2	0.2
4	31.3	0.6	13.1	23.0	26.5	0.5
8	46.0	2.0	18.0	5.7	28.0	0.3
12	67.0	0.0	2.0	0.8	30.1	0.1
16	82.0	0.0	1.5	0.0	16.5	0.0

[a] Expressed as percentage of total number of organisms isolated at each interval.

Source: Campbell and Williams, 1952. Reprinted from Food Technology, Vol. 6, p. 125, 1952. Copyright © by Institute of Food Technologists.

insoluble protein, content of B vitamin, picric acid turbidity or increase in the volatile reducing substances (Cobb and Vanderzant, 1975). White shrimp *(Penaeus setiferus)* inoculated with nonfluorescent pseudomonads and bacilli produce ornithine and ammonia; inoculated with the coryneform bacteria they produce ammonia and trimethylamine.

Bacteria grow freely between the loosely attached carapace and the body of the shrimp. Most are removed during removal of the carapace and deveining. The shrimp are then refrigerated for immediate consumption or are breaded and frozen.

Fluids on shrimp surfaces are a fertile medium for bacterial growth. Shrimp cannot be pasteurized, cooked, or given any heat treatment without destroying the delicate flavor and texture. Guidelines for the processing of breaded shrimp stipulate that at no time can the processing temperature permit staphylococci to grow, and the processed shrimp shall have no more than 100 staphylococci per gram. There is no increase in the total count of coliform bacteria or staphylococci in the production of breaded, raw frozen shrimp when plants exercise good sanitary practices, but the bacteria grow where sanitation and handling practices leave much to be desired (Surkiewicz et al., 1967). The quality of the raw product varies, and little can be done to improve the product that is received by the processor. During the brief period (30 seconds at <200°C) that breaded shrimp are heated to set the batters, the heat-sensitive gram-negative bacteria on the surface of the battered product may be killed, but the more heat-resistant spherical enterococci and staphylococci are able to survive. A typical batter mix may contain relatively few thousands of bacteria per gram at the time of preparation, but during continuous operation at room temperature the counts, chiefly of *Microbacterium lactis,* may rise to 100 million or more, with a concomitant slight lowering of the pH. If the batter mix is kept at 4.5°C or less, as is now required by good manufacturing practices, bacteria do not proliferate (Box 18-5).

Surveys of the bacteriological quality of frozen breaded raw shrimp emphasize the need for better methods of handling (Surkiewicz et al., 1967). Of the samples of fresh or frozen shrimp received for breading, 30% had bacterial counts in excess of 10^6/g, as did 56% of those already breaded and frozen

---

Box 18-5

All batters used in battering and breading are made from a slurry of flour mixed with water. Battered and breaded products are heated simply to set the batter; the coated product receives no significant cooking. The pH ranges between 5.6 and 5.8. During battering, the product usually passes under a continuous flow of batter. The drainage from the battering stage is returned to the reservoir and fresh flour and water are added as needed. In a typical oil cooker, the product is subjected to a temperature of around 197°C for 30 seconds.

(Vanderzant et al., 1973b). Plate counts of samples picked up at retail markets vary from 24×10^6 to 60×10^6/g; enterococci from TFTC to 5500/g; coliforms from an MPN of 0.3 to 8850/g (Nickerson and Pollak, 1972). *Bacillus, Microbacterium, Micrococcus* spp. and coryneform bacteria were predominant in retail samples (Vanderzant et al., 1973b).

Lobster

Lobsters are as highly perishable as are shrimp and prawns. Bacteria are found both on the outer surfaces and in the digestive glands and the intestinal tract. After the lobster dies pseudomonads and achromobacters cause rapid proteolytic souring.

Lobster meat contains an autolytic enzyme system that becomes very active as soon as the lobster leaves its natural environment. When lobsters are caught and packed in ice, the enzymes are activated and the meat undergoes marked changes. Because the activity is much more pronounced in dead lobster meat, lobsters are kept alive in seawater until they are to be cooked. Most lobsters are taken in traps called lobster pots. Tropical lobsters, e.g., *Panulirus versicolor,* are difficult to capture alive because they seldom enter the pots; they are speared. If immediately beheaded, gutted, and cooled, they remain edible for a week.

Chemical composition of the meat differs markedly among lobsters harvested at different times of the year. Hydrogen sulfide, liberated by bacteria, causes a chemical discoloration of the canned meat, called "black salt." Changes in pH values are also correlated with the season; a high pH value seems to accelerate the black discoloration.

Lobster meat contains 1.3 to 1.5% arginine. Its loss due to microbial attack produces undesirable odors and a loss of flavor. Dipping shellfish—and also beef or whale muscle—into a solution of 1% arginine may improve the tenderness and preserve the organoleptic properties (Miyake and Tanaka, 1971). The microbial hydrolysis of arginine to ornithine and ammonia is detected in aging lobster (Sidhu et al., 1974). The source of contamination of cooked lobster meat is warm working surfaces (Varga and Sanderson, 1968).

Crab

The main reason for Captain John Cook's visit to Chesapeake Bay in 1607 was a yearning for its native crab, now called Maryland crab (Damon, 1971). The importance of crabmeat in the diet is shown by the processing of 248 million pounds of blue (Eastcoast) and King and Dungeness (Westcoast) crabmeat in 1968.

Crabs decompose quickly after death. To preserve their freshness, they are cooked in boiling water as soon as possible after capture. The cooked flesh is picked from the carapace and claws (Figure 18-3) and is marketed as refrigerated meat, with an approximate keeping time of 10 days, or is frozen or canned. Live crabs may carry *Clostridium botulinum*, staphylococci, salmonellae, *Vibrio parahaemolyticus,* and shigellae. During cooking the vegetative cells of the pathogens are destroyed; however, spores of types B and E of *Clostridium botulinum* and less often Type F survive (Kautter et al., 1974), although their toxin may be destroyed. Live crabs can be purified by immersion in waters containing 5 ppm chlorine (Nickerson et al., 1939).

Modern crabmeat packing plants with generally good sanitation and personnel practices, clean equipment, adequate washing facilities, and a rapid flow of product can produce meat with fewer than 100,000 bacteria per gram, no *Escherichia coli,* and less than 30 cells of *Staphylococcus aureus* (Phillips and Peeler, 1972). It has been shown experimentally that treatment

Figure 18-3
Picking cooked crab flesh. (Courtesy of FDA Consumer, Rockville, MD.)

of the meat with heavy suspensions of *Lactobacillus bulgaricus* rapidly kills the contaminating low-temperature mesophiles. The lactobacilli themselves do not grow, but achieve inhibition by producing hydrogen peroxide (Gilliland and Speck, 1975).

Yeasts seldom exceed 1000 cells per gram of meat. Bacteria overgrow yeasts rapidly on untreated shellfish products and normally yeasts are of little significance. When crabmeat is irradiated at 0.2 and 0.4 megarad and stored at 0.5 to 5.6°C (33 to 42°F), the numbers of yeasts may exceed 100,000 per gram, but ultimately they are overgrown by bacteria (Eklund et al., 1966).

Yeasts have been associated with cooked meat of the King crab *(Paralithodes comischatica)* and Dungeness crab *(Cancer magister)*. Members of the genera *Rhodotorula* and *Trichosporon* occur most frequently, but *Cryptococcus, Torulopsis, Candida* and a yeastlike organism resembling *Aureobasidium pullulans* also have been recovered (Eklund and Groniger, 1965). On freshly processed Dungeness crabmeat yeast counts range from 200 to 1000 per gram, on King crabmeat from 30 to 500 per gram. Yeasts are generally regarded as saccharolytic rather than proteolytic. However, several isolates from crab meat display proteolytic activity. The organism most active in attacking casein and crab proteins is a *Trichosporon* sp. This yeast and two others—*T. pullulans* and *Cryptococcus scottii*—very actively produce lipase.

References

Allen, E. C., and M. Woodburn. 1972. J. Milk Food Technol. *35*, 540.

American Public Health Association. 1975. Standard Methods for Examination of Water and Wastewater, 12th ed. Washington, D.C.

Andrews, W. H., C. D. Diggs, M. W. Presnell, J. J. Miescier, C. R. Wilson, C. P. Goodwin, W. N. Adams, S. A. Furfari, and J. F. Musselman. 1975a. J. Milk Food Technol. *38*, 453.

Andrews, W. H., C. R. Wilson, A. Romero, and P. L. Poelma. 1975b. Appl. Microbiol. *29*, 328.

Avery, A. C. 1950. Fish Processing Handbook for the Philippines. Res. Rept. No. 26, USDI, Washington, D.C.

Beacham, L. M. 1946. J. Assoc. Offic. Agr. Chem. *29*, 89.

Bhadra, R. and Bose, A. N. 1975. J. Appl. Bacteriol. *38*, 107.

Birdseye, C. 1929. Ind. Eng. Chem. *21*, 854.

Bott, T. L., J. S. Deffner, E. M. Foster, and E. McCoy. 1964. In: Botulism: Proceedings of a Symposium, U.S. Public Health Service. Publ. 999-FDA. Cincinnati, Ohio.

Boyd, J. W., and H. L. A. Tarr. 1955. Food Technol. *9*, 411.

Boyd, J. W., H. M. Bluhm, C. R. Muirhead, and H. L. A. Tarr. 1956. Amer. J. Public Health *46*, 1531.

Bradley, H. C., and B. E. Bailey. 1940. Food Res. *5*, 487.

Brown, L. D., and C. R. Dorn. 1977. J. Food Protection. *40*, 712.

Campbell, L. O., and O. B. Williams. 1952. Food Technol. *6*, 125.

Castell, C. H. 1954. Fisheries Res. Bd. Can. Progr. Rep., Atlantic Coast Sta. *58*, 27.

Center for Disease Control. 1974. Foodborne and Waterborne Disease Outbreaks. Ann. Summary 1973.

Center for Disease Control. 1975. Foodborne and Waterborne Disease Outbreaks. Annual Summary. 1974.

Center for Disease Control. 1976. Foodborne and Waterborne Disease Outbreaks. Annual Summary. 1975.

Center for Disease Control. 1977. Foodborne and Waterborne Disease Outbreaks. Annual Summary. 1976.

Clem, J. O. 1969. F.D.A. Papers *3*(4), 8.

Cliver, D. O. 1971. Critical Revs. Environ. Control *1*, 551.

Cobb, B. F., and C. Vanderzant. 1971. J. Milk Food Technol. *34*, 533.

Cobb, B. F., and C. Vanderzant. 1975. J. Food Sci. *40*, 121.

Cobb, B. F., C. Vanderzant, C. A. Thompson, and C. S. Custer. 1973. J. Milk Food Technol. *36*, 463.

Cobb, B. F., C-P. S. Yeh, F. Christopher, and C. Vanderzant. 1977. J. Food Protection *40*, 256.

Colwell, R. R. 1973. J. Milk Food Technol. *36*, 202.

Communicable Disease Center. 1963. MMWR *12*, 329; 358. Atlanta, Ga.

Cooks, G. C., and W. S. Ritchie. 1938. Food Res. *3*, 569.

Crawford, L., and R. Finch. 1968. Food Technol. *22*, 765.

Damon, G. E. 1971. F.D.A. Papers *5*(4), 21.

DelValle, F. R., and J. L. Gonzales-Inigo. 1968. Food Technol. *22*, 1135.

DelValle, F. R., and J. T. R. Nickerson. 1968. Food Technol. *22*, 1036.

Denis, F. A. 1975. Canad. J. Microbiol. *21*, 1055.

Eklund, M. W., J. Spinelli, D. Miyuchi, and J. Dassow. 1966. J. Food Sci. *31*, 421.

Eklund, M. W., and H. Groninger. 1965. Appl. Microbiol. *13*, 985.

Elliott, R. P. 1952. Food Res. *17*, 225.

Evelyn, T. P. T., and L. A. McDermott. 1961. Canad. J. Microbiol. *7*, 375.

Farber, L. 1952. Food Technol. *6*, 319.

Federal Register. 1970. Part 128a, *35*, 17401.

Fieger, E. A. 1950. Food Technol. *4*, 409.

Fieger, E. A., and A. F. Novak. 1961. In: *Fish as Food,* G. Borgstrom, ed., Academic Press, New York.

Finn, D. B. 1941. Canad. Chem. *25*, 175.

Firman, M. C., A. Abbey, M. A. Darken, A. R. Kohler, and S. D. Upham. 1956. Food Technol. *10*, 381.

Foster, J. F., J. L. Fowler, and J. Dacey. 1977. J. Food Protection *40*, 300.

Fox, L. 1953. Mod. Sanit. *5*(6), 17.

Fugate, K. J., D. O. Cliver, and M. T. Hatch. 1975. J. Milk Food Technol. *38*, 100.

Fujino, T. 1953. Med. J. Osaka Univer. *4*, 299.

Gangarosa, E. G., A. L. Bisno, E. R. Eichner, M. D. Treger, M. Goldfield, W. E. De Witt, T. Fodor, S. M. Fish, W. J. Daugherty, J. B. Murphy, J. Feldman, and H. Vogel. 1968. Am. J. Public Health *58*, 114.

Gradner, E. A., and B. M. Watts. 1957. Food Technol. *11*, 6.

Geiger, E. 1944. Food Res. *9*, 293.

Gilliland, S. E., and M. L. Speck. 1975. J. Food Sci. *40*, 903.

Green, M. 1949. Food Res. *14*, 365, 372, 384, 395.

Gulasekharam, T., T. Velardapillal, and G. R. Niles. 1957. H. Hyg. (Camb.) *54*, 581.

Gunnison, J. B., J. R. Cummings, and K. F. Meyer. 1936. Proc. Soc. Exper. Biol. Med. *35*, 278.

Hanzawa, J., and S. Takeda. 1931. Arch. Mikrobiol. *2*, 1.

Harrison, F. C., and M. E. Kennedy. 1922. Trans. Royal Soc. Can. Ser. III. *16*, 101.

Hefferman, W. P., and V. J. Cabelli. 1967. Shellfish Sanitation Research Center, U.S. Public Health Service, Dept. Health, Education, and Welfare, Washington, D.C.

Highland, M. E. and O. B. Williams. 1944. Food Res. *9*, 34.

Horsley, R. W. 1973. J. Appl. Bacteriol. *36*, 377.

Hunter, A. C. 1920. U.S. Dept. Agr. Bull. 819.

Hunter, A. C. 1939. Food Res. *4*, 531.

Johnson, W. G,. Jr., A. C. Salinger, and W. C. King. 1973. Appl. Microbiol. *26*, 112.

Kautter, D. A., T. Lilly, Jr., A. J. Le Blanc, and R. K. Lynt. 1974. Appl. Microbiol. *28*, 772.

Kiser, J. S., and T. D. Beckwith. 1944. Food Res. *9*, 250.

Koburger, J. A., A. R. Norden, and G. M. Kempler. 1975. J. Milk Food Technol. *38*, 747.

Land, C. W., L. Farber, C. Beck, and F. Yerman. 1944. Ind. Eng. Chem. Anal. Ed. *16*, 490.

Lang, A. W. 1935. U. of Cal. Public Health Report *2*, 26.

Lantz, A. W. 1951. Fish. Res. Bd. Can. Progr. Rept. Pacific Coast Sta. *98*, 82.

Lapin, R. M., and J. A. Koburger. 1974. Appl. Microbiol. *27*, 666.

Lartique, D., A. F. Novak, and E. A. Fieger. 1960. Food Technol. *14*, 109.

Lee, C. F., and L. Pepper. 1956. Com. Fisheries Rev. *18*(7), 1.

Lee, J. S., and D. K. Pfeifer. 1973. J. Milk Food Technol. *36*, 143.

Lee, J. S., and D. K. Pfeifer. 1974. J. Milk Food Technol. *37*, 553.

Levin, R. E. 1971. J. Milk Food Technol. *34*, 277.

Liston, J., and J. Baross. 1973. J. Milk Food Technol. *36*, 113.

MacCallum, W. A. 1955. Food Technol. *9*, 251.

Magar, N. G., and F. Shikmahmud. 1956. J. Sci. Ind. Res. *15C*, 174.

Metcalf, T. G., and W. C. Stiles. 1968. Amer. J. Epidemiol. *88*, 379.

Miller, A., R. A. Scanlon, J. S. Lee, L. M. Libbey, and M. E. Morgan. 1973a. Appl. Microbiol. *25*, 257.

Miller, A., R. A. Scanlon, J. S. Lee, and L. M. Libbey. 1973b. Appl. Microbiol. *25*, 952; *26*, 18.

Miyake, M., and A. Tanaka. 1971. J. Food Sci. *36*, 674.

Morris, E. O. 1973. Ant. u. Leeuw. *39*, 331.

Mueller, G. 1965. Zentralbl. f. Bakteriol. u. Parasitenk. I. Abt. *197*, 295.

Nickerson, J. T. R., and G. A. Pollak. 1972. J. Milk Food Technol. *35*, 167.

Niscolo, W., and H. A. Frank. 1966. Food Technol. *20*, 964.

Olsen, J. C. 1969. F.D.A. Papers *3*(1), 3.

Orillo, C. A., and C. S. Pederson. 1968. Appl. Microbiol. *16*, 1669.

Perry, C. A. 1939. Food Res. *4*, 381.

Petrowa, E. K. 1933. Zentralbl. f. Bakteriol. u. Parasitenk, II Abt. *89*, 345.

Phaff, H. J., E. M. Mrak, and O. B. Williams. 1952. Mycologia *44*, 431.

Phillips, F. A., and J. T. Peeler. 1972. Appl. Microbiol. *24*, 958.

Public Health Service. 1965. P.H.S. Publ. No. 33, Washington, D.C.

Raj, H., and J. Liston. 1961. Appl. Microbiol. *9*, 171.

Ray, S. M. 1971. *Current Topics in Comparative Pathobiology* 1, T. C. Cheng, ed., Academic Press, New York.

Reay, G. A., and J. M. Shewan. 1949. *Advances in Food Research.* 2, E. M. Mrak and G. F. Stewart, eds. Academic Press, New York.

Reed, G. B., and C. M. Spence. 1929. Canad. Biol. Fish. N.S. *4*, 257.

Ross, S. S., and E. O. Morris. 1965. J. Appl. Bacteriol. *28*, 224.

Shewan, J. M. 1937. Rept. Food Invest. Bd. (Brit.) *84*.

Shewan, J. M. 1938. Rept. Food Invest. Bd. (Brit.) *113*.

Shewan, J. M. 1971. J. Appl. Bacteriol. *34*, 299.

Shewan, J. M., and G. Hobbs. 1967. In: *Progress in Industrial Microbiology,* D. J. D. Hockenbull, ed., Ibiffer Books Ltd., London.

Sidhu, G. C., W. A. Montgomery, and W. A. Brown. 1974. J. Food Technol. *9*, 371.

Simidu, W., and S. Hibiki. 1957. Bull. Japan Soc. Sci. Fisheries *23*, 255.

Snow, J. E., and P. J. Beard. 1939. Food Res. *4*, 563.

Spencer, R. 1959. J. Appl. Bacteriol. *22*, 73.

Spencer, R. 1961. J. Appl. Bacteriol. *24*, 4.

Spencer, R., and D. L. Georgala. 1957. Proc. 2nd Int'l. Symposium Found. Microbiol., HMSO, London.

Surkiewicz, B. F., J. B. Hyndman, and M. V. Yancey. 1967. Appl. Microbiol. *15*, 1.

Tanikawa, E., T. Motohiro, and M. Akida. 1967. Food Technol. *21*, 439.

Tarr, H. L. A. 1939. J. Soc. Chem. Ind. *58*, 253.

Tarr, H. L. A., and P. W. Ney. 1949. Progr. Rep. of the Pacific Coast Sta., Vancouver.

Thaysen, A. C. 1936. Ann. Appl. Biol. *23*, 99.

Thomson, A. B., A. K. Davis, J. C. Early, and J. R. Burt. 1974. J. Food Technol. *9*, 381.

Trust, T. J. 1972. J. Fish Res. Bd. Canada *28*, 1185.

Trust, T. J. 1975. J. Appl. Bacteriol. *38*, 225.

Van Veen, A. 1953. Adv. in Food Res. *4*, 209.

Vanderzant, C., E. Mroz, and R. Nickelson. 1970. J. Milk Food Technol. *33*, 346.

Vanderzant, C., and R. Nickelson. 1973. J. Milk Food Technol. *36*, 135.

Vanderzant, C., B. F. Cobb, C. A. Thompson, and J. C. Parker. 1973a. J. Milk Food Technol. *36*, 443.

Vanderzant, C., A. W. Matthys, and B. F. Cobb. 1973b. J. Milk Food Technol. *36*, 253.

Vanderzant, C., C. A. Thompson, Jr., and S. M. Ray. 1973c. J. Milk Food Technol. *36*, 447.

Varga, S. W., and G. W. Sanderson. 1968. Appl. Microbiol. *16*, 193.

Varga, S., R. E. Dobson, and R. Earle. 1977. J. Food Protection *40*, 763.

Virgilio, R., C. Gonzales, S. Mundoza, S. Avendano, and N. Munoa. 1970. J. Food Sci. *35*, 842.

Wilson, T. E., and C. S. McCleskey. 1951. Food Res. *16*, 313.

Wise, R. I., J. B. Winston, and G. Gulli, Jr. 1948. J. Amer. Public Health Assoc. *38*, 1109.

Zobell, C. E. 1961. *Proceedings Low Temperature Microbial Symposium*, Campbell Soup Co., Camden, N.J.

19

Poultry

The term poultry includes domesticated birds that supply meat or eggs—mainly chickens, turkeys, ducks, and geese but also semitame fowl such as pigeons, guinea fowl, peafowl, quail, partridge, and pheasants. Some of these birds were domesticated in antiquity. The domestic chicken, for example, is descended from the wild jungle fowl of Southern Asia and Indonesia.

In many countries much poultry is marketed live, but in most industrially developed countries birds are dressed in a commercial processing plant before being marketed. In the early twentieth century the birds were dressed merely by killing, draining the blood, plucking (New York dressed), and cooling the carcasses. In the United States all poultry is now marketed fully dressed. Although commercial processing eliminates much of the drudgery of preparing birds for the table, it has also introduced a number of microbiological considerations that did not demand attention when birds were held live until shortly before being cooked and eaten.

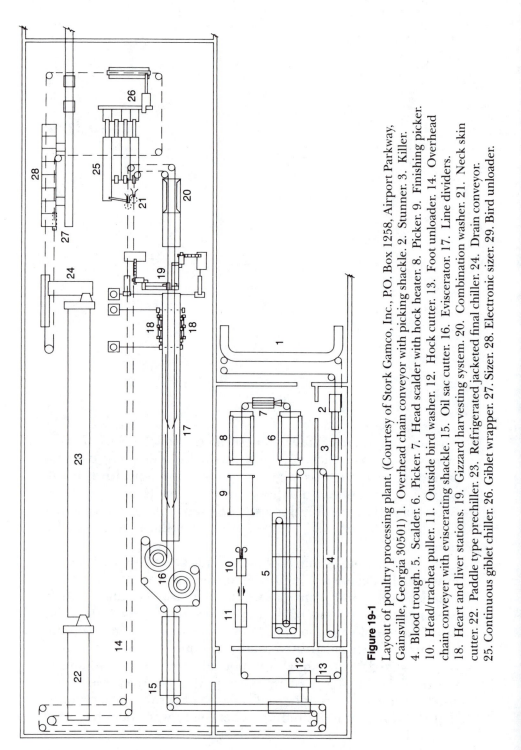

Figure 19-1

Layout of poultry processing plant. (Courtesy of Stork Gamco, Inc., P.O. Box 1258, Airport Parkway, Gainsville, Georgia 30501) 1. Overhead chain conveyor with picking shackle. 2. Stunner. 3. Killer. 4. Blood trough. 5. Scalder. 6. Picker. 7. Head scalder with hock heater. 8. Picker. 9. Finishing picker. 10. Head/trachea puller. 11. Outside bird washer. 12. Hock cutter. 13. Foot unloader. 14. Overhead chain conveyer with eviscerating shackle. 15. Oil sac cutter. 16. Eviscerator. 17. Line dividers. 18. Heart and liver stations. 19. Gizzard harvesting system. 20. Combination washer. 21. Neck skin cutter. 22. Paddle type prechiller. 23. Refrigerated jacketed final chiller. 24. Drain conveyor. 25. Continuous giblet chiller. 26. Giblet wrapper. 27. Sizer. 28. Electronic sizer. 29. Bird unloader.

Box 19-1

In large poultry processing plants 10,000 to 20,000 chickens may be dressed in a single hour (see Figure 19-1). Caged live birds are delivered to the processing plant. Each bird is electrically stunned before its neck is slit for bleeding. This facilitates rapid, humane, and efficient slaughter. The shackled bird is passed through a vat of hot scald water to help loosen the feathers. Water temperature varies from 50 to 60°C for a 30 second exposure, depending upon the age and type of poultry. For broilers, a temperature-time relationship of 50–52°C for 30 seconds is typical. The scalded birds pass through a series of pickers that remove the feathers from the body, tail, wings, and neck. Remaining pin feathers may be removed by singeing. Then the feet and head are removed and the birds are eviscerated. The body cavity is flushed and the lungs are sucked free by a vacuum tube. The edible viscera— heart, liver, and gizzard—are separated from the entrails and, after cleaning, are packaged with the neck and inserted into the body cavity of the dressed bird. By any of various chilling methods the temperature of the meat is lowered to near freezing. The poultry is next packaged and stored at low temperature. In recent years a considerable proportion of carcasses have been dismembered to provide selected parts or segments or further processed products such as poultry rolls, roasts, steaks, or fillets.

The per capita consumption of poultry meat in the United States, 55 pounds per year, exceeds that of any other country. In the past 20 years the amounts of chicken and turkey consumed have doubled and are still increasing. The breeding of a strain of fast-growing broiler chickens has been the major reason. When good management practices are followed, only 7 to 8 pounds of feed are required to produce birds weighing 3¾ to 4 lbs within 7½ weeks. Because of this achievement, in the 35 years from 1935–1970 the number of broilers has increased 70 fold (Figure 19-2). With similar economies in raising turkey poults, their numbers have increased almost sixfold in the same period.

Microbial Loads

Enormous and varied microbial populations are associated with feathers, skin, and feet, with the gut, and with excretions (Ayres et al., 1950; Drewniak et al., 1954). Table 19-1 lists the results of several studies of the generic distribution of microbial contaminants recovered from poultry. Spoilage of processed poultry

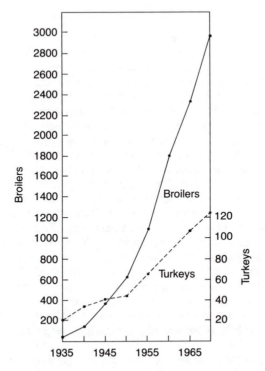

Figure 19-2
Changes in production of broilers and
turkeys over period 1935–1970 (millions of
birds).

is usually the result of bacterial rather than yeast and mold
activity. Yeasts are only a small percentage of the total population occurring on eviscerated poultry. But if conditions eliminate or retard bacteria, yeasts may be an important part of the
spoilage pattern. Yeasts are appreciably more numerous on
carcasses treated with antibiotics of the tetracycline group
(Ayres, 1966).

Microorganisms that eventually cause spoilage of poultry are
either present at the time of slaughter or are introduced by
workers and their cutting tools, or by water and air during
dressing, evisceration, cutting, and cooling. The bacteriological
contamination on the live bird has marked influence on the
number of organisms that can be recovered from the eviscer-

Table 19-1
Genera of bacteria isolated from dressed poultry.

	Gunderson et al., 1947	Ayres et al., 1950	Salzer et al., 1967		Gunderson et al., 1947	Ayres et al., 1950	Salzer et al., 1967
Pseudomonas	+	+	+	Gaffkya	+	+	+
Alcaligenes	+	+	+	Sarcina	+	+	+
Achromobacter	+	+	+	Neisseria	+		
Flavobacterium	+	+		Brevibacterium			+
Escherichia	+	+	+	Streptococcus		+	+
Aerobacter	+	+	+	Lactobacillus			+
Paracolobactrum	+	+		Corynebacterium	+	+	+
Proteus	+	+	+	Microbacterium	+	+	
Salmonella	+	+	+	Arthrobacter			+
Haemophilus	+			Bacillus	+	+	+
Micrococcus	+	+	+	Actinomyces	+		
Staphylococcus	+	+	+	Streptomyces	+	+	

ated product. The number of organisms on the skin surfaces tends to increase during processing; successive steps helps disseminate microorganisms, as samplings at various processing stages indicate. An initial population as low as, say, 1500 organisms/cm² becomes significantly higher after the bird is rough picked—partly through deposit of organisms from material that washes from feet, feathers, feces, and skin while the carcasses are traversing the scald tank, and partly through displacement and redistribution of microorganisms from the feather or pin-hair follicles during picking and pinning. After evisceration and washing, the loads stabilize and, with good sanitation control, are maintained at <50,000/cm². Ordinarily, loads of microorganisms are smaller in the visceral cavity than on the skin. Samplings of the cavities commonly reveal loads from 1400 to 12,000 organisms per cm². Much smaller microbial populations persist in scald tank waters than in chill tank waters. Since the temperature range of scald tank water differs with the type of fowl being processed—i.e., young vs. old birds, chickens vs. turkeys or ducks—the scald temperature is an important consideration. Other factors that must be taken into account are freshness of water, rate of water circulation, and amount of water displaced by the bird. There is considerable difference in numbers of organisms among chill tank waters.

As might be expected, the environment in which birds are processed affects their bacterial loads. For example, in a series of tests, surface contaminants averaged about 4000 organisms per cm² in one plant and 350,000 per cm² in another plant. The birds in the latter plant were received during rainy or snowy weather and were processed before they could become dry. Numbers of enterococci and coliform organisms decreased after immersion in the scald tank and after mechanical picking (Wilkerson et al., 1961).

Much bacterial contamination of the skin surface occurs during manual transfer from the picker to the evisceration line, but very little occurs through evisceration (Drewniak et al., 1954; Keel and Parmelee, 1968). Tumble washing of parts also is effective in reducing the numbers of these organisms (Salzer et al., 1964). Chilling by agitating in ice and water reduced total counts on the skins of fryers significantly more than did air chilling or agitating in chilled water alone (Casale et al., 1965). Although rapid chilling extends product shelf life (Dawson et al., 1963), microbial loads in standard chill tanks increase during the 6-hour chilling of carcasses; however, placing birds in a continuous counterflow-tumble chiller significantly reduces microbial loads (Kotula et al., 1962).

Analysis for Microbial Content

Several methods have been advocated for estimating numbers of organisms on the skin and lean surfaces of poultry. Among them are pressing metal or plastic dishes filled with solidified agar (spot or Rodac plates) against the skin, swabbing known surface areas (Figure 19-3), using cut tissue sections sliced to known depths, smearing material from the underwing area on glass slides (Ziegler and Stadelman, 1955), and shake-rinsing birds of uniform size and weight in containers with equivalent amounts of water. Proponents of each procedure claim it is more expedient or reliable than the others. Procedures using spot or Rodac plates and glass slide smears are quick and con-

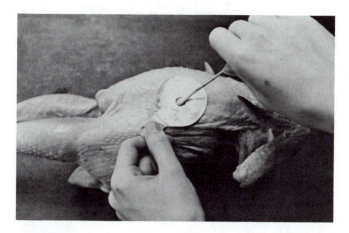

Figure 19-3
A sterile metal template with a cutout of known diameter is
held against the skin surface while a moistened cotton swab
is rolled firmly over the exposed area.

venient for routine purposes, but results of studies indicate that
they are also of limited value, because on some products it is
difficult to find any organisms and on others there are too
many to count. The rinse technique (Mallmann et al., 1958;
Surkiewicz et al., 1969; ICMSF, 1974; Cox and Blankenship,
1975) is unsuited to field use because it requires too many bags
or containers, too much media, and too many of other supplies
and equipment. Results of rinse, cut-section, and swab sampling (Figure 19-4), although not identical, are comparable, and
moist swabs are the easiest, quickest, and most economical to
use. Fromm (1959) advocated using alginate swabs. Higher
numbers of organisms may be recovered from thighs than from
breasts or drumsticks of chicken fryers (Kotula, 1966). The
neck skin, and to a lesser extent, the back and near vent sites
were more heavily contaminated than the leg or underwing
(Patterson, 1972).

For the resazurin reduction test (a procedure commonly used
for determining the quality of milk, as described in Chapter
16), swab samples from poultry may be placed in peptone water
and, after adding resazurin, trypticase soy broth, and skim
milk, incubated at 15 or 30°C until a fluorescent pink color

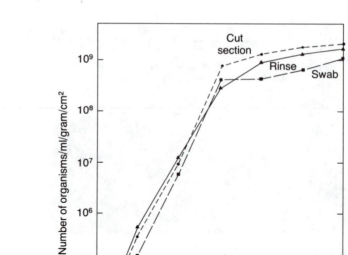

Figure 19-4
Comparison of growth curves obtained by swab, rinse
and cut section methods of sampling. (Ayres, 1959.)

develops. Reduction times often vary, reflecting different
counts, when different areas on the same carcass are sampled
(Table 19-2). Yet reduction times seldom differ by more than 60
minutes, and the differences influence final assessment of poultry condition very little. Good correlation can be obtained between reduction time and numbers of organisms present on
chicken. The following schedule depicts the time needed for
birds of fresh, good, fair, and poor quality to reduce resazurin
to its primary end point, fluorescent pink.

Condition of birds	Reduction time, hours
Fresh	> 8
Good	> 5 but < 8
Fair	> 3.5 but < 5
Poor	< 3.5

Table 19-2
Variation in counts and resazurin reduction time as related to location of sample.

Bird sampled	Sampling site	No. of organisms/cm² ($\times 10^3$)			Resazurin reduction time at 30°C (minutes)
		15°C	30°C	37°C	
1	Leg	17	13	2	480
2	Leg	147	80	87	465
3	Leg	475	170	110	180
1	Thigh	118	85	33	375
2	Thigh	84	42	41	450
3	Thigh	5700	190	1700	180
1	Wing	85	77	4.5	330
2	Wing	120	96	33	450
1	Breast	180	130	100	360
3	Back	1850	194	840	150

Source: Ayres et al., 1956.

Bacterial plate counts and resazurin reduction times also correlate well when these tests are used in poultry sanitation programs. It has been suggested that sanitary practices in poultry
processing plants be periodically checked by the resazurin test,
to eliminate improper practices (Williams et al., 1962).

Off-Odor and Slime

Although a great variety of organisms can be found on live or
freshly slaughtered birds, a much smaller variety is responsible
for the spoilage of refrigerated poultry meat. Immediately
after killing and processing, 75 to 80% of the colonies recovered from chicken parts consist of chromogenic bacteria,
molds, yeasts, and sporeforming microorganisms (Figure 19-5).
The proportion of chromogens and miscellaneous organisms
decreases during storage, and within a few days after processing a rather uniform psychrophilic flora predominates, causing off-odor and slime. Both indications of deterioration are
closely associated with the growth and coalescence of colonies
of several species of *Pseudomonas*. These organisms reproduce
rapidly and cause a characteristic sweetly rancid "dirty dishrag"

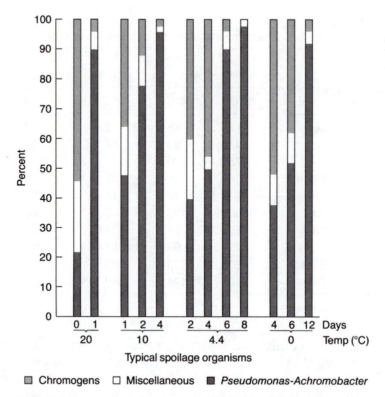

Figure 19-5
Effect of storage at 0, 4.4, 10, and 20°C on organisms associated
with dressed poultry. (Ayres et al., 1950. Reprinted from Food
Technology, Vol. 4, p. 199, 1950. Copyright © by Institute of Food
Technologists.)

odor when dressed chickens are stored long enough at refrigeration temperatures and high humidity. Accompanying the
off-odor, minute, translucent, moist colonies appear in large
numbers on the cut surface and skin. At first the colonies superficially resemble droplets of moisture, but later they enlarge
and become white or creamy in color, finally coalescing into a
more or less uniform, sticky or slimy layer. In this final stage,
poultry meat has a pungent ammoniacal odor in addition to
other characteristic aromas. At the time birds develop off-odor
and slime, *Pseudomonas* and *Achromobacter* spp. account for 90%
or more of the total population (Ayres, 1966). More than 10

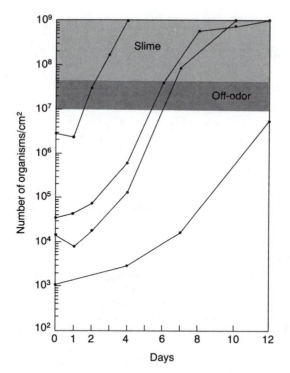

Figure 19-6
Effect of initial microbial load on how long it
takes off-odor and slime to develop at 4°C.
(Ayres, 1959.)

million organisms per cm² (Figure 19-6) are recovered at the
time off-odor is developing; 50 to 60 million cells or more per
cm² are recovered when slime is visible (Ayres, 1959 and 1966).
These levels are comparable to those reported for red meats
(see Chapter 17; Schmid, 1931; Haines and Smith, 1933;
Empey and Vickery, 1933; Ayres, 1959).

Storage

The initial load of microorganisms profoundly affects storage
life. When stored at 4°C, pieces of chicken having high, moderate, and low counts differ widely in bacterial quality after one,
two, or four days (Figure 19-6). The storage temperature of

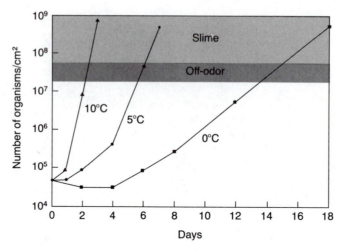

Figure 19-7
Growth curves showing relationships among bacterial
counts, off-odor, and slime formation at 0, 5, and 10°C.
(From Ayres et al., 1950. Reprinted from Food Technology,
Vol. 4, p. 199, 1950. Copyright © by Institute of Food
Technologists.)

poultry greatly influences microbial development and shelf life.
Birds stored at 0°C have a storage life of 14 to 16 days; at 5°C
birds spoil in 6 to 7 days; and at 10°C off-odor and slime may be
observed in 3 days (Figure 19-7) (Ayres et al., 1950).

Contamination of poultry by iron or other minerals in processing water may be partly responsible for development of
spoilage organisms during storage. Solutions containing iron
cause a marked increase in numbers of bacteria (Garibaldi and
Bayne, 1960). Fluorescent pigment does not begin to be excreted until the organisms have used all of the available iron
(Kraft and Ayres, 1961, 1964). Growth of fluorescing pseudomonads on chicken is stimulated by dipping poultry in solutions containing iron. Chickens treated with iron fluoresce two
to three days earlier than do control samples (Figure 19-8).

The shelf life of broilers is directly proportional to the time
carcasses are held in slush ice (Fromm, 1959). Chickens held in
ice for 48 hours before being cut into parts have a longer shelf
life than those cut immediately after cooling (Stadelman et al.,
1957).

Figure 19-8
Fluorescence of chicken wings under ultraviolet light.
Left, control (0 ppm iron); middle, 10 ppm iron;
right, 20 ppm iron. All samples stored five days at
5°C. (Kraft and Ayres, 1964.)

Further Processing

Poultry processed in any manner other than as an eviscerated
carcass is considered by the industry as "further processed."
Such further processing may consist merely of separating the
carcass into parts such as drumsticks, thighs, wings, back,
breast, and neck, or may involve filleting, boning, grinding,
freezing, or converting into products such as rolls, logs, roasts,
steaks, or patties.

Bacterial contamination increases when poultry carcasses are
segmented, filleted, boned-out, or comminuted. A study of
chicken further processed (cut into components and packaged)
in six commercial poultry processing plants and in five retail
stores indicated that counts increased about sixfold in the processing plants and approximately eightfold in the retail stores
(May, 1962). The primary causes were initial microbial loads,
and contact with contaminated work surfaces during separation
of parts and during handling. Meat blocks upon which the
poultry is cut are considered the greatest source of contamination.

Foodborne Poultry-Transmitted Disease

Poultry account for about 10% of reported foodborne disease outbreaks in which the vehicle was identified. Usually the birds were mishandled or stored at improper temperature. In almost 80% of the episodes the poultry was mishandled by food service establishments; most of the remainder occurred in homes; and about 2% were traced to food processors.

Among outbreaks in which an etiological agent was identified, *Salmonella* spp. were responsible for 44%, *Clostridium perfringens* for 26%, and *Staphylococcus aureus* for 26% (Horwitz and Gangarosa, 1976). Various workers reported ranges of contamination by salmonellae of from 2.7 to 34.8%: Galton et al. (1955), 2.7%; Thatcher and Loit (1961), 7.5%; Ladiges and Foster (1974), 8.3%; Wilson et al. (1962), 15%; Sadler et al. (1960), 26%; Woodburn (1964), 27%; Duitschaever (1977), 34.8%.

Salmonellae first contaminate poultry products on the farm, chiefly through feed, feed ingredients, bird droppings, and trough water. When poultry is transported to the processing plant, salmonellae are transferred from droppings in delivery truck cages or from contaminated catching, handling, and feeding equipment. During processing significant percentages of equipment used for picking (76%), eviscerating (29%), and chilling and packaging (16%) are contaminated (Table 19-3). Steam scalding reduces contamination by salmonellae more effectively than conventional hot water scalding. More than twice as many positive samples were recovered from water-scalded birds than from steam-scalded birds (Patrick et al., 1973). Often the same serotype species present on a farm is isolated on turkey carcasses and on processing equipment after the turkeys are dressed (Table 19-4). Although salmonellae can be isolated from two-thirds of the birds after removal of feathers, natural die-off (McDade and Hall, 1964), flushing, and dilution reduce numbers after eviscerating and chilling (10%), but contamination rises again after overnight chilling and grading (17%) (Bryan et al., 1968).

Rapid immunofluorescence (FA) tests detect salmonellae in food products (Georgala and Boothroyd, 1964; Haglund et al., 1964), and are a convenient way of monitoring sanitation in

Table 19-3
Isolations of salmonellae from equipment, listed in order of process flow.

Item	Number of samples	Number positive	Percent positive
Picking room			
Picker 1	14	12	85.7
Picker 2	14	11	78.6
Picker 3	14	7	50.0
Spiral picker	14	12	85.7
Chute	13	9	69.2
Transfer table	13	11	64.6
Totals	82	62	75.6
Eviscerating room			
Evisceration gutter	14	8	57.1
Evisceration knives	12	2	16.7
Head remover	13	4	30.8
Metal shield, cavity washer	7	4	57.1
Spray washer, after evisceration	5	2	40.0
Trussing table	16	10	62.5
Spin chiller 1	14	4	28.6
Slide, chiller 1 to chiller 2	13	4	30.8
Spin chiller 2	14	1	7.1
Spin chiller 3	14	1	7.1
Chill chute and grading table	14	4	28.6
Chill tanks	10	0	0.0
Other equipment	6	0	0.0
Totals	152	44	28.9
Packaging room			
Packaging table	9	3	33.3
Scales	8	0	0.0
Bagger	8	1	12.5
Totals	25	4	16.0
Totals	259	110	42.5

Source: Bryan et al., 1968.

food processing plants (Table 19-5). Nonspecific staining caused by cross-reactions can be minimized by the use of a series of absorbed sera against commonly interfering microorganisms.

Bryan (1971), Gibbs (1971), and Lillard (1971) reported that poultry products often contained *Clostridium perfringens,* which Roberts (1972) also found on 63% of frozen chicken carcasses containing giblets. Several food poisoning outbreaks caused by *C. perfringens* have been cited (Duncan, 1970). Some heat-

Table 19-4

Salmonella serotypes isolated from turkey carcasses and from equipment during successive stages of processing 15 consecutive flocks.

Day	I			II	III		
Flock	A	B	C	D	E	F	(EF)
Fecal droppings, farm	—	e	g	g	i,j	b	
Water troughs, farm	—	—	g	g	i	—	
Fecal droppings, truck	—	—	—	g	—	—	
Picker 1	a	—	a	a,g			c,i
Picker 2	—	e	a,e				b,c
Picker 3	—	—	e	e			g
Spiral picker	b,c	a,c	b,d	d,f			a,b,d
Chute	—	a,e	b,d	g			c
Table (picking)	—	b,e	f	g			b
Carcasses after picking			b,e,f	g	a,b,c,h	b,g	
Carcasses after washing					—	—	
Gutter	—	—	f				b
Knives	—	—	f				—
Head remover			f	b,g			—
Trussing table	b,d	e	f	g			e,i
Spin chill 1	—	—	b	g			—
Slide	a	—	f	f			—
Spin chill 2	—	—	—	—			—
Spin chill 3	—	—	—	—			—
Chute and grade table	—	—	f	g			e
Carcasses before icing	—	—	f	g	a	—	
Chill tank†		—		—	—	—	
Carcasses before packaging†	c	e	f	g	i	f,i	
Grade and packaging table†		—		—			i
Scales†				—		—	
Baggers†				—			—

Source: Bryan et al., 1968.

* Key: a, *S. infantis*; b, *S. anatum*; c, *S. chester*; d, *S. bredeney*; e, *S. typhimurium*; f, *S. cerro*; g, *S. san diego*; h, *S. derby*; i, *S. newington*; j, *S. senftenberg*; k, *S. halmstad*; l, *S. muenchen*; m, *S. stanley*; n, *S. saint paul*; o, *S. blockley*; p, *S. schwarzengrund*; q, *S. montevideo*; —, negative for salmonellae; all blank spaces indicate that no samples were taken.
** After cleanup.
† Isolations made on the following day.
‡ Lapse of one week.

IV	‡	V	VI				VII				VIII	IX	X
**		G	H	I	(HI)	J	K	L	(KL)	M	N	**	O
		k	—	b			b	—			—	—	b
		k	—	—			—	—			—	—	b
		b	l,n	—		—	—	—			—	—	b
g		m			i	m			m,p	b,p	p	—	—
—		m			—	m			b,g	b	p	—	—
—		m			—	—			—				b
b		—			l	—			g	g	a	—	b
—		—			—	—			g	c	q	—	b
—		—			l	b			g	c	c	—	b
		l,m	n,o	g		d,g	d	g		c	a		b
		—	g,l	—		—	e	g		—			b
—		l,m			b	b			—	g	c	—	—
		l											
—		—	n		—	c			—			—	—
—		—	n		—	b			—	g		—	—
—		—			b	—			—			—	—
—		—			—	—			—	g		—	—
—		—			—	—			—	g		—	—
—		—			—	—			—	p		—	—
—		—			—	—			—			—	—
—		—			—	—	—	—				—	—
—					—	—	—	—				—	—
—		—	—	—		—	—				—	—	
		—			b	—				—	—	—	
		—			—	—				—	—	—	
		—			b	—				—	—	—	

Table 19-5

Detection of salmonellae on turkey meat and processing equipment.

Type of sample	Number of samplings	Number positive: FA	Galton (1964)
	Plant A		
Turkey			
whole	70	8	8
cut meat	11	0	0
visceral cavity	1	1	1
roll	9	3	3
roast	31	8	8
Tables	45	11	11
Pans	31	8	8
Cutting boards	12	5	5
Conveyors	23	9	9
Scales	6	2	2
Tanks	9	0	0
Knives, sharpeners, saw	26	6	6
Pincers, tie forms, head remover	9	2	2
Sink, drain, floor, gutter	5	1	1
	Plant B		
Turkey			
whole	23	1	3
leg, breast, wing	25	4	4
roast	11	2	2
roll	2	0	0
Tables	7	1	2
Pans	10	2	2
Cutting board	3	0	1
Conveyor	13	2	4
Scale	7	1	1
Knive, saw	4	0	0
Sharpener	1	1	1
Trough	2	0	0

Source: Haglund, 1965.

resistant spores of the species survive cooking, and at appropriate temperatures grow in cooked chicken (Mead, 1969). Cooking in water at 82°C for 50 minutes did not reduce their numbers, and cooking for 45 minutes at 93°C did not completely destroy them (Craven et al., 1975).

Kraft (1971) indicated that equipment and processing operations were important factors in spreading salmonellae and staphylococci—the latter were often spread by workers' hands. Salmonellae, *Clostridium perfringens,* and staphylococci have been associated with food poisoning that followed the eating of barbequed chicken. Members of these genera grow readily at about 40°C and within eight hours increase approximately 100,000 fold (Pivnick et al., 1968). As more further processed turkey products are made, more information regarding their microbiological quality is needed. From 85 fresh skin and meat samplings, 85% yielded coliforms, 62% yielded *C. perfringens,* 54% *Staphylococcus aureus,* and 15% salmonellae. After freezing for a month at –28.9°C, recovery was 67%, 56%, 53%, and 9%, respectively (Hagberg et al., 1973).

Samplings of fresh ground turkey meat from retail markets in the San Francisco Bay Area provided a mean standard plate count of 8.4×10^7 per gram and a mean MPN for *Escherichia coli* of 19/g. Fecal streptococci were isolated from 95%, *Staphylococcus aureus* from 80%, *C. perfringens* from 42%, and *Salmonella* spp. from 28% of samples, respectively (Guthertz et al., 1976). The mean microbial contamination of preseasoned comminuted turkey meat was 2.2×10^8; the mean coliform plate count was 2.0×10^5; and the *E. coli* count 8.7 (Guthertz et al., 1977).

Regulations of the U.S. Department of Agriculture (1968) require that poultry rolls (other than those cured and smoked) be cooked to an internal temperature of 71.1°C, which is considered sufficient to destroy all pathogenic organisms. No salmonellae survived on Western-type turkey rolls cooked in a waterbath at 74°C for 5.5 hours or at 85°C for 4.5 hours (Bryan et al., 1968). Kinner et al. (1968) had thought that only temperatures as high as 82°C would destroy coliforms and enterococci in juice-spice mixtures used in Eastern-type rolls, but later work from their laboratories (Mercuri et al., 1970) did not demonstrate viable salmonellae in such rolls.

Extending Storage Life

Freezing

Most of the poultry that is not marketed fresh is commercially frozen. In the United States more turkeys are dressed and frozen than are merchandized in any other manner. Freezing is usually done by air blast or by brine immersion followed by air blast. Freezing turkeys by air blast destroys 96 to 99% of the total surface flora (Kraft et al., 1963).

When the birds are immersed in brine before being air blasted, numbers of staphylococci on inoculated turkeys are reduced by about 96%. However, the die-off of microorganisms on the surface of uninoculated birds is less dramatic (84%). Brine immersion and air blast procedures destroy nearly 97% of enterococci, 98% of the total aerobes, and almost 99% of coliforms (Table 19-6). Either freezing method destroys nearly all (99.9%) of the fluorescent bacterial population. Numbers of enterococci and coliform organisms on refrigerated dressed chickens tend to remain constant or decline slowly during 15 to 20 days of storage. Staphylococci and enterococci are more re-

Table 19-6
Effect of freezing methods on numbers of bacteria on turkeys.

	Average number bacteria per cm^2					
	Brine immersion + air blast			Air blast		
	Before freezing	After freezing	% reduction	Before freezing	After freezing	% reduction
Uninoculated						
Total aerobes	1220	195	84.0	3740	160	95.7
Inoculated						
Total aerobes	1,840,000	36,000	98.0	2,883,000	34,500	98.8
Fluorescing						
bacteria	152,700	490	99.9	148,870	475	99.9
Coliforms	43,700	520	98.8	64,500	350	99.9
Enterococci	74,470	2460	96.7	103,700	350	99.9
Staphylococci	16,670	685	95.9	47,525	620	98.7

Source: Kraft et al., 1963.

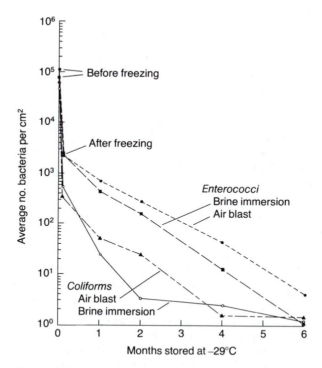

Figure 19-9
Effect of method of freezing and frozen storage on
survival of enterococci and coliforms on turkeys
inoculated with mixed flora. (Kraft et al., 1963.)

sistant than coliforms to destruction by freezing. Freezing inoculated turkey at –30°C and storing at –2 to 10°C drastically
reduces the numbers of coliforms. At this temperature few if
any coliforms can be detected after three months but 2 to 3% of
the enterococci persist for three to four months (Figure 19-9).
Numbers continue to decrease gradually during prolonged
frozen storage (i.e., –29°C); however, a residual viable population of 5 to 10 organisms (total aerobes) can be recovered from
uninoculated birds even after 6 months of frozen storage.

Freezing does not completely free poultry of salmonellae;
low numbers (1/cm²) can be isolated after one month of frozen
storage.

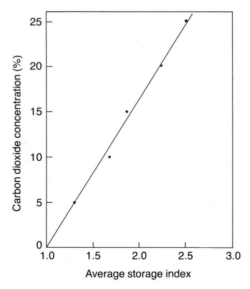

Figure 19-10
Relation of storage days to carbon dioxide
concentration for fresh commercially
dressed chicken (5°C). (Ogilvy and Ayres,
1951. Reprinted from Food Technology,
Vol. 5, p. 97, 1951. Copyright © by
Institute of Food Technologists.)

Carbon Dioxide

Atmospheres containing carbon dioxide improve the storage
life of chicken. The **storage index** of chicken parts—that is, the
ratio of keeping time in the presence of carbon dioxide to keeping time in air—is a linear function of carbon dioxide concentration (Figure 19-10). Chicken parts held in an atmosphere
containing 15% carbon dioxide keep twice as long as pieces
held in air, and parts stored in the presence of 25% carbon
dioxide keep 2.5 times as long. With proper use of CO_2 concentration and holding temperature, poultry meats can be held
for long periods of time. However, a level of 25% carbon
dioxide is considered maximal for general purposes; higher
levels cause discolorations. As storage temperature increases,
the effect of carbon dioxide decreases (refer back to Figure
6-1), and at room temperature it only slightly extends storage
life.

Antibiotics

Fresh poultry is the only food product to which the Food and Drug Administration permitted antibiotics to be added. Poultry flesh was immersed in solutions containing 30 μg/ml of chlortetracycline or oxytetracycline to absorb a concentration of 7 μg/g of flesh at the surface. During the first three days of refrigerated storage, the quantity of antibiotics adsorbed on skin and lean surfaces decreased rapidly; that absorbed by the deep tissues disappeared less rapidly. However, residues on surfaces and in deep tissue remained detectable through 13 days (Figure 19-11). Antibiotics are no longer used for poultry, having been rendered obsolete by improved cleansing and sanitizing practices.

The tetracycline antibiotics were intended to delay microbial spoilage of poultry meats during storage (Ayres, 1966). As many as 7 to 10 days may elapse between slaughter and sale. Figure 19-12 compares flora of control birds and tetracycline-treated birds after various storage times at 4.5°C. The tetra-

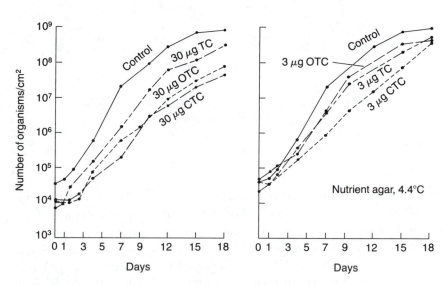

Figure 19-11
Comparison of levels of chlortetracycline (CTC) and oxytetracycline (OTC) found in different tissues (μg/g for deep tissue). (Walker and Ayres, 1958. Reprinted from Food Research, Vol. 23, p. 525, 1958.)

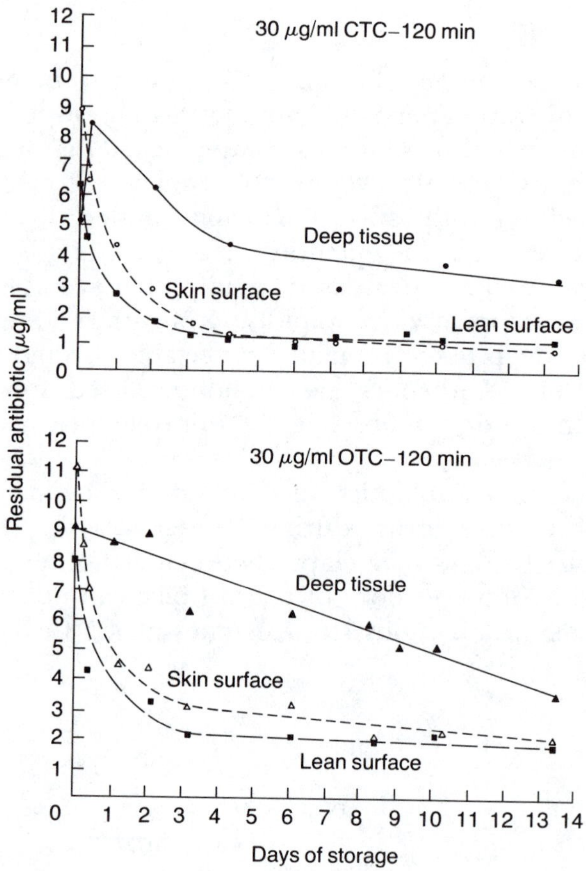

Figure 19-12
Effect of tetracycline (TC), oxytetracycline (OTC), and
chlortetracycline (CTC) on numbers of bacteria. (Ayres
et al., 1966.)

cyclines have little detrimental effect on the growth of
pseudomonads and yeasts. Numbers of yeasts increase on
treated poultry primarily because bacteria that might suppress
them decrease in numbers. Members of genera *Rhodotorula,
Torulopsis, Geotrichum, Candida, Trichosporon,* and *Cryptococcus*
may be present on treated or untreated birds (Njoku-Obi et al.,
1957).

Health officials view such use of antibiotics with suspicion, foreseeing the following problems:

1. Only a portion of the population will be destroyed.

2. Resistant strains will persist.

3. Some of the destroyed natural flora will be replaced by yeasts or other organisms—including potential pathogens—that formerly did not compete favorably.

4. Residual antibiotic may sensitize some consumers.

5. Antibiotic treatment may be substituted for good sanitation practices.

6. The cost of poultry processing will increase.

Packaging

Packaging influences both the bacterial population and the shelf life of the product. Eviscerated chicken packaged in polyethylene and stored at 4°C usually develops undesirable odor and slime within a week. Fryers held in loosely packaged polyethylene bags have the same shelf life as unpackaged birds and a shorter shelf life than those stored in tight-fitting, evacuated packages (Carlin et al., 1957; McVicker et al., 1958).

Examination of poultry under ultraviolet light was considered a convenient method for evaluating shelf life and for demonstrating growth of fluorescing organisms (Cotterill, 1956). Carcasses packaged in polyethylene, pliofilm, and cellophane gave a positive reaction; those wrapped in Cryovac did not. Cotterill reported that chicken packaged by the Cryovac process and held at refrigeration temperatures might be expected to show no fluorescence for as long as 14 days (Barnes and Shrimpton, 1958). However, because off-odors and fluorescence on chicken correlate poorly, the reliability of this technique has been questioned. Nonpigmented pseudomonads were found to grow luxuriantly, although there was no fluorescence.

Since oxygen is required for fluorescence, mere lack of fluorescence indicates not that spoilage is prevented but only that the Cryovac process eliminates most of the free air. Fluorescence is not evident on chicken examined under ultraviolet light until fluorescing bacteria reach the magnitude of 100,000 to 1 million per cm². Usually, fluorescence becomes evident after three to five days of storage, beginning at the edges of the wings, or where packaging materials are not touching the meat, but later becoming widespread over the surface of the skin.

References

Ayres, J. C. 1959. Iowa State J. Sci. *34*, 27.

Ayres, J. C. 1966. Proc. 2nd Intl. Cong. Food Sci. and Technol., Warsaw, Poland.

Ayres, J. C., W. S. Ogilvy, and G. F. Stewart. 1950. Food Technol. *4*, 199.

Barnes, E. M. 1960. Roy. Soc. Health J. *80*, 145.

Barnes, E. M., and D. H. Shrimpton. 1958. J. Appl. Bacteriol. *21*, 313.

Bryan, F. L. 1971. J. Milk Food Technol. *34*, 23.

Bryan, F. L., J. C. Ayres, and A. A. Kraft. 1968. Amer. J. Epidemiol. *87*, 578.

Carlin, A. F., B. E. Holl, and H. W. Walker. 1957. Food Technol. *11*, 573.

Casale, J. O., K. N. May, and J. J. Powers. 1965. Food Technol. *19*, 859.

Cotterill, O. J. 1956. Poultry Sci. *35*, 1138.

Cox, N. A., and L. C. Blankenship. 1975. J. Food Sci. *40*, 1333.

Craven, S. E., H. S. Lillard, and A. J. Mercuri. 1975. J. Milk Food Technol. *38*, 505.

Dawson, L. E., W. L. Mallmann, D. G. Bigbee, R. Walker, and M. E. Zabik. 1963. Food Technol. *17*, 218.

Drewniak, E. E., M. A. Howe, H. E. Goresline, and E. R. Baush. 1954. U.S. Dept. Agric. Circ. 930.

Duitschaever, C. L. 1977. J. Food Protection *40*, 191.

Duncan, C. L. 1970. J. Milk Food Technol. *33*, 35.

Empey, W. A., and J. R. Vickery. 1933. Austr. J. Coun. Sci. Ind. Res. *6*, 233.

Fromm, D. 1959. Poultry Sci. *38*, 887.

Galton, M. M., A. Mackel, A. L. Lewis, W. C. Haire, and A. V. Hardy. 1955. Amer. J. Vet. Res. *16*, 132.

Garibaldi, J. A., and H. G. Bayne. 1960. Poultry Sci. *39*, 1517.

Gibbs, P. A. 1971. Brit Poultry Sci. *12*, 101.

Georgala, D. L., and M. Boothroyd. 1964. J. Hyg. *62*, 319.

Gunderson, M. F., K. D. Rose, and M. G. Henn. 1947. Food Ind. *19*, 1516.

Guthertz, L. S., J. T. Fruin, D. Spicer, and J. L. Fowler. 1976. J. Milk Food Technol. *39*, 823.

Guthertz, L. S., J. T. Fruin, and J. L. Fowler. 1977. J. Food Protection *40*, 322.

Hagberg, M. M., F. F. Busta, E. A. Zottola, and E. A. Arnold. 1973. J. Milk Food Technol. *36*, 625.

Haglund, J. R. 1965. Doctoral dissertation, Iowa State Univ. Library, Ames, Iowa.

Haglund, J. R., J. C. Ayres, A. M. Paton, A. A. Kraft, and L. Y. Quinn. 1964. Appl. Microbiol. *12*, 447.

Haines, R. B., and E. C. Smith. 1933. Dept. Sci. Ind. Res. Food Invest. Bd. (Brit.) Special Rept. No. 43.

Horwitz, M. A., and E. J. Gangarosa. 1976. J. Milk Food Technol. *39,* 859.

ICMSF. 1974. International Commission on Microbiological Specifications for Foods of the International Association of Microbiological Societies. Microorganisms in foods: A series for microbiological analyses: Principles and specific applications. University of Toronto Press, Toronto, Canada, p. 141.

Keel, J. E., and C. E. Parmalee. 1968. J. Milk Food Technol. *31,* 377.

Kinner, J. A., A. W. Kotula, and A. J. Mercuri. 1968. Poultry Sci. *47,* 1442.

Kotula, A. W. 1966. Poultry Sci. *45,* 233.

Kraft, A. A. 1971. J. Milk Food Technol. *34,* 23.

Kraft, A. A., and J. C. Ayres. 1961. Appl. Microbiol. *9,* 549.

Kraft, A. A., and J. C. Ayres. 1964. J. Food Sci. *29,* 218.

Kraft, A. A., J. C. Ayres, K. F. Weiss, W. W. Marion, S. L. Balloun, and R. H. Forsythe. 1963. Poultry Sci. *42,* 128.

Ladiges, W. C., and J. F. Foster. 1974. J. Milk Food Technol. *37,* 213.

Lillard, H. S. 1971. J. Food Sci. *36,* 1008.

McDade, J. J., and L. B. Hall. 1964. Amer. J. Hyg. *80,* 192.

McVicker, R. J., L. E. Dawson, W. L. Mallman, S. Walters, and E. Jones. 1958. Food Technol. *12,* 147.

Mallmann, W. L., L. E. Dawson, B. M. Saltzer, and H. S. Wright. 1958. Food Technol. *16,* 66.

May, K. N. 1962. Food Technol. *16,* 89.

Mead, G. C. 1969. J. Appl. Bacteriol. *32,* 86.

Mercuri, A. J., G. J. Banwart, J. A. Kinner, and A. R. Sessoms. 1970. Appl. Microbiol. *19,* 768.

Njoku-Obi, A. N., J. V. Spencer, E. A. Sauter, and M. W. Eklund. 1957. Appl. Microbiol. *5,* 319.

Ogilvy, W. S., and J. C. Ayres. 1951. Food Technol. *5,* 97.

Patterson, J. T. 1972. J. Appl. Bacteriol. *35,* 569.

Patrick, T. E., J. A. Collins, and T. L. Goodwin. 1973. J. Milk Food Technol. *36,* 34.

Pivnick, H., I. E. Erdman, S. Manzatiuk, and E. Pommier. 1968. J. Milk Food Technol. *31,* 198.

Roberts, D. 1972. J. Hyg. (Camb.) *70,* 565.

Sadler, W. W., R. Yamamoto, H. E. Adler, and G. F. Stewart. 1960. Appl. Microbiol. *9,* 72.

Salzer, R. H., A. A. Kraft, and J. C. Ayres. 1964. Poultry Sci. *43,* 934.

Salzer, R. H., A. A. Kraft, and J. C. Ayres. 1967. Poultry Sci. *46,* 611.

Schmid, W. 1931. Fleisch. z. ges. Kalte Ind. *38,* 1.

Stadelman, W. G., W. W. Marion, and M. L. Eller. 1957. Antibiotics Annual 1956–7.

Surkiewicz, B. F., R. W. Johnston, A. B. Moran, and G. W. Krum. 1969. Food Technol. *23,* 1066.

Thatcher, F. S., and A. Loit. 1961. Appl. Microbiol. *9,* 39.

U.S. Dept. of Agriculture. 1968. Consumer and Marketing Service. Regulations governing the inspection of poultry and poultry products. 7CFR, Part 81, Washington, D.C.

Walker, H. S., and J. C. Ayres. 1958. Food Res. *23,* 525.

Wilkerson, W. B., J. C. Ayres, and A. A. Kraft. 1961. Food Technol. *15,* 286.

Williams, C. V., H. K. Bengsch, and G. Wirksten. 1962. J. Milk Food Technol. *25,* 315.

Wilson, E., R. S. Paffernbarger, M. J. Foter, and K. H. Lewis. 1962. J. Infect. Dis. *109,* 166.

Woodburn, M. 1964. Appl. Microbiol. *12,* 492.

Ziegler, F., and W. J. Stadelman. 1955. Food Technol. *9,* 107.

20

Eggs and Egg Products

Eggs

Domestic fowl the world over produce a total of almost 400 billion eggs annually—70 billion in the United States alone. Most are sold as shell eggs for table use. About 10% of the eggs are broken out for use as frozen or dried egg or egg products.

According to Brooks and Taylor (1955), "roughly 90% of newly laid eggs are free from microorganisms and the true value may be even higher." Nevertheless, poor handling and holding conditions permit microorganisms to penetrate the shell and multiply.

Natural Microbial Inhibitors

The egg possesses both mechanical and chemical defenses against rampant microbial invasion:

Shell. The shell is a calcareous covering composed primarily of calcium carbonate together with 1% magnesium carbonate and 1% calcium phosphate, and held together by about 4% organic

Box 20-1

Eggs are used in many foods for their nutrition, flavor, color, and other functional properties mentioned below. Eggs are excellent sources of essential fats, minerals, and vitamins (A, B, and D); egg protein contains all the essential amino acids in readily digestible form. The bland flavor of eggs allows large-scale use alone or in combination with other ingredients in bakery and confectionery products and in meat and vegetable dishes. The yellow of the yolk adds a much desired characteristic to noodles, salads, and many baked products. In addition, eggs enhance the structure, consistency, and texture of many foods by their thickening, binding, emulsifying and leavening action.

Eggs are graded by size and quality. As to size, current receipt eggs (Figure 20-1) are segregated into one of six weight classes: jumbo (30 oz. per dozen), extra-large (27 oz.), large (24 oz.), medium (21 oz.), small (18 oz.), and peewee (15 oz.). The jumbo and peewee sizes are generally considered specialty items and are not ordinarily sold retail.

Quality of shell eggs depends primarily on the size of the air cell and the shell's cleanliness, texture, and strength. Other factors considered are the firmness, mobility, and clarity of the albumen, definition of the yolk, and freedom from defects as determined by candling. Except for eggs that are dirty or cracked, grade designations used are AA (or fresh fancy), A, B, and C. Eggs with stuck yolks, factory rejects, or eggs containing embryos (as determined by the blood ring condition) are not permitted for sale as food. Similarly, eggs are considered inedible if black, white, green, pink, fungal, or mixed rots are present. Moldy, musty, sour, bloody, or adulterated eggs are likewise not permitted in food channels.

Figure 20-1
Preparation of current receipt eggs for grading and candling.

matter, chiefly protein. The shell serves as a mechanical barrier to prevent microorganisms on dust and on larger soil particles from entering the interior.

Pore Plugs. The calcareous coating is not complete; all over the shell there are some 7000 to 17,000 microscopic openings, or **pore canals,** of various sizes (Tyler, 1953). They are larger and more numerous at the blunt or larger end. A thin transparent coating—or "bloom," or cuticle—of mucin seals the shell and prevents microbial penetration. If the shell becomes wet the mucin dissolves and the pore openings are exposed.

Membranes. There are two **shell membranes:** a thick outer (50–75μm), and a thin inner (17–17μm) barrier. Both are composed of a dense network of interlacing fibers of keratin that act as a filter to prevent microbial penetration; some microorganisms make poor use of egg protein, but spoilage bacteria are supplied sufficient nutrients to grow extensively (Stokes and Osborne, 1956). If the shell of a freshly laid egg is penetrated, 20 days may elapse before microscopic changes are observable or organisms occur in the albumen. Yet the role of the membranes in restraining microbial invasion of the **albumen** may principally reflect the antimicrobial nature of the albumen. A more likely determinant of the lag time is the movement of the yolk towards the shell membrane. If the yolk touches the inner shell membrane, gross infection results (Brooks, 1960; Board and Ayres, 1965).

The contents of the newly laid egg entirely fill the cavity, but as the egg cools from the hen's body temperature (40.5°C) to room temperature, the liquid portion contracts. Also, as the egg ages, the gases trapped in the liquid escape through the pores and the membranes separate to form an "air cell" at the large, or blunt, end. The size of the **air cell** is an index of the egg's age.

Carbon Dioxide. About 10% of the volume of the liquid in freshly laid eggs is dissolved carbon dioxide and water vapor that escape, as the egg ages, through the pore canals and the shell proper. CO_2 has been shown to exert a slight bacteriostatic effect.

Native Proteins. Raw albumen is not readily used by micro-organisms. Some simple nutrients must be present to support growth—even of proteolytic bacteria—until proteinases can be produced. In addition, the albumen within the egg is so arranged that a firm, gel-like layer of thick albumen separates two layers of thinner albumen, thus retarding rapid migration of organisms from the shell and membranes to the yolk. The milky white cords (**chalazae**) that support and protect the yolk also are composed of viscous ropelike albumen that stretches from the inner membrane to yolk.

pH. As the CO_2 and other gases escape, the pH of the albumen increases from 7.6, stabilizing after three days at about 9.4. At pH 9.4 egg albumen is the most alkaline of human food, and few organisms can thrive.

Lysozyme. Egg albumen contains the enzyme lysozyme, which decomposes the cell wall of certain gram-positive cocci and bacilli. The action is especially bacteriolytic for *Micrococcus lysodeikticus*. The enzyme hydrolyzes the mucopeptide constituent of the bacterial cell wall—cleaving the chemical bonds between the N-acetyl muramic acid and N-acetyl glucosamine subunits and releasing the protoplasts. Although the liberated protoplasts retain many vital functions, including continued respiration, increase in size, and even division, prolonged growth is impossible.

Avidin. Raw egg white contains the polypeptide avidin, which forms a stable complex with the vitamin biotin—rendering the latter unavailable to *Saccharomyces cerevisiae* (Eakin et al., 1941). In addition to causing the vitamin to be biologically inactive, avidin inactivates enzymes that contain biotin, such as the enzymes pyruvate carboxylase (Cazzulo and Stoppani, 1967) and carbamylphosphate synthetase (Wellner et al., 1968).

Trypsin Inhibition. Chicken ovomucoid interferes with tryptic digestion; it inhibits bovine, porcine, and ovine trypsins but not human trypsins (Feeney et al., 1969). Cells using this form of enzymatic activity cannot digest egg protein. The role of ovomucoid as a trypsin inhibitor has, however, been questioned (Board, 1966).

Ovoinhibitor. Ovomucoid present in the albumen of the golden pheasant's egg has been shown to inhibit chymotrypsin (Rhodes et al., 1960). Avian ovoinhibitor has also been shown to inhibit subtilisin, a proteolytic enzyme of *Bacillus subtilis* (Matsushima, 1958) greatly resembling bovine α-chymotrypsin (Liu et al., 1971).

Papain Inhibitor. Avian albumen papain inhibitor also inhibits ficin. The molecular weight of this protein is 12,000, and it differs from ovomucoid and ovoinhibitor (Fossum and Whitaker, 1968).

Conalbumin. Raw egg white contains the albuminous protein conalbumin, which inhibits the growth of *Shigella dysenteriae* and certain other bacteria by chelating the iron in a form nutritionally unavailable to them (Alderton et al., 1946). A remarkably stable complex is formed between two iron atoms and one of conalbumin. Conalbumin also binds copper in a similar manner (Fraenkel-Conrat and Feeney, 1950). The presence of conalbumin also enhances pigment production of pseudomonads (Feeney and Nagy, 1952) and reduces glucose utilization and catalase production by *Staphylococcus aureus* (Schade, 1958).

Vitelline Membrane. The vitelline membrane is a thin membrane surrounding the yolk that reduces its penetrability to water and H_2S. The contents of newly laid eggs from physiologically sound birds are essentially free of microbial contamination (Haines, 1939). Although there is disagreement in the literature about how organisms enter the eggs, less than 1% of naturally clean eggs rot during prolonged storage (Brooks and Taylor, 1955). The bird has a common cloaca permitting excretion from the reproductive and digestive tracts. While eggs are being laid the vagina everts through the cloacal opening so that the shell rarely touches the cloaca, the vent lips, or the immediately adjacent feathers.

Most spoilage organisms enter the egg after it has been laid. The shell becomes contaminated as soon as it touches cages, nesting material, the bird's feet or feathers, droppings, or debris. The microflora dominant on the shell are gram-positive (see Table 20-1; Board, 1966), but those causing rots are usually

Table 20-1
Types of microorganisms recovered from the shell of the hen's egg.

	Incidence (%) of organisms on eggs of different conditions from:						
Type of organism	Shops and farms [a]	Egg-breaking plants [b]			Packing stations [c]		
		Clean	Lightly soiled	Heavily soiled	Clean	Lightly soiled	Cracked, etc.
Streptococci	—	8	5	—	—	—	—
Staphylococci	5	30	—	—	9	5	11
Micrococci	18	23	20	—	37	52	42
Sarcinae	2	20	—	—	—	—	—
Arthrobacters	—	—	—	—	5	13	10
Bacilli	30	—	13	5	—	2.5	—
Pseudomonads	6	—	—	—	—	—	—
pigmented	—	—	—	—	1.5	2.5	1
Nonpigmented	—	—	—	—	21	10	23
Achromobacters	19	—	—	—	1.5	2	1
Alcaligenes	—	—	—	—	—	2	—
Flavobacteria	3	—	—	—	—	—	—
Cytophagas	—	—	—	—	—	1	—
Escherichias	4	12	7	2	4.5	7	2
Aerobacters	1	7	—	3	6	0.5	2
Aeromonads	—	—	—	—	1	—	2
Proteus	1	—	20	20	—	—	—
Serratiae	—	—	20	50	—	—	—
Molds	7	—	10	20	—	—	—
Unclassified	—	—	—	—	12 [d]	5 [d]	6 [d]
No. of organisms studied	100	[e]	[e]	[e]	130	164	126

Source: Board, 1966.
[a] Haines, 1938.
[b] Zagaevsky and Lutikova, 1944.
[c] Board et al., 1964.
[d] Aerobic gram-negative bacteria.
[e] Details not given.

gram-negative. As mentioned earlier, the shell is not impervious to microbial penetration, and pseudomonads and other gram-negative bacteria can readily penetrate the shell of the freshly laid egg (Haines and Moran, 1940) and overcome its internal defense mechanisms.

Rots

Green rot, probably the most common microbiological spoilage, is observed by candling with black (ultraviolet) light. Growth of any of several organisms of the genus *Pseudomonas* produces a bright greenish fluorescence in white-shelled eggs candled by UV. But UV candling has only limited use because it cannot detect a variety of other types of spoilage, and because brown-shelled eggs are relatively opaque to black light, green rot of such eggs is not detectable.

During early stages of its growth, *Pseudomonas fluorescens,* which grows readily at refrigeration temperatures (i.e., 1 to 10°C), produces a bright green color in albumen. This defect is difficult to observe with an ordinary incandescent candling lamp. However, when the egg is broken, pseudomonad spoilage generally shows up clearly and such contamination is evidenced by a disagreeable fruity but acrid odor known as a "fluorescent sour." A related species, *P. ovalis,* also produces this type of spoilage (Garibaldi, 1960).

More rarely, species of *Pseudomonas* produce a pink discoloration of the albumen. Some pseudomonads grow in eggs without producing any fluorescence or change in color—producing colorless or white rot. Upon prolonged growth, the albumen becomes thin and watery and the yolk becomes encrusted or disintegrates; colorless rot is then detected by abnormal mobility in the candled egg. Souring of washed cold storage eggs is frequently caused by species of *Pseudomonas* (Lowry and Starr, 1949) or of *Achromobacter* or by coliforms. On rare occasions, contamination by *Serratia* occurs and produces a reddish rot. Although the color is quite distinct and diagnostic, such eggs usually do not have a disagreeable odor. Spoilage of shell eggs has been produced by inoculating with species of *Alcaligenes, Escherichia, Proteus,* and of other genera (Winter, 1942; Funk, 1943).

At least two kinds of black rot are observed in spoiled eggs. One kind, which results from the growth of *Proteus melanovogenes (Aeromonas hydrophila)*, causes the egg contents to appear almost opaque when candled. In the broken-out egg the yolk is hard, black, and solid, and the albumen is entirely liquified, turbid, granular, and murky brown or greenish-brown. Gas pressure may develop, and the egg gives off a fecal or putrid odor (Miles and Halnan, 1937). The second kind of black rot is caused by certain species of *Pseudomonas* (Haines, 1939). When eggs with this type of rot are candled, the yolk appears to be an opaque liquid mass floating freely in the albumen. When the egg is opened, the albumen is fluorescent green or greenish-brown and sometimes viscous and granular, and the yolk is a soft greenish-black mass.

Fungal spoilage often occurs in eggs stored in atmospheres having high relative humidity, i.e., > 80%. *Penicillium, Cladosporium,* and *Sporotrichum* spp. generally produce mold colonies on the outside of the shell and pinspots on its inner surface. In storage atmospheres where the moisture is still more excessive, species of *Mucor, Thamnidium, Alternaria,* and other genera produce a loose network of aerial hyphae, or "whiskers," that eventually may cover the entire surface of the egg. However, according to Vadehra and Baker (1973), sweating has little effect on the spoilage problem.

Mixed rots occur when two or more of these spoilage organisms occur in the same egg. Such eggs are described as "addled."

It has been known for many years that the ovaries can become infected by *Salmonella pullorum*, the only bacterial pathogen to enter the ovum or yolk at the time of ovulation. The bacterium causes the disease bacillary white diarrhea in chicks hatched from infected eggs. In surviving chicks the bacterium resides in the bone marrow until puberty, when it is carried by the blood to the ovaries. In experimentally inoculated birds it is transported by the blood from the alimentary tract to the ovaries (Rettger, 1913; Forsythe et al., 1967). Serospecies of salmonellae other than *S. pullorum* rarely are transmitted to the contents of sound eggs, and then only if these serotypes have produced a gastroenteritis sufficiently severe to give rise to a bacteremia. Gastroenteritis in hens is usually mild and brief, but birds recovering from it may be carriers, and the salmonellae can be recovered from the shells and the fecal material for several weeks after infection (Mundt and Tugwell, 1958). The shell becomes contaminated in the cloaca or by contact with fecal material containing the salmonellae.

Salmonellae on the shell's surface enter the egg meats through cracks in the shell, drop into the malange during the cracking process, and, if the egg is not properly refrigerated, grow and colonize on the shell membrane, which they then digest. They finally pass into the albumen. Exceptions to this route are *Salmonella pullorum* in chickens and *S. anatum* in anserine poultry. The practice of washing eggs in water containing a detergent at 49°C has reduced the probability of eggborne salmonellosis.

Intrinsic *Salmonella pullorum* has a low virulence for the human, although fatality rate among chicks may vary from 5 to 90%. Instances of *S. pullorum* gastroenteritis have been associated with faulty handling of eggs, including those that are thawing. *S. anatum* is somewhat more virulent for humans. At one time it was the common cause of gastroenteritis in parts of Europe where it is customary to eat duck eggs boiled—hence the practice of boiling for nine minutes to ensure safety.

Shortly after the outbreak of World War II large quantities of foods were shipped overseas, by methods that convenience and economy dictated; salmonellae in eggs became an acute public health problem. The number of known serospecies of *Salmonella* in England was said to have more than doubled during the war, after dried and frozen eggs began to be imported from the United States and Canada. At the end of the war dried egg products were incorporated into cake mixes and other convenience foods. Although the problem was speedily recognized, there was no ready, workable solution, such as the pasteurization applied to milk—the eggs became unstable under heat, and material congealed on the sidewalls of heat exchangers, impeding heat transfer. Broken-out eggs are now pasteurized according to the schedule approved by the U.S. Department of Agriculture (1969). The following pasteurization times are suggested for eggs and various egg products:

Whole eggs: heat to not less than 60°C for 3.5 minutes.

Egg whites:
a. neutralize pH, add aluminum to stabilize conalbumin, heat as above; or
b. heat to 52°C for 1 minute, add 0.75% hydrogen peroxide, hold for 2.5 minutes; or
c. heat to 52°C, add 0.1% hydrogen peroxide, hold for 3.5 minutes; or
d. adjust pH to 9, heat to 57°C, and hold for 3.5 minutes.

Dried egg whites with moisture content greater than 6%: incubate at 52 to 54.5°C and hold for 7 to 10 days.

Yolks: heat to 62°C for 3.5 minutes or to 60°C for 7 minutes.

Sugared or salt yolks: heat to 64.5°C for 3.5 minutes or to 62°C for 7 minutes.

Egg nog: There is an admitted risk in the use of egg nog, a nutritious, delicious, and enjoyable food prepared with raw eggs; but the danger appears to be minimal. Egg nog has not been reported as a vehicle of salmonellae in foodborne outbreaks.

Off-Flavors

The musty, fishy, hay, and cabbage-water flavors that sometimes develop in eggs usually are of microbial origin. Mustiness may be caused by *Achromobacter perolens, Pseudomonas graveolens,* or *P. mucidolens.* In fact, the species name *perolens* indicates that the organism is capable not only of penetrating the shell but also of producing opalescence and mustiness within the egg (Turner, 1927). The two pseudomonads also produce strong odors when inoculated into fresh eggs (Levine and Anderson, 1932). Fishy flavor in shell eggs has been attributed to certain strains of *Escherichia coli* (Haines, 1938), while *Aerobacter cloacae* causes the odor of hay in eggs. According to Haines (1939), "cabbage water" flavor precedes pseudomonad green, pink, or black rot.

The bacterial count on the shell surface ranges from 10^2 to 10^8 per shell (see Figure 20-2; Board et al., 1964; Torrey, 1965). Stained and soiled eggs (grade C) have counts of 10^3 to 10^6 organisms/shell, badly soiled eggs 10^4 to 10^7 organisms/shell. Although present on all grades of egg (Table 20-2), coliforms

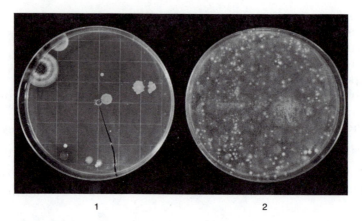

1   2

Figure 20-2
Effect of uncrating on contamination of room air in egg
processing plant. Dish number 1, before uncrating eggs; dish
number 2, after four cases were uncrated. (Board et al., 1964.)

Table 20-2
Recovery of coliforms and enterococci from the shells of eggs.

| | Number of samples | | Percent contaminated with: | | | |
| | | | Coliforms | | Enterococci | |
Grade of egg	Initially	After 1 week at 10°C.	Initially	After 1 week at 10°C.	Initially	After 1 week at 10°C.
A	73	50	10	6	60	a
B	77	50	22	16	69	52
C	74	50	27	26	55	40

Source: Board et al., 1964.

[a] Not analyzed.

and enterococci show higher incidence on eggs of grades B and
C, indicating that the dirt present on their shells is more likely
to include fecal material. Bacterial counts on the shell surface
can be greatly reduced by washing dirty eggs (Forsythe et al.,
1953) (see Figure 20-3). Washing with anionic or cationic
detergent-sanitizing agents has been advocated to minimize mi-

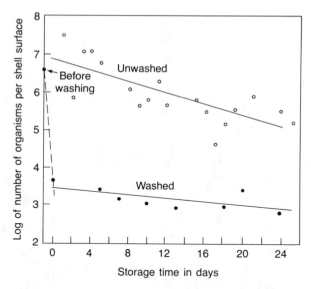

Figure 20-3
Effect of washing on the numbers of microorganisms
during storage at 25°C and 50% relative humidity.
(Forsythe et al., 1953. Reprinted from Food
Technology, Vol. 7, p. 47, 1953. Copyright © by
Institute of Food Technologists.)

crobial loads. While the most effective detergent-sanitizers
combine a quaternary ammonium compound and an alkaline
detergent (Botwright, 1953), these agents do not completely
protect eggs from spoilage. Table 20-3 shows what happens
during storage of eggs cleaned by different methods.

Bacteriological problems with cleaned eggs result primarily
from failure to follow proper practices. Obviously, detergent-
sanitizers cannot protect an egg if microorganisms have already
penetrated the shell. Also, although the shell and membranes
are natural structural defenses, bacteria readily penetrate them
if the pores are exposed while the egg cools. Penetration of
the shell is made easier by washing in cold water or by sanding
or buffing (abrading the egg) while it is cooling. After
pseudomonads and salmonellae have penetrated to the surface
of the shell membrane, some time elapses (three to four weeks
or more) before the microorganisms eventually digest the shell

Table 20-3

Microbiological population changes[a] on the shells of clean, "egg contents" dirty, and light dirty eggs when washed with detergents Vel and Roccal and held in cold storage.[b]

Storage period	Condition and washing treatment								
	Clean			Egg contents dirty			Light dirty		
	Control	Roccal	Vel	Control	Roccal	Vel	Control	Roccal	Vel
	Count per shell surface (× 1000)								
Initial	77	71	140	94,000	93,000	93,000	4500	5900	5900
After treatment	—	0.02	81	—	200	59	—	0.54	3.1
1st month	940	4.0	8.4	4700	49	8.4	9900	3.3	10
2nd–3rd month	250	52	7.2	11000	73	5.4	3400	110	7
4th month	370	3.5	4.7	3000	69	—	6200	13	7
5th–6th month	3000	4.4	26	12,000	1100	4.1	1800	5.4	10
Final[c]	1500	4.6	130	44,000	68	110	800	1100	72

Source: Forsythe et al., 1953.

[a] Standard plate counts—nutrient agar, incubated at 25°C for 72 hours.
[b] 0.5°C, 77% relative humidity, constant air circulation.
[c] Final count made eight days after 5th–6th month; eggs held at 13°C, 77% relative humidity during this time.

membrane and spill over into the albumen (Stokes and Osborne, 1956). If soiled eggs are to be washed to free them of fecal matter and other debris, the water should be warmer than the eggs so that organisms are not forced into and through the shells as the contents contract.

Oiling

Oil processing protects egg quality by reducing evaporation from the eggs during storage and retarding the escape of CO_2. Eggs to be oiled should be dipped in a light paraffin oil (i.e., viscosity of 55 to 60), preferably at 20°C immediately after they have been properly washed and dried. All excess oil should be allowed to drain from the eggs. It is important to keep the oil clean to prevent contamination by fungi, which frequently cause heavy losses of storage eggs. Since these organisms may be transferred to the egg from the oil, oil to be reused should be filtered through a cotton cloth or cheesecloth and sterilized at 82.2°C for 15 minutes. Growth of molds can also be prevented by dipping the eggs in water containing 4% sodium propionate before oil processing.

Thermostabilization

Heat, if properly applied to eggs (i.e., 1.5 minutes at 65.6°C, 3 minutes at 62.7°C, 5 minutes at 60°C, or 15 minutes at 54.4°C), stabilizes the thick albumen (so that the egg maintains a fresh appearance), halts embryonic development of fertile eggs, and destroys many of the spoilage bacteria on the shell by pasteurization (Funk, 1950). Such treatment, called thermostabilization, has proved an effective way of preventing storage losses.

Egg Products

Although most freshly laid eggs are free of microorganisms, nearly all commercial liquid, frozen, and dried egg products contain thousands of bacteria per gram (Redfield, 1920; U.S.

Table 20-4

Changes produced by different genera of bacteria isolated from liquid egg, when inoculated into sterile egg.

Genus	Number of different strains inoculated		Changes produced
Achromobacter	1		No change in 72 hrs
Aerobacter	4		Slight acid odor after 72 hrs
Alcaligenes	17	12	Very sour odor in 60 hrs
		1	Musty odor at 48 hrs
		4	No detectable change
Bacillus	8	6	Coagulation within 18 hrs
		1	Very sour odor after 24 hrs
		1	No change
Chromobacter	2		No change in 72 hrs
Escherichia	15	4	Slight acid odor in 60 hrs
		8	Sour odor after 60 hrs
		3	No detectable change
Flavobacterium	11	2	Coagulation in 120 hrs
		1	Fecal odor at 72 hrs
		4	Slight odor at 120 hrs
		4	No detectable change at 72 hrs
Gram + cocci	3	2	Produced much gas in 18 hrs
		1	No change
Proteus	10	3	Very flat sour odor in 60 hrs
		2	Coagulation in 18 hrs
		4	Very sour odor in 60 hrs
		1	No change
Pseudomonas	7	3	Very sour odor in 60 hrs
		2	Gas within 60 hrs
		1	Sour odor in 60 hrs
		1	No change

Source: Wrinkle et al., 1950.

Dept. of Agriculture, 1969). The principal genera occurring in liquid and frozen egg products are *Alcaligenes, Flavobacterium,* and *Proteus.* Any of these organisms inoculated into sterile liquid whole egg produce spoilage (see Table 20-4; Wrinkle et al., 1950).

Figure 20-4
Commercial production of liquid yolk and albumen from shell eggs.

Pasteurization

Although 87% of all eggs produced in the United States are marketed as shell eggs, some 600 million pounds of egg products are merchandized annually as frozen (350–400 million pounds), dried (165–225 million pounds) or fresh liquid products (40–50 million pounds) (Figure 20-4). In July 1971 pasteurization of egg products in conformity with the terms of the Egg Products Inspection Act became mandatory, and in July 1972 all provisions of the Act relating to eggs, egg products, sanitary practices, and labeling went into effect.

Egg products were first commercially pasteurized by Henningsen Brothers in 1938 (see Figure 20-5) (Goresline et al., 1951). However, pasteurization at temperatures as high as 62.7°C is insufficient to kill all salmonellae. Moreover, although heat readily destroys non-spore-forming pathogens, it also adversely affects many of the functional properties of eggs, such as their ability to foam, thicken, bind, and emulsify. There is at best only a small safety margin between the thermal process

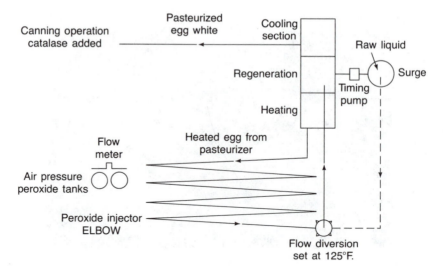

Figure 20-5
Pasteurizing diagram. (Rogers et al., 1966.)

that provides an adequate kill for salmonellae and a heat dosage that could seriously damage the egg product's functional properties.

Liquid Whole Egg and Yolk Products

Regulations governing grading and inspection of egg products state that "strained and filtered liquid whole egg shall be flash heated to not less than 60°C (140°F) and held at this temperature for no less than 3.5 minutes." However, since there is much laminar flow in pasteurizing equipment, this regulation has been criticized as imprecise, for the fastest-moving particles control the effectiveness of the treatment. The *Egg Pasteurization Manual* (U.S. Dept. of Agriculture, 1969) stipulates that *"every particle of strained and filtered liquid whole egg shall be rapidly heated to not less than 60°C (140°F) and held at that temperature for not less than 1.75 minutes."* Except for *Salmonella senftenberg* 775W—which has not been isolated from egg products since 1946—this time-temperature combination appears to be adequate to destroy from 10^5 to 10^7 salmonellae per ml (Winter et al., 1946; Sugihara et al., 1966). However, much more stringent specifications for pasteurizing liquid whole eggs are required in England: 64.4°C for at least 2.5 minutes. The wide safety margin of such treatment appears to be predicated on a negative finding of active enzymatic activity of alpha amylase rather than the need for such a high temperature (Forsythe and Ayres, 1962).

Salmonellae are from 150 to 200% as resistant in liquid yolk as they are in whole egg (Osborne et al., 1954; Garibaldi et al., 1969). Therefore, more severe pasteurizing requirements are essential. This treatment also is suggested for whole eggs with yolks to 33% solids, blends with 27 to 35% solids, and sugared whole eggs. Blends with less than 27% solids and less than 1% sugar and salt should receive the same pasteurization treatment as whole egg. The *Egg Pasteurization Manual* suggests a holding time of 3.5 minutes—fastest particle (fp) 1.75 minutes—at 61.1°C or 6.2 minutes (fp 3.1 minutes) at 60°C. Also, since sugar and salt increase the heat resistance of salmonellae and retard the rate of flow, holding times for these products are 3.5 minutes (fp 1.75 minutes) at 63.3°C or 6.2 minutes (fp 3.1 minutes) at 62.2°C.

Ionizing radiations at a dosage of 0.2 to 0.5 Mrad effectively destroy salmonellae in egg products without the container being opened or the product thawed (Ingram et al., 1961). At these dosages, yolk products show objectionable flavor changes in scrambled eggs and custards, which partially volatilize on spray-drying (Brogle et al., 1957). Egg white products undergo only slight change, but liquid whole eggs are seriously damaged (Mossel, 1960). Ionizing radiations have not been approved for use to date (see Chapter 5).

Liquid Albumen

Among whole egg, yolk, and albumen, raw liquid albumen is the most heat sensitive. Reports consistently show that egg white pasteurized at 57 to 60°C takes unduly long to whip and has low foam stability. Egg white has seven protein components, which are (with their percentage of the total solids content of albumen): ovomucin (1.5%), two globulins (7% each), lysozyme (3.5%), ovalbumin (58%), ovomucoid (10%), and conalbumin (12%). The two globulins and ovomucin and lysozyme are stable to heat, and ovalbumin and ovomucoid are more stable at pH 7 than at 9. Conalbumin, however, is quite heat unstable (U.S. Dept. of Agriculture, 1969). Workers at the Western Utilization Research and Development Division of the U.S. Dept. of Agriculture have shown that although the conalbumin fraction is the most sensitive to heat denaturation, adding small amounts of iron or aluminum makes it more heat stable. They indicate that albumen adjusted to pH 6.6 to 7.0 can be satisfactorily pasteurized by adding 30 ppm aluminum and then heating at 60 to 62.2°C for 3½ to 4 minutes. Under these conditions, the total bacterial count is reduced about 99.9% (3D) and numbers of salmonellae are diminished by 99.999999 (8D) (Cunningham, 1966).

Hydrogen peroxide destroys salmonellae in egg white at a temperature that does no major damage to the functional properties of the proteins (Ayres and Slosberg, 1949). A patented thermal process for egg white (Lloyd and Harriman, 1957), combining hydrogen peroxide (H_2O_2) with heat, permits a lower pasteurization temperature with minimal heat damage to the functional properties of albumen. Catalase, abundant in raw egg white, must be sufficiently heat inactivated that the H_2O_2 can be added and remain effective during pasteurization. Thus, the albumen is heated to 51.2°C for 2 minutes; 0.1% H_2O_2 is then added while the heat treatment is maintained for another 1.5 minutes. Organisms are destroyed and the H_2O_2 decomposes as follows:

$$2 \text{ min } \Delta \text{ } 51.2°C + 1.5 \text{ min } \Delta \text{ } 51.2°C + 2H_2O_2 \xrightarrow{\text{residual catalase}} 2H_2O + O_2$$

This treatment was found to reduce the total plate count by 3 logs (3D) and to eliminate all viable salmonellae from a mixed inoculum containing from 1.9 to 4.1×10^6 cells of *Salmonella typhimurium*, *S. thompson*, *S. montevideo*, *S. heidelberg*, and *S. senftenberg* (see Table 20-5; Figure 20-6; Rogers et al.).

A highly heat-resistant strain—designated *Salmonella senftenberg* 775W—was isolated in 1946 (Winter et al., 1946); being much less heat sensitive than any other species of salmonella tested to date, it is of great value in heat-resistance studies involving vegetative cells of pathogens. Decimal reduction time curves for *S. senftenberg* as well as for several other serotypes of salmonellae indicate F_{140} values of 7 to 9 minutes at pH 5.5 and 2 to 3.5 minutes at pH 8.0 (Anellis et al., 1954).

Table 20-5
Commercial pasteurization of egg white using hydrogen peroxide.

	Raw liquid			Pasteurized [a]		
Total count	Coliforms	Salmonellae		Total count	Coliforms	Salmonellae
144,000	9000	+		220	< 10	—
136,000	2000	−		3360	< 10	—
386,000	65,000	+		610	< 10	—
32,000	10	−		120	< 10	—
140,000	36,000	+		180	< 10	—
70,000	13,000	+		470	< 10	—

Source: Rogers et al., 1966.

[a] Heated at 51.7°C for 3.5 minutes at 0.075% H_2O_2.

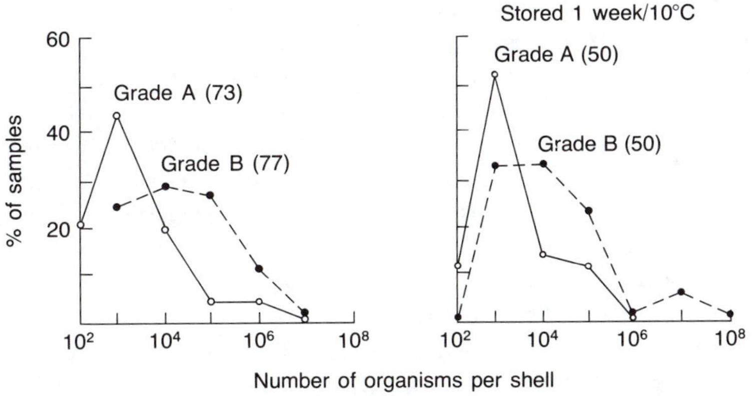

Figure 20-6
Frequency distribution of contamination on the shells of eggs. In parentheses are the number of eggs examined. (Board et al., 1964.)

Dried Albumen

Many years ago the Chinese learned that dried albumen became stable only after undergoing a spontaneous fermentation to free it of reducing sugars. Raw egg white dried before it is free of glucose is not stable; upon prolonged storage, a Mallaird reaction takes place, causing the dried material to darken and lose solubility. On the other hand, if the liquid albumen is stored in a warm place and the casual organisms present are allowed to proliferate, natural fermentation depletes the albumen of its 0.4% content of glucose within two to seven days. But by that time, all too often, *Escherichia coli* or other undesirable microorganisms have become dominant and have produced a fecal or slightly putrefactive odor. In addition, proteolysis may take place and salmonellae thrive in the fermenting albumen. Initially, American processors, offended by the fecal odor of Chinese dried albumen, attempted to dry the product without allowing it to "ferment" beforehand. Although the initial color and appearance of the dried product were quite desirable, they soon learned that its functional characteristics quickly became unsatisfactory. Pure culture inocula utilizing *Saccharomyces* spp. (Hawthorne and Rodenkirchen, 1939; Ayres and Stewart, 1947) and coliforms (Stuart and Goresline, 1942 a and b; Bollenback, 1949) are often used commercially to deplete the glucose and prevent spontaneous fermentation. Owing to the presence of antigrowth factors in egg white, *Saccharomyces cerevisiae* reproduces poorly. Either too many yeast cells must be added or the inoculum must receive an adequate quantity of growth factors (e.g., in yeast extract) to enable those cells present to complete fermentation. The latter condition is preferable, because the resulting albumen has a less breadlike odor and flavor. An alternate procedure is to treat the albumen with an enzyme complex (glucose oxidase) and convert the glucose to gluconic acid (Baldwin et al., 1953).

Albumen with satisfactory functional properties can be prepared by pan drying or spray drying or by blending the two types of dried products. Increasing the temperature of the dried albumen has long been known to decrease the numbers of viable organisms. Since albumen coming from commercial driers contains about 6% moisture, it is practical to use dry heat to destroy salmonellae without impairing the functional properties of the egg (Ayres and Slosberg, 1949; Banwart and Ayres, 1956). A satisfactory treatment requires careful adjustment of the product's moisture content as well as an adequate storage temperature so that the albumen can be whipped satisfactorily, has good solubility, and can be used for producing angel cakes. As Figure 20-7 shows, albumen with 6% moisture held for one

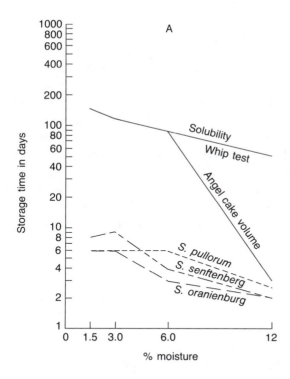

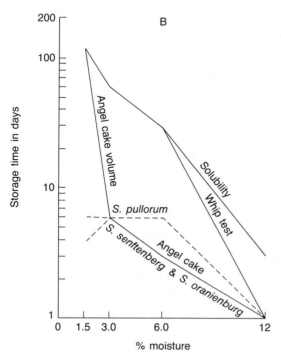

Figure 20-7
Limiting conditions for storing
spray-dried albumen at, A, 50°C;
B, 60°C. Areas above the solid
line indicate storage times for
each of the moisture levels at
which losses in functional
properties were observable.
Areas above the broken lines
indicate time-moisture
relationships in which
populations of the *Salmonella*
species were reduced to fewer
than 0.18 cells per gram.
(Banwart and Ayres, 1956.
Reprinted from Food
Technology Vol. 10, p. 68.
Copyrighted © by Institute of
Food Technologists.)

week at 50°C is freed of salmonellae without impairing its functional properties. However, at 60°C, although viable salmonellae remain, the product will no longer produce a satisfactory angel cake (Figure 20-7 B). Industry experience has shown that powder having approximately 6% moisture needs storage in "hot rooms" (51.2°C) for 7 to 10 days, exclusive of come-up time. For the process to be effective, the dried albumen should be put in storage containers at 51.2°C, at the time that it is discharged from the driers. Otherwise, an additional three to four days storage time is required to attain the same temperature in the product as in the hot room (U.S. Dept. of Agriculture, 1969).

References

Alderton, G., W. H. Ward, and H. L. Fevold. 1946. Arch. Biochem. *11*, 9.

Anellis, A., J. Lusas, and M. N. Raymon. 1954. Food Res. *19*, 377.

Ayres, J. C., and H. M. Slosberg. 1949. Food Technol. *3*, 180.

Ayres, J. C., and G. F. Stewart. 1947. Food Technol. *1*, 579.

Baldwin, R. R., N. A. Campbell, R. Thiessen, Jr., and G. J. Lorant. 1953. Food Technol. *7*, 275.

Banwart, G. J., and J. C. Ayres. 1956. Food Technol. *10*, 68.

Board, R. G. 1966. J. Appl. Bacteriol. *29*, 319.

Board, R. G., and J. C. Ayres. 1965. Appl. Microbiol. *13*, 358.

Board, R. G., J. C. Ayres, A. A. Kraft, and R. H. Forsythe. 1964. Poultry Sci. *18*, 581.

Bollenback, C. H. 1949. Ph.D. Thesis. Iowa State Univ., Ames.

Botwright, W. E. 1953. Amer. Egg Poultry Rev. *15*, 34.

Brogle, R. C., J. T. R. Nickerson, B. E. Proctor, A. Pyne, C. Campbell, S. Charm, and H. Lineweaver. 1957. Food Res. *22*, 572.

Brooks, J. 1960. Appl. Bacteriol. *23*, 499.

Brooks, J., and D. I. Taylor. 1955. *Eggs and Egg Products*. Rept. Food Invest. Bd. No. 60 HMSO, London.

Cazzulo, J. J., and A. O. M. Stoppani. 1967. Arch. Biochem.

Eakin, R. E., E. E. Snell, and R. J. Williams. 1941. J. Biol. Chem. *140*, 535.

Feeney, R. E., G. E. Means, J. C. Bigler. 1969. J. Biol. Chem. *244*, 1957.

Feeney, R. E., and D. A. Nagy. 1952. J. Bacteriol. *64*, 629.

Forsythe, R. H., and J. C. Ayres. 1962. Poultry Processing and Marketing *68*(11), 28.

Forsythe, R. H., J. C. Ayres, and J. L. Radlo. 1953. Food Technol. *7*, 47.

Forsythe, R. H., W. J. Ross, and J. C. Ayres. 1967. Poultry Sci. *46*, 849.

Fossum, K., and J. R. Whitaker. 1968. Arch. Biochem. Biophysics 122.

Fraenkel-Conrat, H., and R. E. Feeney. 1950. Arch. Biochem. *29*, 101.

Funk, E. M. 1943. *Pasteurization of Shell Eggs*. Mo. Agric. Expt. Sta. Tech. Bull. 364.

Funk, E. M. 1950. *Maintenance and Quality in Shell Eggs by Stabilization*. Mo. Agric. Expt. Res. Bull. 467.

Garibaldi, J. A. 1960. Food Res. *25*, 337.

Garibaldi, J. A., R. P. Straka, and K. Ijichi. 1969. Appl. Microbiol. *17*, 491.

Goresline, H. E., K. M. Hager, R. E. Moser, M. A. Howe, and E. E. Drewniak. 1951. USDA Circ. 897.

Haines, R. B. 1938. J. Hyg. *38*, 388.

Haines, R. B. 1939. *Microbiology in the Preservation of the Hen's Egg.* Great Brit. Sci. and Ind. Res. Dept. Food Inv. Spec. Rept. No. 47.

Haines, R. B., and T. Moran. 1940. J. Hyg. *40*, 453.

Hawthorne, E., and J. Rodenkirchen. 1939. Chem. Abstr. *36*, 2942.

Ingram, M., D. N. Rhodes, and E. J. Ley. 1961. Low Temp. Res. Sta. Record Memo No. 365.

Levine, M., and D. Q. Anderson. 1932. J. Bacteriol. *23*, 337.

Liu, W. H., G. E. Means, and R. E. Feeney. 1971. Biochem. Biophys. Acta *229*, 176.

Lloyd and Harriman. 1957. Armour & Co., U.S. Patent No. 2,776,214.

Lowry, F. R., and P. Starr. 1949. Nulaid News *26*(9), 3.

Matsushima, K. 1958. Sci. *127*, 1178.

Miles, A. A., and E. T. Halnan. 1937. J. Hyg. *37*, 79.

Mossel, D. A. A. 1960. Intl. J. Appl. Radn. Isotopes *9*, 109.

Mundt, J. O., and R. L. Tugwell. 1959. Poultry Sci. *37*, 415.

Osborne, W. W., R. P. Straka, and H. Lineweaver. 1954. Food Res. *19*, 451.

Redfield, H. W. 1920. Exam. of Frozen Eggs and Interpretation of Results, U.S. Dept. of Agriculture Bull. 846.

Rhodes, M. B., N. Bennett, and R. E. Feeney. 1960. J. Biol. Chem. *235*, 1686.

Rogers, A. B., M. Sebring, and R. W. Kline. 1966. In: *The Destruction of Salmonellae* ARS 74-37, Albany, Calif.

Schade, A. L. 1958. Abstr. Proc. Intern. Congr. Biochem. 4th Congr., Vienna *127*, 10.

Stokes, J. L., and W. W. Osborne. 1956. Food Res. *21*, 264.

Stuart, L. S., and H. E. Goresline. 1942a. J. Bacteriol. *44*, 451.

Stuart, L. S., and H. E. Goresline. 1942b. J. Bacteriol. *44*, 625.

Sugihara, T. F., R. Ijichi, and L. Linke. 1966. Food Technol. *20*, 1076.

Torrey, G. S. 1965. M.S. Thesis, Iowa State Univ. Library, Ames, Iowa.

Turner, A. W. 1927. Austr. J. Exptl. Biol. and Med. Sci. *4*, 57.

Tyler, C. 1953. J. Food Agric. *4*, 226.

U.S. Dept. of Agriculture. Consumer & Marketing Service, *Grading and Inspection of Egg Products* 1-1-67.

U.S. Dept. of Agriculture. 1969. ARS 74-48, Albany, Calif., *Egg Pasteurization Manual.*

Vadehra, D. V., and R. C. Baker. 1973. J. Milk Food Technol. *36*, 321.

Wellner, V. P., J. I. Santos, and A. Meister. 1968. Biochem. *7*, 2848.

Winter, A. R. 1942. U.S. Egg Poultry Mag. *48*, 506.

Winter, A. R., G. F. Stewart, V. H. McFarlane, and M. Soloway. 1946. Amer. J. Public Health *36*, 451.

Wrinkle, C., H. H. Weiser, and A. R. Winter. 1950. Food Res. *15*, 91.

Zagaevsky, J. S., and P. O. Lutikova. 1944. U.S. Egg Poultry Mag. *50*, 17.

Part IV
Foodborne Illnesses

21

Nonmicrobial Foodborne Illnesses

Foodborne illnesses are conditions of distress following the ingestion of food or drink. Such illnesses may strike one person, or hundreds of persons in a single outbreak, and may be only mildly and temporarily unpleasant, or fatal. They are both microbial and nonmicrobial in origin. Many of the latter display easily recognizable symptoms, but others may mimic illnesses caused by bacteria and viruses.

Opinions differ regarding the classification of foodborne illnesses. In this and the next two chapters these will be discussed as 11 major types:

1. Indigestion;
2. Food intolerance or food sensitivity;
3. Algal toxins;
4. Metal poisons;
5. Phyllotoxins;
6. Manufactured agricultural and household chemicals;
7. Zootoxins;

8. Protozoan disease;

9. Infestations;

10. Microbial infections; and

11. Bacterial and fungal food intoxications (including mycotoxins produced in or on foods after harvest or slaughter).

Many persons are aware of the risks posed by certain foods containing the potent constitutive fungal toxins, phyllotoxins, and zootoxins. Chemical agents, mycotoxins, and agents of infestation, infection, and microbial intoxication usually do not produce changes in foods detectable by color, taste or odor.

Indigestion

Indigestion is acute foodborne distress which follows willful neglect or violation of good eating habits—as, for example, eating too many hotdogs, or the "green apple" syndrome among children. Symptoms of distress, acute abdominal pains and vomiting, occur toward the end of overindulgence or shortly thereafter. The sole remedy appears to be more moderate consumption of the offending food.

Food Intolerance or Sensitivity

Food sensitivities are the food-related reactions termed *allergy, immunological hypersensitivity,* and *anaphyllaxis*—that is, peculiar manifestations of the antigen–antibody reactions following ingestion or contact with foods or inhalation of food dusts. Any food intolerance develops only after one or more instances of exposure to the food, through ingestion or inhalation. In a person sensitive to a food or drugs, sensitization occurs when molecules pass through the intestinal mucosa or, if the food is inhaled as a dust, through the respiratory mucosa. Some molecules of foods or drugs that give rise to hypersensitivity are not proteinaceous, and as such cannot stimulate antibody production. Such molecules serve as *haptens,* which conjugate with proteins and thus become antigenic. Molecules quite often pass through the mucosae; an estimated 20% of the population has antibodies against a variety of common foods. Only occasional individuals develop hypersensitivity. Such individuals are termed *atopic*.

Hypersensitivity to milk, wheat flour, and eggs is common, particularly among children, perhaps because these foods are so abundant in their diet; however, the list of foods that have given rise to hypersensitivity is very long. It is probable that any food frequently or even occasionally consumed, such as celery or green pepper, will stimulate hypersensitivity in some individual. Reactions vary in severity from very mild to exceedingly painful and are occasionally fatal—sometimes within an hour after the food was eaten. Food-specific sensitive persons must either avoid the incriminated food or be desensitized.

Allergic (*all*, changed; *ergic*, reactive) **hypersensitivity** occurs when the antibody and antigen unite and disturb neighboring cells or cells in other parts of the body. Three immediate types, *anaphyllaxis, hives,* and *asthma* are commonly recognized. Delayed food reactions are either less common or less severe, and thus are sometimes not recognized as such.

Anaphyllaxis, the most severe of the reactions, may lead to **shock.** It often is exceedingly painful, and the afflicted person loses consciousness—and sometimes dies—within minutes of consuming the food. **Mast cells,** or granulocytes, which are in the connective tissue and platelets, release histamine, serotonin, and bradykinin into the bloodstream. These three substances stimulate the smooth muscle to contract, damage the liver, and stimulate the blood vessels to dilate, causing **edema** (leakage of fluid into the tissues). Anaphyllaxis can be treated with antihistamine drugs.

In **hives** the antibodies, located in the skin and termed **skin-sensitizing** antibodies, appear promptly upon ingestion of food. Once absorbed in the intestinal tract the offending molecule is transferred to the skin. Or the stimulant may enter skin punctured by the hairs of fruits such as peaches or strawberries. Hives may occur either as a **wheal,** a hard, white swelling that is not infiltrated by leucocytes, or as a **flare,** an extensive, reddened area normally produced by intradermal introduction of the antigen, as in skin testing. Hives may be a mass of flares without visible wheals.

If the antibodies reside in the respiratory mucosa, a reaction gives rise to **hay fever** if the upper respiratory tract is affected, or, if the lungs are affected, to the more serious **asthma** with its attendant difficulty in breathing. Although hives, hay fever, and asthma are distressing to the afflicted person, the binding of the disturbing antigens in the skin or the respiratory surfaces may be beneficial, by preventing the antigen from reaching other tissues with possibly severer consequences.

Delayed hypersensitive reactions occur in sensitized persons hours to days after ingestion of food. The antibodies, which remain attached to the cells that produce them, are termed **reagin** antibodies. The vascular tissue is primarily involved. In contrast to what happens in immediate reactions, in delayed reactions leucocytes infiltrate the affected area, and histamine, serotonin, and bradykinin are not released. A reaction occurring in the skin appears as a red, *indurated* (swollen and hard) area. The positive induration following the intradermal injection of tuberculin is such a delayed reaction.

Algal Toxins

Three divisions of algae, the Pyrrhophyta, Cyanophyta, and Chrysophyta—more commonly known as the dinoflagellates, blue-green algae, and golden-brown algae, respectively—may poison our food supply. Among the several species of the dinoflagellates associated with the poisoning of shellfish are *Gonyaulax catanella, G. acatanella, G. tamarensis, G. monilata, G. polyedra,* and *Gymnodinium breve, G. veneficum, Pyrodinium phoneus,* and *Peridinium polonigium;* among the blue-green algae are *Anabaena flos-aquae, Aphanizomenon flos-aquae,* and *Microcystis aeruginosa;* and among the golden-brown algae are *Prymnesium parvum, Ochramonas danica,* and some strains of *O. malhamensis.* All the organisms occur both in fresh and marine waters, but almost invariably incidents of shellfish poisoning occur in brackish water or seawater. Occasionally species of *Gonyaulax* or *Gymnodinium* become so common in coastal waters that they produce "red tides." The coloration of the water is caused by billions of these algae which produce a "bloom" when they are rapidly reproducing (Gentile, 1971; Shilo, 1971).

Saxitoxin

The toxins produced by the dinoflagellates are sometimes also known as clam poisoning, mussel poisoning, paralytic shellfish poisoning, mytilotoxication, and *saxitoxin*. Various types of shellfish are associated with saxitoxin production—particularly the soft shell clam *(Mya arenaria)*, the Alaska butter clam *(Saxidomus giganteus)*, the Washington clam *(S. nuttalli)*, the Pacific oyster *(Crassostrea gigas)*, the black mussel *(Mytilus edulis)*, the California sea mussel *(M. californianus)*, and the white mussel *(Donax serra)*. This toxin is a low-molecular-weight nitrogenous base having the empirical formula of $C_{10}H_{15}N_7O_3 \cdot 2HCl$. Saxitoxin, which can make shellfish highly toxic within a few days of ingesting the dinoflagellates, is not a postmortem product of decomposition. It is a neurotoxin that has a central effect on cardiovascular and respiration centers and a peripheral effect on the nerve endings, both motor and sensory. It may be a neuromuscular poison (Schantz, 1971).

During toxicosis, which generally develops less than an hour or so after the poisonous shellfish is consumed, the person has a floating or flying sensation but remains mentally alert. Initially there is evidence of peripheral paralysis, such as tingling and numbness about the lips, gums, and tongue, dryness of throat, difficulty swallowing, loss of strength in muscles of the fingertips and toes. Severe cases progress to giddiness and staggering, loss of use of the arms, legs, and neck, respiratory paralysis, and, in extreme cases, death by respiratory failure. The intoxication is treated by purging the stomach with emetics or brisk laxatives. Artificial respiration may also be required.

The toxin is stored in the shellfish digestive gland, liver, gills, and siphon. Saxitoxin is extremely potent: as little as 96 μg seriously poisoned a two year old child. This dosage was the equivalent of 600 mouse units. A mouse unit is the minimum amount of toxin needed to kill a 20 gram mouse in 15 minutes after intraperitoneal injection. In adults, 800μg (5550 mouse units) are known to produce difficulty in respiration, loosening of the teeth, and recurring symptoms for two weeks. From 3000 to 20,000 mouse units are estimated to be fatal to humans.

Removal of the **hepatopancreas** of mussels or the siphon of the butter clam is recommended. About 90% of the toxin is inactivated by cooking for 10 minutes in steam. Cooking with baking soda reduces the toxicity, but the soda impairs flavor. Shellfish should not be harvested in waters in which the algae grow rampantly, and they should not be consumed raw when algal growth is apparent. Waters in which the algae develop are now seasonally posted.

Ciguatera

The green alga *Lyngbia majuscula* produces an unnamed toxin responsible for *ciguatera* poisoning, which occurs after consumption of any of a wide variety of vertebrate fish that feed upon the algae. The toxin passes through the food chain from herbivorous to carnivorous fish—most often implicated are barracuda, amberjack, red snapper, and grouper (Baratta and Tanner, 1970), and some 400 species of marine shore or reef fish inhabiting relatively warm coastal waters. In any species large fish are more likely to be toxic than smaller fish (Halstead, 1967). Ciguatera toxicosis has also occurred among crews of ships capturing fish during oceanic crossings. Intoxication may occur minutes or several hours after toxic fish is eaten. Apparently the incubation period is a function of the amount of toxin ingested (Morbidity and Mortality Weekly Reports, 1974). The victim reacts as though simultaneously suffering a form of metal poisoning, acute gastroenteritis, influenza, mental depression, dilation of the pupils of the eye and blurred vision, possibly temporary blindness, extreme muscular weakness, motor incoordination and paralysis. The fatality rate is estimated at 2 to 7%. Ciguatera is recovered in highest concentration from the fish liver, intestinal tract, testes or ovaries, and muscle. The toxin is heat stable.

Metal Poisons

Both mineral and organic materials toxic to man and animals are widespread in the environment. They occur in foods, often as normal constituents. It is doubtful that any substance exhibits absolute toxicity. For all, there seems to be a threshold level below which the ingested material is either excreted in the urine or feces, or is altered by body metabolism and rendered innocuous. Many minerals such as sodium, potassium, calcium, magnesium, zinc, iron, copper, phosphate, chlorine, and iodine are dietary essentials. Tolerance levels among these are seldom exceeded during the normal preparation and consumption of foods. Sodium intoxication has been known as the "Chinese food syndrome," attributed to use of too much monosodium glutamate in Chinese restaurant cooking. The prominent into-xicating mineral elements are arsenic, lead, mercury, and selenium.

Common Metal Poisons

Arsenic is a common minor element in the crust of the earth and is found in human tissues at approximately 0.3 ppm. It is present in the atmosphere of some smelters, occurs naturally in seafoods and in some drinking waters, but is not normally found in the flesh of food animals, freshwater fish, in a variety of nuts, or in cereals or vegetables. Before the advent of organic pesticides it was commonly used to destroy insects and other pests.

In ocean waters arsenic becomes concentrated in the food chains. Fish and shellfish may have levels of up to 170 ppm, expressed as arsenic trioxide. Fruits, such as apples, that have been sprayed with arsenicals must be washed with dilute hy-drochloric acid. Chemicals used in the food industries must be *food grade,* with the content of toxic metallic elements severely limited.

Threshold levels are difficult to determine. Animals given 0.02% phenylarsenic acid in feed have exhibited no overt symptoms of arsenic poisoning. Persons who drank water containing 45 to 46 mg per gallon exhibited symptoms of chronic arsenic poisoning 2½ years later. Accidental or deliberate introduction of high levels of arsenic into foods has caused death within two to three days, with symptoms of muscular weakness, metallic taste, and internal burning sensations.

Lead occurs naturally in foods at levels exceeding threshold values. For centuries it has been used in making lead dishes and tableware, lead pipes, and pottery glazes. Identification of the contents of jute food bags has been stenciled on in plumber's red lead. Toxic levels of lead have been found—although rarely—in honey made from nectars following spraying of the blossoming crop.

The human daily diet is estimated to contain 0.20 to 0.25 mg of lead, and the average human can excrete about 0.45 mg per day. Lead ingestions above this threshold level are deposited in the bones. Symptoms of lead poisoning include weakness, dental caries, nausea, vague pains, paralysis, and colic.

Selenium occurs in many soils at an average level of 0.25 ppm or 0.5 pounds per 6 inch acre (an acre of soil 6 inches deep; weight 2 million pounds) of soil. It is most commonly used in a spray to control red spider mite and as an ingredient in selective bacteriological culture media. It is found in practically all foods in association with the sulfhydryl ion; wheat contains between 0.1 and 1.9 ppm. Some plants, termed *converter* plants, absorb quantities of selenium from the underlying minerals, and the element is then absorbed by the next succession of plants as the converter plants decay on the surface of the soil.

The toxic level for the human has not been established. Persons living in selenium-rich areas will eliminate nearly 200 μg per 100 ml urine. No human foods produced on seleniferous soils are excluded from the market, because it is believed that no human will acquire a toxic quantity of seleniferous food.

The human diet requires 1 to 2 mg of *copper* per day. Like lead it is easily malleable. Before the advent of glass-lined and stainless steel vessels, most cooking kettles used for preparing a wide variety of food items and in brewing were of copper. The use of copper cookware in the home is not considered hazardous; yet, foods on which a green color develops from prolonged storage in such vessels should not be eaten. Copper is readily dissolved by acids such as lactic, malic and tartaric, and, in plumbing, by water containing oxygen and carbon dioxide.

Copper sulfate was once used as a coloring agent for such common foods as peas, beans, and pickles. Although there were few confirmed cases of copper poisoning, such use is esthetically undesirable, and deliberate addition of copper salts to foods is now prohibited.

Mercury is a common element of seawaters. As solubilized organic (methyl) mercury, it accumulates to high levels in the terminal predators in the food chains (refer back to Figure 18-2).

Mercury poisoning is a relatively rare but well-known, spectacular illness; the reported presence of hazardous mercury levels in seafoods early in the "ecological era" gave rise to anxiety that now appears not to have been justified. However, the discharge of mercurial wastes into waters is not to be condoned. Bacteria residing in waters convert mercury to methyl mercury, which is then deposited in the tissues of the successive predators, terminating in the ultimate predators in the food chain such as the swordfish. The toxic effect of methyl mercury in industrial waters was demonstrated during the early 1950s in Japan when over half of some 100 persons affected died of "Minamata disease," a syndrome that resulted from the ingestion of mercury contaminated fish and shellfish harvested from Minimata Bay.

The massive outbreak of mercury poisoning in Iraq in 1971 probably will remain one of the most bizarre examples of chemical poisoning in history. In this episode, much of the treated seed grain purchased abroad to replenish seed supplies that had diminished through prolonged drought was diverted into flours, resulting in an estimated 6000 deaths and perhaps 100,000 permanently injured.

Other Elements

Aluminum, chromium, nickel, and *tin* commonly come in contact with foods, but with no known instances of toxicity, even though tin in metal cans may become solubilized through reaction with the acids of foods. Aluminum is a common ingredient of baking powder. *Cadmium,* even in small quantities dissolved from plated metals in which foods are prepared or stored, is promptly emetic if ingested. Oysters are especially rich in cadmium. *Zinc* is used as a closure in home canning and also as a plating agent of metal containers. Intoxications attributable to zinc are rare. Probably the amounts normally consumed are far too low to produce harmful effects. Galvanized tubs, pails, and similar containers have been used for pickling cucumbers, and even though the container blackened and the pickles were dark, there were no ill effects from eating the pickles. *Cobalt* salts added to beer to improve its foaming properties have been incriminated in several instances of cardiac failure among heavy beer drinkers. Green, leafy vegetables, which are the richest natural sources of cobalt, contain less than 1% of the levels required to produce signs of toxicity (Underwood, 1973).

Phyllotoxins

Many plants produce substances with pharmacological and toxic effects on humans and animals. Table 21-1 gives a partial list, according to function. The naturally occurring toxicants are discussed in greater detail in several reviews (Ayres et al., 1962; American Chemical Society, 1966; Committee on Food Protection, 1973).

Most substances occur in foods at far below toxic levels—although equivalent amounts injected into the blood stream are exceedingly toxic. Many substances express an adverse effect only if the food is a major and consistent article of the diet, such as cabbages and mustards, the goitrogenic *Brassicae* and *Cruciferae*. Some are toxic only at levels well beyond what are normally consumed, as has been determined by animal experimentation.

Through observation and trial and error, and of course controlled experimentation, humans have learned to recognize many overtly toxigenic foods. Toxigenic foods with rapid effects can be detected by virtually anyone, but this is not always true of toxigenic foods with slowly developing or cumulative effects.

Goitrogenic effects follow the consumption of rutabaga, cabbage, and turnip in abnormal quantities in an unbalanced diet. The Jamaican ackee is toxic only if consumed as the green fruit. The hallucinogenic nutmeg has a pharmacological effect only if 5 grams or more are consumed, a quite remote possibility. Many legumes contain insignificant levels of hemagglutinins, but the castor bean contains the potent intestinal irritant **ricin,** which has been used for this purpose in medicine and also as a poison for moles (Kingsbury, 1964).

Cyanide, fatal to humans at 2 to 3 mg/kg body weight, occurs as the glycoside (sugar and nonsugar moieties) in the seed kernels of stone fruits such as the apricot, peach, cherry and bitter almond, in the leaves of the cherry and in the starchy root manioc. Crushing the kernel (the contents of the pit), leaf, or

Table 21-1
Classes of intoxicants occurring naturally in plants.

Class of intoxicant	Representative plants	General mode of action
Antienzymes	Soy beans and other legumes; potatoes (solanin)	Inhibition of trypsin
Anthroquinone derivatives	Rhubarb leaves; sorrel grass	Vomiting; icterus
Antithyroid compounds	Crucifers; *Laburnum*	Thyroidal adenoma
Carcinogens	*Senecio; Crotalaria;* cycads; sassafras (saffrole)	Hepatocarcinoma; liver damage
Cholinesterase inhibitors	Many common vegetables	Inhibition of nerve impulses
Cyanogenetic glycosides	Stone fruit kernels; lima beans; cherry bark	Cyanosis from liberated HCN
Depressants	Coyotillo; yams; death camas	Decrease in excitability; convulsions
Estrogens	Legumes; anise oil; *Prunus*	Hypertrophy; involvement of estral cycle
Goitrogens	Cabbage family; some fruits	Enlargement of thyroid
Hallucinogens	*Datura;* nutmeg	Derangement of central nervous system
Hemagglutinins	Legumes; castor bean	Depressed growth; hypertrophy of pancreas
Hypoglycemia	Fermented coconut (bongkrek); Jamaican akee	"Vomiting sickness"; hypoglycemia
Lathyrogens	Some members of *Lathyrus; Viccia*	Muscular weakness
Methemoglobinaemia	Greens (chiefly spinach)	Methemoglobin formation
Irritants	Hydroxyphenylisatin in prunes	Purgative
Pressor amines	Foods rich in aromatic amino acids	Pseudoallergenic and anaphyllactic reactions
Stimulants	Coffee, tea	Excitement
Tumorgens	Toxic cycads as emergency foods; saffrole; bracken	Carcinomas

Source: Committee on Food Protection, 1973.

root liberates a hydrolytic enzyme that frees the cyanide. Consuming peach kernels, as some food faddists recommend, could be fatal. It is estimated that 5 kernels contain enough cyanide to kill a 5 year old child, 20 kernels enough to kill an adult. During canning of unpitted cherries, heating to 100°C for 6 minutes inactivates the enzyme. The root manioc, a dietary staple in the tropics, is freed of the cyanogen linamarin by peeling, crushing, washing and leaching in water, and heating or fermenting before drying and eating (Clapp et al., 1966).

Some legume seeds, such as the soy, velvet, horse, and lima bean, contain trypsin and chymotrypsin inhibitors. The common groundnut, or peanut, does not contain inhibitors. The Irish potato contains an anticarboxylase. Most antienzymes are inactivated by thorough cooking (Whitaker and Feeney, 1973).

The carcinogenic safrole, a dioxybenzene, is a major constituent of oil of sassafras and a minor constituent in cocoa, anise, pepper, laurel, wild ginger, mace, and nutmeg. Oil of sassafras is no longer used as a flavoring agent. The proved carcinogenic action of safrole in animal studies suggests that carcinogenesis follows the use of much greater concentrations than are normally encountered in human foods; yet its effect as a cumulative toxicant when consumed as a minor constituent over many years is not known (Hall, 1973). The nut of the cycad, a tropical plant consumed when other foods are scarce, bears the carcinogen cycasin which is hepatocarcinogenic (Miller, 1973).

Pressor Amines

Pressor amines are those decarboxylated amino acids that have stimulating to distressing effects. They are naturally present in flesh and skin of poultry and in fruits such as the banana, plantain, plum, pineapple, and tomato. At normal levels in the diet, they play a necessary role in the hormonal regulation of smooth muscle. They are also produced from the precursor amino acids by the mechanism of decarboxylation during the fermentation of cabbage to sauerkraut, the production of some cheeses and wines, and microbial deterioration of fish of the family Scombroideae. The amines most commonly a problem in the United States are histamine, tyramine, serotonin, and dopamine, and, in parts of the world where the broad bean is eaten, the derivatives of dihydroxyphenylalanine and dopamine cause difficulties (Lovenberg, 1973; Patwardhan and White, 1973).

The human body tolerates approximately 6 mg of pressor amines per meal. Levels up to 40 mg per meal are stimulative but otherwise uneventful in the normal, healthy human; beyond this level, the symptoms of pressor amine intoxication become increasingly severe and may cause death. Symptoms of intoxication include hypertension, headache, flushing, rapid pulse, nausea, vomiting, a burning sensation of the throat, thirst, swelling of the lips, and pruritus. The effects become aggravated if foods with elevated amine content are ingested by tubercular patients being treated with iproniazide, which inhibits monamine oxidase (MAO). At normal dietary levels the amines are oxidized by monoamine oxidases, excreted unchanged with solid wastes, or acetylated and rendered harmless by bacteria such as *Escherichia coli* and *Enterobacter aerogenes*. Illnesses have been noted in patients treated with MAO inhibitors, which block the pathway and inactivate amines after ingestion (Voigt et al., 1974).

Many bacteria can decarboxylate amino acids to produce amines. However, most observed pressor amine intoxication follows consumption of those foods rich in the precursor amino acids in which bacteria function. For example, group D streptococci in cheddar cheese may produce tyramine levels in excess of 100 mg/100 g cheese. Tyramine is found in lesser quantities in cheeses such as Camembert, Brie, Emmentaler and Gruyere, but not in fresh curd cheeses, such as cottage cheese, in which proteolysis of casein has not occurred.

The flesh of members of the marine suborder *Scombroidei*— tuna, skipjack, albacore, and mackerel—contains unusually large amounts of histidine. Bacteria growing on the unrefrigerated flesh decarboxylate the histidine to the pressor amine, histamine, which in turn may be degraded to saurine (from the scombroid fish called *saurus,* the Latin word for lizard). Those who eat bacterially degraded flesh suffer scombroid fish poisoning, or fish allergy. *Proteus morgani* and *Achromobacter* spp. are the bacteria usually involved, although species of *Salmonella, Shigella, Clostridium, Escherichia,* and *Vibrio* also decarboxylate histidine. The normal histamine content of the fresh flesh is 6 mg or less per 100 g (Geiger, 1944); 10 mg per 100 g signifies some measure of microbial deterioration, and concentrations of 100 mg or more per 100 g flesh are associated with illness of the consumer.

Scombroid poisoning usually occurs along coastal areas where freshly caught fish are not immediately consumed or are smoked (Center for Disease Control, 1973), and in inland areas when fish have not been properly refrigerated. In a major outbreak in the United States, involving 232 persons after consumption of canned tuna (Table 21-2), the histamine content was as great as 280 mg/100 g tuna. Both histamine and saurine are stable to cooking. Merson et al. (1974) cite literature stating that scombroid fish poisoning kills more persons than any other form of ichthyosarcotoxism.

Symptoms of scombroid poisoning include burning and sometimes blistering of the mouth; headache, urticaria, or burning and stinging of the skin accompanied by red wheals or blotches; cramps; diarrhea; and simple flushing of the skin. Other symptoms may include nausea; dysphagia, or impairment of speech; thirst; chest pains; prostration; and pruritus, or intense itching of the skin without visible lesions. The onset of the symptoms occurs shortly after the food is consumed, usually within one hour, and an uneventful recovery takes eight hours.

Caffeine, theobromine, and theophilline, in the brews of coffee and tea, are universally accepted as both mildly stimulating and relaxing. In excessive quantities, they lead to a temporarily

Table 21-2

Cases of scombroid fish poisoning from commercially canned tuna fish, February 1973.

State	Cases	Date	Lot no.
Minnesota	182	Feb. 19–22	419, 417
South Dakota	30	Feb. 18–20	419
Oregon	15	Feb. 19–22	417, 419
Wisconsin	5	Feb. 13–20	419

Source: Merson et al., 1974.

elevated pulse rate and blood pressure and acute excitation. The detrimental effects in relation to quantities consumed are controversial. Individuals susceptible to excitation should consume the beverages in moderation or not at all.

Greens, such as spinach, grown on heavily fertilized soils accumulate nitrates. Bacterial action in the intestinal tract reduces the nitrate to nitrite, which is then absorbed. The union of hemoglobin with nitrite renders the hemoglobin functionally inactive. For infants the threshold level is between 49 and 70 ppm—150 ppm in the drinking water may be fatal to them.

Nutmeg contains the pleasing spice ingredient **myristin,** a phenolic compound, which produces euphoria, delirium, and stupor when consumed in abnormal quantities (5 grams or more). Turnip, cabbage, and some fruits contain goitrogenic agents that lead to enlarged thyroids when they form a large part of the diet. The pokeweed contains saponin; various greens contain oxalates that prevent the absorption of calcium from the gut; and even the abnormally large consumption of carrots and other vegetables rich in carotene can result in carotinosis, an orange pigmentation of the skin.

Manufactured Agricultural and Household Chemicals

Insecticides, pesticides, growth regulators, fungicides, and growth stimulators are essential in modern agriculture for the production of adequate quantities of sound food. They include mineral insecticides such as cryolite (sodium aluminum fluoride), lead arsenate, and paris green (copper sulfate, lead arsenite, and acetate); a long list of chlorinated hydrocarbons such as chlordane, dieldrin, and dichloro-diphenyl-trichloro-ethane (DDT); organophosphate compounds such as diazinon, dichlorvos, dipterex, and malathion; dinitro compounds; botanical derivatives such as the alkaloid nicotine and the "miracle" intoxicants pyrethrum and rotenone; hexachlorobenzene as an insecticide; synthetic growth promoters and antibiotics. Toxicity to the human ranges from none, for some compounds such as pyrethrum, to extreme, as for the organic phosphorus compounds. The toxicity of some, such as DDT, is in dispute; DDT accumulates in human and animal tissues without observable damage, but it is believed that at least 25 years of use may be required before its potential for chronic toxicity can be assessed.

In principle, most chemicals are looked upon as adulterants of food. Regulatory control dictates the care that must be exercised in their handling and use, time of application, and residues permitted, if any. Foods become contaminated through:

1. misunderstanding or ignorance of the dangers of the chemicals;

2. negligence, through ignorance or familiarity;

3. faulty spray procedures;

4. the breaking of containers during shipment;

5. leaky containers; and

6. the use of containers, such as bagging material, for purposes other than originally intended.

Intoxications have become occupational hazards among agricultural workers (refer back to Chapter 10 for additional comments).

Zootoxins

Zootoxins are associated only with freshwater and marine foods. Although naturally occurring toxins, they are distinct from the algal toxins that accumulate in marine foods, and from the scombroid toxin that results from microbial deterioration of the fish flesh after capture. Intoxications that are either unknown or rare in the United States include *Moray eel* poisoning; fatal cyanosis from eating porpoise liver; and the nonfatal but painful, stinging, edematous toxicity associated with a fluorescent toxin in the liver of the abalone.

Pufferfish Poisoning

Pufferfish poisoning is a common toxicosis of the Orient, there being several hundred cases of illness and about a hundred deaths occurring each year, primarily among Japanese. No microorganism is known to be involved. There are at least 30 different species of *Fugu,* or pufferfishes, most of them along the coasts of Japan and China. Although for thousands of years they have been known to be poisonous, they are eagerly sought as delicacies. The most toxic species belong to the family Tetraodontidae, and therefore the toxic principle has been called *tetrodotoxin.* It is found in highest concentration in the fish ovaries, gonads, liver, intestine, skin, and spleen. Nearly all deaths from excessive intake of tetrodotoxin occur because, from ignorance or carelessness, the poisonous visceral organs of pufferfish were not removed.

Symptoms of pufferfish poisoning begin within a few minutes of eating, starting with a tingling of the lips and tongue, fingers, and toes. There may be nausea, vomiting, diarrhea, and stomach pain. With acute poisoning there is respiratory distress, loss of reflexes, and progressive paralysis. Most deaths are the result of respiratory paralyses.

Tetrodotoxin is an aminoperhydroquinazoline compound having the chemical formula $C_{11}H_{17}N_3O_8$ and a molecular weight of 319. It is sufficiently heat stable to withstand boiling for 10 minutes. Even short heating periods under pressure, such as are used during canning, are unsafe. Mouse testing with the crystallized tetrodotoxin suggests that fatal toxicity occurs upon ingestion of 8 μg/kg body weight. The poisoning occurs more commonly during the spring and summer.

Protozoan Diseases

Amebiasis

Amebiasis or amebic dysentery occurs throughout the world. Its prevalence is influenced by cultural practices—that is, by disposal of human wastes in such ways that food and water are contaminated. In temperate zones the life of an infected person is rarely threatened by amebiasis, but if visceral organs have been affected and extensively damaged, the person is abnormally susceptible to other illnesses.

Symptoms of the disease may be benign and limited to diarrhea, or they may be severe, with ulceration of the colon and other visceral organs and abscesses of the liver, with permanent damage to all these organs. Typical manifestations include tenderness over the colon or the liver, loose morning stools, recurrent diarrhea, loss of weight, and fatigue. The mortality rate ranges from 2 to 3%.

Recovered patients may become asymptomatic carriers, who may, however, suffer relapses. The reported incidence of amebiasis is low. In 1973, 2199 cases were reported to the National Center for Disease Control. Authorities estimate that 10 million persons in the United States harbor the parasite and that 2 million have symptomatic cases of the disease.

Laboratory diagnosis of amebiasis consists of examining stools for the specific trophozoites and cysts. Their characteristic morphology differentiates them from the saprophytes normally inhabiting the intestinal tract. Of the five parasitic amebae that cause dysentery in humans only *Entamoeba histolytica* invades human tissues regularly. The remaining four are *Entamoeba coli, Endolimax nana, Iodamoeba butchlii,* and *Dientamoeba fragilis.* Human carriers are the major source of the infestation, although the amebae are also found in animals such as rats, dogs, and monkeys, which are thus also possible sources of contamination.

Contaminated food undergoes no spoilage, poisoning, or damage, and gives no visible or other evidence of contamination. Foods may be contaminated directly by handlers with active illness or who are carriers, or indirectly by contamination with human wastes in garden and field either through random practice of elimination or the deliberate use of night soil, and irrigation and washing of vegetables with polluted water. The amebae may also be transferred to foods by flies and cockroaches which have access to both human waste and food.

The first major domestic outbreak in a city of the Temperate Zone occurred in 1933 in Chicago during the Century of Progress World's Fair. The outbreak was attributed to improperly installed plumbing in two hotels; to conserve space, the drinking water pipes were placed inside the sewage pipes. Some of the water pipes developed leaks and siphoned sewage into the drinking water. The disease spread quickly across the country as clientele of the hotels returned to their homes. In this outbreak, 1409 cases were reported, with 98 deaths. In 1955 a similar outbreak occurred in South Bend, Indiana; contaminated drinking water was the source of the parasite.

Entamoeba histolytica has three distinct stages, the **trophozoite,** the **precyst,** and the **cyst.** The trophozoites, or vegetative forms, resemble saprophytic amebae in shape and motility; they range from 8 to 60 microns in length. The trophozoite may live and multiply in the crypts of the large intestine, with enteric bacteria as symbiotic associates. Once the organs have been invaded, the trophozoite no longer depends on the bacteria but obtains its metabolic substances from the tissues (Figure 21-1 A; Jeffrey and Leach, 1966).

The trophozoite passes through the precystic stage to the cyst. The cyst is smaller than the trophozoite, is nonmotile, and has a cyst wall. The cyst is the most resistant form, but it remains viable only if kept moist. It is readily destroyed by desiccation, sunlight, and heat, and by chlorination of water. Cysts, which usually are spherical, range in size from 5 to 20 μ in diameter. When the cyst is ingested, it survives passage through the stomach and excystment occurs in the small intestine with the formation of a four-nucleate ameba. Cysts have been found in the intestinal tracts of flies, and viable cysts have been shown experimentally to pass through the intestinal tracts of the fly and the cockroach (Shaffer et al., 1965).

In the intestinal tract the four-nucleate ameba undergoes cytoplasmic division to form four metacystic trophozoites that grow to mature size. They travel along the intestine to colonize in the ceca or the lower intestinal area. It is not known if the invasion of the host tissue is by physical or by lytic means; there is no evidence of a lytic substance. It is also possible that the advancing amebae extract some vital substance from host cells and move onward, leaving the host cells to undergo lysis. If passage through the intestinal tract is not rapid, the trophozoites transform into the precystic and cystic stages before being excreted. Ulcers in amebiasis vary in size from 2 mm to exceedingly large areas of the intestinal wall.

Effective sanitation is necessary to control amebiasis. Most severe problems arise in areas of crowded living conditions and in rural areas of high population density and poor sanitation and hygiene. Person-to-person contact, filth, flies, food handlers, and contaminated water supplies are involved in the

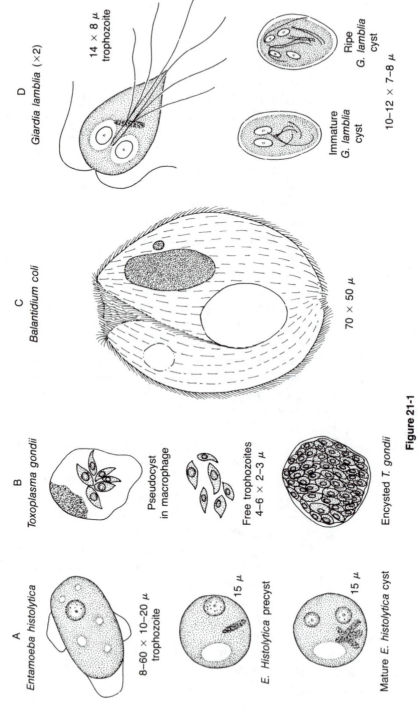

Figure 21-1
Pathogenic protozoa associated with foods.

spread of disease, and all must be considered in control. Treatment makes use of a number of antiamebic drugs, including chlortetracycline and oxytetracycline, diodoquin (64% iodine), chloroquine, emetine hydrochloride, and arsenicals.

Toxoplasmosis

Since its discovery in a small North African rodent in 1908, *Toxoplasma gondii* has been found in a wide range of mammals, in some birds, (Figure 21-2) and, possibly, in reptiles. It is endemic among wild cats, rodents, mice and birds, all of which seem to be unaffected carriers. It may also be spread in raw meats, and is associated with infested pork, lamb, poultry, and beef.

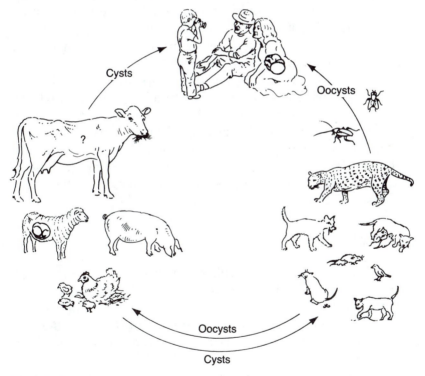

Figure 21-2
Transmission of *Toxoplasma gondii*. (Schwartzberg and Remington, 1975. American Journal of Diseases of Children, 129, 777–779. Copyright 1975, American Medical Association.)

The disease appears to be associated with domesticated cats that have come into contact with feces of homeless cats or those that have eaten wild animals. Oocysts develop in the feces after defecation, and may be transmitted (1) to humans via the mouth after cats or cat litter have been handled or (2) to children playing in sandboxes in which cats have buried their feces, or (3) by cockroaches that contaminate raw meats.

The cysts survive enzymatic action by pepsin or trypsin and germinate in the digestive tract, forming the active trophozoite, which propagates within various cells, penetrates the intestinal wall, and is carried throughout the body. Manifestations include retinal disturbances, encephalitis, and lymphadenopathy (swelling and tenderness of the lymph glands). The most severe forms of the disease, hydrocephalus and blindness, occur in infants and result from intrauterine infection. In the United States infection rates in newborns vary in different geographic areas between 1:1000 and 1:20,000.

While only a small percentage of meat samples may contain *Toxoplasma gondii,* the frequency with which these meats are consumed represents an appreciable opportunity for infection. The encysted parasites in muscle or brain of meat animals are scattered and too small (i.e., < 100 μm) to be seen with the naked eye (Figure 21-1 B). Serological or skin tests for inspection purposes are impractical. The organisms are readily destroyed by freezing or by heating to 60°C. Proper washing of hands and sufficient cooking of foods minimizes the hazard of toxoplasmosis.

Balantidiasis

Balantidium coli is the only ciliated protozoan that parasitizes humans; it is also the largest (50 to 100 μ) of the human protozoan parasites. Only the cyst is infective. It is a tissue invader and produces intestinal pathology similar to that of *Entamoeba histolytica,* although it produces no extra-intestinal lesions (Figure 21-1 C).

Balantidium coli is a common parasite of pigs, which are considered to be a possible source of human infections—although their role in human balantidiasis is controversial. It is likely, however, that an outbreak in the Truk District of Micronesia in 1971, involving 110 cases, occurred because ground or surface water supplies were contaminated with pig feces (Center for Disease Control, 1972).

Human balantidiasis can be transferred from person to person. In the United States it is found occasionally in patients of mental institutions, where poor personal hygiene and the habit of coprophagy among patients implicates person-to-person transmission (Young, 1939).

Giardiasis

The flagellated protozoan *Giardia lamblia* is worldwide in distribution, incidences totaling 400 million. In the United States carriers are estimated to number between 1.5 to 20% of the population. It exists as a trophozoite and as a cyst (Figure 21-1 D). Onset of illness occurs one to three weeks after ingestion. Symptoms include chronic diarrhea, weight loss, abdominal distension, flatulence, greasy stools, and sometimes anorexia, cramps, nausea, and vomiting (Schultz, 1975; Walzer et al., 1971). The illness persists from one to 30 weeks. Asymptomatic carriers are common. The protozoa are restricted to the lumen of the digestive tract, and do not invade the visceral area. Giardiasis is most often associated with contaminated water; foods contaminated with human wastes are presumed to transmit the protozoan to humans.

Infestations

Helminthic infestations are illnesses caused by cestodes (parasitic, highly segmented flatworms), trematodes (parasitic, unsegmented flatworms), and nematodes (long, cylindrical, unsegmented worms). Many infestations are associated with foods characteristic of specific geographic areas, common there and nowhere else. Some are associated with the life style and the culture of the peoples. Some infestations have the potential to be widespread and are; others either affect fewer persons or occur less frequently.

All foodborne infestations enter the human through food or water. The infesting agents are divided into several groups according to life patterns:

1. those transmissible from human to human via the anal-oral route;

2. those associated with warm-blooded meat animals and fish and transferred during the handling of the meat or by ingestion of raw or undercooked foods; and

3. those associated with water and marine life as part of the diet of coastal dwellers.

Once established, helminths may remain in a human for years, shedding eggs continually in intestinal waste. Adequate toilet hygiene, confinement and treatment of human wastes, washing hands after handling potentially affected meats, and proper cooking are recommended measures for preventing the several cestode, trematode, and nematode infestations.

Cestodes

Cestodes are flatworms of the class Cestoda of the phylum Platyhelminthes. *Taenia saginata,* the tapeworm of beef, *T. solium,* the tapeworm of swine, and *Diphyllobothrium latum,* the broad tapeworm of fish, live in the intestinal tracts as parasites of these respective animals. *Echinococcus granulosus,* the hydatid worm, preferentially invades the heart and lungs. In cattle and swine the disease caused by the larval stages of the tapeworm is known as cystercercosis. Symptoms of both tapeworm infestation and cystercercosis are often absent or are quite mild and indefinite. In the early nineteenth century "cleansed" eggs of beef tapeworm were sold in apothecaries as a remedy for obesity.

Taeniasis

Beef and swine tapeworms are essentially alike. Both enter humans when beef or pork infested with the cysts of bladderworms is eaten raw or cooked rare. Infested beef is often known as "measly beef." The bladderworms of beef are known as *Cysticercus bovis,* those of swine *C. cellulosae*—names assigned before they were known to be the intermediate stages of beef and pork tapeworms, respectively. Bladderworms are grayish white sacs about 5 to 10 mm in size. Each sac contains an invaginated head or scolex. In the human intestinal tract the scolex evaginates and attaches itself to the wall of the small intestine by four cup-chaped suckers. The scolex of *C. cellulosae* also has two rows of large and small hooks (Faust et al., 1970).

Segments, or proglottids, develop from the neck of the scolex, each new segment arising immediately behind the neck. The body, or strobila, consists of one to 2000 segments. The strobila of *Taenia saginata* grows to a length of 5 to 10 meters and worms 20 meters in length have been recorded. The strobila of *T. solium* grows to a length of 2 to 7 meters. The proglottids mature in about 10 weeks. The terminal proglottids contain both male and female organs. The gravid uterus of *T. saginata* comprises 15 to 30 lateral branches containing 8000 to 12,000 eggs. The uteri become separated from the strobila and are usually excreted from the bowel with the feces, or they may work their way through the anus. Once voided, motile segments rupture and release eggs (Figure 21-3). The uterus of *T. solium* has 5 to 12 branches. The eggs of the two species are microscopically indistinguishable.

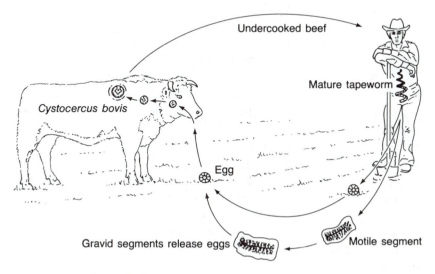

Figure 21-3
Life cycle of *Taenia saginata* in humans and cattle.

Distribution of both tapeworms is worldwide. In some localities they may affect 60% of the population. In Abyssinia, where raw beef is considered a luxury and where possession of a tapeworm is said to aid digestion, as much as 90% of the population is affected. Beef tapeworm once was common in the United States, but the incidence has waned through modern sanitary practices, the confinement of human wastes, and sound cookery.

The eggs of *Taenia saginata* remain infective for 11 days in liquid manure and for half a year in pasture. They cannot develop without passing through alternate hosts. After ingestion, the eggs hatch in the duodenum, penetrate into the lymph system or the bloodstream, and are carried to other organs. In cattle the preferred sites are the internal and external masseter muscles, the heart, muscles of the diaphragm, neck, and shoulders, the intercostal muscles, and the adductors of the hind legs. In these sites they grow into cysticerci, or bladderworms. In cattle, the typical cysticerci are formed in two to four months after ingestion of the egg.

The scolex of the cysticerci is easily recognized under the microscope as a spot in the wall of the cyst. Usually the cyst is surrounded by a cystic membrane formed as a defensive action of the surrounding tissue. The cyst of *Cysticercus bovis* often undergoes degenerative changes such as caseation and calcification.

Symptoms during most human infestations are vague, but occasionally the disease is accompanied by alimentary upset, lethargy, weight loss, and occasionally eosinophilia. There may be a slight or temporary diarrhea with or without abdominal pains and moderate weight loss. Chronic diarrhea may lead to complete exhaustion and the illness may be fatal. Occasionally proglottids may be located in the appendiceal lumen and cause appendicitis. In rare instances the mass of tangled worms causes an intestinal obstruction. *Taenia saginata*—and probably also *T. solium*—causes damage to humans by mechanically irritating the mucosa and by absorbing metabolites by the host who may develop liver damage. *T. solium* may localize in the brain to produce symptoms like those of epilepsy and brain tumor.

Atabrine and mecaprine (acridine drugs) are among the drugs used successfully in treatment. Human taeniasis may also be treated with the extract of the male fern, hexylresorcinol, and tetrachlorethylene.

Economic loss results from the condemnation of infected carcasses. Federal regulations require that carcasses be condemned if most of the cut surfaces reveal two or more cysts within a 100 cm² area. Lightly infected meats are passed after removal of the infected part and storage of the carcasses for 10 days at –10°C or heating to 60°C. In the United States less than 0.1% of cattle and calves are infested, and swine infestation is exceedingly rare.

Hymenolepsis nana, the dwarf tapeworm, occurs in many of the world's warm climates and often is recovered from young children in the Southeastern United States. Transmission of this small worm, only 1 mm in breadth by 25 to 40 mm in length, is usually by the anal-oral route.

Diphyllobothriosis

Diphyllobothrium latum, the broad or fish tapeworm, is a relatively common parasite of humans, dogs, bears and other fish-eating mammals in the fresh-water regions throughout northern and central Europe and Asia and of north central United States, Canada, and Alaska. The endemic areas usually have a cool, moist climate. In the United States the infection is well established in the lake regions of Michigan, Wisconsin, and Minnesota. Other endemic foci are in southern South America, Africa, and Australia.

The life cycle of the fish tapeworm is characterized by two intermediate hosts and takes 8 to 15 weeks to complete. The eggs are liberated from the gravid proglottids through a special uterine pore. A single mature worm at the height of productivity may produce up to 15,000 eggs per gram of human feces. The gravid proglottids frequently break off from the parent worm in chains and are passed in the feces. The eggs are unembryonated at the time they are evacuated in the host's stools.

The eggs embryonate in 10 to 15 days and after reaching cool fresh water (15 to 25°C); a free-swimming ciliated **coracidium** hatches from the egg. If its development is to proceed it must be ingested within 12 hours by a "water flea" of the appropriate species *(Cyclops* or *Diaptomus).* The coracidium loses its cilia while in the cyclops's intestine, burrows into the body cavity of the crustacean, and transforms into a *proceroid* larva in 10 to 12 days. If the infected copepod is consumed by a plankton-feeding fish, the larva migrates into the flesh, viscera, or connective tissue and, in from one to four weeks, transforms into a **plerocercoid.** The plerocercoid larva can be transferred from one fish to a larger, predatory fish. Humans become infected by eating raw or insufficiently cooked, infected fish flesh. The fishes most frequently responsible for human infection in the United States are the barred, wall-eyed, and sand pike, turbot, and carp.

The plerocercoid larva attaches to the intestinal wall and grows at a rate of about 30 proglottids per day, becoming mature in about one month. The fully developed strobila of *Diphyllobothrium latum* is ivory colored, 3 to 10 meters or more long, with as many as 3000 to 4000 proglottids. The individual segments are two to five times as broad as long; hence the name of this tapeworm. It attaches to the small intestine by means of an elliptical (1 × 2 mm) scolex with 2 suctorial grooves. The incubation period is from three to six weeks. The adult worm may live for as long as 20 years in the host. Humans are universally susceptible, and develop no apparent resistance after infection.

Diphyllobothrium latum may produce no symptoms, but in approximately 50% of the infestations diarrhea, heartburn, sense of fullness, hunger pains, loss of appetite (anorexia), nausea, and vomiting occur. A few patients develop severe macrocytic anemia. Toxic symptoms may be associated with massive infections. Diagnosis is made by recovery of eggs from the stools. Treatment consists of the administration of Atabrin (quinacrine hydrochloride).

Control is based on sanitary disposal of human excreta so that viable eggs do not reach fresh water and by cooking of all fish obtained from endemic areas at a minimum temperature of 50 to 55°C for at least 10 min. It has been reported that the eggs are killed by a temperature of –10°C sustained for 48 hours. However, fish caught in endemic areas and shipped on ice to distant cities have infected consumers. Inspection of fish does not appear to be practical, nor does cold smoking or light salting eliminate this parasite.

Hydatid Disease

Hydatid disease is caused by *Echinococcus granulosus,* a cosmopolitan tapeworm. The adult worm, which inhabits the intestines of dogs, wolves, and some other wild canine species, is very small (0.25 to 0.5 cm in length) and consists of three or four proglottids, the first of which is immature while the last is gravid with several hundreds of eggs. A large number of adults may occur in the intestine of an infected dog. The scolex is provided with 4 suckers and with a double crown of about 30 hooks. The adult worm is formed 5 to 7 weeks after ingestion by the dog.

The eggs, 30 to 40 μ in diameter, hatch in the intestine of sheep, cattle, and other intermediate hosts, penetrate the intestinal wall, and usually are carried by the blood to the liver and lungs, or sometimes to any other organ. Although the natural intermediate hosts are ruminants, a wide variety of animals, including humans, are susceptible. The liver and lungs are most often infected. Human disease is most frequent in children, through contact with dog feces.

Echinococcosis is endemic in Canada and Alaska, with an incidence of 10% to 45% among the Canadian Indians and Eskimos. Human infestation is also common in sheep-raising areas, especially in Argentina and Uruguay, but is rare in the United States. Control and prevention are achieved by adequate meat inspection, education of the population (especially farmers and butchers), and, in endemic regions, giving dogs anthelminthic treatments at least twice a year.

Trematodes

Flukes, or trematodes (members of the class Trematoda), are nonsegmented flatworms that have a mouth and an oral sucker located near the anterior end of their body; some of them parasitize humans. They also have a more complicated life cycle than do cestodes. After the fertilized egg is discharged, it hatches into a free-swiming, ciliated larva or *miracidium* that must find and penetrate the tissues of an appropriate snail or other mollusc and transform into a *sporocyst,* which produces daughter *rediae* or cercariae. When mature, these second-generation organisms escape from the mollusc as free-living, tailed *cercariae.* Eventually the cercaria sheds its tail, becoming a *metacercaria.* It then finds an aquatic plant or penetrates the tissues of an aquatic animal, and encysts. The vertebrate host, human included, becomes infected upon ingesting the plant or animal tissue containing the encysted stage, and the metacercariae develop into adult trematodes.

Fasciolopsiasis

The giant intestinal fluke *Fasciolopsis buski* produces inflammation, hemorrhages, abscesses, and eosinophilia in the human intestine. The adult fluke ranges from 8 to 20 mm wide, from 20 to 75 mm long, and 0.5 to 3.0 mm thick. At the time of evacuation in the feces of the host, the eggs are large (i.e., 80 to 130 μ), oval, and unembryonated. After three to seven weeks, in warm (27 to 32°C) fresh water they embryonate; an active miracidium emerges, swimming about until it contacts a planorbid snail, and, after transferring into a sporocyst, produces two redial generations. In the Orient, where the hulls or skins of raw water buffalo nuts, water chestnuts, water caltrops and water bamboo are peeled by placing the plants between the worker's teeth, the ingested metacercarial larvae excyst in the duodenum and in about three months, develop into adult worms attached to the duodenal or jejunal wall. At sites of attachment the worms may produce deep inflamed ulceration of the mucosa. Diarrhea becomes persistent and the absorption of toxic parasitic metabolites produces edema of the cheeks and around the eyes.

Heterophyid Flukes

Infection by two minute (1 to 2 mm by 0.3 to 0.5 mm) pyriform worms results from eating raw, salted, or dried fresh- or brackish-water fish. Infections of humans and other fish-eating mammals by *Heterophyes heterophyes* is often associated with raw, salted, or dried mullet. The same hosts become infected by *Metagonimus yokogawai*, which infests Oriental fresh-water trout and other fresh-water fishes. Mild inflammatory reactions develop at sites where the worms are attached and a mucous diarrhea is usually evident.

Fascioliasis

Fasciola hepatica, or the sheep liver fluke, is a relatively large (i.e., 30 mm by 13 mm) flattened or leaflike fluke; the anterior end bears a characteristic cephalic cone, whereas the midsection is fleshy and the thin posterior end is broadly rounded. Its eggs are operculate and almost identical in size and appearance to those of *Fasciolopsis buski.* Immature when laid, the egg embryonates after 9 to 15 days in water at 22 to 25°C; the miracidium escapes and within a few hours sheds its ciliated coat when the sporocyst invades one of various snails belonging to the family Lymnaeidae (Figure 21-4). Several weeks after penetrating the snail, mature cercariae leave their lymneid host,

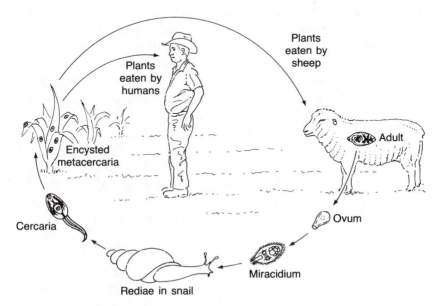

Figure 21-4
Life cycle of *Fasciola hepatica* in sheep and humans.

swim to moist vegetation, shed their tails, and encyst as metacercariae. Sheep and other mammals that consume such aquatic vegetation contract the infection. Human infection usually results from eating raw watercress to which metacercariae are attached. *Fascioliasis* is rare in the United States but common in several Latin American and Mediterranean countries. Lesions are produced by the larval flukes during their migration through the liver parenchyma. As the flukes are developing into adults they bring about fibrosis, inflammation, and cystic enlargement of the bile ducts. Other pathologic changes include jaundice, abscesses, fibrotic lesions, leucocytosis, eosinophilia, persistent diarrhea, and anemia. A daily dosage of 40 mg of emetine hydrochloride until 5 mg per kilo of body weight has been given provides favorable results (Arenas et al., 1948).

Clonorchiasis

The Chinese liver fluke *Clonorchis sinensis* infests the food supply in Asiatic countries, principally Korea and Japan. In countries such as the United States, where fish is seldom eaten raw or poorly cooked, *Clonorchis* does not present a problem. More than 40 species of freshwater fish can serve as secondary hosts for *Clonorchis*. Some of these, the cyprinoids, are eaten in the Orient. If improperly cooked fish is eaten by humans or other mammals, the digestive enzymes break down the cyst wall. Once the trematode has been set free, it migrates through the mucosa and invades the biliary tract, localizing in the bile ducts, where the adult begins shedding as many as 50,000 eggs a day. The eggs progress down the bile duct and are excreted in the feces. At this point, there may be reinfestation and difficulties arise as a result of obstruction of the bile duct, blockage of the tubules in the liver, or cirrhosis (Jacobs, 1962). Heating the fish to 50°C for 15 minutes will kill the encysted metacercariae; drying, salting, marinating, or refrigeration will not.

Paragonimiasis

The Oriental lung fluke *Paragonimus westermani* is endemic in the Far East, where it infects humans and several other crab-eating mammals. The life cycle of this medium-sized ovoidal trematode (8 to 12 mm by 4 to 6 mm), is similar to that of other flukes: the eggs embryonate in water within several weeks and hatch; the free-swimming miracidia enter snails, transform into sporocysts, produce rediae, which then produce cercariae. The cercariae, however, invade and encyst in the musculature of crabs and crayfish. When ingested by humans *Paragonimus westermani* excysts in the digestive tract, migrates through the intestinal wall, and ultimately reaches the bronchioles of the lung. In reaction the host tissue develops a thick fibrous capsule that surrounds the worm. As the worm continues to metabolize and discharge eggs, this irritation results in coughing, pain, epithelial proliferation, inflammation, bloody sputum, eosimophilia, hypertension, and cirrhosis.

Nematodes

The true roundworms, or nematodes (members of the class Nematoda), are nonsegmented, cylindrical, elongate, bilaterally symmetrical, and usually *dioecious* animals. Those inhabiting the human digestive tract range in length from 1.5 mm to 3.0 mm if male trichina and pinworms, to 30 cm or more if the female large intestinal roundworm.

Gongylonemiasis

Gongylonema spp. are found throughout the world, especially in tropical or subtropical areas. Typical hosts are ruminants and other mammals, poultry, and occasionally humans. The vectors of this parasite are roaches and dung beetles that ingest eggs excreted with feces of the vertebrate host. In humans the irritation caused by the nematode is primarily restricted to the area where the worm resides. Although the parasite has been reported to be harmless, it has been known to cause sensitization, illness, and death.

Because the parasite has an immunological action in the host, skin tests are useful in detecting its presence. Immunological response continues even after the host is freed of the parasite. Examinations of meat animals in packing plants in Poland reveal that one-third or more of cattle and sheep and a smaller fraction of the hogs examined harbored the parasite.

Angiostrongyliasis

The rat lungworm *Angiostrongylus cantonensis* is associated with cases of parasitic meningocephalitis, eosinophilic meningitis, lungworm encephalitis, or angiostrongyliasis. The lungworm was first reported as a parasite of humans by Nomura and Lin in 1944 in Formosa. Various species of rats (genus *Rattus*) are the only known natural hosts of adult *A. cantonensis*. Rats from Tahiti, Formosa, Australia, Guam, Ponape, the Truk Islands, the Marshall Islands, and China (Rosen et al., 1962) are infected, as is 12% of the rats in Hawaii (Ash, 1962). Molluscs serve as intermediate host to the nematode; the rat becomes infected by ingesting infected snails and slugs.

The death of the infected rat is believed to result from hemorrhages (caused by young adult worms as they penetrate the small vessels on the surface of the brain) and blockage of the normal blood flow through the pulmonary artery and heart. Another complication is necrosis of lung tissue if death is sufficiently delayed (Jacobs, 1962).

The evidence strongly suggests that this nematode is the major etiological agent of human eosinophilic meningoencephalitis in the Pacific area. In many areas where human infection occurs, sources identified were snails and slugs, fresh-water prawns, and fresh vegetables with infected slugs. In many Pacific areas, accidental ingestion of larvae through infected food or water is quite likely. The nematode infection in humans is characterized by stiffness, progressive loss of memory, high percentage of eosinophils in the white blood cell count, and high numbers of leukocytes in cerebrospinal fluid. Epidemiologic data indicates the infection is painful but seldom lethal (Cheng, 1964).

Trichinosis

The cosmopolitan nematode *Trichinella spiralis* is most often a human parasite where raw or inadequately cooked pork is consumed. The worm was first observed encysted in human musculature by Peacock in London in 1828, but not until Leidy of Philadelphia obtained it from hog flesh in 1846 was it suspected that trichinosis is caused by the ingestion of infected pork.

Trichinella spiralis is too small to be seen in tissue without magnification: diameters of the adult male and female are about 40 to 60 μ and 60 to 80 μ, respectively; lengths are 1.4 to 1.6 mm and 3.0 to 3.8 mm, respectively. Consequently, infection is not detected by visual inspection of animals. Sources of human infection are pork, bear, and walrus meat. In the United States pork prepared as sausage accounts for almost three-fourths of all cases for which a source is identified. Ground beef intentionally or inadvertently adulterated with pork is another important source of infection.

At one time, trichinosis in humans was traced to feeding raw garbage to hogs. The practice is now illegal, and all garbage to be used in animal feeds must be cooked. This precautionary step has reduced the number of infected animals per 1000 from 110 to 1950 to 5.1 between 1966 and 1970 (Zimmermann and Zinter, 1971). All carnivores, including cats, dogs, and rodents may be infested—in fact, when only one hog in a herd became infested, it was assumed that the animal had eaten a rodent.

To complete the life cycle of *Trichinella spiralis* (Figure 21-5), raw or undercooked flesh containing encysted larvae must be consumed by a host. The larvae, released from the ingested muscle by the action of digestive juices, undergo four molts in rapid succession and mature into adults. They localize and mate in the duodenum and jejunum. After copulation the male dies, but the gravid female burrows into the *lamina propria* of the intestinal villi and after four to seven days begins depositing, in the mucosal or lymphatic spaces, large numbers of larvae daily ($>$ 1000) for about six weeks. Some larvae may escape into the intestinal lumen, but most are carried into the bloodstream and reach the skeletal musculature where they encyst and remain throughout the life of the host. The principal locations of trichinellae in the musculature are the intercostal muscles and pillars of the diaphragm where they are found in greatest numbers. Next in order of frequency are tongue, masticatory, and neck (laryngeal) muscles, and muscles of the shoulders, loins, abdomen, and groins.

Since 1950 from 100 to 400 cases of trichinosis have been reported annually in the United States (see Figure 21-6), but these are gross underestimates of the true incidence. In 1944

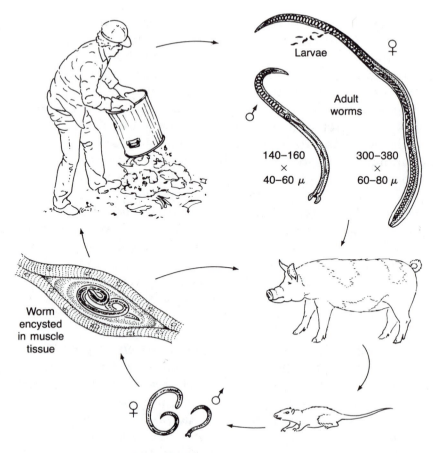

Figure 21-5
Life cycle of *Trichinella spiralis.*

Wright et al. (1944) reported an average incidence of trichinosis of 16% of the entire population, as based on surveys of 5313 persons coming to autopsy from hospitals and coroner's services in 37 states. However, the present adult infection rate in the United States has been estimated to be about 4% (Most, 1965; Zimmermann et al., 1973).

Incidence varies widely among areas of the United States. Surveys show the disease to be high in Eastern Europe and moderately high in Mexico, Chile, and other Latin American states. Epidemics in Sweden, Greenland, and Syria have also

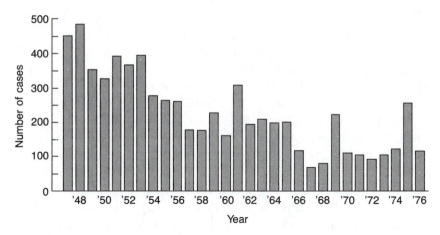

Figure 21-6
Reported cases of trichinosis by year, United States, 1947–1976. (Center for Disease Control, 1977.)

been reported. However, trichinosis is practically unknown in people of the Mohammedan and Jewish religions.

Trichinosis is considered to be a febrile disease, producing gastrointestinal symptoms, periorbital edema, myalgia, petechial hemorrhages, and eosinophilia. Early laboratory diagnosis is extremely difficult and is rarely made. During the first few days after infection, adult worms may be found occasionally by sedimentation of the feces. After the sixth day and until the end of migration, larvae may be found in the blood. They may also be recovered from the spinal fluid. Biopsy for encysted forms is most practically done upon the deltoid, biceps, or gastrocnemius muscles. Compression and digestion methods are used; care must be exercised so as not to be misled by old infections.

Serological and allied tests are not sufficiently specific to be more than an adjunct to clinical diagnosis. The skin test becomes positive earliest but is of no value before the twenty-first day; very light and very heavy infections may not be detected by this method. False positive tests occasionally occur. After the fourth week precipitation and complement-fixation tests become useful. The specificity of precipitation is about the same as that of the skin test. However, the increased accuracy of complement fixation seldom justifies the complexity of the procedure, and the test is seldom used. Postmortem examination methods include digestion, compression, and sectioning. Small bits of diaphragm give the highest percentage of positives, but other striated muscles should also be examined.

The Food Safety and Quality Service of the U.S. Department of Agriculture does not require that raw pork products be inspected for trichina larvae. The federal inspection stamp indicates that the products are processed in accordance with USDA specifications and that "ready-to-eat" pork is heated to an internal temperature of at least 58°C, which is sufficient to kill trichina larvae. Federal inspection regulations also require that all pork being used in uncooked or uncured pork products that are likely to be eaten uncooked by the consumer be continuously subjected to a temperature not higher than one of the following:

Temperature	Duration (days)	
	Group 1	Group 2
−15°C (5°F)	20	30
−23.3°C (−10°F)	10	20
−29°C (−20°F)	6	12

Group 1: pork in separate pieces not to exceed 6 inches in thickness; layers of pork not to exceed 6 inches in depth; or frozen block not to exceed 6 inches in thickness.

Group 2: pork in pieces or layers, or within containers the thickness of which exceeds 6 inches but not 27 inches.

Freezing pork cuts thinner than 6 inches at −15°C for at least 20 days renders trichinae unviable. The pork must be arranged in the freezer to permit free air circulation, to insure prompt reduction of meat temperature as specified in the table. Trichinae survive for 30 days in cured, salted, and smoked meats. The following precautions should be taken to avoid trichinosis:

1. Cook all pork cuts to "well-done."

2. Refuse to eat rare hanburger sandwiches (often hamburger has some pork ground with it).

3. Refrain from eating so-called tenderized hams until they have been cooked to an internal temperature of at least 58.3°C.

Gamma irradiation of pork as a method of control indicates that sexual sterilization of the developing adult worms occurs when trichinous pork is exposed to 10,000 to 12,000 rads of cobalt-60. A dose of 25,000 rads is sufficient to destroy larvae of *Trichinella spiralis* in uncooked pork.

Trichuriasis

Trichuris trichiura, the human whipworm, is much larger than *Trichinella spiralis.* The thin adult male averages about 35 mm in length; the more fleshy female is about 5 to 10 mm longer. While the anterior portions of both sexes are capillary in size and appearance, the posterior end of the male is coiled, that of the female is a straight fleshy club shape. The worms attach by boring into the cecal mucosa; in heavily infected individuals there may be prolapse of the rectum, exposing the wiggling nematodes to view. Trichuriasis is a cosmopolitan digestive-tract nematode in the Southern United States and in moist tropical and semitropical countries throughout the world. Transmission is most common among children who may unwittingly swallow contaminated food or egg-infested soil adhering to toys and playthings. However, adult humans with poor sanitary habits may also become infected. Whipworm infection can be treated orally with thiabendazole (Franz et al., 1965) or with enemas of 0.2% hexylresorcinol.

Capillariasis

This rare but severe intestinal disease is caused by *Capillaria philippinensis,* a small relative of the whipworm. The nematode is a tissue parasite in the liver of domestic and wild animals. When infected livers are consumed, the eggs are liberated from the tissue during digestion, are passed in the feces, and embryonate in soil. The occurrence of capillariasis among humans in certain areas of the Philippines has been associated with the consumption of raw animal foods (Chitwood et al., 1968).

Enterobiasis

Enterobius vermicularis, a small intestinal roundworm known as the pinworm, is restricted to humans, with no alternate or intermediate host. The adult male worm has a diameter of 0.1 to 0.2 mm and is 2 to 5 mm long; the female is considerably larger, having typical dimensions of 0.3 to 0.5 mm and 8 to 12 mm, respectively. Females discharge about 10,000 eggs. The eggs are ovoidal but flattened on one side, measure 20 to 30 μ by 50 to 60 μ, and within a few hours contain a fully developed infective-stage larva. Eggs are readily transferred when contaminated skin is scratched and the fingers are then placed in the mouth (anal-oral transfer) or on foods that, in turn, are eaten directly. Worms are also transferred by common use of unclean toilets, beds, and bedding, bathing too infrequently, and prolonged use of soiled or unwashed underclothing. Eggs pass from person to person by anal-oral transfer and hatch in the intestine; the larvae migrate to the cecum where they mature within two to four weeks. It has been estimated that 10% of the population of the United States, primarily children, are affected. *Pruritus ani,* or rectal itching, is the most common symptom—especially among children. Pinworm infestation can be checked if strict attention is paid to adequate cleansing of hands after each pubic contact.

Ascariasis

The eggs of *Ascaris lumbricoides* enter the human body with raw vegetables and other foods prepared for consumption without further heating. The ingested eggs hatch in the small intestine, and the larvae penetrate the wall and migrate to the liver and lungs, where they develop in one to two weeks and, after molting, they return to the intestine through the bloodstream. Larvae may also be found in the body cavity. Larvae are coughed from the lungs into the mouth, are swallowed, and mate in the intestinal tract. The sexually mature males measure 2 to 4 mm in greatest diameter and 17 to 30 cm in length, and the females measure 3 to 6 mm in diameter and are 20 to 40 cm in length. For a period of 6 to 12 months gravid females produce 200,000 eggs daily. Like the eggs of other infesting agents, ascarids are readily destroyed by heat. They survive putrefaction and can withstand desiccation and cold weather for a long time.

In areas known to be infested, thorough washing of all vegetables with mild chlorine solution is recommended. Piperazine citrate is an inexpensive and effective chemotherapeutic agent.

Anisakiasis

Anisakiasis is the general term for nematodal infestations by larvae of the genera *Anisakis* and *Phocanema,* and possibly others. The illness is characterized by abdominal distress, irritation, inflammation, and sometimes ulceration. Victims may cough up larvae several centimeters in length.

Several hundred cases have been recognized in Holland and in Japan since the disease was first identified in 1955 (Van Thiel, 1966). The illness follows the consumption of raw or mildly heated or lightly marinated infected fish foods such as the green herring of Holland, the Japanese sushi, and the Latin American ceviche. Although only a few scattered cases have been reported in the United States (Jackson, 1975), their numbers may possibly increase as Americans consume more exotic foreign foods in restaurants, and as they become more interested in ethnic and "natural" foods.

The larvae do not survive freezing, full cooking, heavy salting, or marination with strong acid. In Holland the disease became more common once it was no longer the practice to gut herring at sea; it is now mandatory that fish be frozen before use as green herring.

References

American Chemical Society. 1966. Symposium on microbial toxins. In: *Biochemistry of Some Foodborne Microbial Toxins,* R. I. Mateles and G. N. Wogan, eds., MIT Press, Cambridge, Mass.

Arenas, R., A. Espinosa, E. Padron, and R. M. Andrew. 1948. Rev. Kuba Med. Trop. and Parasitol. *4,* 92.

Ash, L. R. 1962. J. Parasitol. *48,* 66.

Ayres, J. C., A. A. Kraft, H. E. Snyder, and H. W. Walker, eds. 1962. *Chemical and Biological Hazards in Food.* Iowa State University Press, Ames.

Baratta, R. O., and P. A. Tanner. 1970. J. Fla. Med. Assoc. *57,* 39.

Center for Disease Control. 1972. Morbidity and Mortality Weekly Reports. *21,* 59.

Center for Disease Control. 1973. Morbidity and Mortality Weekly Reports. *22,* 69, 78.

Center for Disease Control. 1974. Morbidity and Mortality Weekly Reports. *23,* 201.

Cheng, T. C. 1964. The Biology of Animal Parasites. Saunders Co.

Chitwood, M. B., C. Valesquez, and N. G. Salazar. 1968. J. Parasitol. *54,* 368.

Clapp, R. C., F. H. Bissett, R. A. Coburn, and L. Long, Jr. 1966. Phytochem. *5,* 1323.

Committee on Food Protection. 1973. *Toxicants Occurring Naturally in Foods.* Food and Nutr. Bd., Natl. Res. Coun., Natl. Acad. Sci. Washington, D.C.

Faust, E. C., P. F. Russell, and R. C. Jung. 1970. *Clinical Parasitology.* 8th ed. Lea and Febiger, Philadelphia, Pa.

Franz, K. H., W. J. Schneider, and M. H. Pohlman. 1965. Amer. J. Trop. Med. Hyg. *14,* 383.

Geiger, E. 1944. Food Res. *9,* 293.

Gentile, J. H. 1971. In: *Microbiology of Toxins. Vol. 7. Algal and Fungal Toxins.* S. Kadis, A. Ciegler, and S. J. Ajl, eds. Academic Press, New York.

Hall, R. L. 1973. In: *Toxicants Occurring Naturally in Foods.* Committee on Food Protection, Food and Nutr. Bd., Natl. Res. Coun., Natl. Acad. Sci., Washington, D.C.

Halstead, B. W. 1967. Vertebrates. In: *Poisonous and Venomous Marine Animals of the World.* U.S. Gov't. Printing Office, Washington, D.C.

Jackson, G. J. 1975. J. Milk Food Technol. *38,* 769.

Jacobs, L. 1962. In: *Chemical and Biological Hazards in Food.* J. C. Ayres, A. A. Kraft, H. E. Snyder, and H. W. Walker, eds. Iowa State University Press, Ames.

Jeffrey, H. C., and R. M. Leach. 1966. *Atlas of Medical Helmintolology and Protozoology.* E. & S. Livingstone Ltd., Edinburgh, Scotland.

Kingsbury, J. M. 1964. *Poisonous Plants of the U.S. and Canada.* Prentice Hall, Inc., Englewood Cliffs, N.J.

Lovenberg, W. 1973. In: *Toxicants Occurring Naturally in Foods.* Committee on Food Protection, Food and Nutr. Bd., Natl. Res. Coun., Natl. Acad. Sci., Washington, D.C.

Merson, H. M., W. B. Baine, J. Gangarosa, and R. C. Swanson. 1974. JAMA *228,* 1268.

Miller, J. A. 1973. In: *Toxicants Occurring Naturally in Foods.* Committee on Food Protection, Food and Nutr. Bd., Natl. Res. Coun., Natl. Acad. Sci., Washington, D.C.

Most, H. 1965. J. Amer. Med. Assoc. *193,* 871.

Patwardhan, V. N., and J. W. White, Jr. 1973. In: *Toxicants Occurring Naturally in Foods.* Committee on Food Protection, Food and Nutr. Bd., Natl. Res. Coun., Natl. Acad. Sci., Washington, D.C.

Rosen, L., R. Chappel, G. L. Wallace, and P. O. Weinstein. 1962. J. Amer. Med. Assoc. *179,* 620.

Schantz, E. J. 1971. In: *Microbial Toxins. Vol. 7. Algal and Fungal Toxins.* S. Kadis, A. Ciegler, and S. J. Ajl, eds. Academic Press, New York, N.Y.

Schultz, M. G. 1975. J. Amer. Med. Assoc. *233,* 1383.

Shaffer, J. G., W. H. Shlaes, and R. A. Radke. 1965. *Amebiasis: A Biomedical Problem.* C. C. Thomas, Springfield, Ill.

Shilo, M. 1971. In: *Microbial Toxins. Vol. 7. Algal and Fungal Toxins.* S. Kadis, A. Ciegler, and S. J. Ajl, eds. Academic Press, New York.

Swartzberg, J. E., and J. S. Remington. 1975. Amer. J. Dis. Child. *129,* 777.

Underwood, E. J. 1973. In: *Toxicants Occurring Naturally in Foods.* Committee on Food Protection, Food and Nutr. Bd., Natl. Res. Coun., Natl. Acad. Sci., Washington, D.C.

Van Thiel, P. H. 1966. Trop. Geograph. Med. *18,* 310.

Voigt, M. N., R. R. Eitenmiller, P. E. Koehler, and M. K. Hamdy. 1974. J. Milk Food Technol. *37,* 377.

Walzer, P. D., M. S. Wolfe, and M. G. Schultz. 1971. J. Infect. Dis. *124,* 235.

Whitaker, J. R., and R. E. Feeney. 1973. In: *Toxicants Occurring Naturally in Foods.* Committee on Food Protection, Food and Nutr. Bd., Natl. Res. Coun., Natl. Acad. Sci., Washington, D.C.

Wright, W. H., L. Jacobs, and A. C. Walton. 1944. Publ. Health Reports *59,* 669,

Young, M. D. 1939. J. Amer. Med. Assoc. *113,* 580.

Zimmermann, W. J., J. H. Steele, and I. G. Kagan. 1973. Health Service Rep. *88,* 606.

Zimmermann, W. J., and D. E. Zinter. 1971. HSMHA Health Rep. *86,* 9737.

Microbial Foodborne Infections

Foodborne infections are diseases caused by bacteria and viruses that enter the body through the mouth with food. Infections caused by filamentous fungi are unknown. Fermenting yeasts become established in the upper respiratory tract without apparent ill effects, presumably after homemade beer has been drunk. In rare instances, chronic alcoholic intoxication may result from the implantation of a fermenting yeast in the intestinal tract of an individual on a diet rich in starches.

Relatively few of the many thousands of species of fungi, bacteria, and viruses known today can infect or intoxicate humans when consumed with foods. Numbers of bacteria *per se* per unit quantity of food have no significance in determining the hazard of a food. Foods such as ground and cut meats, doughs, batters, fresh and dried fruits and vegetables, milk and natural cheeses, sauerkraut, pickles and fermented sausages may, and do, contain millions of bacteria per gram as determined by standard plate counts.

Most foodborne infecting and intoxicating agents can be classified into several major groups according to their origins, modes of transmission, and conditions for growth. Reducing the incidence of foodborne illnesses requires a knowledge of the specific sources of the pathogens, the foods generally involved, and the modes of control for each of the several groups.

Some agents of foodborne infection are associated only with specific types of food; *Francisella tularensis,* for example, is limited to rabbits and related animals of the *Rodentiae.* Other agents, such as the salmonellae and shigellae—by far the most common bacterial agents of foodborne infections—can potentially be found in a wide variety of nonsterile foods. Disease-producing members of the genera *Proteus, Arizona, Providencia,* the *Bethesda-Ballerup* group, and enteropathogenic *Escherichia coli* have nearly identical vehicles of transmission and modes of control, which also resemble those for members of the genus *Salmonella.*

Gastrointestinal Illnesses

The gastrointestinal illnesses caused by the bacteria named above and by the enteroviruses follow the **anal-oral** or the **anal-fomite-oral** pathway. The fomite can be a food that has been directly or indirectly contaminated with human or animal waste, in which some growth of the bacteria has occurred.

Methods of Limiting Gastrointestinal Diseases

In addition to the broad principles of environmental sanitation, which include insuring that waters used in food processing are potable and that rodents, insects, and birds are under control, four practices are effective in limiting the spread of gastrointestinal disease:

1. exercising meticulous personal sanitation;
2. recognizing raw food sources of infecting agents;
3. preventing contamination of foods under preparation on working surfaces on which other foods, specifically raw meats, have been prepared; and
4. practicing good environmental sanitation.

Meticulous Personal Sanitation. Supervisory personnel must recognize that employees who are new to food handling often have not yet learned practices of food hygiene, and must therefore maintain a continuous program of education, stressing essential rules: to avoid touching pubic areas of the body, to wear only clean under- and outer-clothes, to control expectoration, perspiration, and nasal exudates, and to thoroughly wash hands with soap before returning to the food handling area, after leaving for any reason. Personnel should refrain from scratching their heads, placing fingers in or around the mouth, sneezing or coughing on product, allowing hair to come into contact with foods, or using tobacco in areas where edible products or ingredients are stored, handled, or prepared.

Many employees find it difficult to conceive of the inherent danger of living things as small as bacteria—and many may never have seen a bacterium under a microscope. Without proper training, employees continue the same toilet and sanitary practices common to their home environment—which may be inadequate to protect the public health in the handling of susceptible foods.

Recognizing Raw Food Sources of Infecting Agents. The obvious raw sources of infection are the red meat animals, poultry, fish, shellfish, eggs, and milk. The stated principle applies especially to food service establishments and to industries making use of meats and eggs in dishes to be consumed without further heating. Also, consideration must be given in cases where subsequent heating of foods may be inadequate, such as the underheating of cakes during baking or the cooking of ready-to-eat dinners in a defective oven or by a hurried housewife. Nearly all milk used in the United States is pasteurized, after which it is no longer considered a raw food.

Preventing Food Contamination. As with the first principle, management must be aware of the problem and educate employees. Large commercial food-preparation establishments will have flow lines that eliminate the probability of cross contamination. In smaller establishments such as restaurants, cafeterias, mess halls, delicatessens, the kitchens of schools, churches, camps—

and homes—the same working surfaces may be used for the preparation of raw meats and then of other foods with no intervening sanitization of the surfaces. Hands, working surfaces, and cutlery should be washed after handling raw foods, and the use of clothes and aprons as towels should be discouraged.

Good Environmental Sanitation. Rodents, insects, and birds are instrumental in spreading infectious agents; food plant practices must control such agents. The grounds immediately adjacent to buildings should be free of discarded equipment and refuse of all types, to discourage rodents and insects from taking up permanent or temporary residence or feeding thereabouts. For the same reason weeds and grass surrounding the buildings should be mowed regularly. Buildings must be constructed in ways that deny entrance to these pests.

Internal sanitation involves properly storing all ingredients—for example, by using closed containers for materials dispensed in small quantities or intermittently, and by frequently gathering, confining, and removing accumulated wastes. Raw product and processing areas should be separated physically, and personnel other than supervisors should be discouraged from traveling between these areas. All pieces of equipment, stationary and mobile, with their working surfaces, undersides, flanges, handles, bracing, and supports, as well as cutlery and all receptacles must be cleansed and sanitized as frequently as experience with the specific commodity dictates. Buildings should have easily sanitized walls and floors, and overhead plumbing and other structural and functional facilities should be properly installed to prevent condensation and drip. Handwashing facilities, with an adequate supply of soap, towels and waste containers, should be so placed that employee practices can be monitored.

Enterobacteriaceae in Foods

Foods produced under microbiological control and foods in the marketplace are monitored for the presence of two groups and a genus within one, in the family Enterobacteriaceae: salmonellae, total numbers of coliform bacteria, and *fecal* coliform bac-

teria, or *Escherichia coli.* The abundance of these bacteria on raw
and processed foods is often proportionate to their abundance
in nature. It is unrealistic to expect nonsterile foods to be free
of all such organisms. A rationale can be established to consider
the origins and their modes of ingress into marketed foods and
to establish objective tolerances based on historical knowledge
of the food type. The presence or absence of members of any
one group may bear no relationship to the presence or absence
of other groups.

The widespread distribution of the salmonellae among
domestic and wild animals and in nature has been noted.
Agents of dissemination such as wind, rain, dust, insects, birds,
rodents, polluted irrigation waters, and floods convey the salmonellae to plants and to plant foods during growth in the
fields, storage, or processing. The source of contamination is
controllable only to a limited extent. The processor or the
housewife must recognize the possibility that salmonellae are
present in meats and normally must render such foods safe for
eating by cooking them or by other means of disinfection. Legal
regulations controlling the handling of processed foods are intended to preclude further reproduction of the bacteria.

Heat processing frees foods of salmonellae; if they reappear
it is through negligence by personnel or an unsuspected flaw in
processing. In either case, remedies are instituted immediately,
or as soon as the engineering problems are solved. Examples of
processed foods that once became recontaminated during processing include dried egg products, dried milk, coconut and
candy.

The very low levels of the salmonellae in foods do not generally constitute a direct health hazard. In general, salmonellae
may grow sufficiently to achieve infection if abuse of temperature occurs when the food is being prepared for consumption.
Some foods therefore may present an *indirect* health hazard
and constitute a *legal* hazard to the processor. In the marketplace the consumer purchases heat-processed foods on faith,
relying on the integrity and the ability of the processor to render those foods safe.

Coliform bacteria are those gram-negative bacteria that ferment lactose to acid and gas at 37°C. The limit for potable water

is set at 2.2 coliform bacteria per 100 ml of water, on the premise that polluting material thus diluted represents no threat to the consumer. In the microbiology of foods it becomes necessary to distinguish between *fecal* coliform bacteria and the *nonfecal* coliform bacteria.

The most commonly encountered nonfecal coliform bacterium is *Enterobacter aerogenes*, while the properties of closely related *Hafnia* and *Klebsiella*, *Citrobacter* spp. and *Aeromonas hydrophila* and some species of *Erwinia* result in their inclusion in undifferentiated counts of total coliform bacteria in foods. *Enterobacter*, *Klebsiella*, *Citrobacter*, and *Erwinia* spp. are present on plants and in water. *Citrobacter* spp. resemble *E. coli* very closely in properties and type of colony formation on eosin-methylene blue agar. *Erwinia* spp., plant pathogens, are commonly found as a saprophyte on plant surfaces.

Nonfecal coliform bacteria enter foods during processing as constituents of the normal aerial microflora, or through surface or dust contamination of packaging materials, personnel clothing, and hand transfer through handling. Members of *Enterobacter-Klebsiella* reproduce rapidly in the lower mesophilic range of 10 to 20°C, especially if growth is initiated at higher temperature during cooling. Only *Aeromonas* spp. and some strains of *Klebsiella* grow at 43.3°C and above. Most nonfecal coliform bacteria are eliminated if the fecal coliform bacteria are enumerated at temperatures between 44.5 and 45.5°C.

It is difficult to establish limits on the numbers of nonfecal coliform bacteria in or on unheated raw foods such as fresh vegetables, because their numbers fluctuate naturally. Enforceable limits in terms of numbers per unit quantity of heated foods can be established reliably at levels shown by experience to be reasonable. Excessive numbers in or on moist, heated foods signify inadequate attention to preventing recontamination and to controlling storage temperature.

Escherichia coli is an inhabitant of the intestinal tract and, by contamination, gets on the surfaces of the animal body prior to slaughter. During slaughter and dressing of the carcass, the bacterium is transferred to the cut surfaces. It also enters pecans, almonds, and walnuts when these come in contact with the soil under the trees during harvest (Meyer and Vaughn, 1969;

King et al., 1970; Marcus and Amling, 1973); meats of nuts with soft or cracked shells or open sutures are readily contaminated. Thus, the significance of *E. coli* in foods must include a knowledge of the history of the food and possibly acceptance until mechanisms are developed for specific foods which ensure its destruction.

Escherichia coli is readily destroyed by mild heat. Its presence in heated foods results from direct, personal contamination or indirect contamination once it has been established as an inert or actively growing resident on equipment, utensils, walls, scoops, floors, walls, wash tanks, and other areas. Continuing contamination indicates inefficient and ineffective plant sanitation and sanitizing practices.

Miscellaneous Free-Living Gram-Negative Bacteria

Proteus *Species and the Paracolon Bacteria*. *Proteus vulgaris,* a common resident of the human intestinal tract and the skin, is sometimes found in large numbers in moist foods. It is not a known agent of gastroenteritis. *Proteus morgani* and the closely related *Paracolobactrum* and *Arizona* spp. are implicated as causes of infant diarrhea. *Arizona* spp. ferment glucose to produce acid and gas, but fermentation of lactose is delayed for from 48 hours to as much as 14 days. It is responsible for severe and sometimes fatal enteritis. Also isolated from illnesses are species of *Providencia,* which is nearly anaerogenic, and the members of the *Bethesda-Ballerup* group, which are delayed lactose-fermenting relatives of *Citrobacter.* None of this group of pathogens, except *P. vulgaris,* is a normal resident of the human or animal intestinal tract. It is strongly suspected that foods are a source of infections.

Klebsiella pneumoniae, a common resident of the human intestinal tract, is also a common causative agent of pneumonia in humans. In recent years, it has become increasingly prominent in infant diarrheas, abscesses, and infections of wounds and of the urinary tract, particularly in hospital settings. Low numbers are fairly common on plants, often less than 1×10^3 per gram of seeds or vegetable tissue (Brown and Seidler, 1973). The ability of *K. pneumoniae* to utilize atmospheric nitrogen is not often found among microorganisms.

Pseudomonas *Species.* Most species of this genus are low-temperature mesophiles that either cannot grow at 37°C or grow poorly. *P. aeruginosa, P. maltophila,* and *P. alcaligenes* produce primary and secondary suppurative infections in humans, often after a skin injury such as a burn or after a primary infection by another bacterium has been treated with antibiotics. Each species is found in fresh and stagnant waters, soils, and foods. Only *P. aeruginosa* has been tentatively incriminated as a foodborne pathogen. Raw leafy vegetables serve as fomites, introducing the bacterium into the hospital environment. Identified strains obtained from foods have been isolated from stools of patients.

Pseudomonas cocovenenans produces a potent, often fatal toxin on coconut preparations in parts of Polynesia. The illness is termed bongkrek disease (see Chapter 10).

Enteroviruses

Enteroviruses are viruses that enter the gut. They reproduce in the intestinal tract of the infected person, large numbers of them leaving in the intestinal wastes (Joseph, 1965). According to authorities, carriers and asymptomatic cases far exceed clinical cases. Foodborne viral illnesses are associated with the faulty hygiene of persons actively secreting the viruses while handling foods to be consumed without further heat treatment. These illnesses are also spread when shellfish harvested in polluted waters are eaten raw or undercooked (Fugate et al., 1975).

The *Echovirus* and *Coxsackie* viruses, which produce respiratory diseases, are spread by droplet infection of food—that is, by utensils, dishes, or hands that have touched or received buccal and nasal discharges. The *Poliovirus, Echovirus,* and *Coxsackie* virus are members of the *Picornoviruses (pico,* small; *rn,* RNA). The viruses of hepatitis are placed tentatively in the *Herpesvirus* group.

The relation of *Poliovirus* to foods has become a popular subject of study although there appears to be little concrete epidemiological evidence that any food other than milk is involved in its spread. The virus reproduces in the intestinal tract, from which it invades the motor cells of the central nervous system. Initial symptoms are both gastrointestinal and febrile. The paralytic symptoms range from subclinical to fatal.

It is estimated that the virus of **hepatitis A** (infectious hepatitis) is responsible for more than a half million cases of illness per year. Outbreaks of illness are seasonal and often epidemic. The more common mild form of the disease among children occurs as a sudden fever, malaise, nausea, and abdominal discomfort lasting for several days. The liver may become inflamed and jaundice may occur. Humans are the only known host. The incubation period varies from 10 to 50 days. The virus is closely related to, but distinguishable from, the virus of **hepatitis B** (serum hepatitis), which is not spread through foods.

The common mouse is an unaffected lifetime carrier of the virus causing **lymphocytic choriomeningitis** (Benenson, 1970). The mouse sheds the virus in the feces and urine to contaminate foods with which it comes into contact. The illness, seldom fatal, has an incubation period of 7 to 13 days. Its symptoms resemble those of influenza.

Now that viruses can be recovered and cultivated in culture employing susceptible tissue cells, precise data about the presence of viruses in foods, and their control, can be obtained. Several methods have been developed to liberate viral particles from foods. Among them are homogenization and suspension with suitable suspending agents; and serial filtration through a succession of filters to free the ultrafiltrate from coarse and fine particles of food and bacteria, and to enumerate quantitatively in tissue culture in which the viruses form plaques (Figure 22-1). Plaques appear as vacant areas in the field of growth where the host cells have been affected by the virus. The numbers of the virus are expressed as plaque-forming units. In experimental work, 50% or more of the viral particles introduced to foods has been recovered.

The *Poliovirus* and *Echovirus* spp. are less sensitive to peroxide than are the salmonellae. They are destroyed in

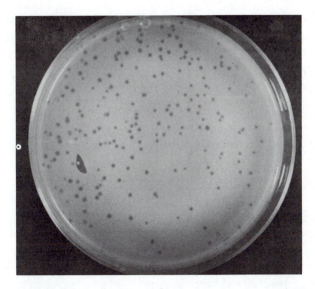

Figure 22-1
Viral plaques: clear areas in an opaque lawn of tissue
culture cells.

commercial egg preparation upon heating to 56.7°C for 20
minutes or to 58.9°C for 3.5 minutes (Strock and Potter, 1972).
In bacon squares, a low-moisture food, stored at 5°C, the
Poliovirus has persisted for 240 days with no loss in numbers of
viable particles (Cliver et al., 1970). The amount of irradiation
required to inactivate the virus in shellfish renders the food
organoleptically unacceptable (Girolamo et al., 1972). The
Poliovirus, Coxsackie virus, and *Echovirus* are markedly more resistant to chemical disinfecting agents than are the *Adeno*-Type
2, *Herpes simplex,* vaccinia, or influenza viruses, possibly because
the latter are lipophilic, whereas the enteroviruses are hydrophilic. The enteroviruses are not affected within 24 hours by
12% concentrations of 0-phenylphenol or Zephiran; they have
a 10 fold greater resistance to bichloride of mercury, and a 2 to
100 fold greater resistance to glutaraldehyde. They are not
affected by isopropyl alcohol (Klein and Deforest, 1973).

A most unusual "foodborne" viral infection is caused by the
Kuru viroid (a particle smaller than a virus). This is a "slow
virus," requiring years to develop in the host. In certain areas of

Polynesia where cannibalism was practiced, the brain of the cannibalized person was a delicacy reserved for women. If the brain was infected, years later the woman suffered degeneration of the tissue of the central nervous system. Males suffered the disease only as if as babies they have been fed some of the brains by their mothers.

Specific Foodborne Infections

Streptococci

Among the members of the genus *Streptococcus, S. pyogenes* is the only agent causing foodborne infections. It is responsible for acute, pus-forming infections (hence the name: *pyo,* pus; *genes,* forming). The bacterium produces diseases such as blood poisoning, puerperal sepsis, and erysipelas, but foodborne manifestations are infection of the throat and scarlet fever. Many strains produce hemolysin and capsules, and prior to the advent of the sulfa drugs and antibiotics the mortality rate was high. Common vehicles for spreading the disease are raw milk and cream contaminated by infected farmers and heated foods, such as ham, that are further handled after cooking. The bacterium resides in the upper respiratory tract. The modern practice of pasteurizing milk and cream and treating the disease with sulfa drugs and antibiotics have reduced the incidence of illnesses attributed to this virulent bacterium.

Salmonellosis

Salmonellosis, the collective term for all human and animal illnesses caused by members of the genus *Salmonella,* has been termed the most serious foodborne disease problem in the United States in recent years. Every year between 1963 and 1975 the Public Health Service reported from 20,000 to 25,000 isolations of salmonellae from humans (Figure 22-2). The figures of reported incidence, however, are misleading: intensive regional surveys indicate that only 1 to 10% of all outbreaks come to the attention of physicians, and not all of these are cultured and reported.

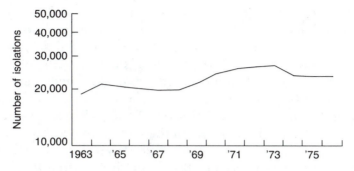

Figure 22-2
Reported human isolations of *Salmonella* spp. in the United States, 1963–1976. (Center for Disease Control, 1977b; Salmonellae Surveillance, 1976.)

Salmonellosis, or enteric fever, has afflicted humans for centuries. The disease was known at least as early as the days of Hippocrates, and typhoid fever was recognized as a specific disease in the mid-nineteenth century. It is probable that many of the serotypes of *Salmonella* known today have been responsible for minor outbreaks of gastroenteritis in the past, but were restricted to a few individuals in a small geographical area. In recent times, a number of factors have contributed to the spread of these bacteria:

1. increasing populations with potential for contaminating the environment,

2. a mobile society,

3. large centrally located food preparatory establishments,

4. large mills processing animal feeds that are then widely distributed,

5. proliferation of nonsterile but susceptible convenience foods,

6. the employment of subclinically ill and asymptomatic workers, and

7. contamination of soils and waterways.

It is a tribute to microbiologists and the food processing industries that once salmonellae were recognized in foods, methods to prevent their entry or persistence were quickly developed. Shredded coconut offers an excellent example of an unsuspected and unexpected occurrence of salmonellae in a food item. During harvest, coconuts are dropped upon soil contaminated by human or animal wastes. During chipping, workers' hands transfer salmonellae from the contaminated husks to the coconut meat. The salmonellae are destroyed by immersing the coconut meat into water at 70 to 80°C for 10 minutes before further processing (Schaffner et al., 1967). Today's foods offer extremely low risk of foodborne infection. Public health authorities have traced most known acquisitions of salmonelloses to sites beyond the control of the microbiologist, such as caterers, restaurants, private and public clubs, homes, group dinners, and camps.

Salmonellosis in its most acute and most often fatal form is particularly a disease of the young, whether human, animal, or bird. Human infants less than one year in age are highly susceptible, but the rate among children less than 4 years of age gradually decreases. The high incidence may indicate a markedly greater susceptibility of young children to salmonellosis, or that parents today more readily consult physicians than did those of previous generations. Another potentiating factor is the intimate association of children with pets such as turtles, which are carriers of salmonellosis.

The incidence of salmonellosis is greatest during the warm months, which the high ambient temperature allows the salmonellae to reproduce in unrefrigerated foods (Figure 22-3). Isolations of salmonellae present a fairly constant cyclic pattern, the fewest recoveries being reported between February and April, the most between July and September. In all probability, outings, picnics, and other social gatherings and a variety of outdoor activities that occur during the pleasant late-summer and early-fall days are equally favorable for microbial activity.

The clinical manifestations of salmonellosis are fever, septicemia, and gastroenteritis. *Fever* is characterized by malaise, commonly by leucopenia (decrease in the numbers of white blood cells), and less commonly by bronchitis and pneumonia.

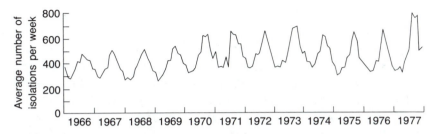

Figure 22-3

Seasonal pattern of human isolations of *Salmonella* spp. in the United States, 1966–1977. (Center for Disease Control, 1978.)

Blood cultures are positive early in the disease, but stool cultures may remain negative for weeks. *Septicemia* implies the presence of the bacteria in the blood, with intermittent fever. Endocarditis, meningitis, pericarditis, osteomyelitis, sinusitis, pneumonia, cholecystitis, pyelonephritis, and abscesses may occur as localized infections. *Salmonella typhi* commonly causes pyelonephritis; its presence in the recovered carrier is attributed to its saprophytic residence in the gallbladder. After surgical removal of the organ or a regimen of antibiotics the person is no longer a carrier. *Gastroenteric symptoms* follow an incubation period of eight hours to several days. The common symptoms are abdominal distress, vomiting, and diarrhea with fever. Usually the salmonellae are confined to the lumen of the intestinal tract, but septic manifestations are frequent in children, and among animals they may establish saprophytic residence in other organs of the body.

The agents of typhoid and paratyphoid fevers, *Salmonella typhi* and *S. paratyphi* A, B, and C (also known as *S. paratyphi, S. schottmuelleri* and *S. hirschfeldii,* respectively) cause illness only in humans and the anthropoid ape. They are among the few salmonellae that are brought under control, because they are restricted to a limited range of hosts. Typhoid fever became a major epidemic disease among both Union and Confederate troops during the War Between the States. Survivors of the disease carried the bacterium to all parts of the country when the troops were disbanded. Thereafter, typhoid fever became a major disease problem of increasing incidence through the contamination of well waters, milk, and foods. Later in that century, measures were instituted to detect and remove human carriers of the causative agent from employment in food preparation. During and after World War I a program to educate the populace was initiated. Large segments of the populace were immunized by a series of three injections at weekly intervals. The serum is a suspension of killed bacteria containing 1 billion cells of *S. typhi* and 250 million each of *S. paratyphi* A and B. These measures have markedly reduced incidence of the disease (Figure 22-4) but have not eradicated it.

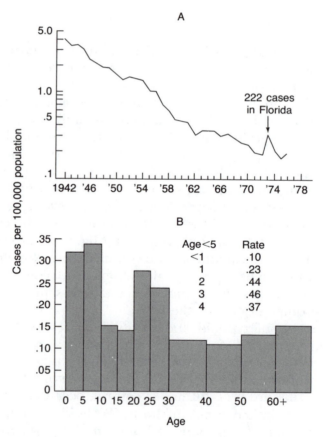

Figure 22-4
Reported cases of typhoid fever per 100,000 population in the United States. A: By year, 1942–1976. B: By age group, in 1976. (Center for Disease Control, 1977a.)

Members of the genus *Salmonella* are freely growing gram-negative rods 1 to 2 μ in length. To acquire energy they oxidize amino acids aerobically and ferment carbohydrates to acid and gas anaerobically. They do not require added growth factors for cultivation or for growth in foods. They are high-temperature mesophiles normally residing in human and animal intestinal tracts. They tolerate a wide range of acidity, temperature, and moisture conditions and therefore survive or grow in foods that do not support more fastidious bacteria.

Many enrichment and selective media have been developed to rapidly recover the salmonellae. Typical complex media contain nutrients such as peptones or protein digests and beef or yeast extracts; the solidifying agent agar; water; and salts either to maintain isotonicity or to serve as buffering agents. The media contain one or more inhibitory agents: bile salts, which reduce the surface tension to that of the intestinal tract, i.e. 37 dynes, so that the vast majority of extraneous bacteria cannot grow; brilliant green, which inhibits the gram-positive bacteria such as the enterococci, lactobacilli, and coliform bacteria *(Escherichia* and *Enterobacter* spp.); sodium thiosulfate to inhibit yeasts and molds; and potassium selenite, which enables rapid outgrowth of salmonellae, even when present in smaller numbers than are coliform and other bacteria. Because the coliform bacteria also grow on many of these solid media, lactose and an indicator are added to impart a distinctive coloration to colonies and media when acid is produced. Since a small number of strains of salmonellae ferment lactose, acid production with this carbohydrate does not provide an absolutely reliable criterion for identifying the coliform bacteria.

Three species of the genus are recognized: *Salmonella typhi* and *S. cholerae suis* are physiologically and serologically distinct, and *S. enteritidis* is a complex consisting of 1800 or more bioserotypes and serospecies. Seldom are more than 200 serotype species encountered in a single year. To identify the serospecies of *S. enteritidis* the cell and the flagella are analyzed for the presence of specific antigens.

Bacterial cells contain specific high-molecular-weight components, termed *antigens,* that when injected into the bloodstream of an animal stimulate the production of antibodies. The antigens are somatic (i.e., cellular), flagellar, or capsular. Among the salmonellae, one to four of the somatic antigens in each serospecies are stable to heating and are employed in the production of antibodies. **Somatic** antigens are composed of lipopolysaccharide, lipid, and glucosamine. Antisera—or more commonly, sera—produced with the whole-cell antigen usually contain several antibodies, each reacting specifically with the type of molecule which stimulated its production. To separate the specific antibody from other antibodies, a serospecies containing a shared antigen is added to the serum; the resulting flocculation of bacterial cells removes the shared antibody. The process is termed **agglutinin absorption,** because the bacteria agglutinate or clump together. Thus, cells containing antigens A and C added to a serum containing antibodies to antigens A and B will remove only antibody A, while antibody B remains in the supernatant serum.

The salmonellae are known to possess 64 somatic antigens (Edwards and Ewing, 1972). These are arranged selectively to divide the many hundreds of serospecies into 13 groups. The term "0" antigen (from the German use of 0 for *ohne,* "without") is frequently applied to these antigens, because they lack a pellicle, a growth characteristic of nonflagellated bacteria.

The **flagellar** antigens are proteinaceous flagellin. The antigenic specificity resides in the amino acid composition and in the sequences of the amino acids in the molecule. Flagellins are thermolabile, and are designated the "H" antigens (from the German use of H for *Hauch,* a film or pellicle, a growth characteristic of flagellated bacteria). Very few of the salmonellae lack flagella. The H antigens are divided into specific antigens—because they were first thought to be distributed within specific somatic groups of the salmonellae—and nonspecific antigens. At present 44 specific and 7 nonspecific antigens are known to exist among the salmonellae.

Each serospecies is characterized with a specific antigenic formula. *Salmonella enteritidis* ser. *Lexington,* or more commonly *S. lexington,* is described as 3, 10 z_{10}, 1, 5, to signify the possession of somatic antigens 3 and 10, the specific flagellar antigen z_{10}, and the nonspecific antigens 1 and 5 (Table 22-1). (Species identified early in the history of microbiology were given distinctively descriptive names, such as *S. typhimurium* (mouse typhoid). Later it became the practice to assign the name of the country, province, state, or city of initial discovery.)

Table 22-1

Antigenic structure of representative Salmonellae according to the schema of Kauffmann-White.

Type	Species of Salmonella	Somatic 0-antigen	Flagellar H-antigen Phase 1	Phase 2
A	*paratyphi*	1,2,12	a	
B	*schottmuelleri*	1,4,5,12	b	1,2
	java	1,4,5,12	b	[1,2]
	saint-paul	1,4,5,12	e,h	1,2
	san-diego	4,5,12	e,h	e,n,z_{15}
	derby	1,4,5,12	f,g	
	agona	4, 12	f,g,s	
	typhimurium	1,4,5,12	i	1,2
	typhimurium var. *copenhagen*	1,4, 12	i	1,2
	heidelberg	1,4,5,12	r	1,2
C_1	*cholerae suis*	6,7	c	1,5
	montevideo	6,7	g,m,s	
	oranienburg	6,7	m,t	
	thompson	6,7	k	1,5
	infantis	6,7	r	1,5
C_2	*newport*	6,8	e,h	1,2
D_1	*typhi*	9,12,Vi	d	
	enteritidis	1,9,12	g,m	
	dublin	1,9,12	g,p	
	panama	1,9,12	l,v	1,5
	javiana	1,9,12	l,z_{28}	1,5
	pullorum	1,9,12		
E_1	*anatum*	3,10	e,h	1,6
	meleagridis	3,10	e,h	1,w
	lexington	3,10	z_{10}	1,5
E_4	*senftenberg*	1,3,19	g,s,t	
F	*rubislaw*	11	r	e,n,x
G_2	*worthington*	1,3,23	z	1,w

Source: Kauffmann, 1966.

In some members of the family Enterobacteriaceae, such as *Escherichia, Citrobacter,* and *Shigella* spp., antigens reside in the exterior capsular coating of the cell. These are designated the "K" antigens. Among the salmonellae, *S. typhi* may contain a capsular antigen (some strains do not) that is designated as the Vi (for virulence) antigen. The presence of the Vi antigen enhances the virulence of the strain.

To locate the source of an outbreak and trace its spread, and to trace the origins of salmonellae in foods, public health workers and epidemiologists find it useful to serotype the salmonellae. For example, before 1971 only six cases of salmonellosis attributed to *S. agona* were known. Thereafter it was reported with increasing frequency and became one of the most commonly reported serotypes during the second quarter of 1972 (Murphy and McConnell, 1973). Its origin and distribution were traced to poultry feed incorporating Peruvian fish meal made from *anchoveta*. It became distributed worldwide. An outbreak in Arkansas was traced to a drive-in restaurant, where food had been prepared on a worktable earlier contaminated with raw poultry produced in Mississippi.

Human ignorance about the biology of the salmonellae has contributed immeasurably to pandemic distribution among wild and domestic, homoiothermal (warm-blooded) and poikilothermal (cold-blooded) vertebrate and invertebrate populations. Snails, caterpillars, wild insects, the common housefly, and cockroaches acquire them as temporary contaminants. It has been estimated that as much as 14% of ocean fish captured offshore harbors salmonellae in the gut. These microorganisms reside in reptiles and turtles, particularly in baby turtles of the pet trade, which have been implicated in dissemination of gastroenteritis. Wild birds acquire salmonellae by drinking water and eating grease at sewage treatment plants and by feeding on discarded, untreated animal wastes. Such incidents have caused epidemics involving thousands of sea swallows in Northern Germany, and in Northern Europe seagulls have distributed serospecies along the line of flight into Sweden. Serospecies have been isolated from duck feathers imported into Europe from the Orient.

In humans outbreaks involving *Salmonella cholerae suis* often are limited to single individuals, and the species appears to be the only member of *Salmonella* other than *S. typhi* and the paratyphoids that is host-specific. The pig is the only natural host of *S. cholerae suis*. It has been suggested that the presence or absence of some yet unknown factor in an organism renders it susceptible or inhospitable to a given microorganism (Taylor, 1965). The remaining salmonellae are not host-specific, and it is a conservative assumption that all serospecies are potentially pathogenic for humans. A preference for specific hosts among domestic and wild animals does exist. *S. pullorum* is commonly associated with chickens, *S. anatum* with anserine poultry, *S. dublin* with cattle, and *S. typhimurium* is unusually virulent for rodents, although it is also a common pathogen for domesticated cattle and turkeys. During the quarter century following its discovery, *S. typhimurium* was considered avirulent for man and it was used as an effective rodenticide in England, France, and parts of Russia; housewives were encouraged to cultivate the bacterium in the home for such use (Cockburn, 1965).

The incidence of salmonellosis is greatest among animals confined in herds, pens, or yards. In England pigs held in contaminated pens for longer than one or two weeks were studied (Hobbs, 1965); these animals had a high degree of salmonella infection (32 to 91%), whereas pigs kept for only two to seven days before slaughter yielded predominantly negative results. If feces are allowed to accumulate, enteric bacteria are encouraged to contaminate feed, water, floors, pens, and body surfaces. Many of the serospecies prominent in causing human illness are also the ones most often isolated from nonhuman sources. Collectively, the 10 most frequently isolated serotypes from human sources represent almost three-fourths of all isolates reported, and the ten most common nonhuman source isolates account for over half of all found in 1976 (Table 22-2).

Typhoid and paratyphoid fevers give rise to *bacteremia* (presence of viable bacteria in the bloodstream). In gastroenteritis, the illness produced by the many serospecies of *Salmonella enteritis,* the bacteria often are restricted to the intestinal tract; however, the severity of salmonellosis is a composite between the virulence of the bacterium and the resistance of the host; and if the illness is severe, bacteremia will occur. Symptoms of gastroenteritis range from acute illness, manifest as diarrhea, abdominal distress, and low mortality, to mild conditions that may be brief and sometimes subclinical.

Individuals recovering from salmonelloses may become carriers. When they do, they shed viable bacteria regularly or intermittently with intestinal wastes. A small percent harbor the bacteria in the urinary and the upper respiratory tracts. The serospecies causing gastroenteritis reside as saprophytes either in the lumen or embedded in the folds of the small intestine.

Table 22-2

Most common serotype species of salmonellae isolated from human and nonhuman sources.

Serotype species of Salmonella	Human sources			Nonhuman sources		
	Number	Rank in 1975	%	Number	Rank in 1975	%
typhimurium	6888	1	28.0	205	1	16.6
newport	1550	2	6.6	45	5	3.6
heidelberg	1474	4	6.3	23	12	1.8
agona	1333	5	5.7	45	5	3.6
infantis	1194	6	5.1	27	11	2.2
saint-paul	883	7	3.8	33	10	2.7
montevideo	308	15	1.3	35	7	2.8
derby	418	11	1.8	34	9	2.8
anatum	256	16	1.1	77	3	6.2
senftenberg	187	20	0.8	143	2	11.5
Total	14,491		61.5%	667		53.8%
All Isolations	23,445			1234		

Source: Center for Disease Control, 1977a.

Years ago it was observed that the percent of carriers following an outbreak of disease dwindles with time. The observation gave rise to the practice of demanding several successive negative stool samples at weekly intervals before permitting workers to return to positions in food handling and preparation. Reliance upon this practice imparts false confidence among those charged with the production of safe foods. Both early studies with typhoid fever (Wilson and Miles, 1946) and more recent work with other serospecies indicate that many carriers profess no knowledge of illness. Those recovered and certified carrier-free are sometimes found, when reexamined months later, to be shedding the initial causative agent. The present state of knowledge indicates that less than full reliance can be placed upon medical examinations and that a maximum of effort must be devoted to the continuing education of the employee in all practices of personal hygiene.

Seemingly recovered animals also become carriers. Commercial animal feeds containing unheated animal by-products have been a major source of the salmonellae to domestic animals in recent years. Pasteurizing tankage, fats, and other susceptible ingredients flowing into compound feeds has reduced the incidence from these sources. Wild animals that frequent farmyards acquire salmonellae from the domestic animals and then bring salmonellae into the herds and flocks. Haddock (1970) attributed extensive carrier salmonellosis in an isolated, large swine herd to several sources: commercial feeds containing the bacteria in numbers too low to be detectable in the small quantities normally sampled; mice and rats in the feed storage areas; flies in the lime pits; cats burying stools in wood shavings; chickens; beetles; and, at the slaughter house, to the low sides of the drinking troughs, which encouraged hogs to wallow in the water.

A persistent carrier state may be limited to only occasional farms (Carpenter et al., 1973). Hogs from such farms transmit the salmonellae to healthy hogs during shipment or while penned together, through the drinking water, floors, soil of pens, and machinery and other equipment used in slaughtering and dressing. The salmonellae reside in the intestinal tract, rectum, caeca, mesenteric lymph glands, liver, spleen, and other animal organs. Upon slaughter, the bacteria are transferred during handling to carcass surfaces. On hog carcasses they are most numerous on the skins of hams near the anal area, from which they spread irregularly over the surface during washing. Some slaughterers now spray dressed carcasses of beef, pork, and lamb with 100 to 200 ppm chlorine to expedite cooling and to reduce the numbers of surface bacteria.

Salmonellae are ubiquitous in the environment, as is shown by a partial listing of areas and foods from which they have been isolated: blankets, hospital bedpans, toilet seats and other bathroom surfaces, bottle warmers, floor polish, toothbrushes, vacuum cleaners, curtains, silverware, dust in the kitchen and window lintels; colby and cheddar cheeses, fresh but not frozen orange juice, cornstarch pudding, various green salads, chicken salad, deviled eggs, lemon meringue pie, semiliquid diet, stews, hollandaise sauce, jelly cakes, custards, cereal powders, barley,

Table 22-3
Confirmed outbreaks of salmonellosis by vehicle of transmission.

Year	Meats	Poultry	Fish	Eggs	Milk and milk products	Fruits and vegetables	Bakery products	Multiple vehicles	Miscellaneous	Etiology unknown	Total
1973	7	3	1	1	5	—	3	7	1	5	33
1974	3	2	1	—	7	1	—	5	3	13	35
1975	9	3	—	—	4	2	2	5	3	10	38
Total	19	8	2	1	16	3	5	17	7	28	106

Source: Center for Disease Control, 1973a, 1974, 1975.

dried yeast, chocolate, coconut cream pie, homemade ice cream, bread dressing, hotdogs, further processed but incompletely cooked meats such as turkey roll, batters, dog food and grilled tomato (Table 22-3). They have not been reported in or on luncheon meats or in meats in casings, where the combination of heat treatment and salt destroys them if present, or in household fats used for cooking. All deliberate search for salmonellae in frozen vegetables has failed to detect them (Splittstoesser and Segen, 1970). Fresh vegetables to be consumed raw, such as lettuce, celery, and radishes may contain coliform and fecal coliform bacteria. Where contamination with salmonellae or other foodborne infectious agents is suspected because of growth on polluted soil or for other reasons, washing in a solution containing (in the United States) 50 ppm chlorine or (in Great Britain) 60 to 80 ppm is recommended. Salmonellae have been found in some foreign canned meats and meat items, but not in domestic items because American processors subject them to higher processing temperatures than do some foreign canners.

Parameters of Growth and Survival. Many studies have been devoted to identifying conditions in which salmonellae are destroyed or merely survive or grow. Although informative and useful, these data do not yield absolute values, for several reasons. For one, each investigator has studied only the few serospecies of interest to him or her. Also, how a serotype behaves in nutrient broth or buffer solution does not reliably suggest how it behaves in foods. Formulated foods are complex, and each manufacturer decides on the proportions among the ingredients. Moreover, the ability of salmonellae to grow or survive is affected by such factors as changes in the acidifying agent, in the quantity of salt or of other soluble or insoluble solids, or in the viscosity of the food or quantity of water. The kind of care that the culture has been given and its conditions of storage can alter resistance or susceptibility. The strains used may be neither typical nor the most resistant of the serospecies tested (Miller et al., 1972).

Salmonellae grow at lower temperatures, at higher acidity—or at lower pH values—and at more extreme values of a_w than those that limit the growth of more fastidious pathogens. While they grow best at 37°C and upward, salmonellae grow quite well at 43°C. Sometimes, to recover salmonellae from foods containing large numbers of extraneous bacteria, the higher incubation temperature is used.

The minimum pH for growth depends on the acidulant: *Salmonella anatum, S. senftenberg,* and *S. tennessee* initiated growth at pH 4.05 in broth acidified with citric or hydrochloric acids; at pH 4.1 to 4.3 in broth acidified with fumaric, gluconic, malic, or tartaric acids; at pH 5.4 in broth acidified with acetic acid; and at pH 5.5 in broth acidified with propionic acid (Chung and Goepfert, 1970). The minimum temperature at which *S. heidelberg* initiated growth at pH 6.0 was 6.5°C; *S. typhimurium,* 7.2°C; and *S. derby,* 9.0°C. Salmonellae did not grow in broth adjusted to pH 5.0 when the temperature was 12°C (Matches & Liston, 1972).

At 8°C *Salmonella typhimurium* tolerates 1% NaCl and *S. heidelberg* no more than 2% NaCl; at 12°C *S. derby* does not initiate growth in 1% NaCl, but at 37°C these salmonellae initiate growth in broth containing 7 to 8% NaCl (Matches and Liston, 1972). Salmonellae have been reported to survive for days or weeks in highly acidic foods such as sauerkraut and pickled fish.

The hardiness of the salmonellae is shown by their ability to survive under adverse conditions. *S. pullorum* has survived 13 months in heavily contaminated soil, a cycle that includes summer heat and dryness and winter cold. *S. typhi* has survived for 300 days in Cheddar cheese stored at 5°C and for 196 days in cheese held at 18°C. It has survived for 105 days in warm soils and for 24 months in cold, wet soils, and for 2 years in ice cream. The salmonellae have been recovered from water originating with a sewage treatment plant that flowed many miles downstream under ice. They have survived for six to nine months in frozen chicken chow mein and three years in dried eggs. They die within 24 hours in refrigerated and frozen orange juice, but freshly prepared juice contaminated by kitchen help has been a vehicle for an outbreak of typhoid fever. Unprotected cells of salmonellae die rapidly on metal surfaces (Enkiri and Alford, 1971).

The maximum reported temperature range at which the salmonellae grow is 44 to 45°C; above this they die. Heating in nutrient broth or phosphate buffer at 60°C for 10 minutes kills most serotypes. However, several strains of *Salmonella senftenberg* that have been isolated from eggs and meat have a heat resistance 10 to 20 times greater than that of most serotypes. But this high resistance to heat is not expressed under all conditions; *S. senftenberg* 775W is less resistant than is *S. typhimurium* when these organisms are heated in milk chocolate having a very low moisture content (Goepfert and Biggie, 1968).

In general, like different serotypes, different strains of the same serotype differ in their survival of heating, although most serotypes are remarkably close to having uniform heat sensitivity. Resistance to heat is increased by the presence of particulate material, thickening agents, denaturable proteins, fats, increased concentrations of salt (NaCl) and sugar, decreased acidity, and reduced moisture content. *Salmonella typhimurium* survived 40 minutes when heated to 85°C in minced chicken muscle (Husseman and Buyske, 1954), and its D value (the time in which 90% of the cells are destroyed at a stated temperature) in milk chocolate was 75 minutes at 90°C. Salmonellae in dry egg white held between 51.7 and 54.5°C were destroyed only after 7 to 10 days.

Spray drying solid and liquid foods at temperatures that yield an economically, physically, and organoleptically suitable product does not ensure death of the salmonellae. Liquid foods must be pasteurized to destroy the salmonellae and *Escherichia coli*.

Drug Resistance among the Salmonellae. In recent years, strains of *Salmonella, Shigella,* and *Escherichia* have emerged that are resistant to one or more of the sulfa drugs and antibiotics (Watanabe, 1971). Single and multiple drug resistance is related to the widespread administration of sublethal levels of antibiotics in animal feeds. Once developed, drug resistance is transferable to other related strains by means of plasmid DNA. Strains of bacteria acquiring the drug-resistant DNA are said to have the R factor, and are described as R$^+$ strains. It is estimated that more than half the domestic animals given feeds containing antibiotics carry R$^+$ (resistant) strains in their intestinal tracts. Among these are *Escherichia coli*, which transfers the R factor to sensitive human pathogens. R$^+$ factors are not often transferred, but in the selective environment of the intestinal tract the continued presence of antibiotics represses the R$^-$ populations. In a free-living, drug-free bacterial society the R$^-$ strains reproduce without inhibition.

Shigellosis

Shigellosis, or bacillary dysentery, is caused by facultatively anaerobic gram-negative, non-spore-forming, rod-shaped organisms belonging to the genus *Shigella* within the family Enterobacteriaceae. There are many points of similarity between shigellae and salmonellae. Shigellae dwell primarily in the intestinal tract, with an optimum temperature of 37°C. They grow both aerobically and anaerobically, producing, under anaerobic conditions, acid but not gas from carbohydrates. They grow freely in warm, bland, moist foods. They are identified through serotyping, but unlike the salmonellae the shigellae have no flagella and thus are nonmotile.

In the United States isolations of these organisms rose markedly in 1973 but has subsided (Figure 22-5) since then. It is believed that the reported incidence represents but a fraction

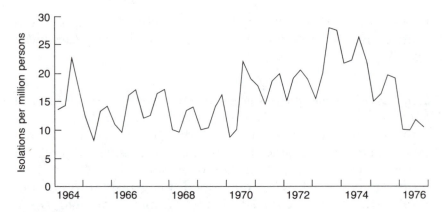

Figure 22-5

Reported isolations of *Shigella* spp., by quarter, United States, 1964–1976.
Only persons in reporting states and territories are included. (Center for
Disease Control, 1977c; *Shigella* Surveillance, 1976.)

of the total number of illnesses and that the severity of the
disease ranges widely. The numbers of reported cases and mortality rates of shigellosis and salmonellosis for 1973 and 1974
are nearly alike. In the developing countries high mortality
rates from bacillary dysentery are related to malnutrition, poor
personal hygiene, and the higher incidence of the disease
among the poor and the ignorant. Outbreaks occur among
groups confined in mental institutions, prisons, children's
homes, in armies in the field and at war, and during recent
years, among persons consuming foods from kitchens serving
large quantities of food. Unlike salmonellosis, most isolations of
shigellae occur in fall and winter rather than in summer.

The genus includes four antigenically unrelated species:
Shigella dysenteriae, S. flexneri, S. boydii, and *S. sonnei. Shigella
dysenteriae* has 10 serotypes; *S. flexneri* has 6 serotypes; *S. boydii*
has 14 serotypes; and *S. sonnei* has one serotype. The serotypes
are not differentiated by cultural procedures but rather by
serological methods. *Shigella dysenteriae* does not ferment mannitol, the other three species do.

Shigella dysenteriae embraces several antigenically unrelated
serotypes. Type 1 was found by Shiga in 1898 to be the etiologi-

cal agent of epidemic dysentery in Japan. Unlike the other dysentery organisms—which act principally in the alimentary tract—the *Shiga* bacillus elaborates a specific thermolabile exotoxin called **Shiga** toxin that is a protein and produces paralysis in various laboratory animals.

All strains of *Shigella* possess potent, heat-stable endotoxins associated with the somatic antigens. These are carbohydrate-protein-lipid complexes. Many shigellae possess capsular or envelope antigens that mask the somatic antigens. The somatic antigens are exposed by heating the cells to 100°C for 15 minutes.

Shigellosis ranges from fairly mild to very severe and fatal. The onset is usually abrupt, requiring from 1 to 7 days of incubation, but sometimes requiring as many as 14 days. The long incubation period can make it difficult to identify the food serving that was the vehicle of infection. Symptoms are abdominal pain and cramps, caused by inflammation of the mucus surface of the large intestine; mucoid, bloody diarrhea; fever to 40°C; and malaise. In severe instances, excessive diarrhea leads to electrolytic imbalance in the bloodstream, erosion of the epithelium of the large bowel, and ulceration leading to formation of scar tissue. There may be kidney failure, jaundice, and persistent internal bleeding. Septicemia does not occur. The infection is localized and organs other than the large intestine are not invaded.

While shigellosis is considered a waterborne disease, it is also foodborne. In the early part of this century, shigellosis was commonly associated with milk and ice cream. In recent years outbreaks involving as many as a thousand persons have been traced to foods such as chocolate pudding, tossed salad, and poi. Because the shigellae occur only in humans, outbreaks can be definitely traced to food handlers.

Methods for controlling shigellosis as a foodborne pathogen are identical with those needed to control salmonellosis. Immunization is possible, but immunity lasts too short a time for the practice to be useful for any persons except those entering areas of high risk. Persons recovering from shigellosis may shed the bacteria for several weeks, but the carrier state seldom exceeds one year.

Shigella dysenteriae is considered the most virulent species. In 1968, after a dormancy of 30 years, *S. dysenteriae* Type 1 reappeared in Guatamala and produced an unusually severe dysentery, with high morbidity and mortality; an estimated 12,500 deaths occurred during the first year (Gangarosa et al., 1972). The Shiga bacillus has since spread to other parts of Central America and Mexico. United States residents traveling in these countries often contract the disease. Contaminated water supplies and person-to-person contact spread the bacteria. As few as 10 cells can cause an infection. Gangarosa et al. (1972) suggest that the pattern of drug sensitivity is mediated by an episome observed in other enterobacteriaceae, and that the epidemic R factor, which mediates multiple drug resistance, may be associated with a factor that enhances virulence.

In the United States *Shigella sonnei* is associated with outbreaks of shigellosis much more often than are the other species. According to the National Center for Disease Control, *S. sonnei* accounts for about 65% of all isolations.

Both the total rate of shigella isolations in the United States and the rate for *S. sonnei* peaked in 1973 (Figure 22-6). The 24% decrease in 1975 reflects a 34.8% decrease in the number

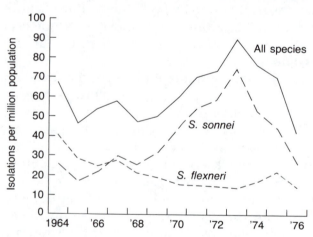

Figure 22-6

Reported isolations of *Shigella* spp., by year, United States, 1964–1976. Only persons in reporting states and territories are included. *S. dysenteriae* and *S. boydii* each account for only 2% of annual total shigella isolates. (Center for Disease Control, 1977c; *Shigella Surveillance*, 1976.)

of *S. sonnei* isolates reported. These decreased from 14,593 in 1974 to 9524 in 1975, whereas the number of *S. flexneri* isolates increased slightly from 4341 to 4740 (a 9.1% increase). Decreases could be illusory: they could be the result of changes in reporting procedures, of fewer stools being cultured, and, as unemployment increased, of fewer persons seeking medical care. However, an actual decrease in incidence cannot be ruled out (Center for Disease Control, 1977d; Shigella Surveillance, 1976).

Enteropathogenic *Escherichia coli*

Escherichia coli is a normal saprophytic inhabitant of the intestinal tract of humans and of warm-blooded animals. Its presence in water and in foods is taken to signify pollution with human or animal wastes. It is an antigenically heterogeneous species, containing 148 somatic, 49 flagellar, and 91 capsular antigens. The antigenic heterogeneity now makes it possible to recognize strains that are associated with the severe and often fatal disease known as scours in piglets, calves, and lambs; that cause outbreaks among newborn; and that are foodborne pathogens. The pathogenic and saprophytic strains are morphologically and physiologically identical.

Among humans enteropathogenic *Escherichia coli* (EEC) causes the illness known as summer diarrhea, or as Montezuma's revenge among travelers to Latin America. It does not occur only during the summer, and is most severe and sometimes fatal among human newborn, with milder manifestations among older children and adults. Adults may be asymptomatic carriers. In severe instances, the bacteria invade the blood, urinary tract, appendix, and peritoneum. EEC is either invasive or toxigenic, but not both. Toxin production is associated with an extrachromosomal episome that is transferable to nonpathogenic serotypes. The infective dose is in the range of 1×10^8 to 1×10^{10} cells. The invasive strains cause epithelial cells to slough in sheets; the bacteria concentrate on the margins of the sheets.

EEC was recognized as a foodborne pathogen in the United States in 1966. It received little attention until 1971, when an outbreak in Washington, D.C., and vicinity was attributed to imported cheese. Undoubtedly the organism caused illness in humans long before these dates. An outbreak of illness in Hungary, in 1937 (Varga, 1938), among persons who had eaten Trappist cheese may have been an EEC infection.

Detecting EEC in foods is cumbersome. EEC differs from nonpathogenic *Escherichia coli* (Yang and Jones, 1969) only by producing rough colonial growth on agar. Serology is the most reliable means of detecting causative groups, but results may be ambiguous because a transferable episome makes hitherto nonpathogenic serotypes pathogenic. In practice, foods suspected of harboring EEC are plated on selective and nonselective media designed to recover *E. coli* and 10 or more colonies are selected for examination.

The same measures restrict EEC in foods as control salmonellosis and shigellosis. Although there are not yet adequate methods for eliminating these pathogens from the food supply (Insalata, 1973), a rigorous sanitation program to eliminate *all* coliforms also is the best way to prevent EEC. Special attention should be given to (1) keeping *Escherichia coli* out of foods, (2) eliminating those present by suitable heat treatment, and (3) refrigeration to prevent them from growing.

Cholera

Asiatic cholera is an ancient disease thought to have originated in India or China, where it has been endemic for hundreds of years. It appeared in Europe early in the nineteenth century. In 1831 an epidemic began in Europe, and Irish immigrants brought cholera to New York in 1832. In 1838 another episode occurred in this country, New Orleans being the port of entry (Smith and Conant, 1960). From 1961 to 1974, El Tor cholera (the El Tor vibrio; now *Vibrio eltor*) swept out of its endemic area in Indonesia and spread into the Middle East across the Soviet Union, Czechoslovakia, Africa, and Western Europe (Figure 22-7). In 1973, 60 cases and 4 deaths occurred in Naples and another 10 cases in Bari, Italy (Center for Disease Control, 1973b). The incidents of cholera in Italy are believed to have originated from infected mussels. Fortunately, only two cases of cholera have occurred in the U.S. since 1911 (Center for Dis-

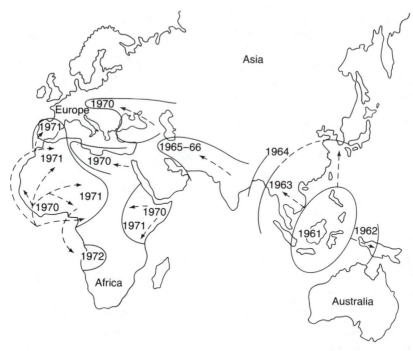

Figure 22-7
Extension of El Tor cholera, 1961–1972. (Principles and Practice of Cholera
Control, Public Health Papers No. 40, WHO, Geneva, 1970. Revised
according to Weekly Epidemiological Record *45* (1970), *46*, (1971), *47* (1972),
nos. 1–20.)

ease Control, 1973b and 1977b); both victims gave a history of
eating large quantities of raw oysters. Shellfish were also associated with an outbreak of epidemic proportions (989 cases) in
Portugal in 1974–1975. In 1974 and 1975 the number of cases
worldwide was 110,152 and 90,415. The number of countries
reporting cholera decreased from 40 in 1974 to 34 in 1975 and
26 in 1976 (Table 22-4). Between 1974 and 1976 African countries reporting cholera decreased from 19 to 14, while almost all
Asian countries reporting cholera in 1974 did so in 1975—
representing 90% of all cholera cases.

Cholera occurs only in humans. It is transmitted by contaminated water, fruits, vegetables, and raw and improperly cooked
fish. Direct person-to-person transmission is rare. High humid-

Table 22-4
Global cholera situation, 1971–1976.

	1971	1972	1973	1974	1975	1976
Number of countries reporting cholera	36	35	36	40	34	26
Number of cases	155,555	67,726	107,856	110,152	90,415	66,804

Sources: Data for 1971, World Health Statistics Report, 1972; for 1972–1976, inclusive data in WHO Weekly Epidemiological Record *52*, 165 and 203, 1977 as cited by Center for Disease Control, 1977a.

ity, warm temperature, and high population density sustain the bacterium's viability and encourage its spread. The disease occurs in epidemic proportions when the sanitary facilities and habits of peoples are disturbed by monsoons, floods, wars, and famine. Because the bacterium does not grow below 15°C, outbreaks in Temperate Zones subside naturally with the onset of cold weather. Human carriers shed the bacterium for three to four weeks after recovery. Immunization with bacterin (injection of a dead cell suspension) is effective for only four to six months.

The organism that causes cholera is *Vibrio cholerae*, sometimes known as *V. comma*. The causal relationship between it and cholera was established by Koch in 1884. The El Tor vibrio *(V. eltor)* was first isolated from travelers at the El Tor quarantine station in the Gulf of Suez in 1905. It is frequently described as being hemolytic, but there occur nonhemolytic strains, which are identified by analysis of the somatic antigens. The Hikojima type predominates in some parts of China; the Inaba type is indigenous in Japan and some parts of India; and the Ogawa strain predominates in other parts of India.

The bacterium is an actively motile, non-spore-forming, small (1 to 3 μm in length and 0.4 to 0.6 μm in width), slightly curved, gram-negative cell with a polar flagellum. It produces catalase and lecithinase, is oxidase-positive, grows freely in simple media, is aerobic, and ferments sugars without producing gas. Six antigenic somatic or 0 groups are known, but sero-group 0:1 causes most occurrences of cholera. It forms indole, reduces nitrates, and liquefies gelatin.

While *V. cholerae* can grow in laboratory media at pH between 6.4 and 9.6, it grows best between pH 7.8 and 8.0. Being alkaline tolerant, it is readily isolated by introduction of stool, water, or food samples into alkaline peptone broth at pH 8.4 to 9.0, which suppresses other bacteria. *V. cholerae* grows rapidly in the upper portion of the tube, and within 6 to 8 hours the cholera red test can be applied. This is a presumptive test in which an intensely red color forms when concentrated sulfuric acid is added to the alkaline broth. The bacterium is also recovered by streaking suspected material on Monsur medium, which contains potassium tellurite, upon which *V. cholerae* produces black-centered colonies. Identification is confirmed if the bacterium is able to ferment sucrose and mannose but not arabinose, and if a positive slide agglutination is obtained with specific group 0:1 serum.

The disease is confined to the lumen of the intestinal tract, the vibrios rarely penetrating to other organs. The bacteria multiply rapidly in the small intestine. The incubation period ranges from as little as six to eight hours to as much as two to three days depending upon the size of the infecting dose. The bacterium produces the enzyme *neuramidase,* a mucinase that hydrolyzes N-acetylneuramic acid, the cell-cementing substance; and an endotoxin is released upon death of the bacterial cells.

When the disease is active the victim may void from 20 to 30 stools per day and lose as much as 15% of body weight or as many as 18 liters of fluid. At first the stools contain fecal matter, but after this reservoir has been evacuated the discharges contain eroded mucosa in yellowish fluid having the characteristic "rice water" appearance. The skin becomes wrinkled and the eyeballs become shrunken by dehydration. Urine excretion is suppressed. The blood pressure falls, severe cramps develop, hypochloremia (loss of electrolytes from the bloodstream) occurs, and collapse and death follow unless the lost fluid and electrolytes are replaced. There is little fever or mental confusion.

Vibrio parahaemolyticus

Vibrio parahaemolyticus is a relative newcomer to the list of microorganisms producing major foodborne diseases. The earliest recognized food poisoning outbreak caused by this bacterium occurred in 1950, in Isaka Prefecture, Japan; it involved 272 patients, of whom 20 died. The victims had eaten partially dried shirazu, a food prepared from the sardine *Engraulis japonica.*

Fujino and associates (1953) recognized the causative agent as a previously undescribed, polarly flagellate, hemolytic rod that resembled *Vibrio cholerae* but did not react with antisera prepared against that species. They assigned it the name *Pasteurella haemolytica.* Sakazaki et al. (1963) placed it in the genus *Vibrio* with the species name *parahaemolyticus,* an appelation confirmed by Colwell et al. (1973) on the basis of numerical taxonomy, base composition, DNA/DNA homology, and sensitivity to bacteriophage.

Illnesses caused by this organism occur mainly throughout the coastal areas of the Far Eastern Pacific Perimeter, where oceanic foods are a staple of the diet. In Japan more than 400 outbreaks occur annually, totaling 5000 to 10,000 cases. In fact, *Vibrio parahaemolyticus* is responsible for over three-fifths of all reported cases of bacterial food poisoning in that country. Elsewhere it is reported to account for 15% of diarrhea in Viet Nam and from 2.9 to 22.6% of all gastroenteritis among hospital admissions in Thailand. The lower incidence throughout Indonesia is attributed to differences in Indonesian and Japanese eating habits: most Indonesian food is well cooked, whereas much Japanese food is eaten raw.

Vibrio parahaemolyticus has been isolated from marine sources in Australia, India, Thailand, Malaysia, the Philippines, Togo, Mexico, and England. It has been detected in countries bordering the Baltic, North, and Mediterranean seas and in Spain. In 1968 it was isolated in the United States (Ward, 1968). Because it is sensitive to low temperature and does not thrive below 15°C, it is isolated from the intercoastal and tidal waters of the United States only between May and October (Liston and Baross, 1973). It has been isolated from the waters of Chesapeake Bay during the summer and from sediments in the bay during the remainder of the year (Colwell et al., 1973).

Vibrio parahaemolyticus has been isolated from 80% of all fresh seafoods examined, in numbers ranging from one to 100 per gram of flesh (Fishbein et al., 1971). The bacterium has been recovered from 70 to 78% of shucked oysters and 50 to 65% of samples of processed crabmeat during September, but none has been isolated from these foods during January through March (Goldmintz, 1974). The bacterium is most abundant in molluscan shellfish and in waters high in organic content. Salted vegetables have been incriminated in the Orient.

Since 1969 at least a dozen outbreaks are known to have occurred in the United States, all involving consumption of crustacean foods. The food involved in several outbreaks had been recontaminated after cooking; in one outbreak undercooked shrimp had been eaten; in another a raw-food delicacy had been eaten. Outbreaks in Maryland (Center for Disease Control, 1971), affected 340 of 580 persons attending picnics; in the delivery trucks steamed crabs had been placed under baskets of live crabs, which drained into them. In another outbreak (Center for Disease Control, 1972b) steamed crabs had been put back into baskets from which *V. parahaemolyticus* had been recovered. After a "shrimp boil" in Louisiana 600 of the 1200 persons attending became ill (Center for Disease Control, 1972b); it is not clear whether they had eaten shrimp that had been recontaminated or simply undercooked. In Honolulu 200 persons attending a luau became ill after eating raw, salted crabs that had been stored overnight at room temperature and chilled shortly before eating (Center for Disease Control, 1972b).

Vibrio parahaemolyticus is a gram-negative, facultatively anaerobic, pleomorphic rod, usually possessing a single polar flagellum. It digests gelatin and ferments sugars to acid. The optimum temperature for growth is 37°C, with little or no growth below 15°C or above 43 to 44°C. The minimal a_w for growth depends upon the incubation temperature and the type and concentration of solute (Beuchat, 1974). In tryptic soy broth the organisms generated fastest in solutions containing 2.9% NaCl—a concentration corresponding to an a_w of 0.992

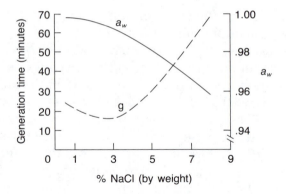

Figure 22-8
Effect of NaCl concentration and
corresponding$_{aw}$ on generation times of *Vibrio
parahaemolyticus* M5250J-2. (Beuchat, 1974.)

(Figure 22-8). The median incubation time for the onset of
symptoms in American outbreaks is 15 to 18 hours, with a
range between 8 and 50 hours; symptoms in Japanese outbreaks appear somewhat more quickly. Symptoms of intoxication, which range from mild to severe and fatal, include abdominal pain, which may be intense; a burning sensation of the
stomach; vomiting and diarrhea with watery stools and sometimes bloody discharges; fever; dyspnea; tachycardia; and
cyanosis. Internal lesions seen at autopsy include catarrh of the
stomach, hyperemia of the mesentery, erosion of the jejunum
and the ileum, congestion of the suprarenal gland, interlobal
hemorrhage of the lungs (Fujino et al., 1974), and conspicuous
edema and infiltration of neutrophiles (Sakazaki et al., 1974).
Sakazaki et al. (1974) believe that the ability of *V. parahaemolyticus* to grow in the intestinal tract may be an essential factor
in its becoming enteropathogenic. *V. parahaemolyticus* causes
wound infections (Twedt et al., 1969) and the death of brown
shrimp (Vanderzant and Nickelson, 1972). It grows in salt
concentrations ranging from 0.5 to 10%, maximum tolerance varying among strains. Ordinarily it grows between pH
5.0 and 8.5, but some strains tolerate pH 9.4. It has a specific

requirement for the sodium ion, which normally is supplied as NaCl, although other sodium salts are equally effective; it does not grow in culture media lacking salt. The bacterium possesses 12 somatic or cellular antigens and approximately 50 K or envelope antigens.

The vast majority of the strains isolated from patients are hemolytic (termed the *Kanagawa* phenomenon), while most isolated from seafoods and waters are nonhemolytic. Accidental ingestions of cultures and tests using human volunteers indicate that several hundred thousand live hemolytic cells are necessary to induce intoxication. Ingesting millions of cells of nonhemolytic culture does not induce illness. Cells killed by heat or chemicals do not affect HeLa cells in tissue culture. Live cells are markedly toxic for mice, but culture filtrates are not. However, Zen Yoju et al. (1974) have purified two nearly identical toxins from highly concentrated broth filtrates that do evoke responses in skin tests and in ligated ileal loops identical to those obtained with live cells.

Nonintestinal Illnesses

Tuberculosis

Tuberculosis is a chronic or protracted, progressive, and fatal disease that has afflicted humans perhaps throughout history. Skeletal remains of early Egyptians are reported to bear tuberculous lesions. Until recently, tuberculosis was among the 10 leading causes of death in the United States. As a respiratory disease, the form it most often takes among primary hosts, it passes through two stages: in the "closed stage" the bacteria are confined in pustules in the lung; in the "open stage" the disease is highly communicable. The bacilli are coughed into the mouth, become airborne, and are inhaled by new hosts. They also leave the mouth through sputum, expectoration, and saliva, contaminating human foods and animals feeds. If swallowed they migrate through the intestinal mucosa to develop secondary foci of infection. They also pass through the body with feces, again becoming airborne as dust from dried feces.

Tuberculosis of the human, bovine, and avian types is caused by *Mycobacterium tuberculosis, M. bovis,* and *M. avium,* respectively. At present *M. tuberculosis* is nearly the exclusive cause of human tuberculosis in the United States. Sputum, nasal exudates, unpasteurized milk, and improperly inspected meats and poultry are the principal sources for humans. Tuberculosis of cattle is attributed largely to *M. bovis;* effective federal herd and meat inspection programs decrease its incidence year by year. In Europe cattle and milk are the prime sources of tuberculosis in children. Food-related sources of tuberculosis were once very common, particularly among children who drank unpasteurized milk. Few children who contracted the bovine type of tuberculosis before reaching 8 to 10 years of age survived beyond 16 to 20 years of age. Bovine tuberculosis is also contracted by humans who eat cheese made with unpasteurized milk or lightly cooked, contaminated meats. Food-related tuberculosis is less often respiratory than gastrointestinal, skeletal, glandular and muscular, because the bacilli enter the body through penetration of the intestinal mucosa and then may be carried to other parts of the body through the bloodstream.

Although *Mycobacterium bovis* is the common cause of tuberculosis among cattle, the *M. avium* complex, common pathogens of birds and poultry, will also produce the disease in cattle. Swine, sheep, and goats also contract the bovine type of tuberculosis, often when confined or grazed near diseased cattle. Most tuberculous swine in America have *M. avium,* which they contract by being near diseased poultry, eating dead birds, or, as was formerly the practice, being fed unprocessed offal. Although susceptible to all three species, swine are most likely to be infected by *M. avium.* Nearly all species of domesticated and wild birds are susceptible to tuberculosis caused by *M. avium.* However, the disease is most common in poultry, probably because such birds are housed in confined areas—a practice conducive to spread of the disease. The incidence of avian tuberculosis varies widely among geographic areas of the United States. Some strains of *M. avium* and *M. intracellulare* can survive temperatures of 71°C for 60 minutes, a time-temperature relationship that may not always be achieved when chicken is roasted, baked, or grilled.

Pasteurization destroys *Mycobacterium tuberculosis* and *M. bovis* in milk. The time-temperature relationship first established for pasteurizing milk—61.7°C for 15 minutes—was based on the heat resistance of these bacteria. To provide a safety factor it was suggested that the duration of heating at that temperature be doubled to 30 minutes.

Federal regulations governing meat inspection require that all food animals with active or generalized tuberculosis be condemned. Carcasses of cattle identified as tuberculin test reactors are approved for human use only if the meats are cooked at 82.3°C for 30 minutes. When tuberculous lesions in swine carcasses are localized and confined to one primary site of infection, the unaffected portion of the carcass may be approved as human food after the affected part is condemned and removed. When the carcass of any swine reveals minimal lesions in more than one primary site, the infected sites are condemned and the unaffected portions of the carcass must be cooked to 82.3°C for 30 minutes. Similar regulations apply to sheep, goats, and horses. Diseased poultry are totally condemned.

Tuberculosis is capable of dramatic resurgence. In Germany and its neighboring allied countries during World War II, because of the war effort, milk was not pasteurized and diseased cattle were not detected. In 1945 the incidence of new cases among children was nearly as high as it had been before control programs were initiated. Once milk was again pasteurized and diseased animals were again detected and removed, the incidence of new cases was again dramatically reduced.

Leptospirosis

Leptospira interrogans, the single species of the genus, consists of 18 named serogroups and approximately 250 serosubgroups. The names of the serogroups are often used as species designations. The spirochaetes are worldwide in distribution, causing leptospirosis in animals and humans.

All members of the genus are morphologically similar. The cells are long, slender, narrowly spirillar, and often with one or both ends curved backward. They are readily visible by dark-field microscopy. The cells stain only with special stains such as

Giemsa and silver stains, not with the customary bacteriological stains. Many serogroups can be cultivated on synthetic media. The spirochaetes survive for long periods of time in water and in soil during moderate weather but are readily killed by heat.

Cattle, swine, sheep, horses, dogs, cats, rats, mice, and a number of wild animals are primary hosts of the spirochaetes, but poultry is not. The bacteria reside in the kidneys in an apparent commensal relationship without harm to the host except during pregnancy. Humans are not a primary host but contract leptospirosis by being around infected animals, by contact with materials contaminated with animal urine, by using contaminated water for drinking, washing, or swimming, or by eating foods contaminated with the urine of infected rodents, often rats. The spirochaete enters human and animal bodies through broken and abraded skin and through the intact conjunctivae of the eye and the membranes of the nose and mouth. Cattle have been experimentally infected by introduction of the specific spirochaete upon the inner surface of the eyelid.

In cattle and swine the disease becomes active with pregnancy and leads to leptospiral abortion, with serious economic losses to the livestock industry. The cattle strain, often called *Leptospira pomona,* is responsible for 13% of human leptospirosis in the United States; it survives very well in hog wallows and soil. Another strain, *L. canicola,* which is responsible for 63% of American leptospirosis, is a mild form of leptospirosis often contracted by children from infected dogs. *L. icterohemorrhagiae,* the most virulent agent for humans, is found in rodents and in cattle. Other serogroups encountered in the United States are *L. autumnalis, L. ballum,* and *L. grippotyphosa.*

Leptospirosis, also called Weil's disease in humans, incubates in from 2 to 20 days. The illness persists from several weeks through several months. The estimated fatality rate in the United States in 1973 was 5%. Symptoms include fever, jaundice, and petechial hemorrhages (very small purplish patches on skin or mucous membranes). The kidneys and the spleen may be involved, and there may be gastrointestinal disturbances, abdominal tenderness and/or pain, and sore throat. The disease has long been recognized as an occupational hazard of veterinarians, farmers, and packing plant employees.

Anthrax

Anthrax is caused by *Bacillus anthracis,* a large, gram-positive, aerobic, spore-forming, nonmotile rod. The organism is widely distributed in nature and is found in soil, dust, water, and decaying vegetation. Anthrax is primarily a disease of herbivorous animals, although all mammals are susceptible in some degree. Sheep, cattle, horses, and swine are the most commonly affected domestic animals; poultry are not infected. Humans contract the disease by (1) coming in contact with infected animals or animal products—the bacteria enter the body through a small wound in the skin; (2) eating diseased animal tissues; or (3) inhaling dust—which most often occurs in the textile industry (and is called woolsorter's disease).

No recorded cases of anthrax in the United States are attributed to consumption of contaminated meat. Slaughterhouse employees are at risk, because swine may suffer from chronic anthrax, which is diagnosed only during postmortem inspection. When swine anthrax is encountered on the kill floor of federally inspected plants, meat inspection regulations require that: all operations stop; the affected carcass be removed; all carcasses from the point of detection to the head dropper be inspected for evidence of anthrax and condemned if positive; all processing equipment and employees' clothing exposed to contamination be thoroughly cleaned and disinfected; the scalding vat water be brought to the boiling point or drained and replaced; all floors, benches, and other equipment contaminated by the affected carcasses be cleaned with water heated to 82.2°C (180°F); employees having contact with infected materials clean and disinfect their arms and hands; and the entire kill floor be cleaned and disinfected.

Brucellosis

Brucellosis is contracted by contact with live animals, by handling meat of infected animals, and by consuming raw milk and milk products. From an incidence of 6321 cases in 1947, cases of human brucellosis in the United States declined to as few as 168 in 1973, but have since undergone a resurgence, i.e., 246 in

1974 and 328 in 1975 (Center for Disease Control, 1976). In California, officially declared free of brucellosis in 1971, the disease returned in 1972 via animals imported from other states to be added to several large dairy herds.

Three species of the genus are virulent to humans. *Brucella melitensis* is named for Sir David Bruce, who first identified the organism, and for the island of Malta, where he did so. The organism, the causative agent of brucellosis (formerly termed Malta fever) occurs in goats. The disease occurs among Americans traveling abroad, when they consume goat or sheep milk or cheese in countries where dairying and cheesemaking are small, often family businesses and milk is seldom pasteurized. *B. abortus* occurs commonly in cattle and *B. suis* in hogs, but the hosts are interchangeable if the animals mingle on farms. Brucellosis of cattle is also known as Bang's disease, in honor of Bang, who discovered *B. abortus* in 1894, and as undulant fever in humans because of characteristic daily fluctuations in fever.

The brucellae are gram-negative, fastidious rods that reside as harmless saprophytes in the mammary tissues of the female animal. They become virulent during pregnancy when they invade the uterus and lead to abortion of the fetus. They leave the animal body in urine, and other cattle, hogs, sheep, and goats ingest them. Brucellae may persist for weeks to months in pond waters and streams, on grasses and in soil, and in the environment of the dairy barn. All parts of the human body can be infected, but brucellosis symptoms may mimic a variety of other infections and abnormalities. Brucellae are among the most heat-sensitive of the gram-negative rods, surviving for only a few seconds at 61°C.

Natural reservoirs among the animal population are eliminated by detecting and removing infected animals from the herd and by vaccinating young animals. In 1975 infected cattle were the main source of human brucellosis. Infected animals are detected by the milk ring test and blood tests. In the milk ring test, a suspension of stained dead cells is added to one ml of milk, mixed and incubated at 37°C for 60 minutes. If antibodies against brucellae are present, they will be in the globulin fraction of the blood, which becomes attached to the surfaces of the butterfat droplets. The bacteria are adsorbed to the surfaces of the butterfat droplets and color the cream layer.

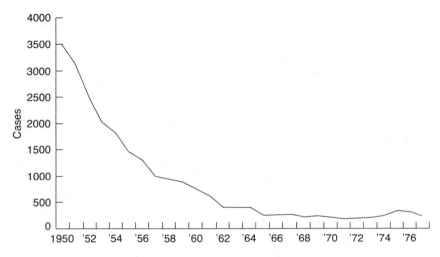

Figure 22-9
Number of cases of brucellosis reported annually in the United States,
1950–1977. (Center for Disease Control, 1978.)

Milk not containing antibodies develops a colored cream layer,
but the milk layer is colored. The blood tests can be quick or
prolonged: in the quick, or two-minute test, measured quantities of serum obtained from a blood sample of the animal are
mixed with measured quantities of stained dead bacterial cells
on a glass plate; in the more time-consuming (overnight) tube
test an unstained antigen is used. Flocculation or agglutination of the bacterial cells in either test signifies the presence
of antibodies to *Brucella* spp. in the blood. Neither test differentiates between a diseased animal, an animal that has recovered from the disease (which rarely happens), or one that has
been vaccinated. In recent years there has been a slight increase
in cases of human brucellosis in the United States (Figure 22-9);
more than half of the cases occurred in people associated with
the meat-processing industry.

Erysipelothrix

Erysipelothrix rhusiopathiae, a very small, slender, gram-positive,
nonsporulating, filamentous rod, is a widely distributed pathogen of poultry and livestock, causing serious economic losses to

growers of swine, turkey, and sheep. The disease may be erysipeloid, arthritic, or respiratory, and among animals herded or penned together may occur in epidemic form. The bacteria are readily transmissible to humans.

The bacterium is markedly resistant to drying, smoking, pickling, and salting. It survives at room temperature for several months but is very susceptible to moist heat; at 55°C it is destroyed within 10 minutes.

As a skin and arthritic disease, erysipeloid is an occupational hazard among farmers and packing plant workers who handle live animals, meat, and poultry. Although the bacterium enters humans usually through wounds on the hands and arms, some persons have become infected by eating pork.

Listeriosis

Listeria monocytogenes, a small, gram-positive, nonsporulating rod, is capable of producing natural infections in humans as well as in chickens, rabbits, sheep, goats, cattle, horses, ferrets, foxes, raccoons, and chinchillas. Many animals harbor the organism without detrimental effect until unfavorable conditions or other infections reduce their resistance. The bacteria may then cause a primary infection or become a secondary invader. The course of the disease varies in different animals.

The infection is believed to be transmitted from animals to humans, either directly, through contact, or indirectly, through milk. Human disease due to consumption of infected meat seems to be rare.

Tularemia

Tularemia is an endemic and epidemic disease of wild rabbits and certain of the *Rodentia* such as ground squirrels. *Francisella tularensis (Pastuerella tularense),* a fastidious gram-negative rod, causes a specific endemic and epidemic disease of rodents, particularly rabbits and squirrels, and is transmitted by a variety of insects. Humans and domesticated animals may also be infected. Incidence of the disease has steadily diminished (Figure 22-10).

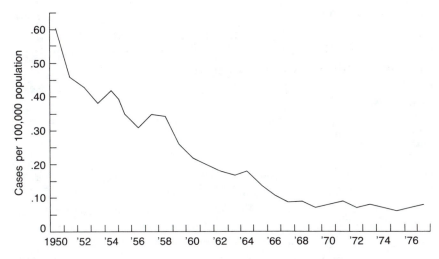

Figure 22-10

Number of cases of tularemia reported annually per 100,000 population by year, United States, 1950–1977. (Center for Disease Control, 1978.)

Humans contract the disease principally by contact with infected rodents, by inhaling dust containing contaminated hair or fur, by drinking contaminated water, or by being bitten by infected flies, ticks, lice, or fleas. Tularemia is an occupational disease among those who handle contaminated carcasses, i.e., commercial rabbit butchers, hunters, and housewives.

Although the organism can be cultivated only on blood agar in an atmosphere of reduced oxyten tension at 37°C, it has been known to survive for months in carcasses or in water. The disease is controlled by avoiding exposure, such as by using protective gloves when handling suspect material and by adequately cooking wild game. The bacteria are susceptible to phenol and other chemicals, antibiotics, and heat (10 minutes at 58°C is lethal).

Q Fever

The causative agent of Q fever is *Coxiella burnetti,* the only known foodborne pathogen in the order Rickettsiales (Canad. J. Public Health, 1949). The microorganism was first isolated in

Australia in 1933 and in the United States in 1938. Several explosive outbreaks have occurred in meat packing plants, usually among workers who handle raw meats. It is also found in the milk of infected cattle, sheep, and goats. It causes a febrile condition and severe pneumonia in humans that lasts for several days.

Yersinia

Yersinia enterocolitica (Pasteurella pseudotuberculosis) is a gram-negative rod that is antigenically related to members of the family Enterobacteriaceae. *Yersinia enterocolitica* has no unusual nutritional requirements. It grows on media used to detect coliform bacteria, salmonellae, and shigellae. It ferments sugars vigorously to produce acid but no gas, in sufficient quantity to impart a positive methyl red reaction. It is motile when incubated at 25°C but not at 37°C. It hydrolyzes urea, but it is distinguished from members of the family Proteeae by its inability to deaminate phenylalanine. It is antigenically heterogeneous, with numerous somatic and flagellar antigens. Many strains ferment sucrose and lactose and produce indol.

In the past this bacterium was considered to be a nonpathogen, but in recent years it has emerged as a virulent pathogen for humans and animals with worldwide distribution. It is endemic in upper New York State, where it is called the "Adirondack disease."

Because white lesions resembling those of tuberculosis appear on the liver, spleen, and lungs of animals the disease is also called pseudotuberculosis. Although such lesions are found in humans, the more common forms of the disease are the highly fatal typhoidlike septicemia and the milder form of enteritis with atypical appendicitis. *Yersinia enterocolitica* is an unusual pathogen because it can produce any of a variety of syndromes. It is the causative agent of gastroenteritis, suppurative arthritis, glomerulitis, acute and sometimes fatal ileitis (which may be mistaken for appendicitis, occasioning appendectomies), mild to severe and often chronic diarrhea (which may last several weeks), pseudo-typhoid fever, erythrema nodosum, cholocystitis, abscesses, meningitis and septicemia (Sonnenwirth, 1974). Because the gastrointestinal tract is so much involved, the pathogen probably takes the oral route of entry.

It has been isolated from the ceca of domestic animals at the time of slaughter, from intestinal wastes of domestic and wild animals, and from molluscs, insects, milk, and ice cream. Human and animal carriers may be asymptomatic. *Yersinia enterocolitica* occurs in surface waters during the cool months, presumably as a result of contamination by wastes from wild animals. Water and food are suspected but unproved vehicles leading to infection. It is found sporadically on refrigerated meats. On experimentally inoculated meats incubated at 7°C, it has increased in numbers by 6 to 7 log units above the original inoculum within 7 days (Hanna et al., 1976). The bacteria require 3 hours of heating to 60°C or 10 minutes at 70 to 80°C for inactivation.

The temperature range for growth extends from 0°C to above 40°C, but the preferred temperature is in the low mesophilic range. Colonies on solid media incubated at 37°C are barely visible. It is perhaps the only truly psychrotrophic bacterium that is pathogenic for humans.

Ornithosis and Psittacosis

Although considered an airborne infection, ornithosis can also be transmitted when diseased bird carcasses are handled. Outbreaks of ornithosis have occurred chiefly among poultry plant workers. The causative organism, *Chlamydia ornithosis,* is distinguished from *C. psittaci* according to the species of bird infected. The latter, causing psittacosis, infects psittacine birds, whereas *C. ornithosis* infects nonpsittacine species including chickens, turkeys, ducks, and pigeons.

References

Benenson, A. S. 1970. *Control of Communicable Diseases in Man.* 11th ed. Amer. Public Health Assoc., Washington, D.C.

Beuchat, L. R. 1974. Appl. Microbiol. *27,* 1075.

Brown, C., and R. J. Seidler. 1973. Appl. Microbiol. *25,* 900.

Bryan, F. L. 1972. J. Milk Food Technol. *35*, 618.

Canad. J. Public Health. 1949. *40*, 398.

Carpenter, J. A., J. G. Elliott, and A. E. Reynolds. 1973. Appl. Microbiol. *25*, 731.

Center for Disease Control. 1971. Morbidity and Mortality Weekly Reports. *20*, 356.

Center for Disease Control. 1972a. Morbidity and Mortality Weekly Reports. *21*, 171.

Center for Disease Control. 1972b. Morbidity and Mortality Weekly Reports. *21*, 245, 282, 341.

Center for Disease Control. 1973a. Foodborne and Waterborne Dis. Outbreaks.

Center for Disease Control. 1973b. Morbidity and Mortality Weekly Reports. *22*.

Center for Disease Control. 1974. Foodborne and Waterborne Dis. Outbreaks.

Center for Disease Control. 1975. Foodborne and Waterborne Dis. Outbreaks.

Center for Disease Control. 1976. Vet. Public Health Notes, June.

Center for Disease Control. 1977a. Morbidity and Mortality Weekly Reports. *25*(53), 65.

Center for Disease Control. 1977b. Salmonella Surveillance Ann. Summary. 1976.

Center for Disease Control. 1977c. Shigella Surveillance Report No. 39. Ann. Summary. 1976.

Center for Disease Control. 1978. Morbidity and Mortality Weekly Reports. *26* (53), 67.

Chung, K. C., and J. M. Goepfert. 1970. J. Food Sci. *35*, 326.

Cliver, D. O., K. D. Kostenbader, and M. R. Vallenas. 1970. J. Milk Food Technol. *33*, 484.

Cockburn, W. C. 1965. *Salmonella, Retrospect and Prospect.* Proceedings, National Conference on Salmonellosis. PHS Publ. No. 1262.

Colwell, R. R., T. E. Lovelace, L. Wan, T. Kaneko, T. Staley, P. K. Chen, and H. Tubiash. 1973. J. Milk Food Technol. *36*, 202.

Edwards, P. R., and W. H. Ewing. 1972. *Identification of the Enterobacteriaceae.* 3rd ed. Burgess Publ. Co., Minneapolis, Minn.

Enkiri, N. K., and J. A. Alford. 1971. Appl. Microbiol. *21*, 381.

Fishbein, M., I. J. Mehlman, and J. Pitcher. 1971. Appl. Microbiol. *20*, 176.

Fugate, K. J., D. O. Cliver, and M. T. Hatch. 1975. J. Milk Food Technol. *38*, 100.

Fujino, T. G., G. Sakaguchi, R. Sakazaki, and Y. Takeda. 1974. *International Symposium on Vibrio parahaemolyticus.* Saiken Publ. Co., Ltd., Tokyo.

Gangarosa, E. J., J. V. Bennett, C. Wyatt, P. E. Pierce, J. Olarte, P. M. Hernandes, V. Vasques, and D. Bessudo. 1972. J. Infect. Dis. *126*, 215.

Girolamo, R. D., J. Liston, and J. Matches. 1972. Appl. Microbiol. *24*, 1005.

Goepfert, J. M., and R. A. Biggie. 1968. Appl. Microbiol. *16*, 1939.

Goldmintz, D. 1974. In: *International Symposium on Vibrio parahaemolyticus.* Saiken Publishing Company, Ltd., Tokyo.

Haddock, R. L. 1970. Amer. J. Public Health *60*, 2345.

Hanna, M. O., D. L. Zink, Z. L. Carpenter, and C. Vanderzant. 1976. J. Food Sci. *41*, 1254.

Hobbs, B. 1965. Proceedings: *National Conference on Salmonellosis,* Public Health Service Publ. #1262.

Husseman, D. L., and J. K. Buyske. 1954. Food Res. *19*, 351.

Insalata, N. F. 1973. Food Technol. *27*(5), 56.

Joseph, J. M. 1965. Assoc. Food and Drug Off. U.S. Quarterly Bull. *29*, 10.

Kauffmann, F. 1966. *The Bacteriology of Enterobacteriaceae.* Williams and Wilkins Company, Baltimore.

King, A. D., Jr., M. J. Miller, and L. C. Eldridge. 1970. Appl. Microbiol. *20*, 208.

Klein, M., and A. Deforest. 1973. Antiviral action of germicides. In: *Chemical Sterilization.* P. M. Borick, ed., Dowden, Hutchisson, Ross, Inc., Stroudsburg.

Liston, J., and J. Baross. 1973. J. Milk Food Technol. *36*, 113.

Marcus, K. A., and H. J. Amling. 1973. Appl. Microbiol. *26*, 279.

Matches, J. R., and J. Liston. 1972. J. Milk Food Technol. *35*, 39, 49.

Meyer, M. T., and R. H. Vaughn. 1969. Appl. Microbiol. *18*, 925.

Miller, D. L., J. M. Goepfert, and C. H. Amundson. 1972. J. Food Sci. *37*, 828.

Sakazaki, R., S. Iwanami, and H. Fukumi. 1963. Jap. J. Med. Sci. Biol. *16*, 161.

Sakazaki, R., K. Tamura, A. Nikamura, T. Kurata, A. Godha, and Z. Kauno. 1974. Jap. J. Med. Sci. Biol. *27*, 45.

Schaffner, C. P., K. Mosbach, V. C. Bibit, and C. H. Watson. 1967. Appl. Microbiol. *15*, 471.

Smith, D. T., and N. F. Conant. 1960. *Zinnser Microbiology*. Appleton-Century-Crofts, Inc., New York.

Sonnenwirth, A. C. 1974. Yersinia. In: *Manual of Clinical Microbiology*. 2nd ed. American Society for Microbiologists, Washington.

Splittstoesser, D. R., and B. Segen. 1970. J. Milk Food Technol. *33*, 111.

Strock, N. R., and N. N. Potter. 1972. J. Milk Food Technol. *35*, 247.

Twedt, R. M., P. L. Spaulding, and H. E. Hall. 1969. J. Bacteriol. *98*, 511.

Vanderzant, C., and R. Nickelson. 1972. Appl. Microbiol. *23*, 26.

Varga, O. 1938. Zentralbl. Bakteriol. Parasitenk. II Abt. *99*, 156.

Watanabe, T. 1971. Infectious Drug Resistance in Bacteria. Current Topics in Microbiol. and Immunol. *56*, 43.

Ward, B. Q. 1968. Appl. Microbiol. *16*, 543.

Wilson, G. S., and A. A. Miles. 1946. *Topley and Wilson's Principles of Bacteriology and Immunity*. 3rd ed. Williams and Wilkins Company, Baltimore.

World Health Org. 1970. Principles and Practices of Cholera Control. Public Health Papers No. 40.

World Health Statistics Report. 1972. *25*(4).

Yang, K., and C. A. Jones. 1969. J. Milk Food Technol. *32*, 102.

Zen Yoju, H., H. Hitokoto, S. Morozumi, and R. A. LeClair. 1971. J. Infect. Dis. *123*, 665.

23

Bacterial and Fungal Foodborne Intoxications

Although many different organisms can grow in foods, only a few elaborate toxins that make the food dangerous to eat (Table 23-1). Among the bacteria the best known are *Staphylococcus aureus, Clostridium perfringens, C. botulinum,* and *Bacillus cereus.*

Various fungi—including several species of mushrooms and *Claviceps purpurea,* which produces ergot—are poisonous when ingested. Certain members of the genera *Fusarium* and *Aspergillus* elaborate toxins that, if they get into foods and feeds, can be carcinogenic for humans or domestic animals. Other species of these same genera—as well as certain species of *Penicillium, Sclerotinia, Pithomyces,* and *Rhizoctonia* that are noted plant pathogens—produce metabolites that have powerful estrogenic, toxic, phototoxic, or cholinergic effects.

Staphylococcal Enterotoxin

Staphylococcal food poisoning is one of the most commonly reported foodborne illnesses in the United States. Each year it is responsible for at least one-fourth of all confirmed outbreaks

Table 23-1
Microbial Toxemias.

Disease	Etiological agent	Principal types	Incubation period	Symptoms
Staph. food poisoning or Staph. enterotoxin	*Staphylococcus aureus*	A, B, C$_1$, C$_2$, D and E	1 to 6 hours Average 2 to 3 hours	abdominal cramps, diarrhea, nausea, vomiting, acute prostration
Botulism	*Clostridium botulinum*	A, B, E and F (others not implicated in foods)	½ day to more than a week Usually 1 to 2 days	nausea, vomiting, diarrhea early, double vision and later, difficulty in swallowing and speech, respiration. Often respiratory paralysis and death
Perfringens food poisoning	*Clostridium perfringens*	A and C (others not implicated in foods)	8 to 24 hours Usually 10 to 12 hrs.	abdominal cramps, diarrhea, nausea, malaise. Vomiting rare.
B. cereus food poisoning	*Bacillus cereus*		8 to 16 hours Usually 11 to 12 hrs.	abdominal cramps, diarrhea, nausea. Vomiting rare.
V. parahaemolyticus food poisoning	*Vibrio para-haemolyticus*		10 to 24 hours Usually 12 to 14 hrs.	abdominal pain, diarrhea, fever, nausea, vomiting

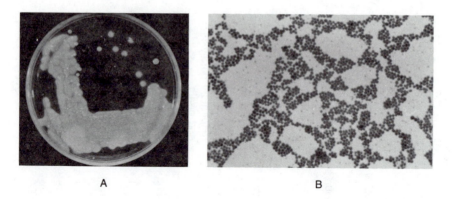

A   B

Figure 23-1

A. Streak culture *Staphylococcus aureus;* B. gram stain of pure culture of
Staphylococcus epidermidis ($\times$ 1000).

reported to the Center for Disease Control (1970, 1971, and
1972) and from one-fifth to over half the cases of foodborne
intoxications.

The specific agent responsible for enterointoxication is
Staphylococcus aureus (from the Greek *staphele,* bunch of grapes;
and the Latin *aureus,* golden—so named because of the characteristic distribution of the cells upon microscopic examination
and the golden color of the colonies on agar surfaces. See Figure 23.1 A). The other commonly occurring species is *S.
epidermidis* (Figure 23.1 B), frequently termed *S. albus* in the
earlier literature. The specific epithet *albus* was applied to
strains of staphylococci lacking pigment and implicated in the
earliest documented report of staphylococcal food poisoning
associated with milk in the Philippines (Barber, 1914). Although *S. aureus* and *S. epidermidis* are the only two recognized
species of the genus, strains producing a rosy pigment *(S.
roseus),* a lemon-yellow pigment *(S. citrinum),* and a halophile *(S.
lactis)* have been reported.

Staphylococcus aureus is a spherical, small (0.5 to 1.5 μ in
diameter), gram-positive, catalase-positive, non-spore-forming
bacterium that usually occurs in clusters but also occurs singly,
in pairs, and in tetrads. Gram-negative cells often are inter-
mingled with the gram-positive cocci.

The bacterium ferments carbohydrates anaerobically, grows at 45°C, resists the lysing action of lysozyme, and possesses the enzymes phosphatase and deoxyribonuclease. It grows very poorly at 10°C. Also, being able to produce acid from glucose facultatively or anaerobically, it can lead a parasitic existence. By these features it can be distinguished from members of the genus *Micrococcus*.

Staphylococcus aureus resides as a saprophyte in the mucus secretions of the nasopharyngeal region; it occurs more often in the posterior nares than in the pharynx. The bacteria leave the nose and mouth in nasal secretions and by expectoration, and are forcibly ejected in droplets formed during coughing and sneezing. They are also transmitted during smoking or nibbling, when fingers touch the lips or nose and during eating, when utensils enter the mouth. The organisms are transient residents of the skin and permanent residents in the sweat glands and in the intestinal tract.

In addition to the food poisoning caused by the enterotoxins, the staphylococci cause localized infections such as pimples, boils, carbuncles, and abscesses, and more generalized infections such as meningitis, osteomyelitis, pneumonia, and mastitis of humans and animals.

Minimal and maximal temperatures for growth under favorable conditions have been established at 6.67 and 45.5°C (Angelotti et al., 1962). The bacterium grows over a range of pH 4.0 to 9.8, doing so at the extremes only under optimal nutritive conditions. Although the cells are most numerous below 20°C, growth is most rapid above that temperature and in media and foods having little acidity. *Staphylococcus aureus* requires nicotinic acid and is favored by the presence of thiamine (Gretler et al., 1955). It requires 11 amino acids for optimal growth: valine, leucine, threonine, phenylalanine, tyrosine, cysteine, methionine, glycine, proline, histidine, and arginine. Uridine is needed if growth is anaerobic. It does not grow in synthetic media and in foods such as synthetic whipped creams, which contain no amino acids or proteins. In sugar-free protein foods it grows aerobically, but carbohydrates favor its anaerobic growth. *S. aureus* reduces nitrate to nitrite and it tolerates, but does not require, NaCl in concentrations of 10% to as high as 15%.

Bacteriophage and Phage Typing

More than 80 strain-specific bacteriophages of *S. aureus* are known. Approximately 75% of all strains isolated are phage typable. Most phage types are related to strains of human origin, but several types are characteristically associated with mastitis in dairy cattle. Thus, phage typing can determine whether intoxications involving milk and its products are of bovine or human origin. In epidemiology the bacterium isolated from foods or from patients' vomita can be traced by phage typing to the human carrier. Recent concern has been expressed that staphylococci and bacteriophages may not be related in reliable ways. Casman (1965) succeeded in conferring type A enterotoxigenicity on a nontoxigenic strain by lysogenizing it with a phage from an enterotoxic culture—a procedure called **transduction.**

Toxins and Toxin Production in Culture

Staphylococcus aureus produces a variety of toxins, including:

1. coagulase, an enzyme that coagulates citrated plasma;
2. the potent *alpha* exotoxin;
3. a hot-cold hemolysin or *beta* toxin;
4. two exocellular leukocytic proteins of the F and S toxins;
5. hyaluronidase, or "spreading factor"; and
6. a group of simple, proteinaceous enterotoxins.

The bacterium produces enterotoxins A, B, C_1, C_2, D, and E. Toxin A, termed a primary metabolite, is produced during the logarithmic phase of growth. This is the most prevalent form of toxin in foodborne illness, probably because many incriminated foods are consumed while the staphylococci are actively growing (Markus and Silverman, 1970). Toxins B and C are produced during the late logarithmic and stationary phases of growth. Toxin A is produced over a wide range of pH, from 4.5

and above (Genigeorgis et al., 1971), whereas toxins B and C are produced over a relatively narrow range of pH near neutrality (Reiser and Weiss, 1969). Although *S. aureus* can grow at an a_w no lower than 0.86, toxin B is produced only at a_w 0.98 upward, and none is produced at a_w 0.96 (Troller, 1971). As concentrations of NaCl increase from 0% to 10%, yields of toxins B and C decrease accordingly, but toxin C is still produced in cured meats having as much as 10% NaCl (Genigeorgis et al., 1971). Toxins are produced only after three days of incubation at 20°C and after longer storage periods at lower temperature. They are produced very rapidly at 37°C. Modern methodology can detect as little as one μg/ml toxin, which equals a population of 10^6 to 10^7 viable cells/ml or gram.

Properties of Enterotoxins

The enterotoxins are simple proteins with relatively low molecular weight (26,000 to 30,000). The actual value for enterotoxin B has been determined to be 28,366 (Dayhoff, 1972). All the enterotoxins are single-polypeptide chains that contain relatively large amounts of lysine, aspartic, and glutamic acids, and tyrosine. Data compiled by Bergdoll et al. (1974) show that there are only two residues of half-cystine (Table 23-2). The enterotoxins are hygroscopic and are soluble in water and salt solutions; they vary in heat resistance and are resistant to the proteolytic enzymes pepsin, trypsin, chymotrypsin, rennin, and papain. Early work (Dack, 1956) indicated that a crude enterotoxin in solution was quite resistant to heat. Filtrates of cultures caused emesis in monkeys, even after having been boiled for an hour. Later data (Read and Bradshaw, 1966) showed that toxin B heated at 121°C was inactivated in 16 minutes. Activity of enterotoxin A decreased by 50% after it was heated at 60°C for 20 minutes (Chu et al., 1966), but less than 50% of toxin B activity was destroyed by heating at 100°C for 5 minutes (Bergdoll, 1967). It should be stressed that few foods in which *Staphylococcus aureus* has grown are subjected to these temperatures, and in fact many foods incriminated in staphylococcal intoxication receive no heat treatment after the point at which the organism may have begun to grow.

Table 23-2

The portion of enterotoxin B molecule which contains the half-cystine residues.

-Lys-Asp-Leu-Ala-Asp-Lys-Tyr-Lys-Asp-Lys-Tyr-Val-Asp-Val-Phe-Gly-Ala-Asn-Tyr-Tyr-
71 72 73 74 75 76 77 78 79 80 81 82 83 84 85 86 87 88 89 90

Gln-Cys-Tyr-Phe-Ser-Lys-Lys-Thr-Asn-Asn-Ile-Asp-Ser-His-Glu-Asn-Thr-Lys-Arg-Lys-
91 92[a] 93 94 95 96 97 98 99 100 101 102 103 104 105 106 107 108 109 110

Thr-Cys-Met-Tyr-Gly-Gly-Val-Thr-Gly-His-Gly-Asn-Asn-Gln-Leu-Asp-Lys-Tyr-Tyr-Arg-
111 112 113 114 115 116 117 118 119 120 121 122 123 124 125 126 127 128 129 130

[a]Bergdoll et al. (1974) consider that the two half-cystines are joined into one cystine residue by folding of the molecule, say as between cystines (amino acids) 92 and 112.

Source: Bergdoll et al., 1974.

Detecting Enterotoxins in Foods

For many years after the elucidation of the role of staphylococ-cal enterotoxin, attempts were made to correlate toxigenesis of suspected cultures with cultural properties and with responses in test animals. It is generally agreed that a toxigenic strain should coagulate citrated human or rabbit plasma and liquefy gelatin, ferment mannitol anaerobically, tolerate potassium tel-lurite, be hemolytic on blood agar, and have a golden pigment when grown on agar; however, many strains fail to fulfil one or more of these criteria. Many test animals, including frogs, pi-geons, nematodes, protozoa, enzyme systems, tissue cultures of monkey kidney cell, monkeys, cats and kittens, and dogs have been used or proposed for use. Cats, monkeys, and dogs are considered the most reliable; mice, rats, and rabbits, having no vomiting mechanism (reverse peristalsis), have little value as test animals.

Primates are the animals most sensitive *per os*. Biological ac-tivity in the rhesus monkey is characterized by emesis or diarrhea within five hours after administration of 5 μg toxin/ 2-3 kg animal through a feeding tube (Chu et al., 1966 and 1969; Borja and Bergdoll, 1967). The cat provides good results when toxin A or B is used but is highly unreactive to toxin C, requiring 50 times as much of it as of A or B to produce emesis.

Subjective animal tests for detecting toxins have been re-placed by direct methods employing immunofluorescence (Stark and Middaugh, 1969; Genigeorgis and Sadler, 1966), hemagglutinin inhibition, reverse passive hemagglutination, immunodiffusion, radioimmunoassay, and the Ouchterlony plate precipitin test (Casman and Bennett, 1964 and 1965; Casman et al., 1969), and an indirect method for detecting thermonuclease (DNase). Tests for the presence of toxin or DNase are of value, because staphylococci die rapidly after reaching the stationary phase and give way to harmless sap-rophytic bacteria, or they may have been killed in heated foods. The Ouchterlony and DNase tests are easily performed; the DNase test detects foods in which staphylococci have grown, but, since some staphylococci producing DNase do not also produce enterotoxin, foods in which DNase is detected should be examined for the toxin (van Schouwenburg-van Focken et

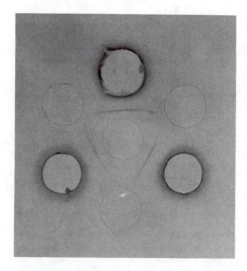

Figure 23-2
Gel diffusion plate showing precipitin
lines between toxic material in center well
and antiserum in three outer wells. The
extract in the other wells did not contain
enterotoxin.

al., 1978). The Ouchterlony test detects toxin by formation of a
zone of precipitation between wells containing antibody in a
food extract containing toxin (Figure 23-2). To detect DNase,
an alkalinized, heated extract of the food is placed in a well in
agar containing deoxyribonucleic acid and an indicator. If
DNase is present, the indicator changes to the acid color
(Lachica et al., 1971).

Intoxicating Levels

Few attempts have been made to test toxin B in human volunteers *per os*. No symptoms of distress were observed when 1 and
10 μg were given to two volunteers (Bergdoll, 1970), but 50 μg
of toxin B estimated to be 50% pure produced typical
symptoms within two to four hours after ingestion (Raj and
Bergdoll, 1969). It is sometimes stated that a concentration of 1
μg/g food will produce obvious symptoms of distress in a consumer.

Symptoms of Intoxication

The enterotoxins cause irritations of the intestinal tract with subsequent response. The severity of the symptoms depends on the dosage and can range from being barely noticeable to causing extreme distress and, although rarely, death. Progressively, the symptoms are a feeling of uneasiness, nausea, diarrhea, vomiting, moderate to severely cramping pains, prostration, coma, and death. The onset is usually within two to four hours after consumption, or earlier if there is severe enterointoxication. During the throes of the disease, the body draws fluid from all tissues to flush the toxin from the intestinal tract and loss of 15 to 18 pounds of body weight has been observed. Disturbance of the salt balance of the bloodstream leads to coma. Even in severe instances, symptoms seldom endure for more than 12 to 15 hours, and the recovery period is uneventful except for a feeling of malaise and general weakness. Affected persons should be kept warm and given lightly salted water to preserve the body fluids, and the episode should be brought to the attention of a physician.

Toxin-Producing Foods

Enterotoxins are produced in moist or fluid foods that are bland or, at most, semiacid, and usually have been heated. Probably no food fitting this description has not at some time been incriminated in staphylococcal food poisoning. The range of foods is broad: cooked meats, particularly ham and poultry and dishes made with these meats; a wide variety of salads; egg products such as custards; natural whipped creams; meringue, cream, custard, and pumpkin pies; creamed vegetables; macaroni; milk and cheese; and soups. With the exception of milk and cheese, where the staphylococci may originate from a mastitic infection of the cow, virtually all incriminated foods have been touched or handled by humans. Most of the foods have been heated or pasteurized to destroy competing bacteria and often are contaminated during cooling.

The ubiquity of the staphylococci in the human nasopharynx makes it virtually impossible to ensure that susceptible foods exposed to handling will be free of staphylococci. Some humans are permanent carriers of toxigenic strains, others transiently so, and some appear seldom to be carriers. The carrier rate varies with the frequency of incidence of respiratory diseases. Limits of tolerance, such as 100 viable cells or clones, as determined by the plate count or other suitable techniques, have been established. The low limit demands that, to prevent multiplication, refrigeration and care of food be given due attention. Numbers in finished, further processed cold foods must not exceed numbers in the raw ingredients. Therefore, foods, such as raw breaded materials, are processed in cool rooms with all components at or near 5°C. The ingredients of foods with high mass and density, such as potato salads and other salads, macaroni, and the like, must be chilled thoroughly before mixing. Potentially contaminated foods such as raw poultry and dressings should be heated to an internal temperature of 74°C to ensure destruction of staphylococci. Large, compact items such as hams, which become contaminated at the surface after heating, should be refrigerated when the surface temperature is no lower than 45°C, to ensure a rapid cooling of the surface to 10°C or less.

Staphylococcal intoxication is almost always associated with foods that have been heated or, like milk, have no competing bacteria to interfere with the growth of the staphylococci. When cold foods become warm the saprophytes that are capable of growing rapidly below 20°C do so, competitively inhibiting the staphylococci so that the saprophytes "provide a built-in safety factor" in warming foods (Peterson et al., 1962). Inhibition may result from depletion of the essential nicotinic acid by the saprophytes (Iandolo et al., 1965).

Perfringens Food Poisoning

Clostridium perfringens, also known as *C. welchii*, the causative agent of human gas gangrene and of stormy fermentation in milk, was recognized as a cause of foodborne illness before the turn of the century by Klein (1895). It became a prominent agent of foodborne intoxication in Europe during World War II (Strong et al., 1963) and was recognized as an agent of intoxication in the United States in 1943 (McClung, 1945). It is now considered to be a major foodborne intoxicant. In 1969 (Center for Disease Control, 1969) nearly 17,000 cases were reported to the National Communicable Disease Center.

Clostridium perfringens occurs as a saprophyte throughout the world in soil, dust, water, human and animal feces, and sewage. Dense populations of the organism are noted in areas where animals congregate, such as grazing areas, feed lots, and watering places. The bacterium is omnipresent in the kitchen (Bryan and Kilpatrick, 1971), where it is a natural and unavoidable contaminant of all raw foods. Fortunately, the numbers present generally are quite low. Small numbers of the vegetative cells and spores of *C. perfringens* are ingested without apparent harm. The organism is present in the human intestinal tract in numbers of approximately 10,000 cells per gram of feces. During an intoxication, one or more million cells per gram are shed.

Perfringens food intoxication is characterized by profuse diarrhea, abdominal cramps, and flatulence. Ordinarily there is no vomiting or nausea. Fever and chills are rare, as are headache, dehydration, and prostration; and fatal cases are extremely rare. Often, the symptoms are fairly mild and of short duration and take the form of an "embarrassing abdominal distress." It is probable that only a small percent of the illnesses are reported. Single episodes may involve one person or several hundreds.

Outbreaks have been associated primarily with beef and poultry, and with foods such as stews and creamed dishes made with these meats. The meats furnish the necessary amino acids for the growth of *Clostridium perfringens*. In typical incidents the foods have been heated and then kept in a warm atmosphere for several hours. The heating depletes the oxygen in the food and brings about chemical reactions that lead to a low Eh, and heat shocks the spores so that they undergo rapid germination. Changes leading to outgrowth of the spores have been observed as early as 30 seconds after heat shock. Growth is initiated at 43°C and continues rapidly during the early portion of the cooling period while the food is still warm. The generation time for *C. perfringens* may be as brief as 9.5 minutes. The bacterium produces copious amounts of gas and, in creamed meats, causes bubbling similar to that of simmering oatmeal.

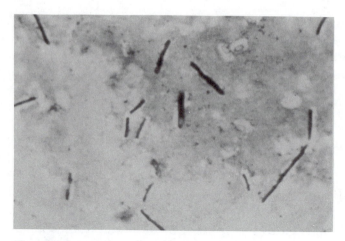

Figure 23-3
Clostridium perfringens (× 1000).

Most reported incidents of perfringens food poisoning occur as the result of failure to prevent cross contamination and to maintain proper control of temperature (Bryan and Kilpatrick, 1971). Outbreaks occur from improperly handled foods in food preparatory establishments such as institutions, schools, restaurants, cafeterias, delicatessens, and catered meals.

The illness has been described as a live-cell intoxication, since it is always associated with the consumption of foods with large numbers of viable cells (Hauschild, 1973). The heating to 63°C of a prepared food which has been subjected to temperature abuse destroys both the vegetative cells and the intoxicating agent.

Clostridium perfringens is a large, gram-positive, spore forming, anaerobic rod occurring singly, in pairs, and in short chains (Figure 23-3). The ovoid spores are centrally located, unswollen, and few in number. Although spores often are seldom seen in microscopic examinations of foods, they grow if suspected material is heated to 80°C for 10 minutes and placed in suitable subculture. The organism grows well in the pH range 5.0 to 8.5, between 15 and 43°C, and at Eh values below −115 millivolts, i.e., above that of venous blood but below that of arterial blood. *C. perfringens* may be inhibited by other bacteria such as enterococci and lactobacilli (Kafel and Ayres, 1969). It tolerates 6% or more of sodium chloride, but less of sodium nitrite. An evident nitrite effect, termed the Perigo effect, is produced in heated foods (Perigo and Roberts, 1968; see Chapter 17).

Reducing temperature to 20°C or less rapidly kills cells (Traci and Duncan, 1974). Storing foods frozen for 24 hours may reduce the population of cells by 99% (Hall, 1962). Very few of the spores of food poisoning strains germinate without heat activation (Barnes et al., 1963; Roberts, 1968). *Clostridium perfringens* seldom produces spores in meat (Mead, 1969) and appears not to sporulate in turkey meat at 20°C (Kim, 1965), in minced beef at pH 5.8 or less, in food products containing fermentable carbohydrates (Despaul, 1964; Mead, 1969), or in products with less than 0.96 a_w (Kim, 1965; Strong et al., 1970).

Five types of *Clostridium perfringens* are known, but only types A and C have been implicated in foodborne intoxications. Type A is the classic bacterium of human gas gangrene; type B is associated with lamb dysentery; type C with struck of sheep and human hemorrhagic enteritis and necrotic enteritis; type D with infectious enterotoxemia or pulpy kidney in lambs and grass sickness in horses; and type E with enterotoxemia in animals. These five types are distinguished from one another by the alpha, beta, epsilon and iota (α, β, ϵ, and ι) exotoxins produced by them. Two of the eight toxins produced by the several types of *C. perfringens* cited are hemolytic. Nygren (1962) attributed the specific action of foodborne intoxication to the hydrolysis of phosphorylcholine by phospholipase C, but this has been challenged by Dack (1964) and Weiss et al. (1966). *C. perfringens* type A food poisoning is much more common than type C and has milder symptomatology. *C. perfringens* type A food intoxications appear to develop as follows (Hauschild, 1973a):

1. ingestion of food contaminated with 10^6–10^7 vegetative *C. perfringens* cells per gram;

2. multiplication and sporulation of *C. perfringens* in the small intestine;

3. production of enterotoxin in the sporulating cells;

4. release of the toxin by cell lysis;

5. accumulation of fluid in the cell lumen caused by the action of enterotoxin;

6. diarrhea resulting from excess fluid accumulation.

Symptoms of illnesses attributable to *Clostridium perfringens* type C are severe abdominal pain, vomiting, diarrhea, necrotic inflammation of the small intestine, and a higher mortality rate than type A. Outbreaks of this illness, as a consequence of consuming canned meat contaminated with heat-resistant *C. perfringens* type C, were reported in Germany after World War II. While several serotypes of *C. perfringens* are present in the human intestinal tract, seldom does a perfringens intoxication involve more than one serotype. Little is known of how the toxin acts *in vivo*. The accumulation of fluid in the intestinal lumen may result from the increased permeability of the blood capillaries in the intestinal wall (Niilo, 1971), but it has been questioned whether permeability changes are the primary cause of fluid accumulation in the intestine (Hauschild, 1973).

The toxicity is inactivated by pronase and by the protease of *Bacillus subtilis,* but not by trypsin, chymotrypsin, or papain (Duncan and Strong, 1969). The UV absorption spectrum has a maximum of 278 to 280 nm and a minimum of 250 nm. The specific toxicity of the purified toxin is about 2000 mouse MLD/mg N; lethal doses generally cause death of mice within 20 minutes (Hauschild and Hilscheimer, 1971). The toxin has a Stokes radius of 2.6 nm (Hauschild, 1973), an isoelectric point of pH 4.3, and an apparent molecular weight of $36,000 \pm 4000$.

Detecting *Clostridium perfringens* Type A Enterotoxin

Methods for detecting type A enterotoxin by animal feeding have yielded to sophisticated biological and immunobiochemical methods, which became feasible as the toxin was separated from the living cell and purified. Animal feeding trials are insensitive to small quantities of toxin, are expensive to conduct, and are not reliable unless numerous precautions are taken.

The rabbit ileal (or ligated) loop technique, used to study the cholera toxin (Burrows and Musteikis, 1966), has been applied to the study of Type A enterotoxin (Duncan et al., 1968). In this procedure the exposed ileum of a fasting rabbit is ligated into segments 12 to 14 cm in length and the ligated sections are injected with the material under test. The ligated section distends if fluid accumulates because capillary permeability increases, vasodilation occurs, and intestinal mobility increases (Nilo, 1971). The ileal loop technique detects 6.5 to 29 μg of toxin (Hauschild, et al., 1971). The technique has been adapted to the study of *Vibrio cholerae, V. parahaemolyticus,* enteropathogenic *Escherichis coli,* salmonellae, and shigellae, all of which are customarily incriminated in live-cell intoxications. An intradermal injection into the rabbit of 0.125 μg or less produces an erythematous, nonnecrolyzing wheal in 30 minutes.

The immunobiochemical tests include the single-gel diffusion test in which toxin and antibody react to form a precipitate at an interface; the Ouchterlony double-diffusion test, in which antigen and antibody in separate wells diffuse into the agar and form a feathery precipitate where they meet; the electroimmunodiffusion test, in which protein in an agar well is stimulated by a high voltage and migrates into agar containing antibody (Duncan and Somers, 1972); and the reverse passive agglutination test, which makes use of erythrocytes coated with antibody in an agglutinating reaction, and which detects as little as 0.00,005 mg of toxin (Genigeorgis et al., 1973).

Botulism

Botulism (from the Latin *botulus,* sausage) is a rare neuroparalytic disease caused by consuming foods containing toxin of *Clostridium botulinum.* Being a saprophyte, the organism seldom grows or produces toxin in the live animal; it can do so only by growing in food. Also, although moderate numbers of spores can be ingested without apparent harm, consuming food contaminated with the preformed toxin is exceedingly dangerous, because it is the most lethal substance known.

Clostridium botulinum is a gram-positive, spore-forming, anaerobic, rod-shaped organism 0.5 to 0.8 by 3.0 to 8.0 μ with rounded ends occurring singly, in pairs, and in short-to-long chains, that ferments glucose and maltose but not lactose or salicin. It produces hydrogen sulfide and liquifies gelatin but does not reduce nitrates or produce indole. In milk *C.*

botulinum produces acid but does not digest casein or produce a stormy fermentation. The species produces eight immunologically distinct heat-labile toxins, all of which appear to be simple proteins.

The Toxins

Strains producing types A, B, and E toxins cause botulism in humans, whereas those producing C_α, C_β and D types primarily cause botulism in birds and nonhuman mammals. Type E botulism is caused mainly by consumption of contaminated marine seafoods, although freshwater fish have also been implicated. Only two outbreaks of type F botulism have occurred and, as far as is known, type G, isolated from soil in Argentina, has not caused cases of botulism in humans or animals.

Type A botulinal toxin was obtained in crystalline form and shown to be a simple protein of high molecular weight (i.e., 900,000) composed of amino acids only (Lamanna et al., 1946; Abrams et al., 1946). Later, purified type E toxin was obtained (Sakaguchi and Sakaguchi, 1961) in Japan by chromatographic procedures. It has a molecular weight of 350,000. Molecular weights of the ultimate toxic unit for type A toxin were reported (Gerwing et al., 1967) to be 12,000, those of type B to be 10,000 and of type E, 18,000. Although no other investigators were able to duplicate these findings, the controversy stimulated further investigations on the molecular size of these toxins (Schantz and Sugiyama, 1974). As a result, crystalline toxin A was found to dissociate into two components—one with hemagglutinating properties and a molecular weight of about 500,000, the other with an average molecular weight of 70,000 but with two to three times the specific toxicity of the parent toxin. Similarly, fractionation of the crystalline type A toxin yielded a hemagglutinating fraction of high molecular weight and a toxic fraction having a molecular weight of 120,000–150,000 but with three to five times the toxicity of the parent toxin (Das Gupta and Boroff, 1967). The smaller protein—about 20% of the crystalline toxin—was considered to be the neurotoxin and was designated α toxin; the hemagglutinating fraction (about 80%) was designated the β fraction. A type B neurotoxin of 167,000 molecular weight was isolated (Das

Gupta et al., 1968) and, when its associated nontoxic protein was removed, the molecular weight of type E neurotoxin was determined to be 150,000 (Kitamura et al., 1969).

Clostridium botulinum does not produce the fully toxic molecule; instead, a progenitor toxin is activated to its full toxicity by enzymes (Duff et al., 1956). The nontoxic component of the complex may form a shield to protect it against digestive processes so that it can enter the bloodstream. Orally ingested toxins, absorbed chiefly through the upper intestinal tract, are transported to the lymphatic system where the progenitor toxin is hydrolyzed to the highly toxic derivative toxin and a nontoxic residue (Sugii et al., 1977). The derivative toxin can be divided into two antigenic, nontoxic fragments with molecular weights of 111,000 and 59,000, respectively.

The toxin is carried by the bloodstream to target nerves, where it inhibits endings to effector elements (Figure 23-4). Synapse blockage is accomplished by interfering with some unknown step in the release of acetylcholine from the nerve endings. Animals and humans experience some relief when treated

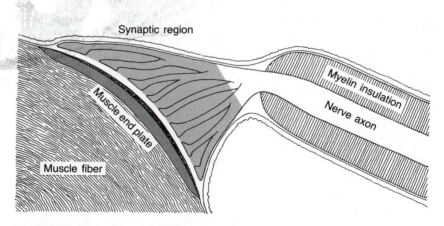

Figure 23-4
Nerve-muscle connection shows the actions of different nerve poison. Botulinal and dinoflagellate toxins (horizontal hatching) prevent the synaptic region from producing acetylcholine, which acts on the end plate to make muscle contract. Curare (vertical hatching) blocks the nerve impulse by making the end plate less sensitive to acetylcholine. (From "Poisonous Tides" by S. H. Hutner and John J. A. McLaughlin. Copyright © 1958 by Scientific American, Inc. All rights reserved.)

with guanadine and homoserine, substances that accelerate the release of acetylcholine at the nerve endings.

The toxins are good antigens, but natural immunity in people rarely occurs because the dose needed to elicit antibody production is larger than a fatal dose. Formaldehyde converts the toxins to a toxoid that has no toxicity but is still an antigen and so can be effectively used to immunize humans and animals. In addition to producing antisera for treatment of botulism in humans, the toxoids are valuable for immunizing those working with the toxin, with livestock, or with fur-bearing animals.

Toxicogenesis

Meats and proteinaceous, low-acid vegetables contaminated by *Clostridium botulinum* often develop especially obnoxious odors. Foods with more acid and those low in proteins may become just as toxic without much evidence of putrefaction. Type A strains and most cultures of type B are proteolytic and sufficiently putrefactive to produce an obnoxious odor in proteinaceous foods, but some strains of B and those of type E are not. Thus, a canned food may look, smell, and taste normal but contain a toxin. Although A and B produce toxin in low-protein foods such as string beans and corn, they give no marked indications of putrefaction. The organism ferments carbohydrates, producing gas, which, however, is also sometimes not evident.

Several combined factors determine whether spore germination occurs or at what rate and to what extent these include the composition of the food or medium, especially its nutritive properties; moisture content; pH; oxidation-reduction potential; salt content; and storage temperature and duration. Thus, the nutritive properties of the food are likely to determine the minimal pH or temperature and the maximal concentration of sodium chloride for growth and toxin production.

Although foods are known to differ as culture media for *Clostridium botulinum*, much of the supporting evidence is empirical. Most of the studies have been on toxin production in various foods. Meats, fish, and low- or medium-acid canned foods have been shown to support toxin production and to differ in the potency of the toxin formed. Toxin production in most foods is prevented by a pH of 4.5 or less, but the lowest pH allowing spore germination is considerably higher—for example, 4.8 in bread and pineapple-rice pudding and 5.0 in veal infusion broth. *C. botulinum* is able to grow and produce toxin in foods that otherwise are too acid if other microorganisms, especially molds, also are growing in the food and presumably raising the pH locally or generally. Conversely, *C. botulinum* fails to produce toxin when growing in the presence of large numbers of *C. sporogenes.* A maximal pH of 8.9 was found for vegetative growth but the toxin is unstable at pH values above 6.8.

Food composition and temperature determine what concentrations of sodium chloride are necessary to prevent growth and toxin production in foods. If sodium nitrate or disodium phosphate is present in cheese spread, less sodium chloride is needed to prevent toxin production. Under favorable cultural conditions 8% or more of salt is needed to inhibit *C. botulinum;* more salt is needed at 37°C than at 15°C.

Temperature is an important factor in determining if toxin production will take place and what the rate of production will be. Different strains of *C. botulinum* types A and B vary in their temperature requirements. A few strains of these two types have been reported to be able to grow at 10 or 11°C, but the lowest temperature for spore germination of most strains is considered to be about 15°C. Their maximal growth temperature is about 48°C. Although 30 to 37°C often is given as the optimal temperature for the organism, usually more toxin is produced at a lower temperature, possibly because the toxin is less stable at the higher temperature. Type E strains produce gas and toxin in 1 to 1½ months at as low as 3.3°C. The heat treatment necessary to destroy the toxin depends upon the type of organism producing the toxin and the medium in which it is heated. In the laboratory, heat treatments of from a few minutes to 30 minutes at 80°C inactivate the toxin; in practice it is recommended that suspected foods be kept at a full boil for at least 5 minutes. The toxin is destroyed in cheese by 7.3 Mrad of gamma rays and in broth by 4.9 Mrad. The toxin has been known to persist in foods for long periods, especially at low storage temperatures.

Table 23-3
Main feature of the different types of *C. botulinum.*

Type	Differentiated by	Year	Species mainly affected	Commonest vehicles	Highest geographic incidence
A	Leuchs Burke	1910 1919	Man; chickens ("limberneck")	Home-canned vegetables and fruits; meat and fish	Western United States, Soviet Ukraine
B	Leuchs Burke	1910 1919	Man, horses, cattle	Prepared meats, especially pork products	France, Norway, Eastern United States
C_α	Bengston	1922	Aquatic wild birds ("western duck sickness")	Fly larvae (*Lucilia caesar*); rotting vegetation of alkaline ponds	Western United States and Canada, South America, South Africa, Australia
C_β	Seddon	1922	Cattle ("Midland cattle disease"), horses ("forage poisoning"), mink	Toxic forage; carrion, pork liver	Australia, South Africa, Europe, North America
D	Theiler and Robinson Meyer and Gunnison	1927 1929	Cattle ("lamziekte")	Carrion	South Africa, Australia
E	Gunnison, Cummings, and Meyer Kushnir	1936 1937	Man	Uncooked products of fish and marine mammals	Northern Japan, British Columbia, Labrador, Alaska, Great Lakes region, Sweden, Denmark, USSR
F	Moeller and Scheibel Dolman and Murakami	1960 1961	Man	Home-made liver paste	Denmark
G		1973		Soil	Argentina

Source: Dolman and Murakami, 1961, with additions.

Spores of *Clostridium botulinum* occur in virtually all agricultural soils and are transferred to vegetable crops by wind, rain, dust, insects, and by handling during harvest. Prevalence of the several *C. botulinum* types differs geographically (Table 23-3).

Compared with the spores of most other species of *Clostridium*, those of some of the putrefactive anaerobes, including *C. botulinum*, have a comparatively high resistance to heat. The heat treatment necessary to destroy all of the spores in a food will depend upon the kind of food, the type and strain of *C. botulinum*, the medium in which the spores were formed, the temperature at which they were produced, the age of the spores, and the number of spores present. The following heat treatments are needed to destroy all spores of *C. botulinum* (Esty and Meyer, 1922):

100°C	6 hours
105°C	2 hours
110°C	36 minutes
115°C	12 minutes
120°C	4 minutes

In general, spores of organisms of types C, D, and E are less heat resistant than those of types A and B; type E spores are inactivated in 15 minutes at 80°C.

Before 1950 a case fatality rate of about 60% obtained but has since gradually decreased—for example, to 23% for the 10 year period 1964–1973 (Figure 23-5) but has since dropped.

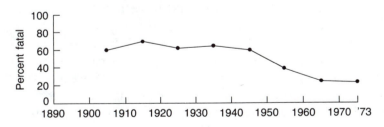

Figure 23-5
Foodborne botulism death-to-case ratios, by 10-year periods, 1899–1973. (Center for Disease Control, 1975.)

Table 23-4

Food products causing botulism outbreaks, 1899–1973.[a]

Botulinum Toxin type	Vegetables	Fish and fish products	Fruits	Condiments[b]	Beef[c]	Milk and milk products	Pork	Poultry	Other[d]	Total
A	96	7	22	16	3	2	2	1	5	154
B	24	3	5	4	1	2	1	2		42
E	1	19								20
F					1					1
A&B	2									2
Total	123	29	27	20	5	4	3	3	5	219

[a] Includes only outbreaks in which the toxin type was determined.
[b] Includes outbreaks traced to tomato relish, chili peppers, chili sauce, and salad dressing.
[c] Includes 1 outbreak of type F in venison, and 1 outbreak of type A in mutton.
[d] Includes outbreaks traced to vichyssoise soup, spaghetti sauce, and corn and chicken mash.
Source: Center for Disease Control, 1975.

Botulism in Humans

Faulty vegetable processing has caused about 60% of all botulism outbreaks in which the toxin type was determined (Table 23-4). Fish and fish products have been responsible for the next largest percent of incidents; fruits and condiments also were important offenders. Occasionally, milk and milk products, pork, poultry, beef, mutton, and venison are implicated.

For almost three quarters of a century, a mean of 5 to 25 cases occurred annually in the U.S. (Figure 23-6). Most incidents in which the toxin type was identified were attributed to type A until the decade 1960–1969, during which type E was the most prevalent. Since then both types A and B have been more common than type E. In 1963 there were 47 cases of botulism, of which, in contrast to the usual pattern, 24 were traced to contaminated commercially processed foods; 22 of the 24 were caused by type E resulting from ingestion of fish products. In 1973, 34 cases were reported.

Several illnesses have been misdiagnosed as botulism. Those most likely to be confused include Guillain-Barre (GB) syn-

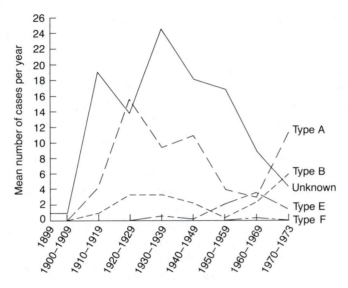

Figure 23-6
Mean number of cases of foodborne botulism per year,
by toxin type. (Center for Disease Control, 1975.)

drome, staphylococcal food poisoning, chemical intoxication,
cerebrovascular accidents involving the midbrain, myasthenia
gravis, tick paralysis, phenothiazine idiosyncratic reaction,
and diphtheria. Of 438 suspect botulism outbreaks reported
to CDC between 1964 and 1973 only 17.1% proved to be
botulism, whereas while GB syndrome accounted for 10.5%.
GB syndrome can simulate botulism, but muscular cramps,
paresthesiasis, and an elevated spinal fluid protein in the absence of bacterial cells help distinguish the GB syndrome.
Botulism can be confirmed by demonstrating that the patient's
serum is toxic for mice. In the human victim the first symptoms—vomiting, nausea, abdominal pain, and diarrhea—
generally occur within 18 to 36 hours after the toxin is ingested.
These symptoms are succeeded by lassitude and vertigo.
Dryness of the mouth, throat, and tongue, and widely dilated
and unreactive eye pupils or double vision follow. Later indicators include difficulty in swallowing and breathing, blurred
vision or diplopia, paralysis, drooping of the face muscles,

weakness of the tongue, diaphragm, neck, and extremities. Often there is constipation. The pulse usually remains normal and fever is absent. The principal cause of death is progressive respiratory distress, but air blockage, pulmonary infection, and cardiac arrest may also be involved. The amount of toxin needed to cause death in humans has been estimated to be from 0.1 to 1 μg, or from 3000 to 30,000 mouse intraperitoneal LD_{50} doses. The latter dosage is calculated by suspending serial dilutions of the toxin in .05 M phosphate buffer at pH 6.8 and intraperitoneally injecting these aliquots into separate groups of six to eight 20 gram mice and plotting the percent kill in each group within a 96 hour period.

Inadequately heated or cured foods prepared in the home are the most frequent cause of botulism (Table 23-5). Before the end of World War II a much larger proportion of foods was canned each year in the home than has been since the war. In fact, the frequent outbreaks of botulism that occurred as a result of improper home processing led the Federal Bureau of Home Economics about 1920 to become interested in launching an educational program concerning the need for pressure cooking when meats and vegetables were preserved by home canning. It was pointed out that increasing the pressure by 10 pounds raised the processing temperature to 115.5°C, greatly increasing the bactericidal effect—in fact, the killing effect at 115.5°C is 14 times that at 100°C. This campaign, together with ever-increasing commercial processing, resulted in fewer and fewer outbreaks. The difference between the numbers of outbreaks for home-processed and commercially processed products was most sharply defined between 1940 and 1949. In that decade a commercial product caused only one outbreak, while home-processed products caused 120—about 90% of all outbreaks in humans resulting from improper sterilization. While it is possible that the toxin may be produced *in vivo* by vegetative cells of C. botulinum (Starin and Dack, 1928) there is little experimental evidence to support this view (Center for Disease Control, 1972).

Occasionally, botulism occurs with no involvement of food. Ten cases of wound botulism were reported between 1943 and 1973.

Table 23-5

Outbreaks of foodborne botulism attributed to commercially processed or home-processed foods, 1899–1973.

Source of food	1899	1900–1909	1910–1919	1920–1929	1930–1939	1940–1949	1950–1959	1960–1969	1970–1973	Total
Home processed	1	1	48	77	135	120	50	42	21[a]	495
Commercially processed	0	1	14	26	6	1	2	10	2	62
Unknown	0	0	8	13	13	13	51	26	7	131
Total	1	2	70	116	154	134	103	78	30	688

[a] Includes one outbreak in which the vehicle was canned by the owner of a restaurant and sold to his customers.
Source: Center for Disease Control, 1975.

Infant Botulism

A recent phenomenon in the United States and England, botulism in infants, has been attributed to the growth of *Clostridium botulinum* in the infant's body, accompanied by formation of toxin. Although honey has been implicated in some episodes, no specific foods have been incriminated, nor has there been any apparent predisposing weakened condition of the infant. Since spores of *C. botulinum* occur in the environment and on foods, particularly fresh fruits and vegetables, it is assumed that the spores gain entry with any of a number of foods.

Bacillus cereus Poisoning

Bacillus cereus (from *Ceres,* the Latin name of Demeter, the Greek goddess of agriculture) occurs commonly in soils. It is readily isolated from nearly all plant foods and dehydrated foods, spices, seasoning mixes, flours, starches, and skim milk powder, and from cooked foods such as rice, potatoes, soup mixes, and spaghetti (Kim and Goepfert, 1971). In any of these foods its counts range from near zero to 10,000 per gram. It has no unique nutritional requirements.

Small numbers of spores and vegetative cells probably are swallowed daily with impunity. When *Bacillus cereus* cells are present in large numbers in foods, however, they produce an intoxication characterized by acute abdominal pain, flatulence, and diarrhea. Headache and dizziness are common; dehydration and prostration may occur; but nausea, vomiting, fever, and chills are rare. The illness appears within 8 to 24 hours after consumption of the food, and the symptoms usually last less than 24 hours.

As early as 1906 European workers recognized *Bacillus cereus* as a causative agent of foodborne illness. It has received attention in the United States only in recent years (Goepfert et al., 1971). In southern Europe it is a summertime disease. Pumpkin greens cooked with immature fruit are a common vehicle. In the United States and the United Kingdom it is often associated with cooked rice in restaurants, which is frequently prepared in large quantities as a matter of convenience, but a variety of other foods and meat dishes have also been incriminated.

Bacillus cereus forms nonswollen, cylindrical spores and vegetative cells 0.9 μ or more in diameter. A pearly iridescence characterizes the surface of the colony. When stained with a simple stain such as methylene blue, the cells appear granulated internally. Only two other species in the genus *Bacillus* form large, granulated cells: *B. anthracis,* which forms a colony resembling a Medusa head but rarely occurs in foods; and *B. mycoides,* which forms across the agar long tendrils resembling tufts of Angora wool.

Bacillus cereus grows in laboratory media at temperatures of 10 to 50°C and at pH 4.9 to 9.3. The spores have a D_{100} value in skim milk of 2.7 to 3.1 minutes and of 5 minutes in low-acid foods (pH >4.5). In phosphate buffer at pH 7.0, D values at 85, 90, 95, and 100°C are 220, 71, 13, and 8 minutes, respectively. Since few meat dishes and vegetables are heated to these temperatures for such periods of time, the potential for survival is great. In pasteurized milk, the spores gradually settle to the bottom of the container and germinate to produce a soft coagulum known as bitty milk or bitty cream. Pasteurized milk has not been incriminated in intoxications caused by *B. cereus.*

The physiological properties of the bacterium are seemingly well described. It is, however, variable in its ability to ferment many sugars and to reduce nitrate (Kim and Goepfert, 1971). It hydrolyzes starch and produces acetylmethylcarbinol. Most strains hydrolyze casein and gelatin, produce penicillinase, and are resistant to lysozyme.

Intoxication occurs in foods in which *Bacillus cereus* can grow rapidly and to high numbers—foods that have been heated to destroy the competing microflora and given prolonged storage at room temperature. The spores quickly germinate at temperatures approaching 50°C. The intoxication invariably is associated with the consumption of foods containing large numbers of live, vegetative cells. The toxin is thermolabile and is destroyed simultaneously with the vegetative cells when foods are reheated to 60°C. Therefore, unused portions of food should be promptly refrigerated or, if held at room temperature for a few hours, reheated to avoid intoxication by *B. cereus.*

Foods in which the presence of *Bacillus cereus* is suspected are cultured by surface plating in serial dilution on the KG medium of Kim and Goepfert (1971), containing mannitol, egg yolk, and polymyxin. Colonies of *B. cereus* are recognized by the pearly surface and by the zone of precipitation brought about by the action of lecithinase C on the egg yolk. An isolated strain is confirmed as an enterotoxic agent by first injecting 0.05 ml culture filtrate into the shaved skin of a rabbit, then injecting a 10% solution of Evans Blue into the ear vein. The entire shaved skin turns gray-blue, but darker areas develop at the site of the injection. The diameter of the dark blue area indicates the potency of the toxin produced by the strain.

When *Bacillus cereus* is administered orally, cats and dogs prove to be quite sensitive to its enteropathogenic activity. If sufficient numbers of cells are given, cats develop watery, thin diarrhea and may die (Nikodemusz and Gonda, 1963). Not all strains of *B. cereus* isolated from foods bring about the enteropathogenic effect (Glatz et al., 1974; Powers et al., 1976).

Mushroom Toxins

Relatively few people in the United States collect wild mushrooms for culinary purposes and only 27 deaths were reported between 1958 and 1968 (Benedict, 1972). In Western Europe, however, where edible mushrooms are eagerly collected, poisonings are frequent. In Switzerland almost 5% cases of mushroom poisonings recorded between 1919 and 1958 (i.e., 96 of the cases) were fatal (Alder, 1961). In Japan fatalities from mushroom poisoning are reported to be even higher (Romagnesi, 1964).

The green or white *Amanita* spp. cause more than 95% of fatal mushroom poisonings (according to Alder's data, 91 of the 96 deaths). Specimens are often confused with *Agaricus campestris,* the mushroom produced commercially, and with the edible *Tricholoma equestre.* White *T. equestre* looks like *A. campestris,* although the former has faintly pink lamellae that become darker red with maturity; this mushroom has a typical aniselike odor (Wieland and Wieland, 1972). The amanita is a white-gilled, white-spored mushroom borne on a sturdy stem arising from the *volva* (a sheath, or cup), with a parasol-shaped cap that

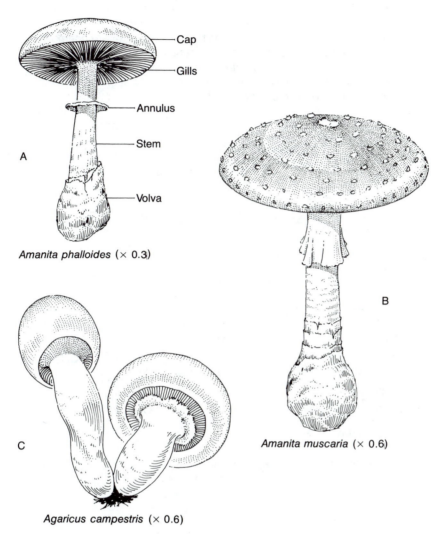

Amanita phalloides (× 0.3)

Amanita muscaria (× 0.6)

Agaricus campestris (× 0.6)

Figure 23-7
Morphology of poisonous mushrooms, A., *Amanita phalloides,* and, B.,
Amanita muscaria; and a nonpoisonous mushroom, *Agaricus campestris.*

readily separates from the stem, and an *annulus,* or ring, high
on the stem. The ragged shaped volva at the base of the stem
remains in the soil (Figure 23-7).

Often mushrooms of unknown vintage eaten accidentally or
deliberately may behave as hallucinogens with no far-reaching
deleterious effect. In many instances, the nature of the chemical

Box 23-1

All of the toxic peptides of the amanitas are cyclic and are composed of only a few amino acids. Each cyclo-peptide has a molecular weight of about a thousand and contains a gamma-hydroxylated amino acid and a sulfur atom derived from a cysteine residue connected to the indole nucleus of a tryptophane side chain.

	R_1	R_2	R_3	R_4	R_5
Phalloidin	OH	H	CH_3	CH_3	OH
Phalloin	H	H	CH_3	CH_3	OH
Phallisin	OH	OH	CH_3	CH_3	OH
Phallacidin	OH	H	$CH(CH_3)_3$	CO_2H	OH
Phallin B (tentatively)	H	H	$CH_2C_6H_5$	CH_3	H

Source: Wieland, 1968.

The common cyclic octapeptide structure of the α, β, γ, and ϵ amanitins and of amanin and amanullin is shown.

Box 23-1 continued

	R_1	R_2	R_3	R_4
α-amanitin	OH	OH	NH_2	OH
β-amanitin	OH	OH	OH	OH
γ-amanitin	OH	H	NH_2	OH
ϵ-amanitin	OH	H	OH	OH
Amanin	OH	OH	OH	H
Amanullin	H	H	NH_2	OH

Source: Wieland, 1968; Wieland and Wieland, 1972.

Amanin differs from β-amanitin only in lacking a phenolic hydroxyl group in position 6 of the indole nucleus. Amanullin, which is totally nontoxic, has a second molecule of isoleucine rather than a γ-hydroxylated sidechain. Apparently the amatoxins require a -hydroxy group for toxic activity.

compound is unknown. Symptoms of intoxication by the poisonous mushroom are complex, resembling those of drugs that act on the central nervous system. Vomiting and diarrhea, the first symptoms of intoxication, do not become apparent until 10 to 24 hours after ingestion. By that time the lethal toxic peptides have begun to act. Two families of poisonous peptides are involved, the phallotoxins and the amatoxins. The *phallotoxins* include (their minimal lethal mouse doses/kg of body weight are in parentheses) *phalloidin* (2.0), *phalloin* (1.5), *phallicidin* (2.5), *phallisin* (and *phallin* B) (10). The *amatoxins* are resolved into *amanitins* of types α– (0.35), β– (014), γ– (0.2), δ–, and ε– (1.0), and into *amanin* (0.5) and the nontoxic *amanullin* (Wieland, 1968). In the amanitas there are about 10 mg phalloidin, 8 mg α-amanitin, 5 mg β-amanitin and 0.5 mg γ-amanitin per 100 g fresh tissue or 5 g dry weight. The slow-acting amatoxins begin their action as soon as they reach the liver, and death is caused by the rapid and irreversible destruction of liver cells. Since the lethal dose of the amanitins is less than 0.1 mg/kg human body weight, the toxin content of a mushroom weighing 50 g (containing about 7 mg amanitins) may be sufficient to cause death.

The phallotoxins act quickly and, when consumed in high dosages, cause death in animals within one to two hours. Because the action of the amatoxins is delayed, even very high doses are not lethal within less than 15 hours. Relative toxicity is, however, just the opposite: the amatoxins are stated to be 10 to 20 times more toxic than the phallotoxins and therefore are considered the major poisonous constituents of the deadly amanita (Wieland, 1968).

Whereas the phallotoxins act specifically on the liver (Siess et al., 1970), the amatoxins attack both the kidney and the liver. Phalloidin has a marked affinity for the microsomal fractions of the liver cell and affects metabolism only if administered to the intact animal. Phalloidin itself may not be toxic, but drug-metabolizing enzymes of the liver convert it into a toxic compound that, in turn, affects the structure of the liver's endoplasmic reticulum.

The amanitins primarily affect the nuclei. The nucleoli of liver cells begin to fragment, causing cytoplasmic lesions that are not entirely restricted to liver cells; cytological changes are also observed in kidney. The RNA content decreases during the first 24 hours of an intoxication of the animals with α-amanitin. Although RNA synthesis may not be impaired, mouse liver cells are less able to incorporate ^{14}C-orotic acid, and α-amanitin added to mouse liver nuclei inhibits the RNA polymerase reaction, which is activated by manganous ion and ammonium sulfate. During the course of intoxication the cytoplasm is so affected that within two to four days death results from cellular necrosis (Wieland and Wieland, 1972).

Amanita muscaria, the fly agaric, is much less toxic to humans than is *A. phalloides,* the green death cap or destroying angel of Central Europe or the deadly agaric of North America. In this country the white *A. verna* and the yellow *A. citrina* are as common as *A. phalloides* and often have been reported to incite fatal poisoning.

No statistics are available concerning muscarine poisoning in the United States. In Europe several cases are reported each year (Eugster, 1957). Extracts of *Amanita muscaria* kill insects, presumably by muscarine chloride, $C_9H_{20}O_2N^+Cl$, an oxo-heterocyclic quaternary salt, 2-methyl-3-hydroxy-5-trimethyl-ammonium methyltetrahydrofuran chloride. The mushroom is reported to be used in Siberia as a hallucinogen. Other mushrooms, especially many species of *Inocybe* and *Clitocybe*, have much larger amounts of muscarine. The inocybe mushrooms cause muscarine toxicosis distinctly different than the toxic state produced by *A. muscaria.*

Muscarine

The active hallucinogenic principle may be muscimol, an amine easily derived by decarboxylation of ibotenic acid, which occurs naturally in a relatively high concentration (0.03 to 0.1%1).

Ibotenic acid   Muscimol

Ibotenic acid and tricholomic acid, its erythro-2,3-dihydro derivative, have great tastiness. These compounds have 20 times the flavor intensity of sodium glutamate and have a synergistic action with nucleotide seasonings.

Ergotism

The overwintering stage of *Claviceps purpurea,* a fungal pathogen of rye—and, less commonly, of barley, wheat, and other grains—produces a toxic product called *ergot,* or St. Anthony's fire. (Further details concerning the life cycle of this fungus are

given in Chapter 12.) Fungal tissue grows on the grain, trans-
forming, enlarging, and replacing the seed tissue with a compact, hard, dark purple sclerotial mass that is collected as the
ergot commercially marketed. The alkaloids contained in ergot
possess profound toxic characteristics, but after purification
and when used in proper dosages, ergonovine, ergometrine,
ergometrinine, ergotamine, ergotaminine, and other active
principles have important pharamacological properties. Trimethylamine gives the drug its unpleasant odor.

Ergot contains at least five isomeric pairs of alkaloids; the
levorotatory isomers have proved to be medicinally useful in
inducing uterine contraction to cause abortion during early
pregnancy or, later, to hasten childbirth and to return the
uterus to normal size afterward. Also, it is used to initiate con-
tractions of bladder, stomach, intestines, and other involuntary
muscles, to control bleeding, to selectively block the sympathet-
ic nervous system, and to alleviate certain vascular disorders
such as migraine headaches. Lysergic acid serves as a nucleus
for the alkaloids of ergot, and so is a natural source of lysergic
acid diethylamide (LSD), a drug that induces hallucinations.

Alimentary Toxic Aleukia (ATA)

Outbreaks of a mycotoxicosis called **alimentary toxic aleukia**
(ATA) were recorded in the nineteenth century in Russia,
where the toxicity of smut and rust fungi for humans and animals was actively investigated in the second half of that century
(Bilay, 1960). Despite these early studies relating mold-
contaminated feeds with disease, it was not until 1943–1944
that a severe widespread epidemic of humans and animals in
the Orenburg district of Russia was observed and definitively
related to the consumption of moldy millet (Joffe, 1963). In
1944 there was many fatalities, and ATA or septic angina was
noted among persons who had consumed grain that had been
cut but, during the siege of Stalingrad, had overwintered un-
harvested in the fields in 1942–1943 (Figure 23-8; Joffe, 1965).
Clinical manifestation of the disease included leukemia, ag-
ranulocytosis, necrotic angina, hemorrhagic diathesis, sepsis,
and aregenerative exhaustion of the bone marrow.

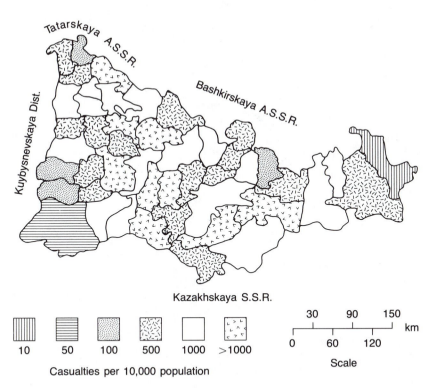

Figure 23-8
Casualties related to consumption of moldy millet in a district in the Soviet Union after the siege of Stalingrad. (Joffe, 1963. Reprinted from *Plant and Soil 18*, 31.)

The production of toxin was associated with fungal growth on the grain at low temperature. Numerous isolates from the genera *Fusarium, Cladosporium, Alternaria, Penicillium,* and *Mucor* were associated with the exposed grain, but it was not until later that Joffe (1960 and 1962) reported on the prevalence of *Fusarium* mycotoxicosis in diseased grain and pointed out the psychrophilic character (i.e., growth at −2 to −10°C) and other biological properties of these fungi.

T-2 Toxin

T-2 toxin, or diacetoxyscirpenol, is isolated from *Fusarium tricinctum* strain T-2 (Figure 23-9 A), reported to be one of the major toxin-producing species on moldy corn in the Midwest.

A

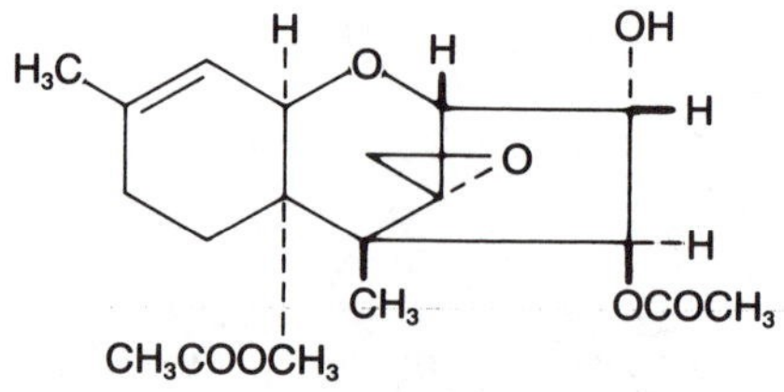

Diacetoxyscirpenol

B

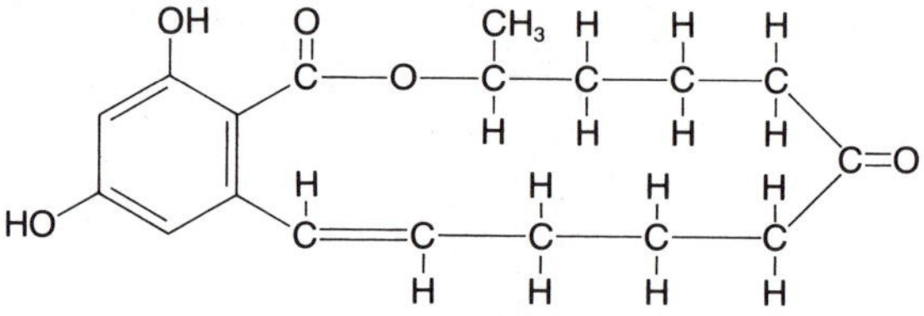

Zearalenone

Figure 23-9
A. T-2 toxin, or diacetoxyscirpenol; B. F-2 toxin, or zearlaenone. (Courtesy F. S. Chu.)

If 0.1 mg of T-2 toxin is applied to the skin of three-week-old rats, the test area becomes swollen and sensitive to pressure. Between the third to sixth day a large scab forms over the treated area, and when a larger dose (i.e., 0.25 mg) is given, the animals become lethargic and ataxic and lose appetite; soon thereafter most of the animals die. Japanese investigators (Tatsumo et al., 1968; Ueno, 1970) reported severe dermal irritation of experimental animals when *F. tricinctum* was applied to their skin. Laboratory workers frequently suffer serious dermal reaction from inadvertent contact with extracts containing these compounds (Grove, 1970).

Mycotoxins of the trichothecene type may have caused the widespread ATA outbreaks that occurred in Russia. Originally Joffe (1962) suggested that ATA was caused by organisms such as *Fusarium sporotrichioides* and *F. poae* which may be identical with *F. tricinctum* (Snyder and Hansen, 1945); the fusarenones and nivalenol might be similarly implicated metabolites. Saito and Tatsumo (1971) believe that these scirpene metabolites are most likely responsible for ATA. Diacetoxyscirpenol causes severe damage to the hematopoietic systems—one of the pathological aspects of ATA.

Zearalenone

Zearalenone, or F-2, a resorcylate—or, more exactly, 6-(10 hydroxy-6-oxo-trans-1-undecenyl)-B-resorcylic acid lactone (Figure 23-9 B) produced by *Fusarium graminearum* or *F. roseum*— is a toxin having an estrogenic effect on swine, poultry, and other animals. When the perfect stage of the causal fungus is recognized, it is then known as *Gibberella zeae*. *F. graminearum* causes ear and stalk rot of corn; the infection is recognized by a pinkish-white mycelial growth at the tip of the ear and even on the husks—although areas of the latter may be darkened. Individual infected kernels may be shrunken and heavily overgrown and have blotches of color ranging from pinkish-white to pinkish-red, particularly on the face (Figure 23-10). It also causes molding of barley and some other grains.

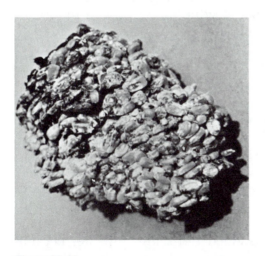 

Figure 23-10
Left: Corn kernels overgrown by *Fusarium graminearum;* Right:. Close view showing fungal development.

Fusarium or *Gibberella* damage becomes a serious problem during cold, wet growing seasons or when harvested grain is permitted to remain damp and cool for extended periods. During normal seasons at harvest time, corn has a moisture content of 14% while activity of *G. zeae* does not occur until a moisture level of > 20% has been attained. During the fall and winter of 1972–1973 much of the corn crop in southern Michigan, northern Indiana, and northwestern Ohio was contaminated by this fungus. Moisture levels for corn in this area ranged from 20 to 35%.

Swine fed such corn respond in one or more of the following ways: (1) they refuse to eat the grain; (2) they vomit; (3) they show poor weight gain; (4) they suffer reproductive disorders; and (5) they abort. Corn that swine refuse to eat may be acceptable to cattle and poultry, but not always, according to isolated reports. Feed with over 5% "Gib-damaged" kernels may be rejected by swine, but at somewhat lower levels the animals may eat the feed and become ill or experience estrogenic imbalances. In prepuberal gilts the vulva becomes swollen and there may even be vaginal prolapse; the uterus becomes enlarged, the ovaries shrunken. Pregnant sows may abort. In young males the testes atrophy and the mammary glands enlarge.

Zearalenone is monitored by federal regulatory agencies because it is feared that a toxic residue might remain in animal tissues and enter the human food chain.

Aflatoxins

In 1960 two separate mycotoxicoses were reported that stimulated much interest in the metabolic activity of various fungi. One mycotoxicosis involved the loss of turkey poults fed moldy peanut meal. Because its true cause was unknown at the time, workers at the Tropical Products Institute in Britain called the disorder "Turkey X Disease." Brazilian groundnut (peanut) meal was identified as the source of the toxic agent. Shortly thereafter a similar disease caused severe losses of ducklings in Kenya and Uganda. Poisoned birds lost appetite, became lethargic, and suffered wing weakness. In young birds that had ingested the suspect meal for a few days, liver parenchyma cells were necrotized, and bile duct cells proliferated excessively.

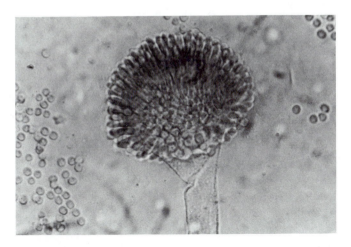

Figure 23-11
Photomicrograph of *Aspergillus flavus* (× 1000).

Examination of the feed revealed that the mold *Aspergillus flavus* Link (Figure 23-11) produced the toxin. A characteristic blue fluorescence was observed in the toxic meal. Further study demonstrated that the toxic material was extractable with chloroform and consisted of four fractions that were separable by chromatography and that fluoresced as individual spots when exposed to long-wave (ultraviolet) light.

At about the same time, 1960, that the two mycotoxicoses just described were reported, hepatomas of hatchery-reared rainbow trout, *Salmo gairdneri*, appeared in epidemic proportions at various localities in the United States and Europe. This trout hepatoma appeared in private, state, and federal hatcheries after dry, pelleted commercial feeds had been introduced as a primary dietary source. When known hepatomagenic diets were fractionated, four lipid soluble carcinogenic agents were isolated. Investigators quickly learned that certain strains of the common mold *Aspergillus flavus* that grew in damp peanut meal or cottonseed products were able to synthesize highly substituted coumarins containing bifuran rings and lactone configurations (Asao et al., 1963, 1965) (Figure 23-12). Owing to their blue or green fluorescence when exposed to ultraviolet light,

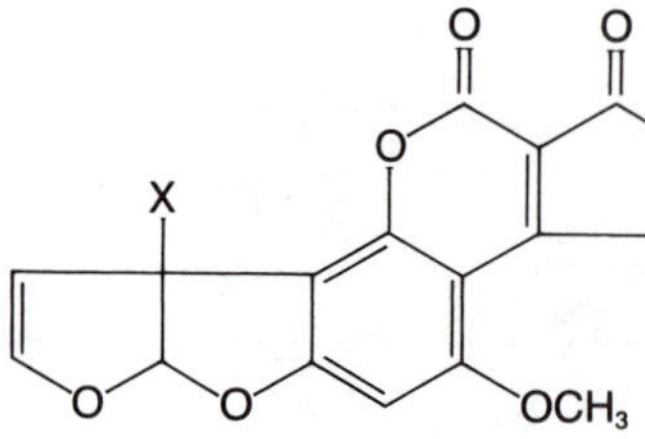

B₁ : X = H
M₁ : X = OH

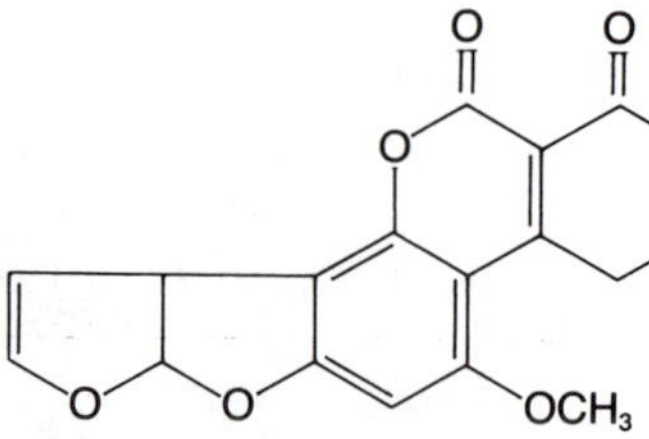

G₁

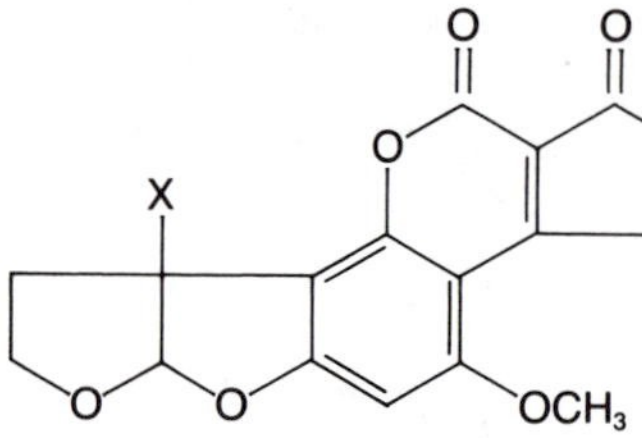

B₂ : X = H
M₂ : X = OH

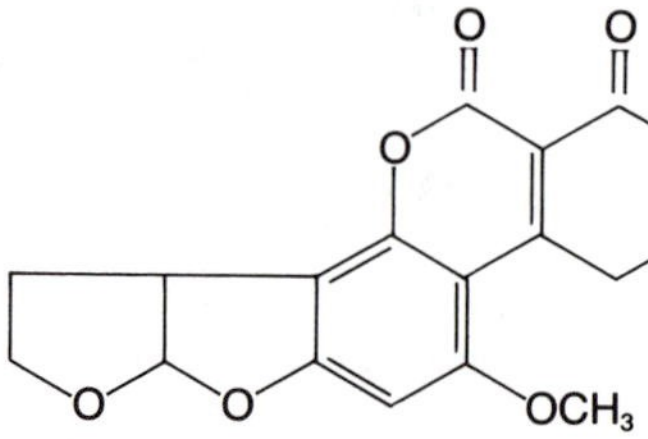

G₂

Aflatoxin

A

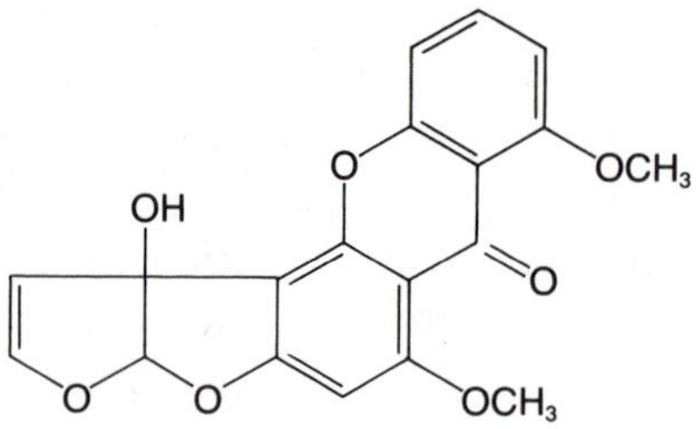

Aspertoxin
(OH-O-methylsterigmatocystin)

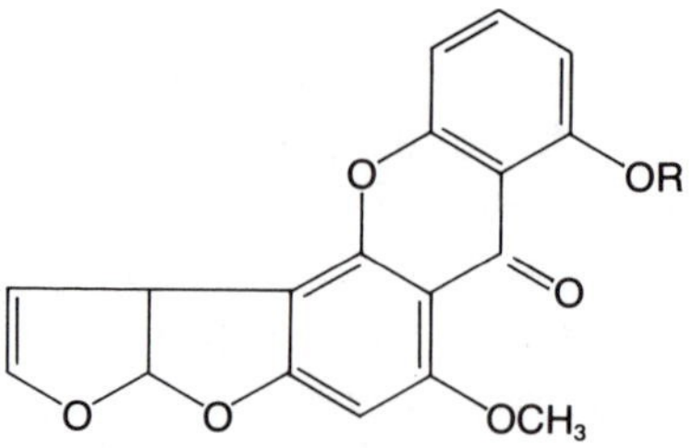

Sterigmatocystin (R = H)
O-methylsterigmatocystin (R = CH₃)

B

Figure 23-12
A. Molecular structures of aflatoxins B₁, B₂, G₁, G₂, M₁, and M₂; B. Structures of aspertoxin and sterigmatocystins. (Courtesy of Dr. F. S. Chu.)

the four isolated compounds were designated aflatoxins B_1, B_2 and G_1, G_2.

The pentanone structure of the coumarin nucleus that fuses with the bifuran in the B series is replaced by a six-membered lactone in the G series. If animals have eaten feed experimentally contaminated by toxic groundnut meal, two hydroxyaflatoxin derivatives, M_1 and M_2, appear in the female animal's milk; these derivatives have been isolated from several samples of market milk in South Africa. Also isolated have been two additional hydroxy derivatives of aflatoxin, i.e., B_{2a} and G_{2a}, and a phenolic derivative of aflatoxin B_1, appearing mainly in conjugated form in the urine of rhesus monkeys. In addition to the aflatoxins, several substituted xanthone furofuran compounds, the sterigmatocystins and aspertoxin, have been implicated as carcinogens. Within this family of mold metabolites, B_1 is the most acutely toxic with an LD_{50} of 0.36 mg/kg. M_1 is almost as active, followed by G_1 (0.79 mg/kg), B_2 (1.8 mg/kg), and M_2 and G_2 (3.45 mg/kg). Aspertoxin is only 1/25 toxic as B_1 and the toxicities of B_{2a}, G_{2a}, and sterigmatocystin are 1/60, 1/100, and 1/250 that of B_1, respectively.

Evidence continues to mount implicating aflatoxins as human carcinogens. In 1963 an anonymous writer speculated (J. Amer. Med. Assoc., 1963) on the relation of moldy diet to the high incidence of hepatoma among African Bantu tribes, where hepatomas represent 68% of carcinomas—an incidence five to eight times higher than that reported in the United States. Extensive liver cancer also occurs in Asiatic countries, where customary dietary patterns include sundry mold-fermented foods.

Regional high incidences of hepatomas are related to the necessity of ingesting contaminated foods because of food scarcity (Alpert et al., 1968). During treatment for kwashiorkor, 3 of 20 children who accidentally consumed some commercial peanut meal having an aflatoxin content of 300 µg/kg of meal developed hepatomegaly after 3 weeks (Parpia and Sreenivasamurthy, 1971). Needle biopsies from the livers of the other 17 showed pathologies progressing from fatty liver infiltration to different grades of "Indian childhood cirrhosis" within a year.

Hundreds of Thai children 1 to 13 years of age die annually of disease of unknown etiology manifested by vomiting, convulsions, coma, and death. Chemical analysis of 23 autopsy specimens from hospitalized children who died with acute encephalopathy and fatty degeneration of the viscera (EFDV) revealed the presence of aflatoxin B_1 in 22 of the 23 EFDV cases. High levels of aflatoxins have been detected in human tissue—93 μg B_1/kg in liver, 123 μg/kg in stool, 127μg/kg in stomach and intestinal contents, 8 μg/ml in bile, and trace amounts in brain, kidney, and urine (Shank et al., 1971). The extent of aflatoxin contamination in food samples from Thai markets seemed to parallel the disease both seasonally and geographically. Also noted were marked similarities between Reye's syndrome (acute aflatoxin B_1 poisoning in young macaques) and EFDV in Thai children; it was suggested that aflatoxins and possible other mycotoxins may be a factor in the etiology of EFDV.

In 1974 a large number of deaths in parts of western India were linked to the consumption of corn contaminated with aflatoxins. Symptoms were jaundice, hepatitis, rapidly developing ascites, portal hypertension; the mortality rate was as high as 20% of those affected. Corn harvested during unseasonally heavy rains became contaminated through a combination of conditions: drought and, through poverty and ignorance among the populace, improper storage. From 6.25 to 15.6 ppm aflatoxin was found in the contaminated corn and it was estimated that the level of aflatoxin consumption ranged from 2 to 6 mg daily for several weeks. Several hundred persons in more than 200 villages were affected. An autopsy liver study indicated bile duct proliferation, which appeared to result from aflatoxicosis (Food Chem. News, 1976).

Aflatoxin is a potent hepatic-carcinogen for ducks, rats, rainbow trout, ferrets, hamsters, guinea pigs, and monkeys. The hepatoma induced by aflatoxin ordinarily is classified as a liver-cell adenoma. In their earliest stage, hepatomas brought about by the ingestion of aflatoxin appear as slightly elevated or subsurface gray-to-yellowish 1 to 2 mm nodules in the liver parenchyma; more advanced tumors are generally firm, yellowish, and more-or-less lobulated nodes. Carcinomas are often observed to metastasize.

A wide variety of commodities have been shown to yield aflatoxin in the presence of *Aspergillis flavus* or *A. parasiticus;*

e.g. almond paste, apricot seed paste, bakery products, barley, beans, brazilnuts, bread, cassava, cocoa beans and products, coconut oil cake, copra, corn, corn grits, cottonseed meal, cowpea, hazel nuts, locust beans, millet, palm kernel, pea, peach seed paste, peanuts, peanut butter, peanut cake, peanut meal, pecans, pepper corns, pistachio nuts, raisins, rice, sesame, sorghum, soybean meal, spaghetti, wheat, and wheat flour.

Peanuts present a special problem because the fruit, after fertilization, penetrates into the soil, where various fungi contaminate the shell, seed, and testa. Much of the damage sustained by peanuts occurs after the crop has been dug but before it is dry. Mechanical damage in harvesting, improper drying and storage, and insect infestation are important factors. In developing countries there are several practices that tend to raise moisture content in crops and thus favor mold development: manual harvesting of peanuts from fields flooded with water to soften the earth; storing crops that are still wet, or storing dry crops (Figure 23-13) where there is poor protection from water seepage; and harvesting in humid weather.

Figure 23-13
Sacks of peanuts contaminated by mold after water had leaked through a hole in the tarpaulin.

In cotton fields in hot humid weather *Aspergillus flavus* often causes the cotton bolls to rot. Under these circumstances aflatoxins often develop on and in cottonseeds by the time of harvest.

In the United States and Canada peanuts and cottonseeds are monitored for contamination by aflatoxin. In 1969 Canada rejected 7.5 million pounds of peanuts because of such contamination. The currently accepted level of 20 ppb of aflatoxin in foods is based on a 50 fold safety factor that some consider too high—a reduction to 15 ppb has been suggested (Federal Register, 1975).

During production, processing, and storage, almonds, brazil nuts, coconut, hazel nuts, pistachios, and pecans must be properly handled. When stored improperly, both unshelled and shelled nuts support extensive mold growth and toxin production. Also, these nutmeats are subjected to further contamination when processed as halves, pieces, granules, meat, and pastes or when introduced into a variety of products in the home or commercially.

Aflatoxin content in crude crushed Indian peanut oil was found to vary from 20 to 100 μg/kg. Not only did the toxin have great stability in this medium, but foodstuffs fried in such oil absorbed much of the toxin. However, refining, hydrogenating, and bleaching vegetable oils reduced their mycotin content. Prolonged heating of unrefined peanut oil partially destroyed the original aflatoxin content. If the oil was heated for 10 minutes at temperatures of 120 or 150°C, 50% of the mycotoxin was destroyed; at 200°C, 69% was destroyed; and at 250°C, 96% was destroyed (Parpia and Sreenovasamurthy, 1971).

A similar reduction (i.e., 60%) in aflatoxin content was noted when pecans were roasted in coconut oil or margarine at 190°C for six minutes. Dry roasting seedling pecan halves at 190°C for 15 minutes reduced the aflatoxin B_1 and G_1 content of these artificially contaminated nutmeats by 80%, but for Stuart pecan halves, over twice the size of the seedlings, there was a 45% reduction. After roasting there were no viable molds and only 3% of the quantity of toxin originally introduced remained in the oil or margarine bath (Escher et al., 1973).

On many foods molds may be present incidentally, or, more seldom, as intentional ingredients of the final product. A large number of different kinds of molds occur on the surfaces of "fermented" sausages and country-cured hams (Leistner and Ayres, 1968). Among those fungi isolated from heavily molded hams or salami, members of the genera *Penicillium* and *Aspergillus* were predominant. On sausages penicillia prevailed during the entire ripening process, but on hams they were succeeded by aspergilli after prolonged aging. Only the isolates of *A. flavus* and *A. parasiticus* produced chloroform-extractable, fluorescent compounds that coincided with aflatoxin standards on thin-layer chromatoplates (TLC) (Ayres, 1973). Aflatoxin G_1 levels were 10 times those of B_1. Small amounts of G_2 were also detected, but only traces of B_2 were found. Given favorable conditions of temperature and humidity, some strains of *A. flavus* and *A. parasiticus* might be able to produce large amounts of aflatoxins in mold-ripened sausages (Leistner and Tauchmann, 1970).

Sterigmatocystins

The sterigmatocystins—which include sterigmatocystin itself, O-methyl sterigmatocystin, 5-methyl sterigmatocystin, and aspertoxin (a hydroxylated derivative of 0-methyl sterigmatocystin)—resemble the aflatoxins structurally but have a xanthone rather than a coumarin nucleus attached to the bifuran moiety (Figure 23-8 B). In addition to being produced by *Aspergillus flavus,* they are elaborated by *A. versicolor, A. nidulans, A. rugulosus,* a species of *Bipolaris,* and *Penicillium luteum.* When these organisms are grown on cornmeal, levels of sterigmatocystin ranging from 0.75–1.2 mg/kg are attained. Sterigmatocystin has been isolated from wheat and has been found on coffee beans.

When injected, the sterigmatocystins are toxic to ducklings, mice, rats, and monkeys, and when fed to them are carcinogenic. Large quantities of the sterigmatocystins permitted to enter the food supply could present a health hazard.

Ochratoxins

Aspergillus ochraceus is widely distributed in nature. Like other mycotoxins, the toxic principle has been given a trivial name: *ochratoxin.* Ochratoxin A is the major toxic compound isolated,

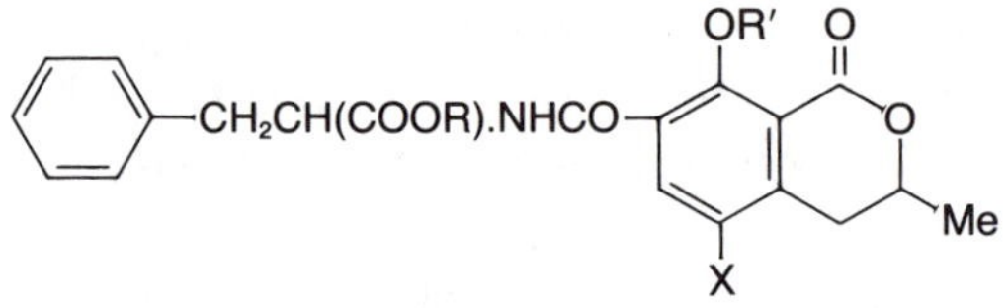

Ochratoxins

	R	R'	X
A	H	H	Cl
B	H	H	H
C	CH₂CH₃	H	Cl

A

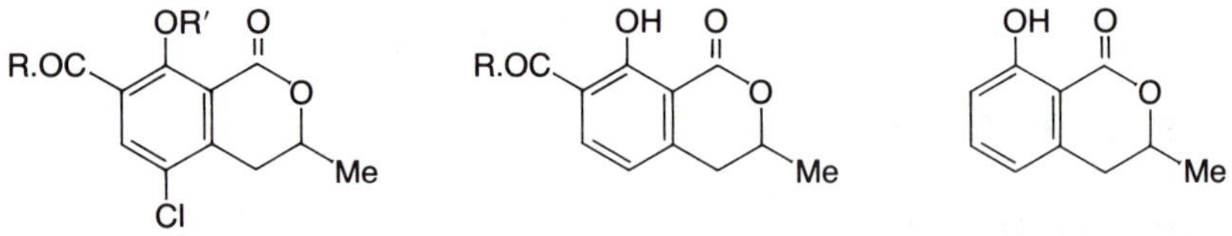

Isocumarins

B

Figure 23-14
A. Structure of ochratoxins; B. Related isocoumarins. The isocoumarin with no ROC side-chain is mellein. (Courtesy F. S. Chu)

but it is not dominant on all substrates on which the mold is grown. Ochratoxin A produces a greenish fluorescence under ultraviolet light. Structurally it is composed of an L-phenylalanine joined to a chlorinated isocoumarin by an amide bond. Ochratoxin B, a nonchlorinated analog, fluoresces blue under ultraviolet light. Methyl and ethyl ester derivatives of these parent compounds also exist (Figure 23-14).

Ochracin (mellein) and 4-hydroxy mellein, compounds that are metabolites of *Aspergillus melleus* and *A. ochraceus,* have a coumarin structure similar to that of the ochratoxins but without the L-phenylalanine moiety. At least two other members of the *A. ochraceus* group, *A. sulphureus* and *A. melleus,* as well as *Penicillium viridicatum,* have been reported to elaborate ochratoxin (Lai et al., 1970).

Several strains of *Aspergillus ochraceus* Wilhelm have been iso-lated from stored cottonseed, citrus fruit, peanuts, and South African stock feeds (Steyn, 1971). A survey of the microflora of South African legume and cereal products showed 46 strains of 22 species of fungi to be toxic for ducklings. Among these, three of five strains of *A. ochraceus* proved to be toxigenic. One of the most toxic strains (K-804) was isolated from sorghum grain. Four of five strains of *A. ochraceus* Wilhelm found in peanuts were toxic to day-old cockerels. A strain of *Penicillium viridicatum* from moldy barley produced ochratoxin A and cit-rinin; cool climates may favor the natural occurrence of these toxigenic penicillia rather than that of their competitors (van Walbeek, 1973). A *Penicillium* sp. isolated from packaged ham produced ochratoxin, and a strain of *A. ochraceus* obtained from Brazil nuts that produced ochratoxin A and small amounts of aflatoxin. However, the production of aflatoxin by organisms other than *A. flavus* and *A. parasiticus* has been questioned.

Male rats metabolized injected ochratoxin A to isocoumarin and an unidentified green fluorescent metabolite. All three compounds were present in urine and feces of rats after a single dosage of toxin (Steyn, 1971). Hens fed 1 to 2 ppm of ochratoxin A laid fewer eggs than those on the control diet. Many of the laying hens receiving the toxin at a level of 4 ppm died and the survivors failed to lay eggs.

Early workers reported that ochratoxin A had about the same order of toxicity as did aflatoxin B_1—i.e., an LD_{50} of 25 μg/duckling—but this has been revised to an LD_{50} of 135 to 170 μg/duckling (Steyn and Holzapfel, 1967). Ochratoxin B is only one-sixteenth as toxic as ochratoxin A to chicks and trout. For weanling rats, the single-dose lethality of ochratoxin A is three times less than that of aflatoxin B_1 but, while it induces acute nephrosis and liver damage in rats, chicks, and rainbow trout, it is apparently noncarcinogenic.

Ochratoxin A was found to occur at a level of about 130 ppb in a sample of low-grade corn and at 100 ppb in samples of rolled wheat associated with the loss of dairy cattle. In the latter, *Penicillium viridicatum* contamination was implicated (Scott et al., 1972).

At 20 to 25°C, *Aspergillus ochraceus* Wilhelm was able to invade grain with a moisture content of more than 16%. *A. ochraceus* and *A. flavus* were observed to be the predominant fungi of red and black peppers. Autoclaved cornmeal inoculated with *A. ochraceus* Wilhelm isolated from the pepper was rendered toxic to ducklings and rats (Christensen et al., 1967).

In the Far East *Aspergillus ochraceus* Wilhelm and related species were reported as constituents characteristic of the microflora of the fermented fish preparations *katsuobushi* and *katsuoshuto*. In the United States Patent No. 1,313,209 covers the use of *A. ochraceus* for inducing a desirable change during the fermentation of coffee (Steyn, 1971). In the light of the information now available, such uses need to be reviewed.

Patulin

Patulin (also named clavacin, clavatin, claviformin) is a toxic and carcinogenic lactone metabolite elaborated by species of the genera *Penicillium, Aspergillus* and *Byssochlamys. P. claviforme, P. divergens, P. equinum, P. expansum, P. griseofulvum, P. lapidosum, P. leucopus, P. novozealandiae, P. patulum, P. urticae, A. clavatus, A. giganteus, A. terreus,* and *B. nivea* are capable of producing patulin. Some of these organisms produce appreciable amounts of the toxin at temperatures below 2°C. *B. nivea* is a heat-resistant fungus that contaminates fruit juices.

Patulin has a molecular weight of 154 and the empirical formula $C_7H_6O_4$ (Figure 23-15 C). It was considered for use as an antibiotic but then rejected because of its toxicity to higher animals. Patulin inhibits cell division, presumably by rendering the sulfhydryl groups of the contractile fibers of the chromosomal spindle inelastic. Reaction with glutathione renders patulin nontoxic.

Patulin has been found on a wide range of agricultural commodities including grain, chick starter, malt feed, flour, molded bread and bakery goods, sausage, and fruit, but its public health significance in any of these except the last is questioned. Apples naturally rotted by *Penicillium expansum* often contain patulin (refer back to Figure 15-6). One might expect to find small amounts of this toxigenic compound in apple juice and other apple products. Levels as high as 1 g/liter of patulin were expressed from rotten tissue. However, in market samples, concentrations of one part per million were more typical. Con-

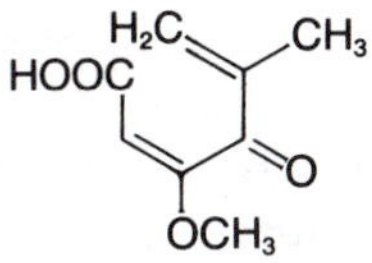

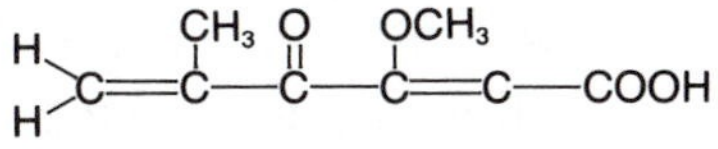

Penicillic acid

A

B

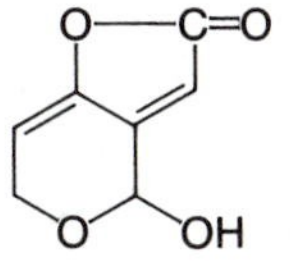

Patulin

C

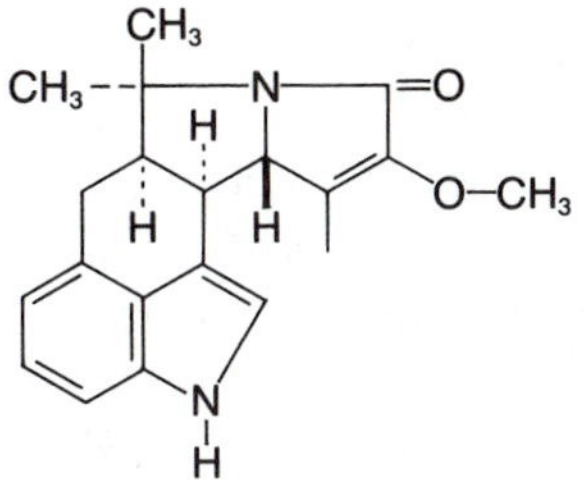

Cyclopiazonic acid

D

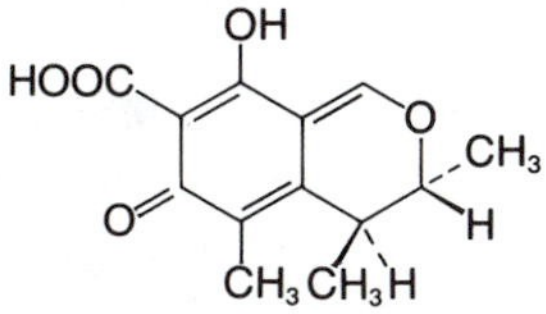

Citrinin

E

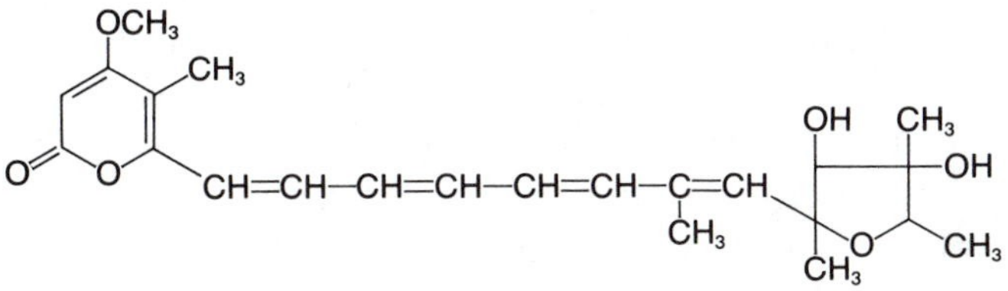

Citreoviridin

F

Figure 23-15
Mycotoxins produced by penicillia.

sequently, one must conclude that primary interest in patulin derives from the fact that it is a proved carcinogen of laboratory animals and therefore cannot be present at any level.

Penicillium Mycotoxins

Species of the genus *Penicillium* produce more than 25 toxic or antibiotic metabolites in food and feed. The organisms are prevalent in agricultural products ranging from stored seeds and milled products to prepared foods such as cured meats and cheese. The *Penicillium* mycotoxins include penicillic acid as well as the yellowed rice toxins—cyclopiazonic acid, citrinin, and citreoviridin, and *P. islandicum* metabolites—and the rubratoxins (Figures 23-15 A–F and 23-16). While the incidence of *Penicillium* mycotoxins in animal feeds is of critical and urgent importance, their significance in human foodstuffs remains to be ascertained.

Penicillium islandicum Metabolites

Penicillium islandicum Sopp causes yellowing or "yellowesis" of domestic and imported rice. The organism produces islanditoxin, a colorless hydrophilic chlorine-containing peptide (cyclochlorotine), $C_{25}H_{36}O_8Cl_2$, and luteoshyrin, $C_{30}H_{22}O_{12}$, a lipophilic 1,1-bis(2,4,5,8-tetrahydroxy)-7-methyl-1-2,3-dihydranthriquinone (Figure 23-16 A, B). Compounds related to luteoskyrin, rubroskyrin (Figure 23-16 C), and *rugulosin*, have been isolated from *P. ruburum* and *P. rugulosum* respectively.

Islanditoxin is extremely hepatotoxic to laboratory animals, causing severe liver damage, hemorrhaging, and death. LD_{50} values for 10 g mice range from 4.75 μg subcutaneously to 65.5 μg orally. It is suggested that this toxin interferes with carbohydrate metabolism by causing glycogen granules to disappear from the injured liver. By contrast, luteoskyrin is slow acting and larger dosages are necessary to elicit pathological response. The LD_{50} for 10 g mice is 1.47 mg subcutaneously and 2.21 mg orally. Liver damage caused by luteoskyrin is of the so-called acinus central cytotoxic pattern characterized by centrilobular necrosis of the liver and diffuse fatty metamorphosis of the liver cells. Luteoskyrin is thought to bind to DNA and cause pigment changes and alter activity of DNA-dependent RNA polymerase (Saito and Tatsuno, 1971).

Citreoviridin

Citreoviridin, with the empirical formula $C_{23}H_{30}O_6 \cdot CH_3OH$ (Figure 23-15 F) is composed of a pyrone chromophore conjugated with a polyene moiety. Several species of the genus *Penicillium* produce this metabolite in moldy or "yellowed" rice. One of these, originally named *Penicillium toxicarium* and since determined to be the same as *P. citreoviride* Biourge, causes in animals paralysis as well as respiratory and circulatory disturbances that resemble symptoms of beriberi in man. Apparently, citreoviridin localizes in the central nervous system and causes respiratory paralysis. It is mildly toxic—the minimal lethal dose for rats being 8 to 30 mg/kg depending upon the route of administration (Saito and Tatsumo, 1971).

Citrinin

Citrinin, $C_{13}H_{14}O_5$, is a slow-acting nephrotoxin produced by *Penicillium citrinum* Thom and at least seven other species of penicillia (Figure 23-15 C). This fungus, found in all of the world's major rice-producing countries, has been frequently isolated from polished yellowed rice. It fluoresces when viewed under ultraviolet light. Since several other species of penicillia and aspergilli synthesize citrinin, its occurrence on various food products is expected. Seven strains of *P. viridicatum* isolated from country-cured ham were able to produce citrinin when inoculated onto that meat, which was then stored at room temperature (25 to 30°C). At temperatures of 15°C or lower little citrinin was synthesized (Wu et al., 1974).

Citrinin has severe renal toxicity capabilities. The kidneys of rats fed 5 mg citrinin evidence turbid swelling and dilatation. However, the true toxicity of citrinin is unknown (Saito and Tatsumo, 1971).

Rubratoxins

Penicillium rubrum produces two toxic metabolites: rubratoxins A and B. Rubratoxin B (Figure 23-16) is the bisanhydride, $C_{26}H_{30}11$, while rubratoxin A has the same structure with one of the anhydride groups reduced to the lactol. Rubratoxin B is the main metabolite and is a higher homolog of byssochlamic acid and probably is produced by the head-to-tail, head-to-tail coupling of two C_{13} units.

Penicillium rubrum was isolated from corn toxic to pigs, mice, horses, and a male goat (Burnside et al., 1957) and has repeatedly been isolated from moldy feeds. Yet the role that the organism plays in naturally occurring disease is still in doubt

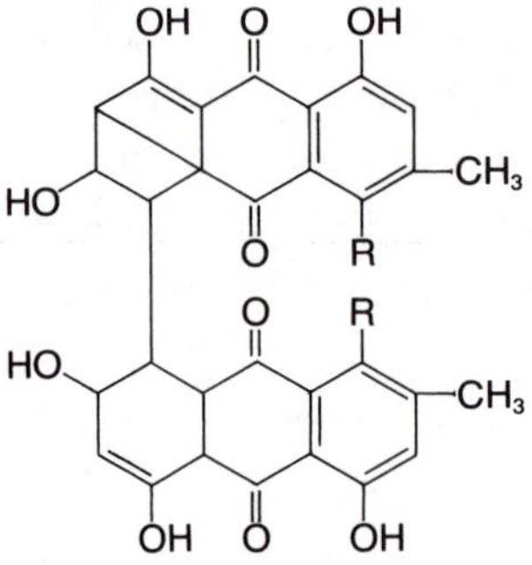

Luteoskyrin (R = OH)

A

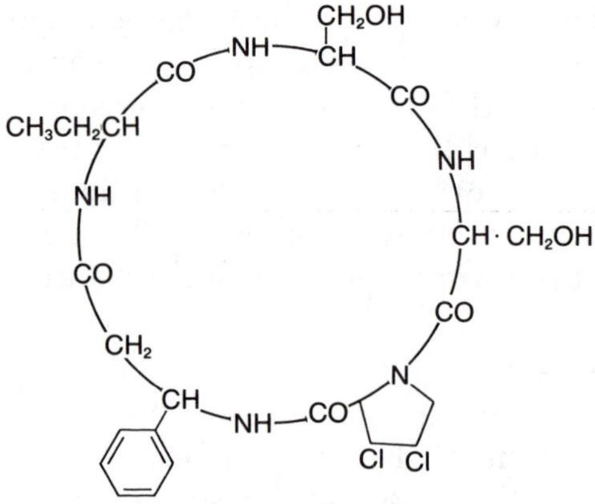

Islanditoxin

B

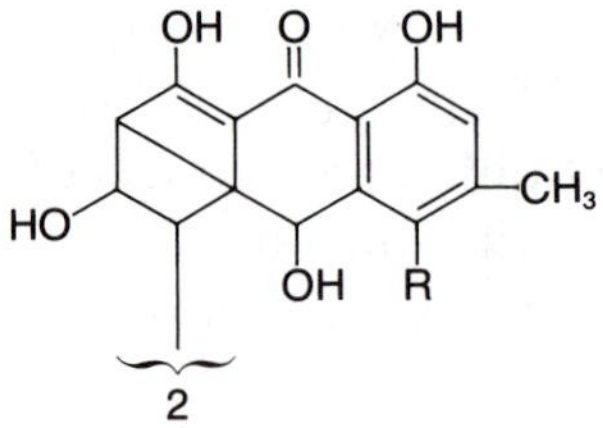

Rubroskyrin

C

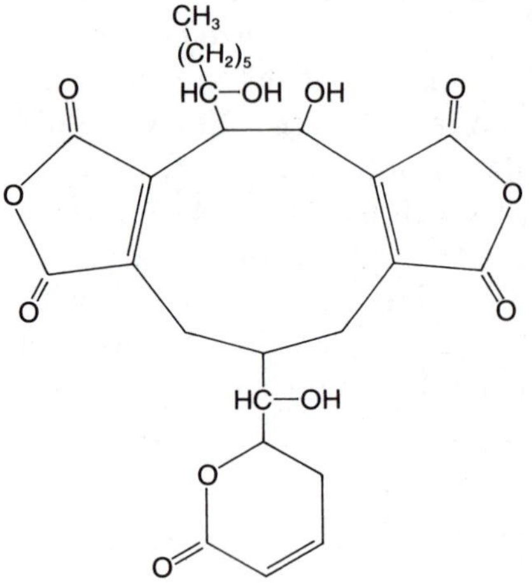

Rubratoxin B

D

Figure 23-16

Toxins produced by, A and B, *Penicillium islandicum* and by, C and D, *P. rubrum*. (Courtesy of Dr. F. S. Chu.)

and the nature of the toxicity of rubratoxin is not clear. There is the possibility that the rubratoxins may potentiate the activity of more dangerous compounds, such as the aflatoxins (Wogan and Mateles, 1968).

Penicillic Acid

Penicillic acid, $C_8H_{10}O_4$, exists as a substituted γ-keto acid—e.g., γ-keto-β-methoxy-α-methylene-A-hexenoic acid—or as a γ-hydroxy lactone (Figure 23-15 A, B). It was first isolated in 1913 from *Penicillium puberulum* Bainier (Alsberg and Black, 1913) and its toxicity for various laboratory animals established. This acid-lactone is supposed to be carcinogenic (Ciegler et al., 1971).

At least 10 species of penicillia and aspergilli synthesize penicillic acid. Several strains of *A. ochraceus* growing on a sucrose-glutamic acid-salts medium produced 0.7 g/liter penicillic acid (Udagawa et al., 1970). Kurtzman and Ciegler (1970) thought this compound responsible for the toxicity of "blue-eye" diseased (Figure 23-17) high-moisture corn that had been stored for six months at 1°C. At dosages as low as 0.1 mg, tumors began growing in rats, and 1.0 mg/doses subcutaneously injected twice weekly induced transplantable tumors in

Figure 23-17
"Blue-eye" disease of corn.

rats (Dickens and Jones, 1963). However, data are needed to substantiate its toxicity by the oral rather than by the subcutaneous route.

About 10% of 346 cultures of *Penicillium* isolated from mold-fermented sausage synthesized penicillic acid on liquid media (Ciegler et al., 1972). In view of this metabolite's toxicity for livestock and poultry, it is strange that a patent was issued allowing use as a feed supplement.

Cyclopiazonic Acid

Penicillium cyclopium Westling has been associated with moldy corn toxicosis in cattle and other farm animals in the United States (Albright et al., 1964; Wilson et al., 1968). Cyclopiazonic acid, empirically $C_{20}H_{20}N_2O_3$ (Figure 23-15 D), is produced by this organism and a closely related species, *P. puberulum* Bainier. While it has been encountered frequently on stored grain products intended for human consumption, and has produced a tremorgenic-diuretic toxin able to kill sheep and horses and to cause acute toxicosis in ducklings and rats, it has not to date been incriminated with mycotoxicoses affecting humans.

Psoralens

Psoralens, or furocoumarins, representing a fused structure of furan and coumarin rings, are phototoxins produced by *Sclerotinia sclerotiorum* (Figure 23-18). Xanthotoxin (8-methoxypsoralen), 4,5,8-trimethyl psoralen, and bergapten (5-methoxypsoralen) have all been associated with the pink rot of celery. These psoralens occur in the leaves, stems, roots, and fruit of a variety of plants. A blistering skin disorder, or dermatitis, has been reported to be common among celery workers coming into contact with plants infected by *S. sclerotiorum*. An unique characteristic for lesion production is that, following mold contact, the skin must also be exposed to sunlight or to ultraviolet radiation in the range of 280 to 360 nm. Unless such exposure occurs within a few minutes of contamination, no lesions are produced.

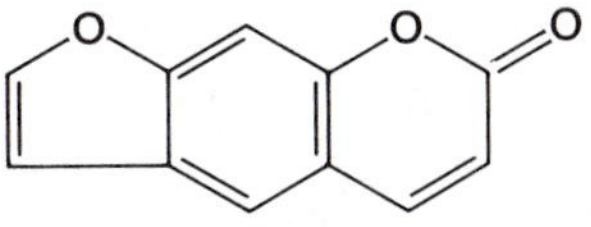

Psoralen

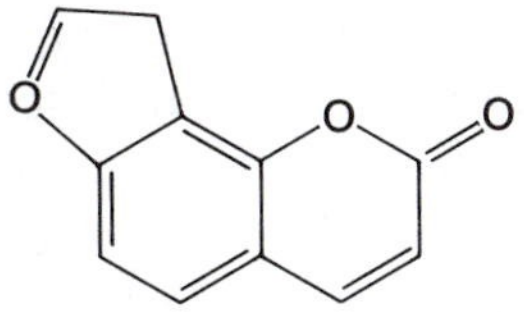

Isopsoralen

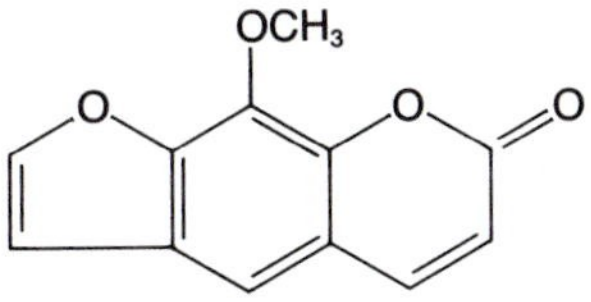

Xanthotoxin

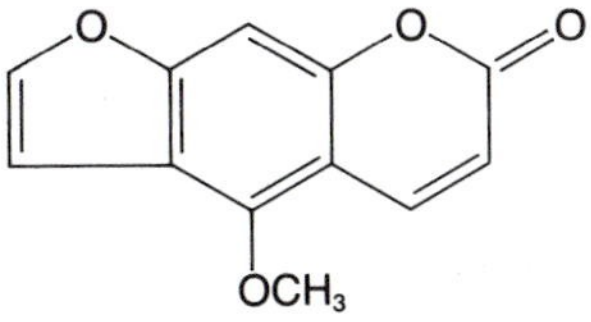

Bergapten

Figure 23-18
Molecular structures of psoralens.

The fungus produces xanthotoxin and bergapten in the rotted areas of celery only under experimental conditions, and only if the celery and *Sclerotinia sclerotiorum* are metabolized concurrently (Wu et al., 1972). These compounds are not produced at a temperature lower than 15°C on celery or on artificial media, nor are they formed on the nonmetabolizing vegetable. Similarly, although the mold grew well on carrots, sweet potatoes, artichokes, oranges, cucumbers, and turnips, no psoralen production could be detected in these commodities.

Psoralens appear to be germination and growth inhibitors, antifungal agents, and mutagenic agents. The minimal effective concentration (MEC) of 4,5,8-trimethosypsoralen by skin test is 0.1 μg per square inch for rabbits, 2.5 μg per square inch for guinea pigs, and 10 μg/square inch for humans. The oral MEC is 4.6 mg/kg for the guinea pig and for humans; 10 mg/day accompanied by exposure to sunlight results in erythema.

The mold *Sclerotinia sclerotiorum* can attack almost any kind of horticultural crop and is one of the most widely distributed pathogens of the vegetable world. The imperfect stage of this fungus, *Botrytis cinerea,* infects grapes and causes rapid evaporation of water from juice. Since gray mold increases the concentration of sugar without adversely affecting the flavor, it is used in France in the production of dry sherry of high alcoholic content. In the light of the phototoxic action of *S. sclerotiorum,* use of this mold for concentrating grape juices seems questionable.

References

Abrams, A., G. Regeles, and G. A. Hattle. 1946. J. Biol. Chem. *164,* 63.

Albright, J. L., S. D. Aust, J. M. Byers, T. E. Fritz, B. O. Brodie, R. E. Olson, R. P. Link, J. Simons, H. E. Rhodes, and R. L. Brewer. 1964. J. Amer. Vet. Med. Assoc. *144,* 1013.

Alder, A. E. 1961. Deut. Med. Wochenschr. *86,* 1121.

Alpert, M. E., G. N. Wogan, and C. S. D. Davidson. 1968. Gastroenterology *54,* 149.

Alsberg, C. L., and O. F. Black. 1913. U.S.D.A. Bur. Plant Ind. Bull. 270.

Angelotti, R., H. E. Hall, M. J. Foter, and K. H. Lewis. 1962. Appl. Microbiol. *10,* 193.

Asao, T., G. Buchi, M. M. Abdel-Kader, S. B. Chang, E. L. Wick, and G. N. Wogan. 1963. J. Amer. Chem. Soc. *85,* 1706.

Asao, T., G. Buchi, M. M. Abdel-Kader, S. B. Chang, E. L. Wick, and G. N. Wogan. 1965. J. Amer. Chem. Soc. *87,* 882.

Ayres, J. C. 1973. Acta Alimentaria (Budapest) *2*, 285.

Barber, M. A. 1914. Philippine J. Sci. *9,* 515.

Barnes, E. F., J. E. Despaul, and M. Ingram. 1963. J. Appl. Bacteriol. *26,* 415.

Benedict, R. G. 1972. Mushroom Toxins Other Than *Amanita.* In: *Microbial Toxins VIII.* S. Kadis, A. Ciegler, and S. J. Ajl, eds. Academic Press, New York.

Bergdoll, M. S. 1967. The Staphylococcal Enterotoxins. In: *Some Foodborne Microbial Toxins.* G. N. Wogan and R. I. Matches, eds. MIT Press, Cambridge, Mass.

Bergdoll, M. S. 1970. Enterotoxins. In: *Microbial Toxins III. Bacterial Protein Toxins.* J. C. Montie, S. Kadis, and S. J. Ajl, eds. Academic Press, New York.

Bergdoll, M. S., I-Y Huang, and E. J. Schantz. 1974. J. Agric. Food Chem. *22,* 9.

Bilay, V. I. 1960. *Mycotoxicoses of Man and Agricultural Animals.* U.S.S.R. (Eng. Transl.). Off. Techn. Serv., U.S. Dept. Commerce, Washington, D.C.

Borja, C. R., and M. S. Bergdoll. 1967. Biochem. *6,* 1467.

Bryan, F. L. 1971. J. Milk Food Technol. *32,* 381.

Bryan, F. L., and E. G. Kilpatrick. 1971. Am. J. Public Health *61,* 1869.

Burnside, J. E., W. L. Sippel, J. Forgacs, W. T. Carll, M. B. Atwood, and E. R. Doll. 1957. Amer. J. Vet. Res. *18,* 817.

Burrows, W., and M. Musteikis. 1966. J. Infectious Diseases *116,* 183.

Casman, E. P. 1965. Ann. N. Y. Acad. Sci. *128,* 124.

Casman, E. P., and R. W. Bennett. 1964. Appl. Microbiol. *12,* 363.

Casman, E. P., and R. W. Bennett. 1965. Appl. Microbiol. *13,* 181.

Casman, E. P., R. W. Bennett, A. E. Dorsey, and J. E. Stone. 1969. Health Lab. Sci. *6,* 185.

Center for Disease Control. 1969. Morbidity and Mortality Weekly Report. Ann. Supplement Summary.

Center for Disease Control. 1970. Morbidity and Mortality Weekly Report. Ann. Supplement Summary.

Center for Disease Control. 1971. Morbidity and Mortality Weekly Report. Ann. Supplement Summary.

Center for Disease Control. 1972. Morbidity and Mortality Weekly Report. Ann. Supplement Summary.

Center for Disease Control. 1975. Botulism in the United States, 1899–1973.

Center for Disease Control. 1976. Morbidity and Mortality Weekly Report. *25*, 75, 269.

Christensen, C. M., H. A. Fanse, G. H. Nelson, F. Bates, and C. J. Mirocha. 1967. Appl. Microbiol. *15*, 622.

Chu, F. S., F. Crary, and M. S. Bergdoll. 1969. Biochem. *8*, 2890.

Chu, F. S., E. Thadhami, E. J. Schantz, and M. S. Bergdoll. 1966. Biochem. *5*, 2381.

Ciegler, A., R. W. Detroy, and E. B. Lillehoj. 1971. Patulins, Penicillic Acid, and Other Carcinogenic Lactones. In: *Microbial Toxins VI. Fungal Toxins*. Academic Press, New York.

Ciegler, A., H. J. Muntzlaff, D. Weisleder, and L. Leistner. 1972. Appl. Microbiol. *24*, 114.

Dack, G. M. 1956. *Food Poisoning*. 3rd ed. Univ. Chicago Press, Chicago, Il.

Dack, G. M. 1964. Food Technol. *18*, 1904.

Das Gupta, B. R., and D. A. Boroff. 1967. Biochem. Biophys, Acta *147*, 603.

Das Gupta, B. R., D. A. Boroff, and E. Rothstein. 1968. Biochem. Biophys. Res. Commun. *32*, 1057.

Dayhoff, M. 1972. Atlas Protein Sequence Struc. *5*, D227.

Despaul, J. E. 1964. *Food Poisoning Microorganisms*. Defense Subsistence Supply Center, Chicago, Il.

Dickens, F., and H. E. H. Jones. 1963. Brit. J. Cancer *17*, 100.

Dolman, C. E., and L. Murakami. 1961. J. Infect. Dis. *109*, 113.

Duff, J. T., G. G. Wright, and A. Yarinsky. 1956. J. Bacteriol. *92*, 455.

Duncan, C. L., and E. B. Somers. 1972. Appl. Microbiol. *24*, 801.

Duncan, C. L., H. Sugiyama, and D. H. Strong. 1968. J. Bacteriol. *95*, 1560.

Duncan, C. L., and D. J. Strong. 1969. J. Bacteriol. *100*, 86.

Escher, F. E., P. E. Koehler, and J. C. Ayres, 1973. J. Food Sci. *38*, 889.

Esty, J. R., and K. F. Meyer. 1922. J. Infect. Dis. *31*, 650.

Eugster, C. H. 1957. Helv. Chim. Acta *40*, 880.

Federal Register. 1975.

Food Chem. News. 1976. *18*(29), 25.

Genigeorgis, C., S. Martin, C. E. Franti, and H. Riemann. 1971. Appl. Microbiol. *21*, 862.

Genigeorgis, C., and W. W. Sadler. 1966. J. Food Sci. *31*, 441.

Genigeorgis, C., G. Sakaguchi, and H. Rieman. 1973. Appl. Microbiol. *26*, 111.

Gerwing, J., B. Mitchell, and D. Van Alystyne. 1967. Biochem. Biophys. Acta *140*, 363.

Glatz, B. S., W. M. Spira, and J. M. Goepfert. 1974. Infect. Immunity *10*, 299.

Goepfert, J. M., W. M. Spira, and H. U. Kim. 1971. J. Milk Food Technol. *34*, 12.

Gretler, A. C., P. Mucciolo, J. B. Evans, and C. F. Niven. 1955. J. Bacteriol. *70*, 44.

Grove, J. F. 1970. J. Chem. Soc. *Sec. C*, 375.

Hall, H. E. 1962. U.S. Public Health Service *1142*, 50.

Hauschild, A. H. W. 1973. *Microbial Foodborne Infections and Intoxications*, Health Protection Branch, Dept. Natl. Health and Welfare, Ottawa, Canada.

Hauschild, A. H. W., and R. Hilscheimer. 1971. Canad. J. Microbiol. *17*, 1425.

Hauschild, A. H. W., R. Hilscheimer, and C. G. Rogers. 1971. Canad. J. Microbiol. *17*, 1475.

Iandolo, J. J., C. W. Clark, L. Bluhm, and Z. J. Ordal. 1965. Appl. Microbiol. *13*, 646.

J. Amer. Med. Assoc. 1963. *184*, 57.

Joffe, A. Z. 1960. Bull. Res. Coun. Israel *9D*, 101.

Joffe, A. Z. 1962. Mycopath et Mycol. Appl. XVI, (3), 201.

Joffe, A. Z. 1963. Plant and Soil XVII (1), 31.

Joffe, A. Z. 1965. In: *Mycotoxins in Foodstuffs*. G. N. Wogan, ed. MIT Press, Cambridge, Mass.

Kafel, S., and J. C. Ayres. 1970. Proc. 3rd Intl. Congress Food Sci. and Technol., Warsaw.

Kim, C. H. 1965. Purdue Univ. Dissertation Abstracts *26*(3).

Kim, H. U., and J. M. Goepfert. 1971. J. Milk Food Technol. *34*, 12.

Kitamura, M. S., S. Sakaguchi, and G. Sakaguchi. 1969. J. Bacteriol. *98*, 1173.

Klein, E. 1895. Zentralbl. für Bakteriol. u. Parasitenk *18*, 737.

Kozaki, S., S. Miyagaki, and G. Sakaguchi. 1977. Infect. and Immunol. *18*, 761.

Kurtzman, C. P., and A. Ciegler. 1970. Appl. Microbiol. *20*, 204.

Lachica, R. V. F., C. Genigeorgis, and P. D. Hoeprich. 1971. Appl. Microbiol. *21:*585.

Lai, M., G. Semeniuk, and C. W. Hesseltine. 1970. Appl. Microbiol. *19*, 542.

Lamanna, C., H. W. Eklund, and D. E. McElroy. 1946. Science *103*, 613.

Leistner, L., and J. C. Ayres. 1968. Die Fleischwirtschaft *48*, 62.

Leistner, L., and F. Tauchmann. 1970. Die Fleischwirtschaft *50*, 965.

McClung, L. S. 1945. J. Bacteriol. *50*, 229.

Mead, G. C. 1969. J. Appl. Bacteriol. *32*, 86.

Nilo, L. 1971. Infect. Immun. *3*, 100.

Nygren, B. 1962. Acta Path. Microbiol. Scand. Suppl. *160*.

Parpia, H. A. B., and V. Sreenivasamurthy. 1971. Importance of Aflaxtoxins in Foods with Reference to India. Proc. SOS/70, Third Intl. Congress Food Sc. and Technol. G. F. Stewart and C. L. Willey, eds. Institute of Food Technology, Chicago.

Perigo, J. A., and T. A. Roberts. 1968. J. Food Technol. *3*, 91.

Peterson, A. L., J. J. Black, and M. F. Gunderson. 1962. Appl. Microbiol. *10*, 16.

Powers, E. M., T. G. Latt, and T. Brown. 1976. J. Milk Food Technol. *39*, 668.

Raj, H., and M. S. Bergdoll. 1969. J. Bacteriol. *98*, 833.

Read, R. B., and J. C. Bradshaw. 1966. Appl. Microbiol. *14*, 130.

Reiser, R. F., and K. F. Weiss. 1969. Appl. Microbiol. *18*, 1041.

Roberts, T. A. 1968. J. Appl. Bacteriol. *31*, 133.

Romagnesi, M. H. 1964. Bull. Soc. Mycol. France *80*, IV.

Saito, M., and T. Tatsumo. 1971. Toxins of *Fusarium nivale*. In: *Microbial Toxins VII. Algal and Fungal Toxins*. S. Kadis, A. Ciegler, and S. J. Ajl, eds. Academic Press, New York.

Sakaguchi, G., and S. Sakaguchi. 1961. Jap. J. Med. Sci. Biol. *14*, 242.

Schantz, E. J., and H. Sugiyama. 1974. J. Agric. Food Chem. *22*, 26.

Scott, P. M., W. van Walbeek, B. Kennedy, and D. Angeti. 1972. J. Agric. Food Chem. *20*, 1103.

Shank, R. C., C. H. Bougeois, N. Keschamros, and P. Chandavimol. 1971. Food Cosmet. Toxicol. *9*, 501.

Siess, E., O. Wieland, and F. Miller. 1970. Arch. Pathol. Anat. Physiol. Abt. B *6*, 151.

Snyder, N. C., and H. N. Hansen. 1945. Amer. J. Botany *32*, 659.

Starin, W. A., and G. M. Dack. 1928. J. Infect. Dis. *36*, 383.

Stark, R. L., and P. R. Middaugh. 1969. Appl. Microbiol. *18*, 361.

Steyn, P. S. 1971. *Microbial Toxins. VI. Fungal Toxins*. A. Ciegler, S. Kadis, and S. J. Ajl, eds. Academic Press, New York.

Steyn, P. S., and C. W. Holzapfel. 1967. Tetrahedron *23*, 4449.

Strong, D. H., J. C. Canada, and B. B. Griffiths. 1963. Appl. Microbiol. *11*, 42.

Strong, D. H., E. M. Foster, and C. L. Duncan. 1970. Appl. Microbiol. *19*, 980.

Sugii, S., I. Ohishi, and G. Sakaguchi. 1977. Infect. and Immunol. *17*, 491.

Tatsumo, T., M. Saito, M. Enomoto, and H. Tsunoda. 1968. Chem. Pharm. Bull. (Tokyo) *16*, 2519.

Traci, P. A., and C. L. Duncan. 1974. Appl. Microbiol. *28*, 815.

Troller, J. A. 1971. Appl. Microbiol. *21*, 435.

Udagawa, S., M. Ichinoe, and H. Kurata. 1970. In: Proc. 1st U.S.-Japan Conf. Toxic Microorgs., Honolulu, Hawaii, 1968. U.S.D.I.-U.J.N.R. Panels on Toxic Microorgs., Washington, D.C.

Ueno, Y. 1970. In: Proc. 1st U.S.-Japan Conf. Toxic Microorgs., Honolulu, Hawaii, 1968. U.S.D.I.-U.J.N.R. Panels on Toxic Microorgs., Washington, D.C.

van Schouwenburg-van Focken, A. W., J. Stadhouders, and J. A. Jans. 1978. Neth. Milk Dairy J. *32*, 217.

van Walbeek, W. 1973. Canad. Inst. Food Sci. Technol. J. *6*, 96.

Weiss, K. F., D. H. Strong, and R. A. Groom. 1966. Appl. Microbiol. *14*, 479.

Wieland, T. 1968. Science *159*, 946.

Wieland, T., and O. Wieland. 1972. In: *Microbial Toxins VIII*. S. Kadis, A. Ciegler, and S. J. Ajl, eds. Academic Press, New York.

Wilson, B. J., C. H. Wilson, and A. W. Hayes. 1968. Nature *220*, 77.

Wogan, G. N., and R. I. Mateles. 1968. Prog. Ind. Microbiol. *7*, 149.

Wu, C. M., P. E. Koehler, and J. C. Ayres. 1972. Appl. Microbiol. *23*, 852.

Wu, M. T., J. C. Ayres, and P. E. Koehler. 1974. Appl. Microbiol. *27*, 427.

Index